Chemogenomics in Drug Discovery

Edited by
H. Kubinyi, G. Müller

Methods and Principles in Medicinal Chemistry

Edited by R. Mannhold, H. Kubinyi, G. Folkers

Recently Published Volumes:

G. Molema, D. K. F. Meijer (eds.)

Drug Targeting
Vol. 12

2001, ISBN 3-527-29989-0

D. Smith, D. Walker, H. van de Waterbeemd

Pharmacokinetics and Metabolism in Drug Design
Vol. 13

2001, ISBN 3-527-30197-6

T. Lengauer (ed.)

Bioinformatics – From Genomes to Drugs
Vol. 14

2001, ISBN 3-527-29988-2

J. K. Seydel, M. Wiese

Drug-Membrane Interactions
Vol. 15

2002, ISBN 3-527-30427-4

O. Zerbe (ed.)

BioNMR in Drug Research
Vol. 16

2002, ISBN 3-527-30465-7

P. Carloni, F. Alber (eds.)

Quantum Medicinal Chemistry
Vol. 17

2003, ISBN 3-527-30456-8

H. van de Waterbeemd, H. Lennernäs, P. Artursson (eds.)

Drug Bioavailability
Vol. 18

2003, ISBN 3-527-30438-X

H.-J. Böhm, G. Schneider (eds.)

Protein-Ligand Interactions
Vol. 19

2003, ISBN 3-527-30521-1

R. E. Babine, S. S. Abdel-Meguid (eds.)

Protein Crystallography in Drug Discovery
Vol. 20

2004, ISBN 3-527-30678-1

Th. Dingermann, D. Steinhilber, G. Folkers (eds.)

Molecular Biology in Medicinal Chemistry
Vol. 21

2004, ISBN 3-527-30431-2

Chemogenomics in Drug Discovery

A Medicinal Chemistry Perspective

Edited by
Hugo Kubinyi and Gerhard Müller

WILEY-VCH Verlag GmbH & Co. KGaA

Series Editors:

Prof. Dr. Raimund Mannhold
Biomedical Research Center
Molecular Drug Research Group
Heinrich-Heine-Universität
Universitätsstrasse 1
40225 Düsseldorf
Germany
raimund.mannhold@uni-duesseldorf.de

Prof. Dr. Hugo Kubinyi
Donnersbergstrasse 9
67256 Weisenheim am Sand
Germany
kubinyi@t-online.de

Prof. Dr. Gerd Folkers
Department of Applied Biosciences
ETH Zürich
Winterthurerstrasse 19
8057 Zürich
Switzerland
folkers@pharma.anbi.ethz.ch

Volume Editors:

Prof. Dr. Hugo Kubinyi
Donnersbergstrasse 9
67256 Weisenheim am Sand
Germany
kubinyi@t-online.de

Dr. Gerhard Müller
Axxima Pharmaceuticals AG
Max-Lebsche-Platz 32
81377 München
Germany
gerhard.mueller@axxima.com

Library of Congress Card No.: Applied for
British Library Cataloging-in-Publication Data
A catalogue record for this book is available from the British Library.

Bibliographic information published by Die Deutsche Bibliothek
Die Deutsche Bibliothek lists this publication in the Deutsche Nationalbibliografie; detailed bibliographic data is available in the internet at http://dnb.ddb.de.

Printed in the Federal Republic of Germany.
Printed on acid-free paper.

Cover design 4t GmbH, Darmstadt
Composition Manuela Treindl, Laaber
Printing betz-druck GmbH, Darmstadt
Bookbinding Buchbinderei J. Schäffer GmbH & Co. KG, Grünstadt

ISBN 3-527-30987-X

Dedicated to the memory of the great medicinal chemist
Dr. Paul Janssen (1926–2003),
the discoverer of many breakthrough medicines.

Contents

Chemogenomics in Drug Discovery: A Medicinal Chemistry Perspective.
Edited by Hugo Kubinyi and Gerhard Müller

ISBN: 3-527-30987-X

Preface

The term chemogenomics is applied to a diversity of approaches that use chemical compounds to probe biological systems. While all of the approaches have at least some relevance to drug discovery, the methods can be differentiated according to the extent to which they employ stochastic versus directed approaches. Stochastic chemogenomics approaches probe the global response of a biological system on exposure to chemical compounds. Focused chemogenomics approaches use chemicals as detailed probes of biochemical pathways that can play a key role in target identification and validation. An integrated chemogenomics platform uses affinity-based screening, directed combinatorial chemistry, and structure-based drug design to rapidly develop drug-like tool compounds that can validate a target-based therapeutic hypothesis *in vivo*.

Chemogenomics approaches are evolving to overcome key problems limiting the efficiency of drug discovery in the postgenomic era. Many of these limits stem from the low success rates in finding drugs for novel genomics targets whose biochemical properties and therapeutic relevance is poorly understood. The fundamental objective of chemogenomics is to find and optimize chemical compounds that can be used to directly test the therapeutic relevance of new targets revealed through genome sequencing. The chemogenomics approach defers investment in biological target validation to a later stage in the discovery cycle, where resources can be deployed more efficiently and with a higher probability of success, thus providing a more direct route to finding new drugs.

The present volume on "Chemogenomics in Drug Discovery" is organized in three main sections. General aspects in the first section are dedicated to privileged structures as target family-directed masterkeys (G. Müller), drug discovery from side effects (H. Kubinyi), the value of chemical genetics in drug discovery (K. Russell) and structural aspects of binding site similarity (A. Bergner and J. Günther).

The second section focuses on target families such as kinases (R. Buijsman), ion channel modulators (K.-H. Baringhaus and G. Hessler), and phosphodiesterases (M. Hendrix and C. Kallus). In addition, the contribution of molecular informatics for chemogenomics (E. Jacoby et al.), chemical kinomics (B. Klebl), as well as proteochemometrics (J. Wikberg et al.) are discussed.

Chemical libraries are the topic of the final section and cover chemogenomics in compound library and template design for GPCRs (T. R. Webb), computational filters in lead generation (W. Guba), navigation in chemical space (G. Schneider

Chemogenomics in Drug Discovery: A Medicinal Chemistry Perspective.
Edited by Hugo Kubinyi and Gerhard Müller

ISBN: 3-527-30987-X

and P. Schneider), natural product derived combinatorial libraries (M. A. Koch and H. Waldmann), and combinatorial chemistry in chemical genomics age (R. Joseph and P. Arya).

We are grateful to the Volume Editors for their enthusiasm to organize this volume and to work with such a fine selection of authors. We also want to express our gratitude to Frank Weinreich from Wiley-VCH for his valuable contributions to this project.

Dr. Paul A. J. Janssen, former Director of Janssen Pharmaceutica N. V., Beerse Belgium, and founder of the Center for Molecular Design, Vosselaar, Belgium, unexpectedly died on November 11, 2003. As he was one of the most prominent medicinal chemists and discoverer of many breakthrough medicines, the Volume and Series Editors would like to dedicate this book to the memory of this great man.

March 2004

Raimund Mannhold, Düsseldorf
Hugo Kubinyi, Weisenheim am Sand
Gerd Folkers, Zürich

A Personal Foreword

Chemical Genomics versus Orthodox Drug Development is the title of an essay published in the February issue 2003 of Drug Discovery Today (Drug Discovery Today **8**, 157–159, 2003), discriminating between two pharmaceutical research approaches; the chemical genomics-based approach on one hand, as opposed to the classical way of drug development, adhering to the accepted traditional strategies on the other hand. Embedded in this apparent contradiction, defined by established medicinal chemistry and the post-genomic approaches characterized by -omes and -omics tags, this volume of *Methods and Principles in Medicinal Chemistry* attempts to re-position the core discipline of Medicinal Chemistry right into the centre of chemogenomics. Since chemogenomics is widely claimed to address key issues posed by the sharp decrease in pharmaceutical industry's productivity, the role and relevance of modern medicinal chemistry has to be re-emphasised in this context.

All contributions of this issue focus on aspects of the systematic investigation of molecular recognition phenomena that underlie drug–target interactions, and subsequent extrapolation either within compound classes or within target families with the ultimate aim to enhance efficiency of the drug discovery process.

G. Müller, H. Kubinyi, and K. Russell elaborate in their contributions on different aspects of classification and systematisation. The target family-directed masterkey concept conveyed by G. Müller intentionally takes advantage of privileged structures that are tailor-made to explore entire gene families, thus accounting for the required scalability of a once established chemistry concept in a chemogenomics framework. The systematic exploitation of observed side-effects associated to known drugs is described by H. Kubinyi as an efficient approach towards high-content leads for novel targets and respective diseases. In more general terms, K. Russell introduces into the manifold conceptual interfaces between biology and chemistry on a chemical genetics platform. Apart from the aspects of target identification and validation, the chemogenomics idea is developed out of the chemical genetics realm by extrapolating compounds from tools to high-quality leads.

Predominantly, the book covers systematic elaborations on pharmaceutically relevant target families with clear focus centred around systematic medicinal chemistry access routes towards the distinct members of those target clusters. Contributions by R. Buijsman and B. Klebl and colleagues provide detailed insights into the world of protein kinase inhibitors. While R. Buijsman systematically focuses on the detailed structural requirements of protein kinase binding sites that

Chemogenomics in Drug Discovery: A Medicinal Chemistry Perspective.
Edited by Hugo Kubinyi and Gerhard Müller

ISBN: 3-527-30987-X

determine small molecule design strategies, B. Klebl and co-workers provide detailed insights into chemical kinomics, highlighting chemical genomics, chemical validation strategies, chemical genetics approaches, and a chemical proteomics technology, always emphasising the multiple purposes of specifically developed kinase inhibitors.

Medicinal Chemistry approaches towards the target family of phosphodiesterases, ion channels, and G protein-coupled receptors under a chemogenomics paradigm are introduced in three distinct contributions. M. Hendrix and C. Kallus elaborate the element of systematic strategies within medicinal chemistry for phosphodiesterase inhibitors where common substructures are described to address conserved features of an entire target family. Privileged chemotypes that qualify for a target family-directed library design concept form the basis for a chemogenomics-based discovery strategy pursued for ion channels, as described by K.-H. Baringhaus and G. Hessler. T. Webb refers to the area of G protein-coupled receptors, where ligand-derived information is systematically used to design target family-directed scaffolds that, upon further chemical variation, allow for rapid lead generation.

Contributions by R. Joseph and P. Arya as well as M. A. Koch and H. Waldmann focus on synthetic aspects towards lead structures originating from natural product-derived scaffolds. R. Joseph and P. Arya refer to two complementary approaches, the synthetic access to focussed libraries around bioactive natural product cores, and diversity-oriented synthesis aiming at 3D scaffold diversity for hit generation, respectively. On the other hand, M. A. Koch and H. Waldmann emphasise the correlation of natural product-based library concepts with structural features of targeted protein domains, thus strengthening the privileged structure concept from a bioorganic viewpoint.

Systematic application and conceptual combination of chemoinformatics, bioinformatics, and structural genomics approaches are covered by a variety of contributions in this book. E. Jacoby and colleagues report on design strategies for combinatorial compound libraries pursuing a system-based chemoproteomics approach that is exemplified on the target family of G protein-coupled receptors. Numerous aspects of ligand based *in-silico* design techniques are reviewed in detail by G. Schneider and P. Schneider, touching upon algorithms and applications of e.g. similarity searching, or pharmacophore models. W. Guba and O. Roche highlight pragmatic applications of computational strategies for addressing drug-like characteristics of chemotypes within the framework of lead finding and optimisation. A. Bergner and J. Günther propose a systematic approach towards a deeper understanding of target binding site characteristics and corresponding similarities, thus integrating unique and precious protein structure knowledge into the chemogenomics discussion. Finally, J. Wikberg and co-workers report on a novel bioinformatics approach, termed proteochemometrics, to develop detailed insights into molecular interaction space, by scrutinising binding data of different compound series targeted towards different receptor systems.

As the field of chemogenomics is still maturating, this book is an attempt to highlight the role of medicinal chemistry in the multi-disciplinary set-up that is required for a successful drug discovery environment. Careful consideration of all aspects discussed within this book will undoubtedly facilitate the development of a clear definition of chemogenomics. In this context, the book will be helpful for numerous researchers in the life science community, currently addressing any aspect of drug discovery and development in pharmaceutical industry, as well as in academia.

All chapter authors are very much acknowledged for their great enthusiasm, their preparation of the manuscripts within a tough time frame and the high quality of their contributions. The Editors would also like to thank Dr. Frank Weinreich and the staff of Wiley-VCH for their engagement in the production of this monograph.

April 2004

Hugo Kubinyi
Gerhard Müller

List of Contributors

Pierre Acklin
Novartis Institutes for Biomedical Research
Novartis Pharma AG
Lichtstrasse 35
4056 Basel
Switzerland
pierre.acklin@pharma.novartis.com

Prabhat Arya
Steacie Institute for Molecular Sciences
National Research Council of Canada
100 Sussex Drive
Ottawa, Ontario K1A 0R6
Canada
prabhat.arya@nrc-cnrc.gc.ca

Karl-Heinz Baringhaus
Aventis Pharma Deutschland GmbH
Chemistry/Computational Chemistry
Building G 878
65926 Frankfurt/Main
Germany
karl-heinz.baringhaus@aventis.com

Andreas Bergner
Cambridge Crystallographic Data Centre
12 Union Road
Cambridge CB2 1EZ
United Kingdom
current address:
Discovery Partners International AG
Gewerbestrasse 16
4123 Allschwil
Switzerland
abergner@discoverypartners.com

Rogier Buijsman
N. V. Organon
Lead Discovery Unit – Chemistry
Molenstrat 110
5340 BH Oss
The Netherlands
rogier.buijsman@organon.com

Henrik Daub
Axxima Pharmaceuticals AG
Max-Lebsche-Platz 32
81377 München
Germany
henrik.daub@axxima.com

Chemogenomics in Drug Discovery: A Medicinal Chemistry Perspective.
Edited by Hugo Kubinyi and Gerhard Müller

ISBN: 3-527-30987-X

Wolfgang Guba
F. Hoffmann-La Roche Ltd.
Pharmaceuticals Division
4070 Basel
Switzerland
wolfgang.guba@roche.com

Judith Günther
Schering AG
CDCC/Computational Chemistry
13342 Berlin
Germany
judith.guenther@schering.de

Martin Hendrix
Bayer Healthcare AG
Pharma Research
Forschungszentrum Aprather Weg
42096 Wuppertal
Germany
martin.hendrix.mh@bayer-ag.de

Gerhard Hessler
Aventis Pharma Deutschland GmbH
Chemistry/Computational Chemistry
Building G 878
65926 Frankfurt am Main
Germany
gerhard.hessler@aventis.com

Edgar Jacoby
Novartis Institutes for Biomedical Research
Novartis Pharma AG
Lichtstrasse 35
4056 Basel
Switzerland
edgar.jacoby@pharma.novartis.com

Reni Joseph
MDS Pharma Services
2350 Cohen Street
Montréal, Québec H4R 2N6
Canada
reni.joseph@mdsps.com

Christopher Kallus
Bayer Healthcare AG
current address:
Aventis Pharma Deutschland GmbH
Industriepark Höchst
Building G878
65926 Frankfurt/Main
Germany
christopher.kallus@aventis.com

György Kéri
Vichem Chemie Ltd.
Herman Ottó u. 15
1022 Budapest
Hungary
keri@vichem.hu

Bert M. Klebl
Axxima Pharmaceuticals AG
Max-Lebsche-Platz 32
81377 München
Germany
bert.klebl@axxima.com

Marcus A. Koch
Max Planck Institute of Molecular Physiology
Department of Chemical Biology
Otto-Hahn-Strasse 11
44227 Dortmund
Germany
marcus.koch@mpi-dortmund.mpg.de

Hugo Kubinyi
Donnersbergstrasse 9
67256 Weisenheim am Sand
Germany
kubinyi@t-online.de

Maris Lapinsh
Department of Pharmaceutical
Biosciences
Uppsala University
Box 591, Biomedicum
751 24 Uppsala
Sweden
maris.lapinsh@farmbio.uu.se

William F. Michne
AstraZeneca Pharmaceuticals
Department of Chemistry
1800 Concord Pike
Wilmingtion, DE 19850-5437
USA
william.michne2@astrazeneca.com

Gerhard Müller
Axxima Pharmaceuticals AG
Max-Lebsche-Platz 32
81377 München
Germany
gerhard.mueller@axxima.com

Peteris Prusis
Department of Pharmaceutical
Biosciences
Uppsala University
Box 591, Biomedicum
751 24 Uppsala
Sweden
peteris.prusis@farmbio.uu.se

Olivier Roche
F. Hoffmann-La Roche Ltd.
Pharmaceuticals Division
4070 Basel
Switzerland
olivier.roche@roche.com

Keith Russell
AstraZeneca Pharmaceuticals
Department of Chemistry
1800 Concord Pike
Wilmington, DE 19850-5437
USA
keith.russell1@astrazeneca.com

Gisbert Schneider
Institute of Organic Chemistry and
Chemical Biology
Johann Wolfgang Goethe-Universität
Marie-Curie-Strasse 11
60439 Frankfurt am Main
Germany
g.schneider@chemie.uni-frankfurt.de

Petra Schneider
Schneider Consulting GbR
George-C.-Marshall-Ring 33
D-61440 Oberursel
Germany
petra.schneider@moleculardesign.de

Ansgar Schuffenhauer
Novartis Institutes for Biomedical
Research
Novartis Pharma AG
Lichtstrasse 35
4056 Basel
Switzerland
ansgar.schuffenhauer@pharma.
novartis.com

Herbert Waldmann
Max Planck Institute of Molecular
Physiology
Department of Chemical Biology
Otto-Hahn-Strasse 11
44227 Dortmund
Germany
herbert.waldmann@mpi-
dortmund.mpg.de

Thomas R. Webb
Chembridge Research Laboratories
16981 Via Tazon, Suite G
San Diego, CA 92127
USA
twebb@chembridge.com

Jarl E. S. Wikberg
Department of Pharmaceutical
Biosciences
Uppsala University
Box 591, Biomedicum
751 24 Uppsala
Sweden
jarl.wikberg@farmbio.uu.se

Introduction

Gerhard Müller and Hugo Kubinyi

The term "chemogenomics" evolved from the merger of chemistry and genomics. Since this chapter introduces into a volume of *Methods and Principles in Medicinal Chemistry* that is entitled "Chemogenomics in Drug Discovery: A Medicinal Chemistry Perspective", the attempt is made to provide a valid definition of chemogenomics in the context of drug discovery, more specifically in the context of medicinal chemistry. Prior to a more precise definition of chemo-related aspects of chemogenomics, the parent discipline of genomics should be highlighted first.

The starting definition of genomics derives from Tom Roderick, as first cited in print by Victor A. McKusick and Frank H. Ruddle in the inaugural edition of the new journal *Genomics* [1]. At that time, genomics distinguished large-scale mapping and sequencing efforts from molecular studies of only a few genes. Over time, genomics has shifted in meaning to any studies that involve the analysis of DNA sequence, and even to the study of how genes affect biological mechanism and phenotype. This still includes the original meaning of genomics, but extends well beyond that. Alongside with the biotechnology boom, genomics became a meaning that is broader still and it was adopted as a buzzword, also to attract venture capital, particularly in the period of 1998 to 2001, when many new companies emphasised their involvment in proteomics and bioinformatics, categories that clearly overlap with genomics. By the end of 2001, the term was considerably broader in meaning and had become purely arbitrary in some cases.

Even the 2003 report of the WHO entitled "Genomics and world health: report of the Advisory Committee on Health Research" (Geneva, WHO 2003) begins with an overoptimistic statement, clearly supporting the idea of a direct gene-to-clinic fast-track:

> *"The complete sequencing of the human genome, announced in 2001, marked the culmination of unprecedented advances in the science of genomics, the study of the genome and its function. The availability of genome sequences for many living organisms clearly has important implications for health improvement, and it has been widely predicted that elucidation of the sequences will lead to a revolution in medical research and patient care."*

Chemogenomics in Drug Discovery: A Medicinal Chemistry Perspective.
Edited by Hugo Kubinyi and Gerhard Müller

ISBN: 3-527-30987-X

Later in the report, a more adequate and realistic assessment is given, correcting the rocketing expectations widely raised:

> *"An overoptimistic picture of the applications and benefits of genetic research has been drawn. The potential medical applications of genomics are considerable and will lead to major advances in clinical practice but the time-scale is difficult to predict."*

Up to now, chemogenomics has been applied to a diversity of approaches that use chemical compounds to interrogate biological systems [2–4], but since some of these approaches have only peripheral relevance to drug discovery, we felt it worthwhile to focus on those aspects that address key issues posed by the sharp decrease in pharmaceutical productivity that has occurred in the post-genome era.

And still, pharmaceutical industry severely suffers from a productivity gap, even though a plethora of new technologies were implemented in the R&D structure of virtually all pharmaceutical and biotechnology companies. The majority of those innovative technologies have sent drug development costs soaring, unfortunately with no measurable rise as yet in number of new chemical entities reaching the market. As a generic conclusion, medicinal chemistry is viewed as a still limiting factor in the creation of new drugs. The immense flow of gene and protein data at the turn of the millennium led to the irresistible idea that once all of the disease targets were characterized, drugs for each would eventually follow straightforwardly.

Alongside with the appearance of more and more gene and protein data, a whole suite of "-omes" and "-omics" emerged. Generally, an "-omics" describes a technology toolbox that is developed to study a specific object of interest at the largest possible scale with highest degree of systematisation. Consequently, the object of interest is the corresponding "-omes" that is associated to the respective "-omics". Over the last five years, for virtually any classical process step of the traditional drug discovery and development value chain, a distinct -omics technology was born, no longer working on single defined objects, but on the associated -omes. The spectrum ranges from bibliomics, biomics, cellomics, chromosonomics, degradomics over genomics, glycomics, immunomics, interactomics, lipidomics, metabolomics, methylomics up to peptidomics, physiomics, regulomics, transportomics, and vaccinomics, just to mention a selection of those technology toolboxes [5].

Chemogenomics, in most general terms, has been defined as the discovery and description of all possible drugs to all possible targets [6]. Whereas such an attempt would undoubtedly be the most systematic approach towards chemogenomics, it remains impossible to ever achieve this goal.

Today, the most widely used definition of chemogenomics refers to the perturbation of biological systems with the help of small molecules, thus gaining a holistic understanding of the interaction of such molecules with complex molecular systems. In this context, chemogenomics is simply a subset of genomics in which the focus is on small molecules [7, 8].

The ability to study certain subjects systematically on a large scale, better to the largest possible extent, rather than on a one-by-one and on a case-by-case basis, is what finally renders a specific technology to mature towards an associated "-omics". Just as genomics is the extension of genetics to a genome-wide scale, chemogenomics has been defined as being the extension of chemical genetics to a genome-wide scale. Chemical genetics is exactly the study of biological processes using small molecule intervention, as opposed to genetic intervention. Just as genetics offers a way to study biology by modulating gene function through mutation, chemical genetics seeks to study biology my modulating protein function with low-molecular weight compounds. For application of low-molecular weight compounds to perturb complex biological systems not a great deal of medicinal chemistry is required.

However, chemogenomics also has been predicted to produce chemical ligands for all important proteins which should enable chemical modulation of their activities, both positively or negatively, and completely of selectively [7]. With this definition a claim is being made that directly reaches through into the innermost core of medicinal chemistry. Unfortunately, statements of that type raise expectations that by no means account for the complexity of any drug discovery attempt. Projecting this ambitious goal of chemogenomics as outlined above [7] into the reality of e.g. generating selective agonists for protein-binding G protein-coupled receptors, then one can get the feeling that this sounds more like science fiction than science. Also the design of protease activators, instead of inhibitors, will emerge as a major undertaking. In essence, the majority of today's definitions of chemogenomics refer to the process by which small molecules are used to gain insight into the function of novel biological targets. Whenever chemogenomics is also seen as a parallel approach to target validation and drug discovery, not too much of medicinal chemistry-directed know-how is included. This insight was one of the major driving forces to edit the current book and to lay the emphasis on medicinal chemistry aspects that help to re-define chemogenomics.

Today, the majority of scientists involved in drug discovery and more and more senior executives realize that the pay-offs of the automation and miniaturization attempts associated with most of the "-omes" and "-omics" seem farther away then originally hoped. It is obvious that the technological revolution of the last decade has altered the way we pursue drug discovery in general and organic chemistry, one of the underlying core disciplines of medicinal chemistry in particular. Success in lead finding and optimisation still requires skilled scientists making the correct choices on e.g. which hits are likely to play out as tractable leads that, upon optimisation will finally take the numerous hurdles that any pre-clinical candidate must surmount.

In addition, there is increasing evidence that implementation of those technologies especially in small and mid-sized pharmaceutical companies resulted in a fascination in technology that finally led to a defocusing of R&D efforts. In this context, the editors believe that chemogenomics as advocated in this book will define a new suite of tools and strategies that will primarily support the medicinal chemist scaling his or her lead generation and optimisation capabilities from distinct single experiences towards a broader and more systematic understanding and subsequent application within drug discovery and development.

The scale-up of lead discovery from a case-by-case to a "genome-wide" effort requires general guidelines that can be applied e.g. throughout entire target families, or allow to systematically explore tailored compound classes. In most general terms, chemogenomics seeks for a correlation of tailored compound collections with well-chosen target classes, thus utilizing a systematic effort aligned along two dimensions, the compound dimension as well as the target dimension.

As genomics is concerned with taking large-scale sequence information to the next higher level, thus annotating a functional understanding, chemogenomics attempts to promote chemical structure-encoded information on a higher level in order to correlate compound space with target space.

The editors do hope to have succeeded with this volume to position the scientific discipline of Medicinal Chemistry into the focus of chemogenomics, thus making it less of a buzzword by shifting the content of this "-omics" significantly towards the chemisty-related aspects. On purpose, the emphasis is laid on chemistry to remind that one of the major bottlenecks is the chemical aspect of chemogenomics, which was and still is often underestimated within the today's chemogenomics discussion.

References

1 V. A. McKusick, F. H. Ruddle, *Genomics* **1987**, *1*, 1–2.
2 F. R. Salemme, *Pharmacogenomics* **2003**, *4*, 1–11.
3 S. L. Schreiber, *Chem. Eng. News* **2003**, *81*, 51–61.
4 S. L. Schreiber, *Bioorg. Med. Chem.* **1998**, *6*, 1127–1152.
5 see: http://www.genomicglossaries.com/content/omes.asp
6 P. R. Caron, M. D. Mullican, R. D. Mashal, K. P. Wilson, M. S. Su, M. A. Murcko, *Curr. Opin. Chem. Biol.* **2001**, *5*, 464–470.
7 X. F. S. Zheng, T.-F. Chan, *Drug Discov. Today* **2002**, *7*, 197–205.
8 D. E. Szymkowski, *Drug Discov. Today* **2003**, *8*, 157–159.

I
General Aspects

Chemogenomics in Drug Discovery: A Medicinal Chemistry Perspective.
Edited by Hugo Kubinyi and Gerhard Müller

ISBN: 3-527-30987-X

1
Target Family-directed Masterkeys in Chemogenomics

Gerhard Müller

1.1
Introduction

Chemogenomics aims at providing a small molecule for every protein encoded by the human genome for use as a molecular probe of cellular function and, in parallel, as a possible lead candidate for drug development [1, 2]. This contribution attempts to present a novel medicinal chemistry concept, namely the target family-directed masterkey concept, which is based on tailor-made privileged structures [3] as a key element of chemogenomics. In this context, chemogenomics is envisioned as a still evolving and maturing paradigm in drug discovery, rendering the entire preclinical discovery process more efficient through the systematic application of new medicinal chemistry concepts on a 'genomic' scale. Rather than following the classical approach, in which a single target protein is tackled at a time within a distinct disease area, the masterkey concept [3] offers the opportunity to process multiple related members of a target family simultaneously across numerous therapeutic areas. The masterkey concept is considered as a chemogenomics platform, since it allows one to deal with a large number of potential protein targets with increased efficiency in lead generation, delivering target-specific molecules amenable to parallel optimization toward progressible preclinical candidates. This novel medicinal chemistry concept will contribute to an urgently required renaissance of chemistry within the multidisciplinary area of drug discovery and development. To position the masterkey concept into today's landscape of drug discovery, a general overview of the pharmaceutical industry's current situation and performance over the last decade is given.

The pharmaceutical industry is one of the largest industries worldwide still exhibiting a strong growth potential. The associated market volume reached US $ 365 billion in 2001, representing a 12% growth over the preceding year [4]. The top 10 leading products alone accounted for more than 10% of that market (US $ 40 billion), reaching 22% growth over 2000. Apart from a single recombinant protein, notably erythropoietin, the business predominantly deals with classical small-molecule drugs (Table 1.1 and Figure 1.1) [4].

Chemogenomics in Drug Discovery: A Medicinal Chemistry Perspective.
Edited by Hugo Kubinyi and Gerhard Müller

ISBN: 3-527-30987-X

Table 1.1 Top 10 best-selling pharmaceutical products of 2001 [4].

Brand name	*Ingredient*	*Company*	*Indication*	*Sales (growth) [$ billion]*
Lipitor®	Atorvastatin	Pfizer	hypoercholesterolemia	7.0 (31%)
Prilosec®	Omeprazole	AstraZeneca	ulcers	6.1 (0%)
Zocor®	Simvastatin	Merck & Co.	hypercholesterolemia	5.3 (25%)
Norvasc®	Amlodipine	Pfizer	hypertension	3.7 (14%)
Prevacid®	Lansoprazole	Takeda/Abbott	ulcers	3.5 (13%)
Zyprexa®	Olanzapine	Eli Lilly	schizophrenia	3.2 (35%)
Celebrex®	Celecoxib	Pharmacia/Pfizer	pain, arthritis	3.1 (32%)
Procrit®	Erythropoietin	J&J/Amgen	anemia	2.9 (35%)
Paxil®	Paroxetine	GlaxoSmithKline	depression	2.8 (19%)
Vioxx®	Rofecoxib	Merck & Co.	pain, arthritis	2.6 (44%)
all 10 products				**$ 40 200 000 000**

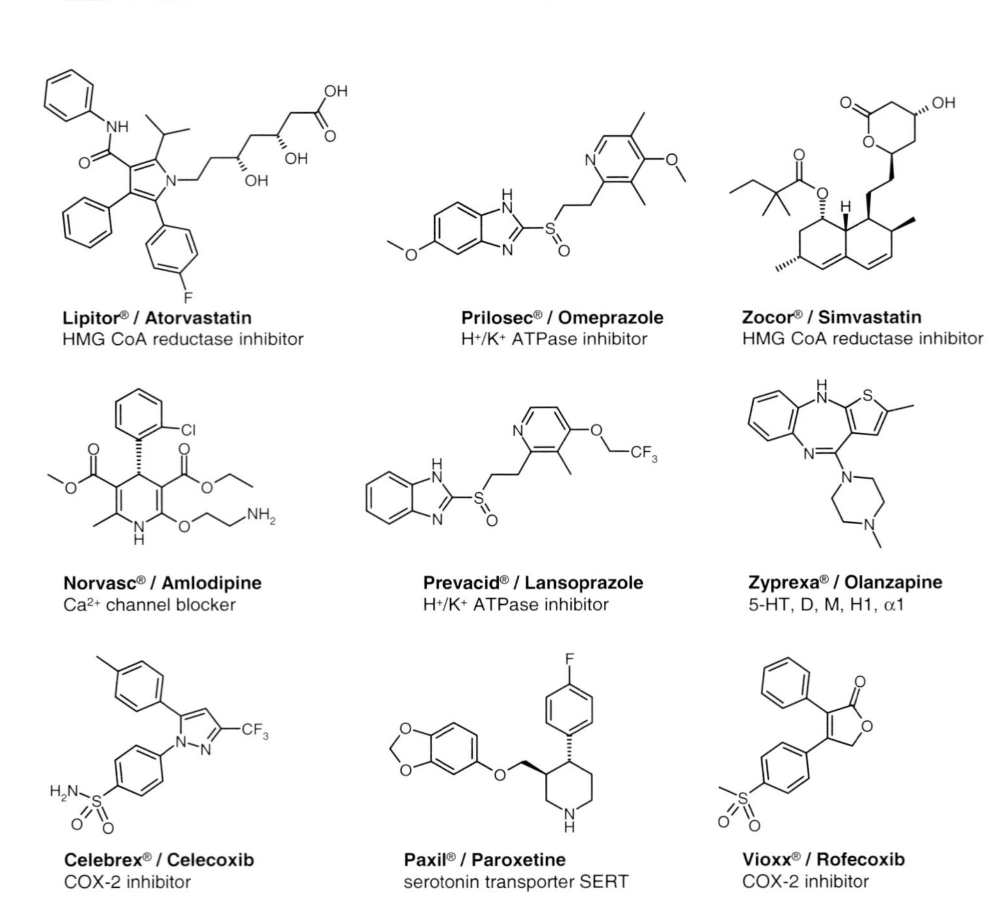

Figure 1.1 Chemical structures of the best-selling drugs of 2001 [4]. Nine of the top ten drugs are low-molecular-weight compounds.

Table 1.2 Fastest-growing products of 2001. Growth is given relative to sales in 2000 [4].

Brand name	***Ingredient***	***Company***	***Indication*** *targets*	***Sales (growth) [$ million]***
Nexium®	Esomeprazole	AstraZenec	ulcers *proton pump inhibitor*	623 (999%)
Protonix®	Pantoprazole	Altana/Wyeth	ulcers *proton pump inhibitor*	695 (426%)
Advair®	Albuterol & Fluticasone	GlaxoSmithKline	asthma *β2 agonist & corticosteroid*	1103 (351%)
Remicade®	Infliximab	J&J/Schering-Pl.	Crohn's, arthritis *IgG mAB*	753 (168%)
Aciphex®	Rabeprazole	J&J/Eisai	ulcers *proton pump inhibitor*	1017 (99%)
Rituxan®	Rituximab	Idec/ Genentech Roche	Non-Hodgkin's lymphoma *IgG mAB*	743 (88%)
Seroquel®	Quetiapine	AstraZeneca	schizophrenia *(5-HT's, D's, H1, α2)*	793 (82%)
Actos®	Pioglitazone	Takeda/Eli Lilly	diabetes *PPARγ agonist*	1151 (79%)
Avandia®	Rosiglitazone	GlaxoSmithKline	diabetes *PPARγ agonist*	1128 (65%)
Diovan®	Valsartan	Novartis	hypertension *angiotensin antagonist*	736 (63%)
Celexa®	Citalopram	Lundbeck/Forest	depression *SSRI's (SERT)*	1107 (61%)

The current therapy leaders are the anti-ulcerants such as Omeprazole (Prilosec®) or Lansoprazole (Prevacid®), which accounted for US $ 19.5 billion in 2001, thus claiming a market share of 6%. The anti-ulcerants are closely followed by the cholesterol reducers, e.g., Atorvastatin (Lipitor®) and Simvastatin (Zocor®), totaling US $ 18.9 billion, the antidepressants (US $ 15.9 billion), and the NSAIDs (nonsteroidal anti-inflammatory drugs; US $ 10.9 billion). Calcium antagonists, antipsychotics, oral antidiabetics, ACE inhibitors, cephalosporins, and antihistamines complete the list of the best-selling therapy classes, with sales between US $ 9.9 and 6.7 billion. It is interesting to note that four classes grew more than 20% in 2001, namely the cholesterol reducers (22%), the antipsychotics (30%), the oral antidiabetics (30%), and the antihistamines (22%) [4].

Since the sales numbers taken from the 2001 analysis only represent a snapshot in time of a considerably dynamic field, a trend analysis with more predictive value can be obtained from the growth performance of products in conjunction with the respective market share (Table 1.2) [4].

Again, proton pump inhibitors such as Esomeprazole (Nexium®), Pantoprazole (Protonix®), and Rabeprazole (Aciphex®) dominate the list of fastest-growing products. The two PPARγ (peroxisome proliferator-activated receptor) agonists Pioglitazone (Actos®) and Rosiglitazone (Avandia®) have already had sales of more than US $ 1 billion each and still show dramatic growth performance. Among the fastest-growing products in 2001 (Table 1.2), two immunoglobulin G monoclonal antibodies are found, namely Infliximab (Remicade®) for the treatment of Crohn's disease and arthritis, and Rituximab (Rituxan®) against non-Hodgkin's lymphoma. It is also noteworthy to mention that Valsartan (Diovan®), one of the first nonpeptide angiotensin II receptor antagonists, goes against the dominant role of the ACE inhibitors in antihypertensive therapy [4].

At first glance, an industry with an associated market worth of more than several hundred billion US dollars, based on products that achieve more than US $ 5 billion annual sales with fast-follower products that display an annual growth performance of several hundred percent might be in good shape with a bright future perspective. However, to maintain the healthcare industry's prospects for sustained growth and to meet the changing needs of a global and aging society, increases in productivity on the order of two- to four-fold are urgently required [5]. Productivity in this context refers to research-intensive innovative drugs for numerous unmet medical needs. But the discovery and development of new medicines is an expensive and time-consuming effort. The Tufts University Centre for the Study of Drug Development found that the time from synthesis of a new drug to US marketing approval has increased dramatically [6]. The Tufts data indicate that this period has increased from ~8 years for approvals in the 1960s to more than 14 years in the 1990s (Figure 1.2).

Lengthening development times dramatically increase the costs of bringing a new drug to market by increasing the capital needed for research and development activities. According to the PhRMA (Pharmaceutical Research and Manufacturers

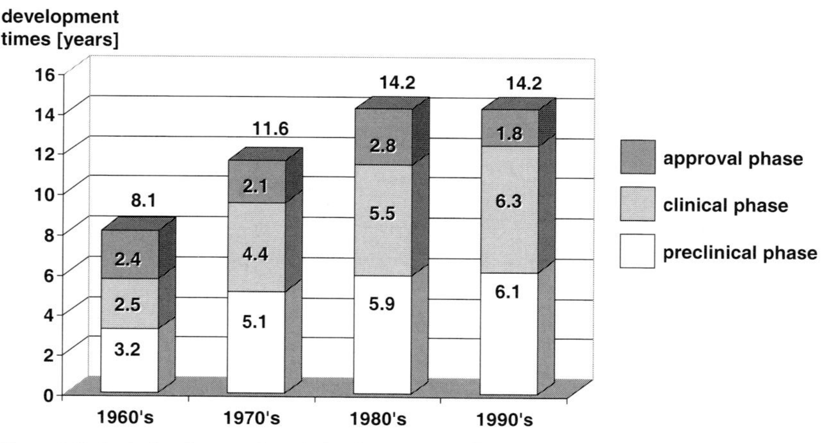

Figure 1.2 Analysis of research and development timelines in recent years [6]. The preclinical research, clinical development, and approval times are explicitly given.

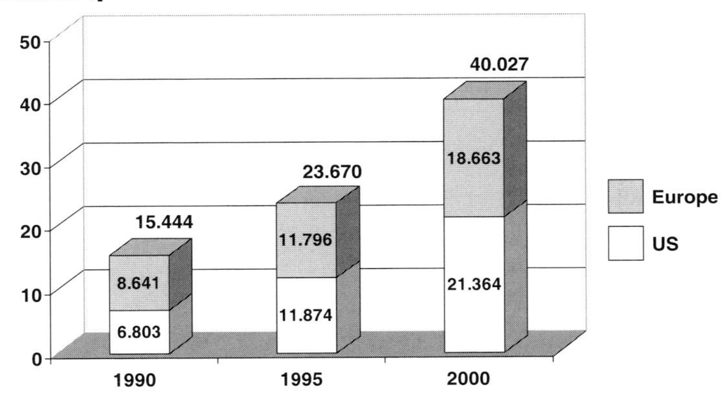

Figure 1.3 Development of R&D expenditures in recent years [7]. Total R&D costs are derived from investments made by US and by European companies, respectively.

of America) Annual Membership Survey 2002 and data from the EFPIA (European Federation of Pharmaceutical Industries and Associations), research and development expenditures in the pharmaceutical industry were greater than US $ 40 billion in 2000 (Figure 1.3) [7].

Despite a steady increase in research and development funding within the pharmaceutical industry, the number of NCEs (new chemical entities) reaching the market has failed to increase over the past decade [8]. Pollack reports in the Business/Financial Desk section of the 19 April 2002 edition of the *New York Times* in an article entitled "Despite Billions for Discoveries, Pipeline of Drugs Is Far From Full", that "Fewer new drugs are being discovered despite constant increase in spending on research and development, causing some to worry that [the] new product pipeline may be running dry; industry's output of new drugs has risen only moderately in [the] last two decades despite [a] more than six-fold increase, after adjusting for inflation, in research and development spending, to more than $ 30 billion annually in [the] last few years, output has actually declined."

Apart from the increasing research and development costs and longer development times, shorter exclusivity times for key products impose novel challenges, if not threats to the overall healthcare industry. Viable Intellectual Property (IP) strategies ensure that invention-based research and development investments are protected. Strong IP protection is essential for the preservation and growth of a research-based enterprise, and thus for the sustained development of new and better medicines. Competition among research-based pharmaceutical companies is continually increasing. One company's patent on a specific drug does not preclude other innovator companies from producing rival medicines to treat the same disease. Increased competition in the search for new and improved drugs has led to a shortening period during which, e.g., a novel blockbuster drug is without challenge on the market. The anti-ulcer drug Tagamet®, introduced in 1977, had 6 years on

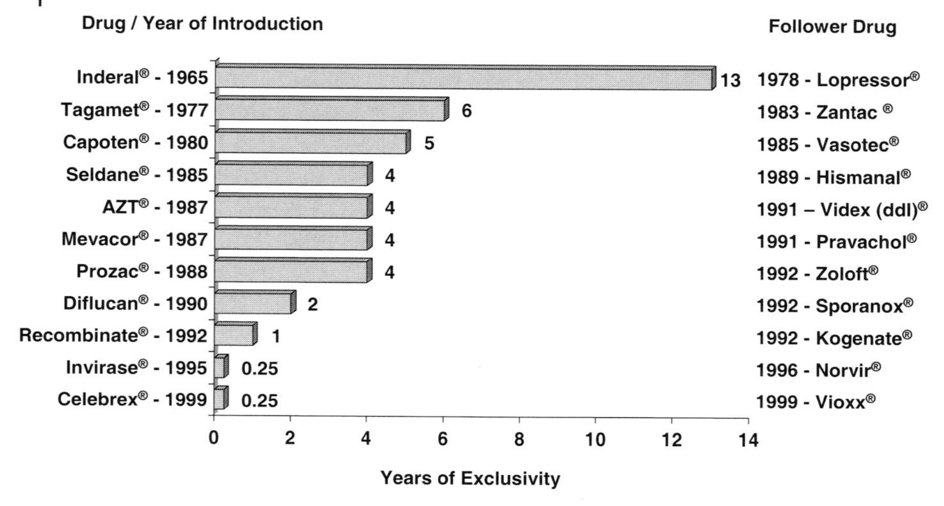

Figure 1.4 Development of exclusivity timelines in recent years [9]. The first-in-class compounds are given on the left, and the corresponding me-too drugs are given on the right.

the market before the follower drug Zantac® was introduced. In contrast, the HIV protease inhibitor Invirase®, released in 1995, was on the market for only 3 months before the first competitor drug, Norvir® was approved (Figure 1.4) [9].

This dramatically shrinking period of market exclusivity clearly represents an often-underestimated challenge for the future of research-based pharmaceutical companies.

To withstand the threats on the record of pharmaceutical research and development productivity, numerous so-called paradigm changes have been announced over the past 10–15 years, all aimed at resolving major bottlenecks along the value chain of drug discovery [10–12].

From the mid 1990s onwards it was widely claimed that the trend toward structure-based molecular design seemed to have fallen well short of expected productivity gains. Concomitantly, molecular biology gained full impact in the early discovery phases of preclinical research, and the accompanying development of high-throughput screening (HTS) and combinatorial chemistry was believed to efficiently remove one of the most persisting bottlenecks in drug discovery, notably the generation of progressible lead compounds [13]. Consequently, the predominant drug discovery process pursued by most pharmaceutical companies was based on large and growing collections of compounds for use in HTS assays. The high degree of sophisticated automation that emerged in the chemical laboratories in pharmaceutical research enabled the chemists to construct large screening libraries in a relatively short time. Organic chemists involved in combinatorial chemistry or automated synthesis were tempted to generate libraries that spanned as large a volume of principally accessible space as possible, corroborated by the introduction of the 'maximum chemical diversity' concept [14]. In this context, assessments of the size of the virtual organic chemistry space containing reasonably sized compounds broadly range from 10^{40} to even greater than 10^{100} [15], with 10^{62} as a commonly quoted middle-range estimate [16]. Given these unimaginably huge numbers, any attempt to systematically scan this virtual molecular diversity space by synthetic means is clearly doomed to failure.

In contrast to the assumption that the number of identifiable leads is related to the degree of molecular diversity encoded in a multimillion compound library, current experience clearly suggests that clinically useful candidates exist as small tight clusters within the molecular diversity space [16, 17]. The number of therapeutically relevant protein targets within the human genome was analyzed to be in the range of 600 to 1500 [18], so aiming at maximal chemical diversity is an inefficient molecular design strategy unless we expect a vast number of yet undiscovered clinically useful targets to be out there.

Despite all advances and investments made in developing HTS technology and combinatorial chemistry concepts, it was recently concluded that HTS has not lived up to what it was hoped, since there is not a significant number of HTS-related INDs (investigational new drugs) [19]. Obviously, the technical compromises and loss of precision that occurred on adaptation of many assays for high-throughput platforms generated high error rates that made hit follow-up by chemistry teams a very laborious, inefficient, and frustrating process. According to a worldwide survey among screening departments, only 43% of targets processed through HTS initiatives generated progressible leads, thus emphasizing that certain target families are HTS-resistant and that success in HTS is a direct correlate to the quality of a compound library [19]. The average success rate of 43% was obtained from an amazing variability ranging from 5% success rate to up to 100%. From 44 HTS laboratories, 326 leads were reported to be found in 2001. However, a 'lead' in this study was defined as a hit, confirmed by more than one assay *in vitro* in a manner that shows biologically relevant activity that correlates to the target of interest. To be a lead, the compound must further show evidence that a structure–activity relationship can be built around it [19]. According to industry-wide accepted standards, a lead is generally characterized more stringently. In most cases, a lead emerges from a chemical optimization program, displaying efficacy in a disease-relevant animal model. Further, a structure–activity relationship should already be elaborated around the respective compound class with promise for achieving a balanced ADME (absorption, distribution, metabolism, excretion) profile before a lead decision is made.

In brief, the stochastic medicinal chemistry approaches pursued over the past 10 years by focusing on HTS in combination with combinatorial chemistry have not met the expectations that were raised at the beginning of this 'big numbers game'. Consequently, the focus of HTS laboratories is now gradually changing toward a concerted effort to improve the relevance and quality of assay operations, in that high-information-content screens are becoming part of the screening philosophy [19]. A steadily increasing number of HTS departments are involved in conducting secondary screens and numerous *in vitro* ADMET (absorption, distribution, metabolism, excretion, toxicology) assays. Also, the compound selection criteria are changing. In contrast to screening a few distinct libraries with tens of thousands of compounds based on, e.g., an identical scaffold, front-loading of drug-likeness and structure- or mechanism-related design approaches are pursued. The concept of 'fewer of many' as opposed to 'many of fewer' offers a substantially increased likelihood of discovering viable lead compounds [17]. The current disillusion with screening huge, diverse libraries [20–23] has led to the tendency to screen libraries that are biased toward our current medicinal chemistry know-how [24, 25]. This reflects both the nonuniform distribution of drugs in chemistry space and the realization that ADMET properties are as relevant as, or even more relevant than, pure target affinities in the search for candidate compounds with realistic therapeutic potential [16, 26, 27]. The current situation in the medicinal chemistry-driven drug discovery disciplines is clearly characterized by a 'knowledge vs. diversity paradox' [28]. The lifeblood of medicinal chemistry is still the combined understanding and improvement of structure–activity and structure–property relationships so as to gear optimization programs toward valuable clinical candidates that possess a balanced activity, selectivity, and ADMET profile [29]. Once this process gains strength, it automatically plays against the molecular diversity paradigm, since the concept of 'similarity' becomes a more successful design principle than the maximal diversity strategies.

After it was realized that neither computer-aided drug design nor the HTS/combinatorial chemistry approaches alone significantly improved the record of pharmaceutical industry productivity as measured by high-quality clinical candidates, the progress made in genome sequencing by the end of the last decade was widely believed to have revolutionizing impact into the research and development area. Although the entire pharmaceutical industry concentrated its efforts on fewer than 500 protein targets with fairly well-elucidated biological functions and disease relevance during the 'pre-genome' era [30, 31], a thorough application of genomics and related technologies was considered to be generating a tidal wave of novel drug targets that would sweep over the pharmaceutical industry with numerous benefits for drug discovery and development in its wake [32]. The knowledge of all genes encoded in the human genome was initially envisioned to provide unprecedented opportunities for the discovery of new drugs with novel modes of action. Headlines such as "Bioinformatics Battles Breast Cancer" [33] suggested an yet undiscovered shortcut on the pathway from a therapeutic hypothesis past the corresponding assays, hits, leads, and pre-clinical candidates, to innovative therapeutic products, and thus to immediate profit. It was only in 2001 that a renowned pharma-consulting company stated that the impact of genomics, if applied rigorously, would help to halve the cost and time it takes to develop a new drug [32]. Surprisingly, in the same year a different consulting firm arrived at a completely contrary conclusion that, rather than improving research and development productivity, the impact of genomics on drug discovery and development is primarily reflected by an increase in costs [34]. The quality of target validation in genomics efforts was identified as a major obstacle. The flood of new targets for discovery programs is not, by far, matched by the required information content about these targets [35]. This lack of knowledge at the outset leads to numerous problems downstream in the drug discovery process and to even worse attrition rates. The analysis further reveals that until 1995, large pharmaceutical companies had up to 50 targets under scrutiny annually. Of these, 70% were preceded with a wealth of associated information known from the literature. Most likely, drugs were already on the market or in late-stage clinical trails against these targets. With the emergence of large-scale genomics approaches, it is estimated that, in 2005, the number of targets per company will increase to ~200 annually, with only 30% of these precedented by any aspects of target validation. In consequence, a significant increase in the absolute number of preclinical development projects will occur. Due to the insufficient validation state of the targets under investigation, a drop in preclinical research success by 25% and a decline in clinical phase II success by 20% are predicted. The result of this costly attrition is that the net present value per NCE will drop dramatically [34].

After a couple of years of progressing into the 'omes' and 'omics' era, one has to recognize that the expected productivity gain has not been achieved. Undoubtedly, genomics approaches produced a vast increase in biologically relevant information. However, the translation of this information into efficient discovery strategies, as opposed to interesting science, has proven elusive.

1.2 Medicinal Chemistry-based Chemogenomics Approach

Although several diverse factors affect pharmaceutical productivity, it is clear that much of the lost productivity can be attributed to the failure of any new technology, by itself, to make the complex discovery process more efficient. As with the stringently applied computer-aided drug design and the combined HTS/combinatorial chemistry approaches, also for all genomics strategies the key hurdle in the way of increased productivity lies not within the associated technology itself, but with how to efficiently organize the implementation of real new paradigms. Only new conceptual thinking, novel assemblies of well-validated technologies, like the ones described in the previous paragraph, within different organizational architectures will be best suited to creating meaningful improvements in research and development productivity.

The highly synchronized orchestration of all the above-mentioned technologies derived from structure-based molecular design – protein modeling, bioinformatics, high-throughput screening, combinatorial chemistry, automated synthesis, genomics, and proteomics – defines the framework for a revised definition of 'chemogenomics', taking advantage of a classification principle within pharmaceutical research that is aligned according to target families, rather than to disease areas.

The emphasis in this contribution is laid on an emerging paradigm in drug discovery, notably the systematic classification of therapeutically relevant target classes according to structure and function. These are subsequently correlated with family-wide recognition motifs that can be translated into lead-like low-molecular-weight ligands, ultimately placing the molecular compound at the heart of a chemogenomics approach. It is the systematic exploration of these densely populated target families with a proven potential to yield drugs that opens up a yet unexploited opportunity to increase preclinical research productivity based on the genomics and proteomics advances of the last years.

This opportunity of systematization of drug discovery strategies follows from accepting that biology as well as chemistry knowledge gained from one target can be transferred to 'adjacent' targets in the same gene family. Even though systematization might require significant commitment of time and resources, it allows enormous efficiencies to be gained through economies of scale, provided that the target families are of significant size, richness, and diversity in therapeutic value. Most importantly, an accumulation of target class-specific know-how is created over time, whereby past experiences allow rapid attack on new members of the target cluster of interest [36]. Comparative analyses over, e.g., inhibitory capabilities, mechanism of action, and even binding modes of whole series of compounds against dozens of members of a distinct enzyme family occur more frequently only in the recent literature, giving first indications of a paradigm change in the way we pursue lead finding and optimization under a chemogenomics perspective [37–39].

From a molecular design point of view, a precious knowledge base including target structure, mechanism, and viable medicinal chemistry approaches toward distinct representatives of densely populated target families is available. For almost any given new emerging target from, e.g., the G protein-coupled receptor, ion channel, nuclear receptor, kinase, or protease family, a medicinal chemistry strategy could be devised even prior to completing a corresponding high-throughput screening campaign. In this context, the concept of target family-directed molecular masterkeys provides an alternative to both blind screening attempts and stringently applied structure-based design approaches. Available knowledge on the structure and/or function of target families is encoded in low-molecular-weight substructures that, upon decoration, deliver high-quality lead structures for further expansion toward viable preclinical candidates. The medicinal chemistry aspects of classification of target family-wide commonalities in ligand recognition for the design of privileged structures and the subsequent application of the masterkey concept to broadly launch into the target cluster represent a chemogenomics approach. Genomics information is directly utilized to drive lead discovery processes.

1.3 Densely Populated Target Families

According to a thorough analysis of drugs listed in the pharmacopoeia in 1996, the total number of proteins within the human organism for which pharmaceutical research had produced drugs is less than 500 [30, 31]. This analysis from the pre-genome era further speculated on the existence of up to 10 000 potentially relevant proteinogenic drug targets within the human genome, albeit irrespective of their biochemical nature. Also in the post-genome era, hit generation and subsequent optimization work based on applied medicinal chemistry are still on the critical path toward a viable preclinical candidate. Consequently, a drug target survey 'from a compound's point of view' seemed to be a more reasonable approach toward the often-mentioned druggable genome. Researchers at Pfizer (UK) systematically mined the human genome for rule-of-five [40] compliant putative drug targets and produced a list of approximately 400 nonredundant proteins that have been shown to bind low-molecular-weight compounds with binding affinities below 10 μM [18]. Not surprisingly, a great percentage of those targets cluster into target families such as G protein-coupled receptors (GPCRs), kinases, proteases, ion channels, and nuclear hormone receptors. Based on the assumption that, once a member of a target family is amenable to rule-of-five compliant compounds, the entire gene family is druggable, a theoretical number of approximately 3000 putative protein targets was derived by systematic extrapolation within the corresponding gene families (Figure 1.5).

Most importantly, the analysis revealed the potentially interesting target classes for which medicinal chemistry is obviously capable of producing low-molecular-weight compounds, even though irrespective of any proven disease relation. The fact that these are multimember gene families allows an enormous opportunity for systematization within the discipline of medicinal chemistry. Privileged structures can be designed that account for a family-wide commonality in terms of enzymatic mechanisms and/or molecular recognition elements. The most densely populated target families, notably kinases, GPCRs, ion channels, proteases, nuclear hormone receptors, and phosphatases, represent attractive fields of activity for medicinal chemistry. Even though approximately 30% of all marketed drugs target GPCRs, approximately 7% of drugs address ion channels, and approximately 4% of marketed drugs bind to nuclear hormone receptors, only 2 drugs address kinase targets (Imatinib (Gleevec®) and Gefitinib (Iressa®)), only 1 drug addresses a Ser-protease (Argatroban (Acova®)), and only 1 metalloprotease is inhibited by marketed drugs (angiotensin converting enzyme (ACE)). From these results it is obvious that there is definitely sufficient if not tremendous room for innovation within the realm of hit generation and optimization.

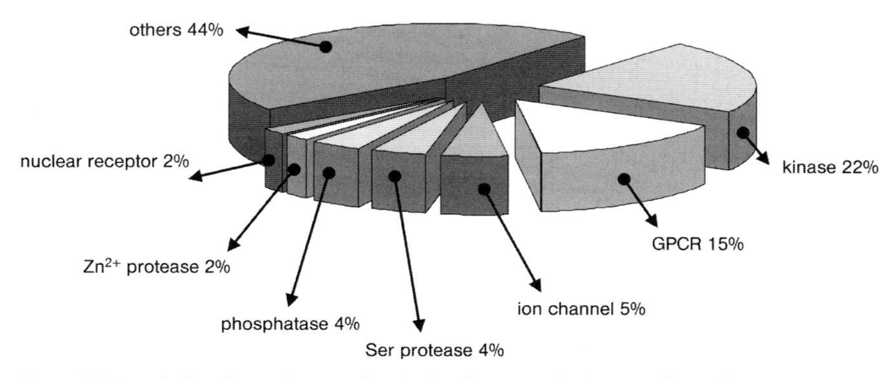

Figure 1.5 Detailed analysis of protein family distribution in the human 'druggable genome' [18].

However, it should be emphasized again that the number of targets that are, in principle, druggable (app. 3000) by far exceeds the number of drug targets, since no aspects of disease relation were considered in this analysis [18]. Our current view of the number of 'druggable drug targets' is far more modest and, thus, realistic as compared with the very optimistic expectations that were spread within the scientific community a few years ago. Within the human genome, the number of pharmaceutically relevant target proteins is estimated to be in the range of 600 to 1500 [18].

1.4 Privileged Structures: A Brief Historical Assessment

The term 'privileged structure' originates from Evans and coworkers at the Merck Sharp and Dohme Research Laboratories, who focused in 1988 on the design of benzodiazepine-based CCK-A antagonists [41], and was later updated by Patchett and Nargund [42] of the same company. Evans et al. [41] refer in their publication to a finding from 1986 by Chang and colleagues [43], who discovered that the previously described analgesic κ-opioid agonist tifluadom [44] also acts as a peripheral CCK receptor antagonist. This documented activity of a single compound at two different target proteins of the same gene family (GPCRs) led Evans and colleagues to conclude that a single molecular framework can provide ligands for diverse receptors. They stated, "What is clear is that certain 'privileged structures' are capable of providing useful ligands for more than one receptor", and later in their paper "judicious modification of such structures could be a viable alternative in the search for new receptor agonists and antagonists" [41].

The concept of privileged structures was occasionally referred to during the mid 1990s, when the use of solid-phase synthesis for generating combinatorial libraries was extended to nonpeptide structures. In a review published in 1996 by Ellman's group, privileged structures were defined as templates that have previously provided potent compounds against a number of different receptor or enzyme targets [45]. It is interesting to note that in Ellman's concept, a clear differentiation is made between privileged structures and so-called designed templates, the latter being based on key recognition motifs for specific protein targets. In this definition, a privileged structure just needs to have a proven record in delivering therapeutically relevant compounds, irrespective of any understanding of what makes a structural element privileged, while a designed template bears encoded information of a specific molecular recognition principle [45]. The syntheses of a dozen library designs are summarized under the privileged structure–templated approach (Figure 1.6) [45].

It is evident that all described privileged structures are heterocyclic scaffolds with various ring sizes, heteroatom distributions, and substitution patterns. The synthesis of these scaffolds was considered to be achievable by means of solid-phase chemistry, thus yielding high-dimensional libraries. No emphasis is laid on whether the template itself contains pharmacophoric elements or is just a scaffold with

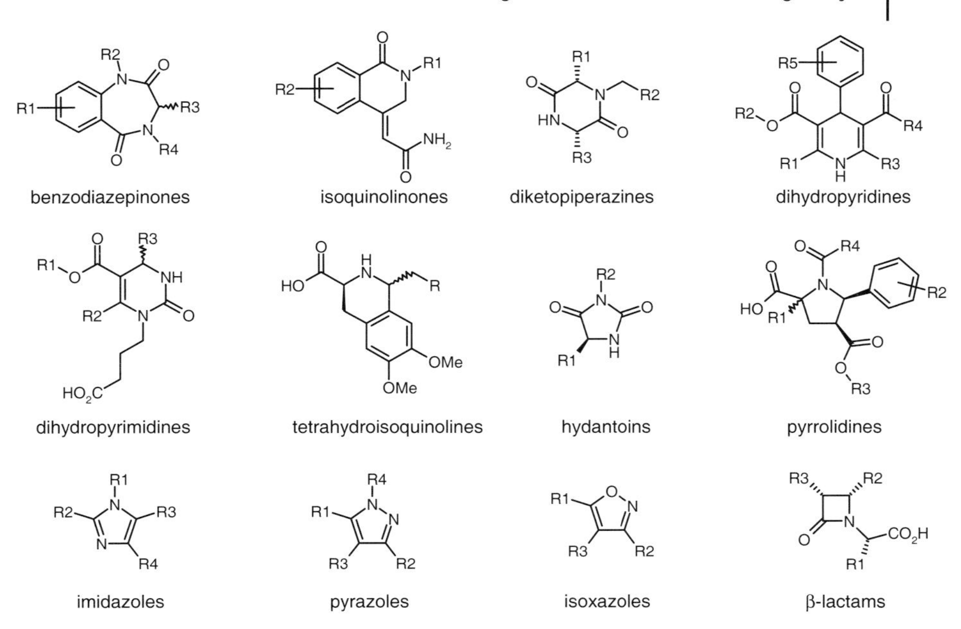

Figure 1.6 Structural scaffolds that served as templates for so-called privileged structure-based combinatorial compound libraries. R-groups are depicted at positions of combinatorial variations [45].

appropriate orientational characteristics for peripheral groups that contain the pharmacophoric elements required for target binding. It is obvious that in these days the opportunity to enrich the toolbox of solid-phase synthesis for library generation dominated the conceptual considerations, rather than a detailed understanding of structural determinants governing, e.g., details of family-wide conserved ligand recognition principles or improving lead- or drug-like properties of the library constituents.

1.5 Privileged Structures with Undesired Target Profiles

The most generic definition of a privileged structure refers to substructural elements emerging in compounds that showed effects on more than one target protein [41], irrespective of the corresponding target families they might belong to. This specific characteristic of compounds that are discovered, e.g., as hits in numerous different biological assays covering a broad range of protein targets, is not necessarily a desired profile. The elimination of so-called 'frequent hitters' from compound libraries was recently described by a group at Hoffmann-La Roche [46], since those compounds were shown either to bind nonspecifically to a variety of targets or to interfere with the utilized assay read-out methods. These compounds were clearly considered as undesirable starting points for optimization programs in medicinal chemistry. Obviously, a differentiation between these promiscuous binders and privileged structures is required, thus refining the original definition of Evans et al. [41]. This is further supported by recent studies on underlying mechanistic phenomena from Shoichet's group at Northwestern University in Chicago [47]. An in-depth study of screening hits that appear to be not drug-like with a noncompetitive mode of action and contradictory structure–activity relationships revealed a common mechanism that accounts for that undesired compound profile. The investigated compounds tend to form molecular aggregates, as determined by dynamic light scattering and electron microscopy, with particle sizes of 30 to 400 nm in diameter. It is noteworthy that this phenomenon is not restricted to compounds that a trained medicinal chemist would classify as not drug-like, but also occurs for drug-like molecules such as steroids and kinase inhibitors [48] and even for known drugs (Figure 1.7) [49].

In a comparative study, 50 unrelated drugs were assayed for inhibition of β-lactamase, chymotrypsin, and malate dehydrogenase, although none of these enzymes were considered targets of the selected drugs. Out of these 50 drugs, 7 compounds were identified as behaving as aggregation-based inhibitors (Figure 1.7). Further mechanistic studies revealed that the observed nonspecific inhibition resulted from

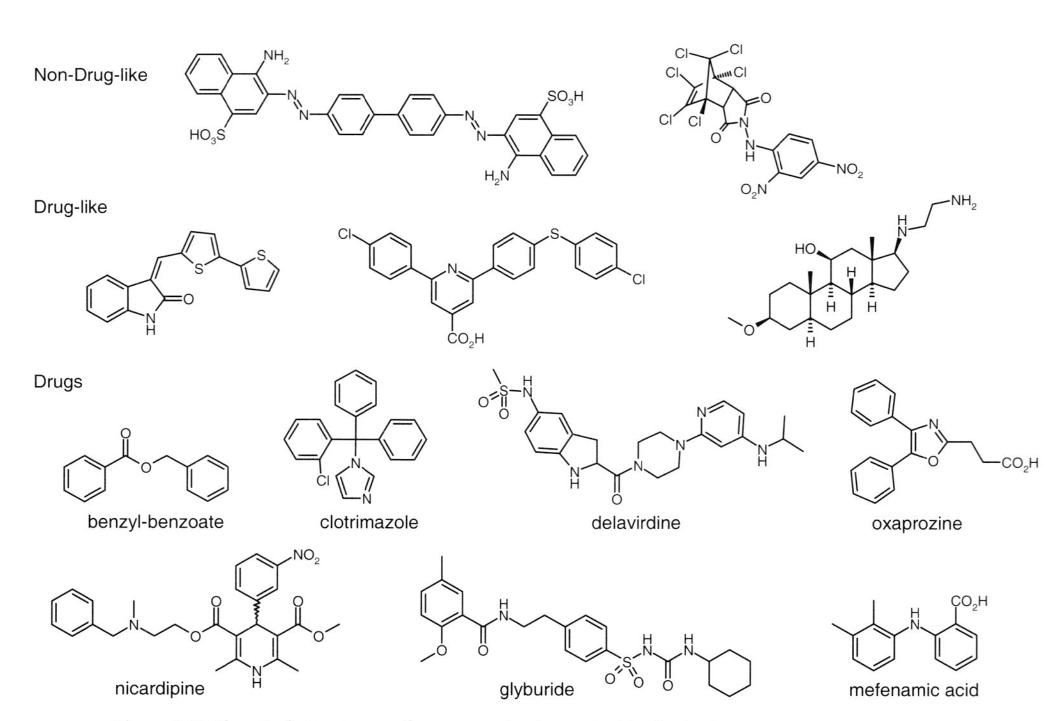

Figure 1.7 Chemical structures of compounds shown by biophysical investigations to form molecular aggregates, thus eliciting false-positive biological activities in a variety of biochemical assays [47–50].

the reversible adsorption of enzymes to the surface of the aggregates, formed by the promiscuous inhibitors, as shown by electron microscopy [50]. By this specific mechanism, the aggregate-forming inhibitors reversibly sequester enzyme from the assay system, resulting in apparent inhibition. It was further shown that this enzyme–aggregate adsorption could be reversed and even prevented by the addition of detergents [50, 51]. The findings that high-quality compounds also tend to elicit nonspecific biological activities by forming aggregates not only renders numerous screening hits and associated optimization programs highly questionable, but also defines a principally new de-selection criterion for hit and lead assessment, which requires biophysical investigations before any significant medicinal chemistry resource assignment.

1.6 File Enrichment Strategies with Recurring Substructures

Although the studies mentioned above revealed undesired types of 'privileged structures', numerous investigations have focused on the identification of desirable privileged structural elements. Even though this contribution is not aimed at reviewing those studies, two early approaches should be mentioned because they can be seen as pioneering studies that provided guidelines for the medicinal chemist for the fragmentation of compounds into core structures and peripheral decoration and for how this approach might drive chemistry programs based on privileged molecular fragments. In this context, the interested reader is referred to the excellent work of Bemis and Murcko from Vertex [52, 53] and of Lewell and colleagues from Glaxo Wellcome [54, 55]. Both studies reveal frequently recurring substructural elements that can be employed for proactively enriching as well as focusing chemistry efforts toward the more productive regions of multidimensional molecular diversity space. A variety of further in-silico tools have been developed since then, as exemplified by a recent study of Sheridan of the Merck Research Laboratories, who developed a method to identify molecular substructures that are associated with, e.g., a therapeutic area or a mechanism-based biological activity [56].

Fesik and his group at Abbott Laboratories pursued an experimentally based procedure of identifying fragments with generally high propensity for protein binding [57] that was conceptually modeled after the Vertex [52, 53] and Glaxo Wellcome approaches [54, 55] that relied on in-silico database mining methodologies. NMR-based screening of more than 10 000 selected fragment-type compounds against 11 target proteins revealed 12 privileged substructures that appeared with statistical significance in compounds that were shown to bind to the selected targets.

The main conclusion from this study refers to the preferential utilization of those substructures in combinatorial libraries with the aim of qualitatively enriching an in-house screening compound collection. The quality criterion, of those structural elements of being privileged, relates only to the observation that compounds containing one of the identified fragments might bind to an unspecified target with higher than average probability. A clear rationale as to why those fragments tend to bind and a fragment-target family relation are not apparent.

1.7
Recurring Structures Devoid of Target Family Correlations

Once the focus of privileged structures is laid on conformationally constrained core structures, any trained medicinal chemist can easily identify recurring structural motifs from his own work or from literature studies. The benzodiazepine scaffold [58], for example, is believed to mimic a rigid β-turn peptide conformation that might exhibit prominent orientational characteristics for a pharmacophore-encompassing molecular periphery. Such structural features, combined with the plausibility of introducing three to four points of diversity onto the corresponding scaffold during its construction, have made numerous scaffolds appealing substructures for, e.g., combinatorial library synthesis, and thus are easily called privileged structures (see Figure 1.6) [45, 58, 59]. Indeed, these types of templated libraries emerged as prolific sources of hits against a broad range of enzyme and receptor targets, but the primary rationale for the generation of those focused

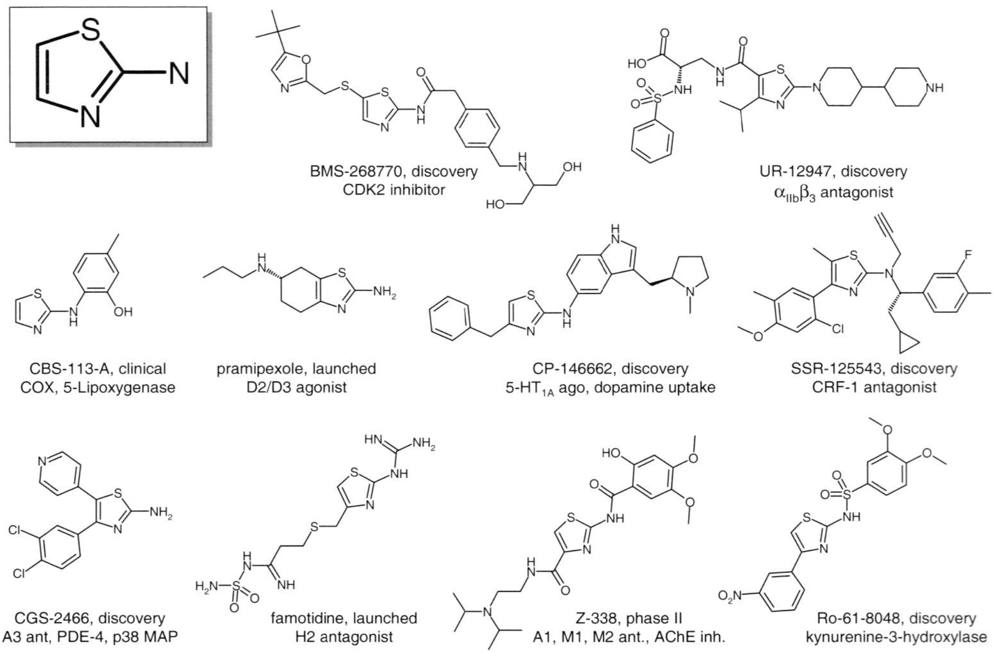

Figure 1.8 Chemical structures of derivatized 2-aminothiazoles displaying biological activities on numerous target proteins.

compound files always was a common synthetic route, rather than an explicit pharmacophore pattern comparison across the target types to be screened.

This is exemplified by the 2-aminothiazole core that is found in numerous drugs, as well as clinical and preclinical candidates, addressing a broad spectrum of targets [3]. No target family correlation is evident; instead, the compounds bind to enzymes such as cyclooxygenases, phosphodiesterases, kinases, acetylcholinesterase, and numerous members of the GPCR family and integrins (Figure 1.8) [3]. Most likely, the versatile chemistry approaches delivering a decorated 2-aminothiazole-derived compound are the main reason for this scaffold to appear as a recurring structural motif in compounds targeting members of numerous different gene families.

As outlined above (Figure 1.6) [45], oligo-substituted five-membered heterocycles with a conserved vicinal (1,2) di-phenyl substitution pattern are ideal representatives of a recurring core structure that is found in numerous biologically active compounds, including cyclooxygenase inhibitors, kinase inhibitors, GPCR antagonists and even agonists, phosphatase inhibitors, and dopamine transporter inhibitors (Figure 1.9) [3].

Since the spatial extent of the common underlying 1,2-di-phenyl substituted heterocycle by far exceeds that of a generic scaffold, and the nature of the underlying

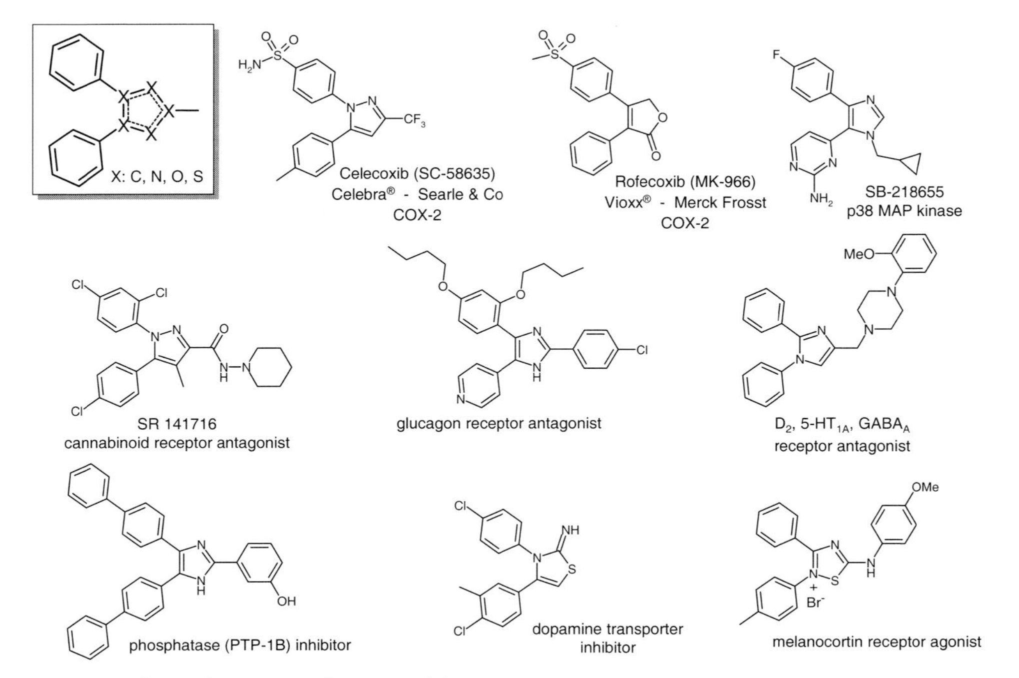

Figure 1.9 Chemical structures of enzyme inhibitors, receptor agonists, and antagonists that all refer to a common underlying molecular topology (box in upper left) consisting of a central five-membered heterocycle with at least two aromatic rings attached in a vicinal arrangement.

heterocycle is quite diverse, a versatile chemistry can be ruled out as the main reason why this structure type frequently occurs in biologically interesting low-molecular-weight compounds. Even though crystallographically derived structures for a variety of these compounds in complex with their target proteins are known, a structural interpretation on the privileged status of the common fragment remains unclear. The heterocycle bearing two adjacent phenyl rings in a vicinal relation clearly prevents the two aromatic rings from hydrophobic collapse, thus representing an orientational variation of the frequently occurring diphenylmethane moiety [60] within biologically active compounds. This nicely indicates that chemical similarity does not necessarily correspond to biological similarity.

The tricyclic neuroleptics and antidepressants, classical pharmaceutical textbook compounds, represent an amazing example of compounds with a high degree of chemical similarity displaying a bewildering array of biological activities. A search in drug databases based on the generic structure depicted in Figure 1.10 reveals more than 150 released drugs, with Zyprexa® (Table 1.1 and Figure 1.1), number six on the list of best-selling drugs in 2001, among them. None of these drugs seems to display a 'clean' target profile; instead, the desired neuroleptic and antidepressant activity is predominantly achieved by antagonistic activity against an array of biogenic monoamine receptors from the GPCR family, in addition with, e.g., serotonin-uptake inhibitory activity (Figure 1.10).

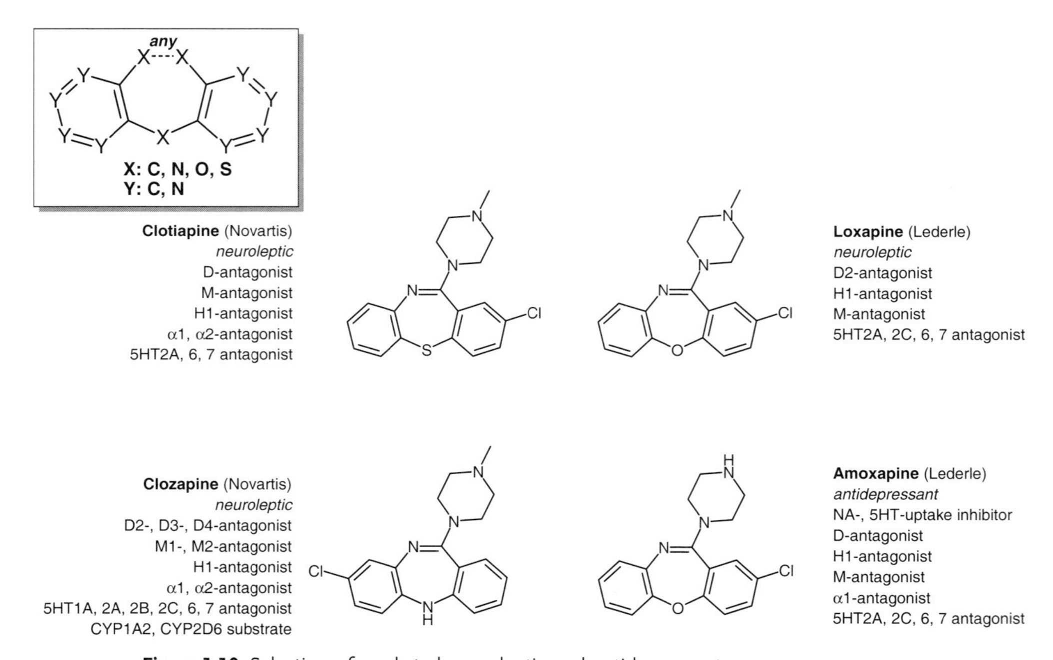

Figure 1.10 Selection of marketed neuroleptic and antidepressant drugs belonging to the class of 'tricyclics'. All depicted drugs have a common underlying core structure (box in the upper left). The receptor and enzyme targets for each compound are given explicitly.

Anti-inflammatory and antihistaminic activity is encoded by a more selective receptor antagonist profile, as exemplified by Olopatadine and Loratadine, respectively (Figure 1.10). Based on the compounds shown in Figure 1.10, a target family correlation might emerge, in that these compounds mainly address the classical neurotransmitter-binding GPCRs, apart from a few other target proteins. However, the database search mentioned above not only revealed compounds depicted in Figure 1.10, but also a broad range of related analogs, individuals of which bind quite selectively to a whole spectrum of enzymes and receptors (Figure 1.11).

Based on these findings, the question remains as to how a medicinal chemistry setup could take advantage of these results in future programs. Since there is no detailed insight in, e.g., a highly conserved compound–target interaction mode for the structural elements described above, these findings can only serve to guide, e.g., combinatorial chemistry initiatives or compound acquisition so that more emphasis is laid on similarity to these recurring fragments, instead of undertaking the attempt to systematically scan the molecular diversity universe.

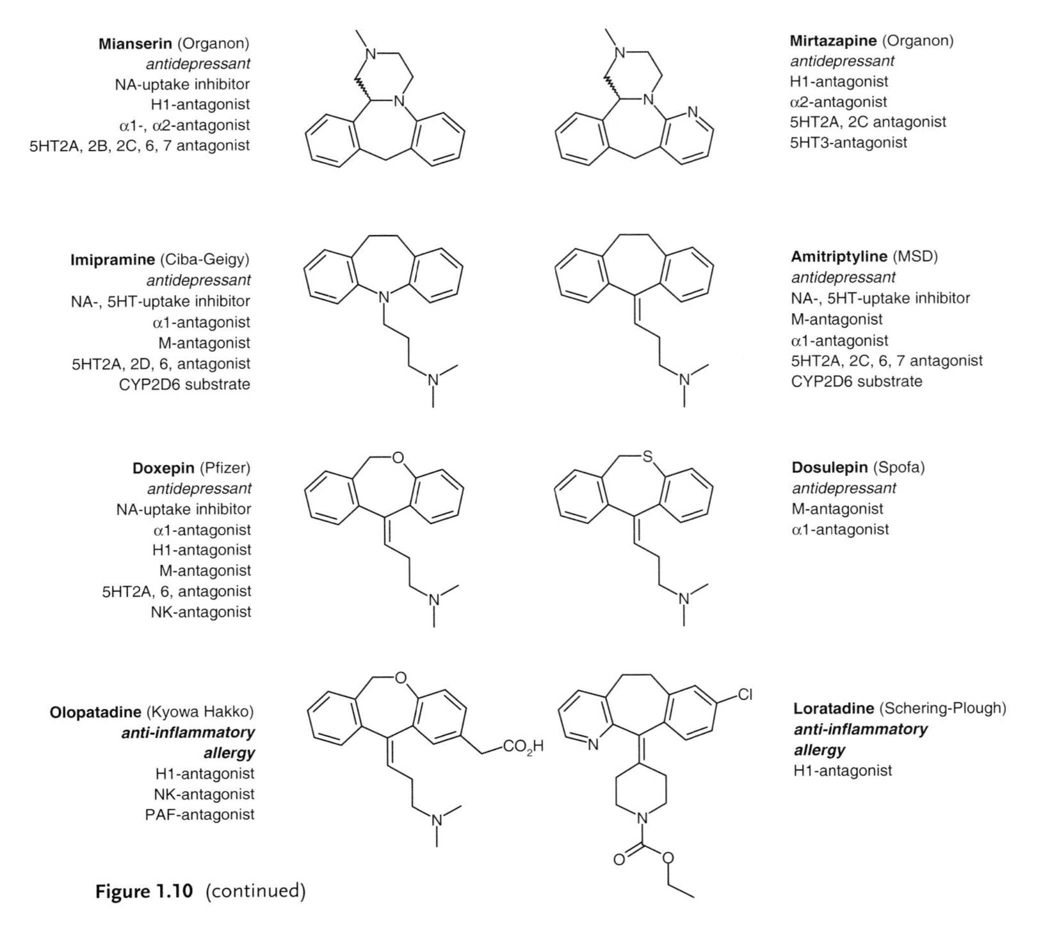

Figure 1.10 (continued)

GPCRs

Epinastine (Boehringer Ing.)
anti-inflammatory
allergy
H1-antagonist
"launched"

(Merck & Co.)
hypertension
ET-A antagonist
"discovery"

L-640,035 (Merck Frosst)
anti-thrombotic
prostanoid receptor antagonist
"discovery"

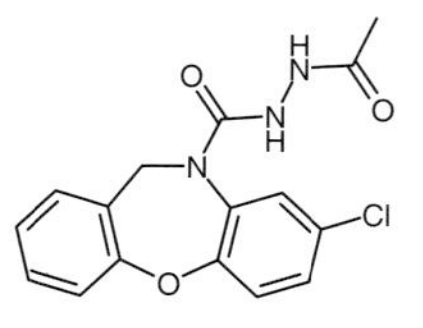

SC-19220 (Searle)
analgesic
PGE1 antagonist
"discovery"

channels & receptors

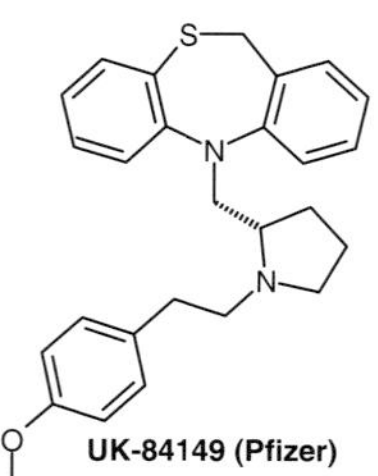

UK-84149 (Pfizer)
gastric motility disorder
Ca^{2+} channel blocker
"discovery"

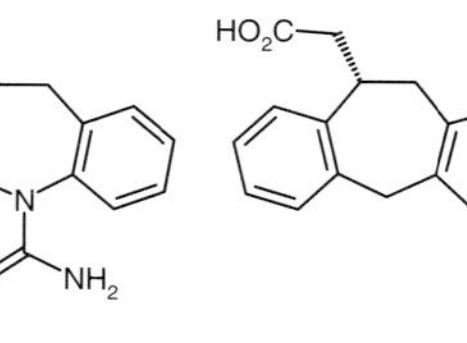

Oxcarbazepine (Novartis)
analgesic
Na^{+} channel blocker

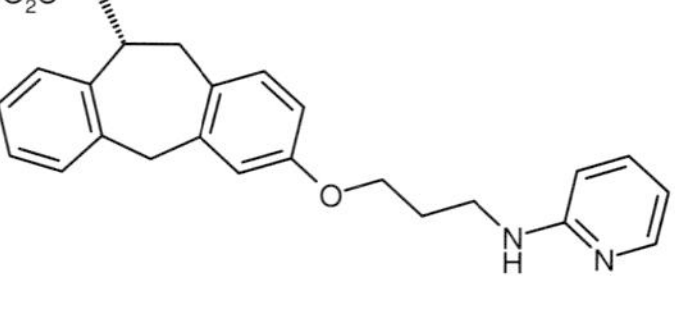

SB-265123 (GlaxoSmithkline)
cancer, osteoporosis
$\alpha_v\beta_3$ antagonist
"preclinic"

HX-600 (Nikken)
cancer
retinoic acid receptor
(RAR) antagonist
"discovery"

enzymes

Nevirapine (Boehringer Ing.)
anti-HIV
RT inhibitor

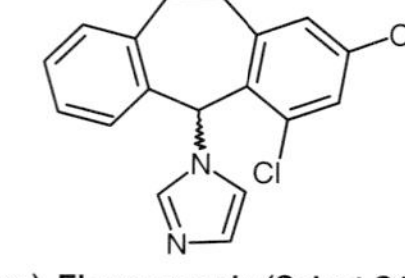

Eberconazole (Salvat SA)
fungicide
ergosterol synthesis inhibitor
"pre-registration"

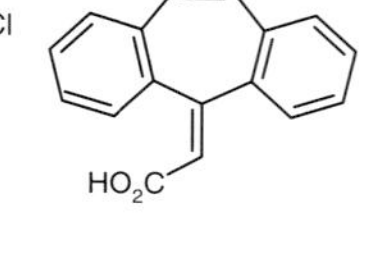

WY-41770 (Wyeth)
pain, inflammation
COX inhibitor
"discovery"

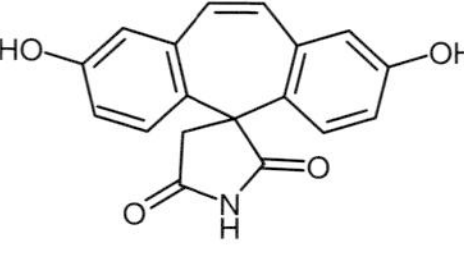

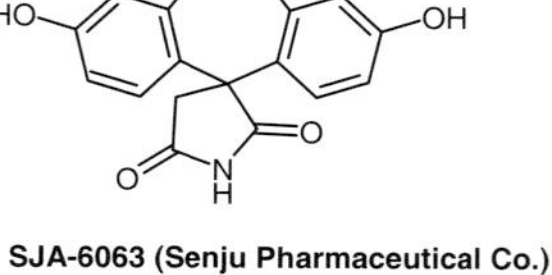

SJA-6063 (Senju Pharmaceutical Co.)
diabetes
aldose reductase inhibitor
"discovery"

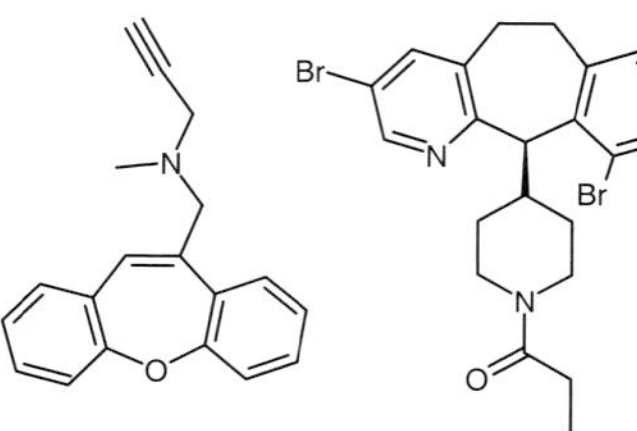

TCH-346 (Novartis)
Alzheimers, Parkinson
GAPDH inhibitor
"phase 2 clinics"

Lonafarnib (Schering-Plough)
cancer
RAS FTase inhibitor
"phase 2 clinics"

KF-17828
(Kyowa Hakko Kogyo Co.)
hyperlipidemia
ACAT inhibitor
"discovery"

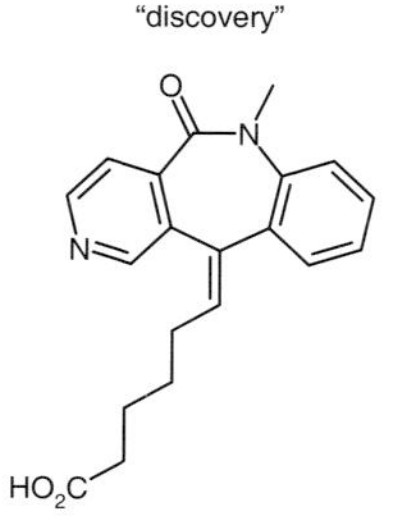

KF-13218
(Kyowa Hakko Kogyo Co.)
anti-inflammatory
thromboxane
synthetase inhibitor
"discovery"

Figure 1.11 Chemical structures of GPCR antagonists (top), channel blockers and antagonists for receptors other than GPCRs (middle), and enzyme inhibitors (bottom) that contain the same tricyclic skeleton as shown in the structures in Figure 1.10.

1.8
Convergent Pharmacophores for Target-hopping

The idea of proactively reusing already-established inhibitor, agonist, or antagonist concepts for a second lead-finding initiative aimed at another target is occasionally described in the literature. At the end of the 1990s, researchers from Rhône-Poulenc Rorer reported a novel arylsulfonylhydroxamic acid template (Figure 1.12) as an underlying structure for a scaffolded combinatorial library [61, 62]. Subtle changes in functional decoration displayed the necessary pharmacophoric patterns for inhibition of members of either of two different target families, namely the matrix metalloproteases (MMPs) and the phosphodiesterases (PDEs), respectively.

Even though there is little if any apparent structural similarity between the natural substrates of the two enzyme classes – peptide sequences for MMPs vs. cyclic nucleotide monophosphates for PDEs – there is an obvious element of convergence of pharmacophoric arrangements. Minor changes in the backbone and aromatic ring substituents yielded class-specific compounds (Figure 1.12). Apart from the initial achievement of generating compounds that discriminate between the target classes, nondiscriminating molecules were considered to constitute an intriguing opportunity for developing dual inhibitors of MMPs and PDE4, based on the

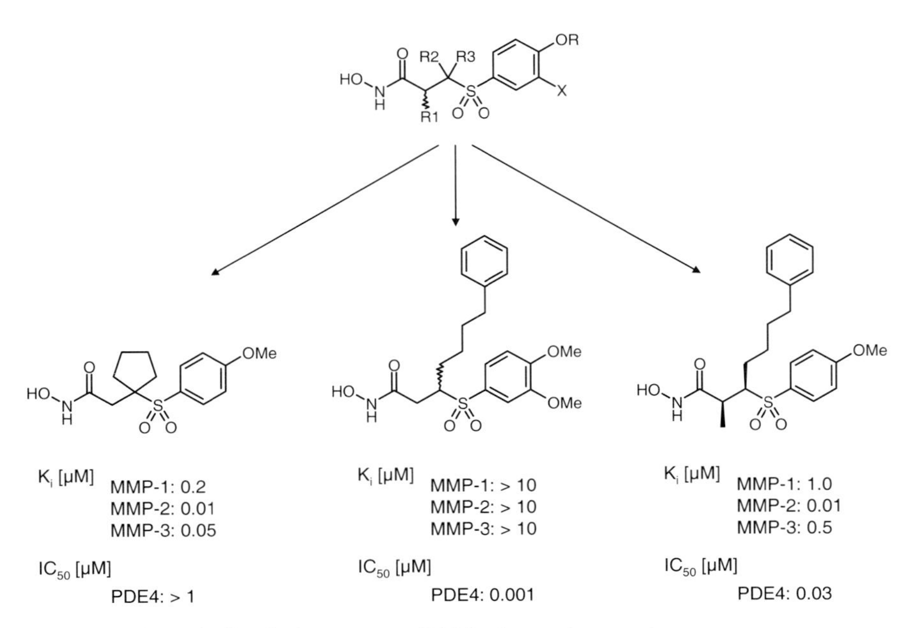

Figure 1.12 The arylsulfonylhydroxamate scaffold (top) served as template structure for compounds active on various members of the matrix metalloprotease family, as well as of phosphodiesterase 4. By variation of the molecular periphery (R-groups, X; top), discriminating compounds (bottom left, bottom middle), as well as dual inhibitors (bottom right) were obtained [61, 62].

rationale that MMPs and PDE4 are key intervention points in a variety of inflammatory diseases [62]. By fine-tuning the decoration pattern of the common arylsulfonyl-hydroxamate template, compounds with well-balanced inhibitory profiles for MMP-1, MMP-2, MMP-3, and PDE4 were identified (Figure 1.12) [62]. The emphasis in these studies was primarily on convergence of a template design and utilized combinatorial chemistry principles to maximize the impact of a single established synthetic route in delivering compounds with more than a single application. Although the pharmacophore relation between the two target families addressed in these studies remains unclear, a certain element of convergence emerges with the strategy of generating dual inhibitors against different targets for the treatment of the same pathologies.

A research team at Sterling Winthrop aiming to develop low-molecular-weight bradykinin B2 receptor antagonists used a well-defined element of convergence, notably a proteolytic enzyme cleaving two different oligopeptides that specifically bind to distinct receptors [63]. The metalloprotease angiotensin-converting enzyme (ACE) cleaves angiotensin I to produce the vasoconstrictive angiotensin II within the blood pressure controlling renin–angiotensin-system. But the hypotensively active nonapeptide bradykinin is also cleaved by ACE to yield inactive products. Since the enzyme ACE obviously recognizes both peptide sequences, a convergent conformational relation exists between the two peptides. Consequently, ACE inhibitors may also display structural features similar to those of bradykinin at its ACE cleavage site. Based on this hypothesis, ACE inhibitors should also show binding potential to the bradykinin B2 receptor, a member of the GPCR superfamily. This last hypothesis is based on the assumption that the bioactive conformations of bradykinin in the active site of ACE and the binding pocket of the B2 receptor are identical, or at least highly similar. Based on preliminary structure–activity relationships obtained from bradykinin-derived peptide analogs, an aromatic moiety is required for B2 receptor binding at amino acid position 8 (P1′ as ACE substrate), together with two terminal arginine sidechain-borne positive charges (Arg1 and Arg9) that span a distance of roughly 10 Å. The authors projected these pharmacophoric groups onto a classical ACE inhibitor, i.e., Quinalapril, to obtain a hybrid compound consisting of the ACE inhibitor skeleton as core, with bradykinin-specific decoration elements attached to it (Figure 1.13) [63].

The Tic moiety was retained and the Zn^{2+}-coordinated carboxylate group of the phenylbutanoyl fragment was removed to facilitate the incorporation of one of the terminal charged groups. The resulting dipeptide analog showed a submicromolar affinity as a B2 receptor antagonist. In this study, the convergent pharmacophore of two apparently different bioactive compounds was ascribed to conformational similarity, derived from the finding that both peptides served as substrates for the same enzyme.

A group at Fujisawa Pharmaceutical aiming at the discovery of nonpeptide bradykinin B2 receptor antagonists also exploited the suggestive structural correspondence between the same two peptides [64]. The applied lead-finding strategy focused on a primary screen of only 300 compounds carefully selected from a previously pursued angiotensin II antagonist program. From that, a weak hit was identified bearing a cyanopyrrolyl-phenyl substituent, an isoster of the biphenyltetrazole moiety that became almost the trademark of nearly all marketed nonpeptide angiotensin II antagonists (Figure 1.14).

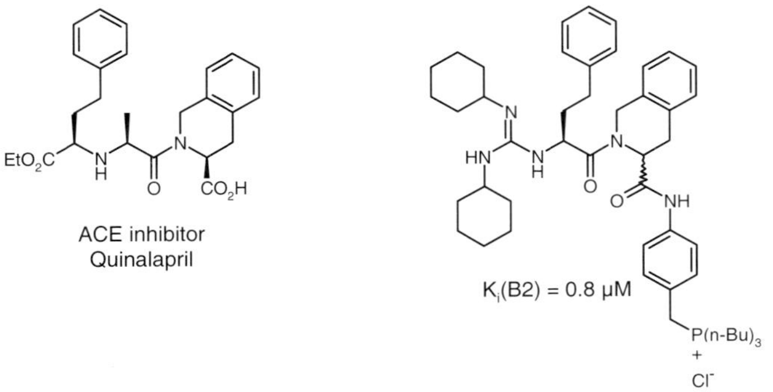

Figure 1.13 A suggestive structural correspondence between the ACE inhibitor Quinalapril (left) and bradykinin was used to design novel B2 receptor antagonists, one of them shown on the right [63]. For details of this convergent pharmacophore strategy, see the text.

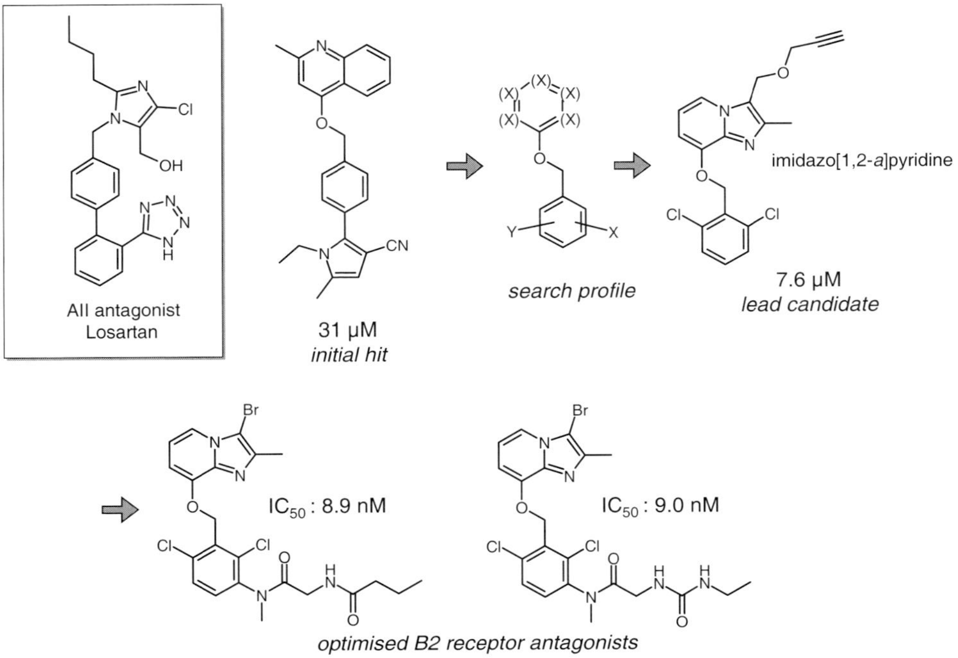

Figure 1.14 Schematic illustration of the B2 receptor antagonist design approach utilized by Fujisawa Pharmaceutical [64–66].

Based on that high-micromolar hit, the benzyloxy-heteroaromatic substructure was defined as a constant core for a virtual database screening that revealed 400 compounds containing the desired structural element of the search profile. From screening of these compounds, the benzyloxy-substituted imidazo[1,2-a]pyridine core (Figure 1.14) emerged as a promising candidate that was synthetically expanded into a lead finding and optimization program [65]. Highly active B2 receptor antagonists were identified possessing promising pharmacokinetic properties that were clearly superior to those of peptide-based candidates [66]. This work represents a textbook example of how a target similarity (AII receptor vs. B2 receptor) and a convergence point of two distinct proteolytic cascades of two different peptides were conceptually overlaid on ligand similarity. Once established, the concept provided the basis for rapid lead finding and optimization in a follow-up project, reusing molecular scaffolds and associated chemistries. Starting with a hit structure encoding the convergent pharmacophore, specific compounds were obtained by subsequent rounds of optimization. From a conceptual viewpoint, these studies represent a rudimentary interpretation of the privileged structure-based masterkey philosophy [3], since two members of the same target family were addressed specifically by series of compounds that can be traced back to common precursor chemotypes. However, those structural elements that encode the family-wide commonality in ligand binding, i.e., the ultimate privileged structure, and those that account for the final target selectivity still remain unclear.

Although in the previous case studies the convergent pharmacophore could be ascribed to a target similarity and a common binding event to exactly the same enzyme, in the following example common co-substrates serve as the cross-relating entities defining common pharmacophoric patterns and associated chemistries among different targets. The enzymes of interest in these studies are carbohydrate sulfotransferases as potential anti-inflammatory targets [67, 68] that catalyze the transfer of a sulfuryl group from the sulfate donor 3′-phosphoadenosine-5′-phosphosulfate (PAPS) to a hydroxy or amino group of an acceptor saccharide. This co-substrate very closely resembles ATP, which is the phosphate donor for all kinase-catalyzed phosphorylation reactions, thus establishing a relation of a sulfotransferase inhibitor project to known kinase inhibitor concepts (Figure 1.15).

The fact that the hydrophobic adenine binding pocket of crystallized estrogen sulfotransferase [69] and of the heparin N-sulfotransferase [70] are similar to those of kinases was also taken into consideration. On the basis of these parallels, kinase inhibitors were screened for cross-reactivity with carbohydrate sulfotransferases. Compounds from these purine-based libraries (Figure 1.15) displayed, at 2.4 μM PAPS ($0.5 \times K_M$), inhibitory activities in the range of 20 to 40 μM against the GlcNAc-6-sulfotransferase NodH [71]. Following the same concept, active inhibitors against the estrogen sulfotransferase were also identified in purine-based libraries [72] (Figure 1.15).

These few examples were chosen to illustrate different elements of convergence that can be exploited to utilize suggestive pharmacophore relations for cross-fertilizing medicinal chemistry projects on targets in the same or different protein families. However, the applied pharmacophoric relations did not reveal generic

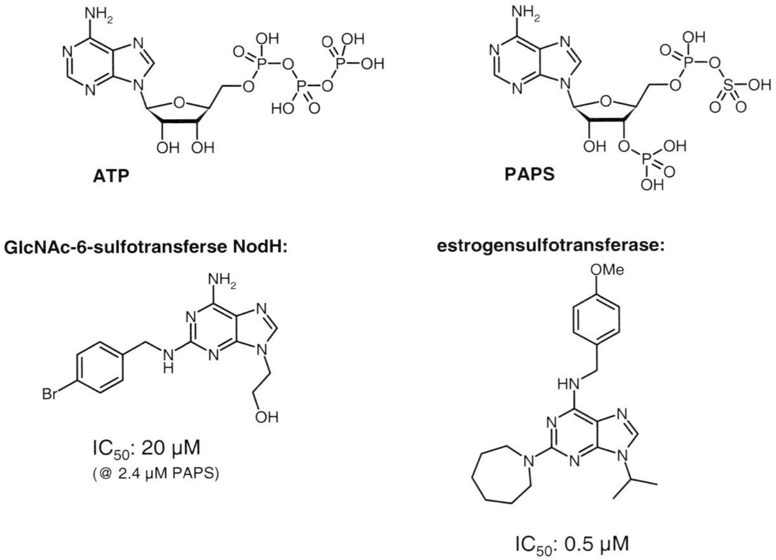

ATP

PAPS

GlcNAc-6-sulfotransferse NodH:

IC_{50}: 20 μM
(@ 2.4 μM PAPS)

estrogensulfotransferase:

IC_{50}: 0.5 μM

Figure 1.15 Co-substrate similarity between ATP (top left) and PAPS (top right) initiated a directed search for carbohydrate sulfo-transferase inhibitors based on kinase inhibitor compound collections [67, 68, 72]. The resulting sulfotransferase inhibitors together with their inhibitory activities are shown at the bottom.

privileged structures with appropriate functional decoration patterns that would allow systematic exploration of a target family in a chemogenomics perspective.

1.9 Target Family-directed Masterkey Concept

The target family-directed masterkey concept [3] represents the most rational and stringent application of privileged structures and is tailor-made to a systematic exploration of entire target classes by a once-established medicinal chemistry concept. A privileged structure in this context is a substructural element with a proven correlation to a target family. It encodes a single or a variety of key structural elements that account for a target family-wide commonality in ligand binding. Illustrative examples of this quality of privileged structures are reverse-turn mimics with appropriate functional groups for decoration that simulate a specific peptide-derived backbone conformation required for a peptide sequence to be recognized by a variety of receptors [73, 74]. Although the skeleton of the turn mimic itself is the molecular imprint of a common underlying recognition principle, thus being the privileged structure, the peripheral decoration with selected pharmacophoric groups ensures finally achieving the required selectivity for distinct peptide-binding receptors. Since the recognition of peptide-encoded pharmacophores in reverse-turn conformations is a recognition principle widely spread over several target families (e.g., integrins [75], GPCRs [76], SH2 domains [77]), the target family bias of a corresponding masterkey concept is less pronounced.

The protein superfamily of proteases [78, 79], however, is an ideal framework for a directed privileged structure-based masterkey concept. It has already been reported that the 5,5-*trans*-fused lactam moiety was systematically optimized and explored as a serine protease-directed scaffold by GlaxoSmithKline and has delivered progressible lead compounds for various members of that target class [3], such as thrombin [80, 81], elastase [82, 83], HCMV protease [84, 85], and the hepatitis C virus-encoded NS3-4A protease [86, 87]. Here, the initially identified scaffold was engineered toward the serine protease-wide commonality in substrate binding and processing [3].

Proteases in general, and cysteine proteases [88] in particular, still represent a major challenge for lead identification approaches, since these enzymes have turned out to be resistant to, e.g., HTS-based lead-finding initiatives [78]. Consequently, the family of cysteine proteases provides an ideal framework for the elaboration of a masterkey concept, also because extensive family-wide characteristics in substrate binding and processing are known. In general, a minimal fragment of 4–6 residues of a peptide sequence is bound in an extended conformation, while all cysteine proteases work with a direct nucleophilic attack on the scissile peptide bond with the thiol sidechain of the active-site cysteine residue [89]. Based on this enzymatic mechanism, the majority of inhibitor design principles established in the past years employ irreversible alkylation reactions of the catalytically active thiol group with, e.g., α-haloketones, α-diazoketones, epoxides, or vinyl sulfones [89]. Since irreversible alkylation has tremendous toxic potential, due to nonspecific alkylation of other biomolecule-encoded nucleophiles, reversibly modifying compounds are considered as inhibitors of choice. Aldehydes, ketones, nitriles, and α-ketoesters also bind covalently to the cysteine sidechain, but in a reversible reaction [78, 89]. Due to the chemical nature of these classical protease-directed warheads, the corresponding, mostly peptidomimetic, inhibitors allow only either the left-side (unprimed) or the right-side (primed) areas flanking the cleavage site of the peptide-binding canyon within a protease structure to be addressed [90]. To achieve sufficient binding affinity and, more importantly, the required target selectivity, active-site-spanning cysteine protease inhibitors, as already developed and marketed for aspartic protease inhibitors, are most preferred. They allow one to address binding pockets at will on both sides of the catalytic centre. For that purpose, a bifunctional building block including the mechanism-directed warhead in the central part of the privileged structure allows not only addressing the thiol sidechain by reversible complex formation, but also expanding the structure in both the unprimed and the primed direction of the recognition pocket. In this context, structural analogs of the 1,3-diaminopropanone moiety have emerged as versatile cysteine protease inhibitor designs that reversibly form a hemithioketal when bound to the target enzyme (Figure 1.16). Further, they can be appropriately modified on both amino groups to explore binding epitopes all along the binding channel. Researchers at Merck Research Laboratories first described the concept in 1994 for the synthesis

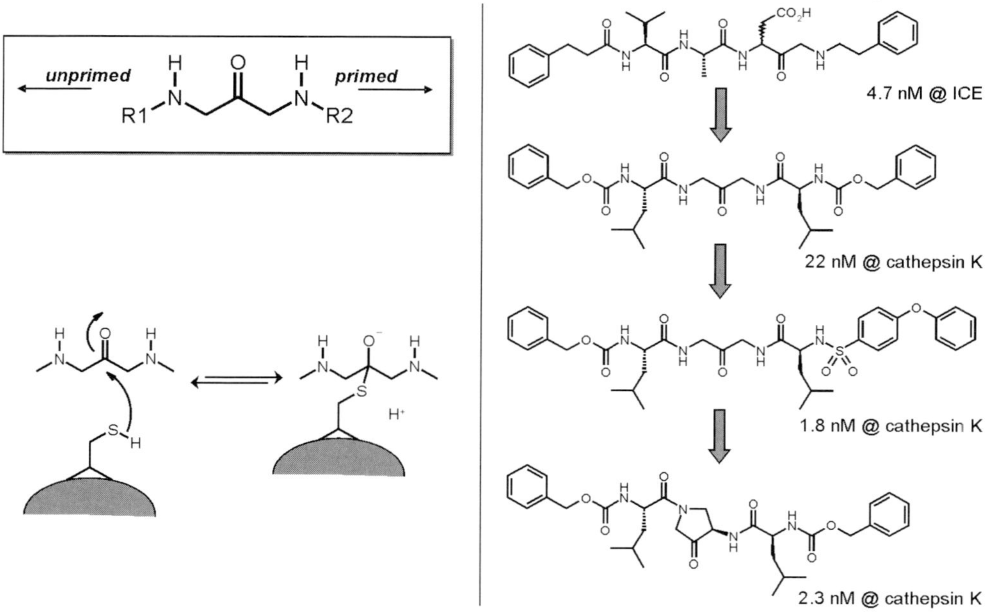

Figure 1.16 Left: Schematic presentation of the 1,3-diaminopropanone core moiety as cysteine protease-directed and active site-spanning inhibitor principle (top). Upon reaction with the enzyme nucleophile, the ketone is reversibly converted to a hemithioketal (bottom). Right: Peptidomimetic cysteine protease inhibitors of subsequent generations are depicted together with their inhibitory activity and primary targets.

of interleukin 1β converting enzyme (ICE) inhibitors, yielding a peptidomimetic compound with nanomolar inhibitory activities [91]. No evidence of irreversible or time-dependent inhibition was observed for the bis-acylated diaminopropanone moiety-containing compounds (Figure 1.16).

The same concept was reintroduced 3 years later by a medicinal chemistry team at SmithKline Beecham Pharmaceuticals designing cathepsin K inhibitors, initially yielding a C_2-symmetric 22 nM compound (Figure 1.16) [92]. Upon optimization of the primed-side substituent of the diaminopropanone core, the inhibitory potency was further increased to 1.8 nM [92] (Figure 1.16). Based on analysis of crystal structures of enzyme–inhibitor complexes (PDB code: 1AU0 [92], 1AU2 [92]), conformational constraints were introduced to optimize the steric fit, yielding a series of cyclic diaminoketone derivatives as novel privileged structure cores [93]. An x-ray cocrystal structure of one of the cyclic diaminoketone-based inhibitors showed that the inhibitor spans both sides of the active centre [93]. The versatility of the diaminopropanone core as a cysteine protease-directed privileged structure was recognized, and combinatorial chemistry concepts based on that inhibitor principle were established in industry [94] as well as in academia [95]. At SmithKline Beecham, a four-dimensional library comprising only 18 compounds was reported to deliver compounds that were assayed against a variety of cysteine proteases with reported activities against cathepsin K, cathepsin L, and cathepsin B (Figure 1.17) [94].

Ellman's group expanded the scope of the difunctionalized ketone core, in that not only 1,3 diaminoketones were prepared combinatorially, but also 1-amino-propanones with acyloxy and alkylated mercapto substituents in the 3-position (Figure 1.17) [95]. In this model study, no biological data on target proteases were reported [95].

Seto and coworkers at Brown University further expanded the idea of a bifunctional cyclic ketone as a central building block for active site-spanning cysteine protease inhibitors, to a cyclohexanone nucleus yielding a design for a two-dimensional combinatorial library (Figure 1.17) [96]. This strategy is the extension of previous work by that group on monofunctionalized heterocyclohexanones (Figure 1.17) that were found to be active serine protease inhibitors [97], as well as cysteine protease inhibitors [98]. These amino-substituted heterocyclohexanones allowed binding-mediating entities to be positioned only on the unprimed side of the substrate-recognition pocket. In contrast, the disubstituted and active site-spanning core

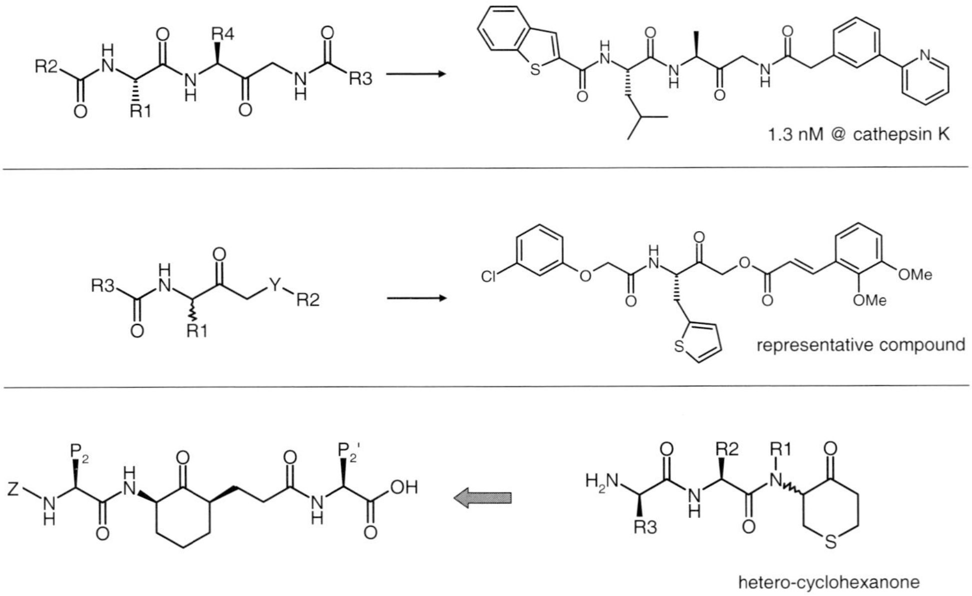

1.3 nM @ cathepsin K

representative compound

hetero-cyclohexanone

Figure 1.17 Combinatorial library designs for compound collections that are based on cysteine and serine protease-directed scaffolds. Top: Four-dimensional library design [94] (left) utilized by SmithKline Beecham for generating cathepsin K inhibitors (right). Middle: Three dimensional library design (left) employed by the Ellman group [95] to generate viable protease inhibitors (right). Bottom: The active site-spanning cyclohexanone core (left) was conceptually derived from a non-active site-spanning mono-substituted heterocyclohexanone derivative [96–98] (right).

bridges over the catalytic centre and allows combining primed- and unprimed-side binding epitopes. Synthesis and biological evaluation of a 400-member library revealed detailed structure–activity and structure–selectivity relationships against cathepsin B, plasmin, urokinase, kallikrein, and papain [97].

The difunctionalized linear or cyclic molecular skeletons encompassing a quiescent warhead targeted against active-site nucleophiles of proteases emerged as a validated privileged structure with proven target family correlation. Additionally, the privileged structure offers sufficient opportunities to engineer the required peripheral diversity into the inhibitor compounds, since two to four diversification points allow for tailoring selectivity, thereby fully exploiting the conceptual advantage of active site-spanning inhibitors over the classical serine or cysteine protease inhibitor concepts. In this context, combinatorial chemistry approaches were established that aided in the generation of target family-biased compound libraries with front-loaded rationales based on experimental validation of the privileged core entity.

Interestingly, researchers at Hoffmann-La Roche reported the foundations for a similar concept applicable to aspartic proteases in 1999, identifying disubstituted piperidines as renin inhibitors (Figure 1.18) [99, 100]. Also here, a cyclic disub-

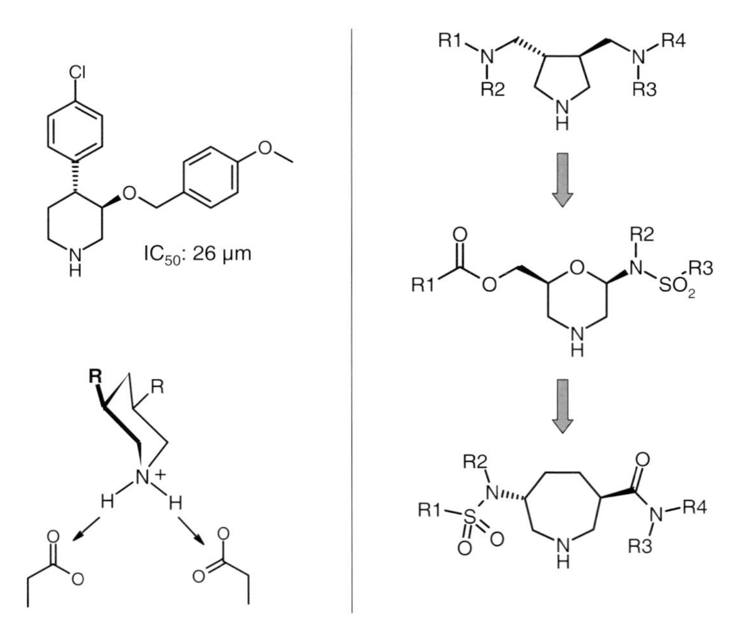

Figure 1.18 Left: Piperidine-derived inhibitors (top) were shown to inhibit aspartic acid proteases by bridging the two catalytically active aspartate residues (bottom). Based on this finding [99, 100], together with the knowledge of the cyclohexanone-based active site-spanning cysteine protease inhibitors [96–98], novel cyclic warheads against aspartic acid proteases can be designed as target family-directed privileged structures [3, 101].

stituted structure contains the protease family-specific warhead, in this instance, a protonated secondary amine forming two charge-enforced hydrogen bonds to the catalytically active aspartate residues, while the peripheral substituents reach into binding pockets on both sides of the catalytic centre.

This clearly resembles the inhibition mode of the serine and cysteine protease inhibitors described above. Iterative refinement (Figure 1.18), e.g., by variation of ring size and symmetrization of the functional decoration pattern, combined with subsequent extrapolation of the renin-specific finding to the entire aspartate protease family, is the apparent basis for a new generation of nonpeptide, lead-like inhibitors with multiple therapeutically relevant endpoint opportunities. The five-membered 3,4-di(aminomethyl)-pyrrolidine (Figure 1.18) served as the core structure for highly active aspartate protease inhibitors [101].

Additional tailor-made privileged structures that allow for systematic exploration of pharmaceutically relevant target families are given elsewhere [3].

1.10 Conclusions and Perspective

The concept of target family-directed masterkeys based on privileged structures in its most stringent definition [3] contributes to the repertoire of chemogenomics, since it provides a powerful means for lead generation that is systematically applicable to entire gene families. This establishes the required forward-integration of genomic and proteomic data into the realm of synthetic chemistry. Genomics- and proteomics-derived technologies, originally envisioned to provide multiple opportunities for the discovery of new drugs with first-in-class and ideally best-in-class characteristics, has undoubtedly delivered novel insights into the genetic and mechanistic basis of several diseases, also reemphasizing the central importance of target validation for the success of a drug discovery project. Together with the rapid explosion of heterogeneous data that have become available to pharmaceutical research teams, novel technologies and a high degree of automation have emerged in all drug discovery-related disciplines. Despite all these revolutionary technologies that have emerged in recent years, solid indications for a sustained decrease in the pharmaceutical industry's productivity are apparent. Recent technologies have not yet succeeded in increasing the output of NCEs or in reducing the costs and timelines for developing new drugs. Those technologies creating just new data have by far outpaced the ability to contextualize these data into the framework of drug research. The entire biotechnology revolution, mainly based on the allure of 'omes' and 'omics', seems to have made only a peripheral contribution to therapeutic intervention in challenging diseases. Obviously, overly optimistic predictions contributed to the development of totally unreasonable promises and expectations as to the immediate impact of any novel technology on the discovery of new therapeutics. Within the complex and challenging business of drug discovery, experience will always win over expectations. This old insight increasingly dominates today's drug discovery in the pharmaceutical industry and is having growing impact also on the still young biotechnology industry, which is currently undergoing a difficult period of consolidation. Since the majority of pharmaceutical companies are organized around therapeutic areas with research teams sequentially scrutinizing targets from various receptor and enzyme families, the gene family-directed masterkey approach presented in this contribution is predicted to have maximal effect in smaller and more flexible medicinal chemistry oriented biotechnology companies.

The major gain in lead discovery and optimization efficiency is achieved by repeatedly using established and steadily growing knowledge on a target family in all involved areas of biology and chemistry. Once optimized, technical procedures for, e.g., protein production, purification, assay development, and screening, can be used for numerous members of a target family of interest. A considerable percentage of a compound collection with built-in target family bias, preferentially based on tailor-made privileged structures [3], will show activity against distinct members of the target enzyme or receptor cluster, with emerging structure–activity and structure–selectivity relationships, respectively. Multiple use of target family-directed biology and chemistry resources is definitely more efficient than starting from scratch for each new discovery project, and so will accelerate lead finding and optimization campaigns considerably.

In the chemogenomics approach outlined in this contribution, increased productivity and shorter timelines are achieved by a strict reuse of well-designed chemistry concepts, based on the mutual overlap between privileged structure-based pharmacophore space and the structural and physicochemical requirements of the ligand binding site of target family members. This overlap is the privileged structure-encoded information content that is optimized for complementarity toward the target family-wide commonality in molecular recognition. A clear structural understanding of this relation is required to leverage the intrinsic potential of a target family approach with associated multiple therapeutic scopes.

In conclusion, this contribution aims at strengthening the 'chemo' aspect of chemogenomics. Over the long-term, chemogenomics will further mature into a synthesis of genomics- and proteomics-derived approaches and modern medicinal chemistry, resulting in a fully integrated approach to drug discovery.

Acknowledgements

Bert Klebl (Axxima Pharmaceuticals AG), Hugo Kubinyi (University of Heidelberg), Stan van Boeckel (N. V. Organon, Oss), and Marion Gurrath are acknowledged for numerous fruitful discussions on various topics of drug discovery in general, and on refining the definition of chemogenomics in particular. Further, the technical support provided by Elvira Wessner (Axxima Pharmaceuticals AG) is greatly acknowledged.

References

1 B. R. Stockwell, *Nat. Rev. Genet.* **2000**, *1,* 116–125.

2 G. McBeath, *Genome Biol.* **2001**, *2,* 1–6.

3 G. Müller, *Drug Discov. Today* **2003**, *8,* 681–691.

4 S. Class, *Chem. Eng. News* **2002**, *80,* 39–49.

5 G. M. Mine, Annu. *Rep. Med. Chem.* **2003**, *38,* 383–396.

6 J. A. DiMasi, R. W. Hansen, H. G. Grabowski, *J. Health Econ.* **2003**, *22,* 151–185.

7 Pharmaceutical Research and Manufacturers of America (PhRMA), Pharmaceutical Industry Profile 2002 (Washington, DC, PhRMA **2002**).

8 F. S. Smith, *Nat. Rev. Drug Disc.* **2003**, *2,* 95–96.

9 Pharmaceutical Research and Manufacturers of America (PhRMA), Pharmaceutical Industry Profile 2002 (Washington, DC, PhRMA **2003**), see also: http://www.phrma.org/publications/publications/profile02/index.cfm.

10 J. Drews, *Science* **2000**, *287,* 1960–1963.

11 J. Drews, *Drug Discov. Today* **2000**, *5,* 2–4.

12 J. Drews, *Drug Discov. Today* **2003**, *8,* 411–420.

13 D. Hunter, *J. Cell. Biochem.* **2001**, Suppl. *37,* 22–27.

14 E. J. Martin, J. M. Blaney, M. A. Sinai, D. C. Spellmeyer, A. K. Wong, W. H. Moss, *J. Med. Chem.* **1995**, *38,* 1431–1436.

15 W. P. Walters, *Drug Discov. Today* **1998**, *3,* 160–178.

16 C. A. Lipinski, *J. Pharmacol. Toxicol. Methods* **2000**, *44,* 235–249.

17 R. A. Goodnow, *J. Cell. Biochem.* **2001**, Suppl. *37,* 13–21.

18 A. L. Hopkins, C. R. Groom, *Nat. Rev. Drug Discov.* **2002**, *1,* 727–730.

19 S. Fox, H. Wang, L. Sopchak, S. Farr-Jones, *J. Biomol. Screening* **2002**, *7,* 313–316.

20 L. Weber, *Curr. Opin. Chem. Biol.* **2000**, *4,* 295–302.

21 D. F. Veber, F. H. Drake, M. Gowen, *Curr. Opin. Chem. Biol.* **1997**, *1,* 151–156.

22 G. Dorman, P. Krajcs, F. Darvas, *Curr. Drug Disc.* **2001**, 21–24.

23 M. Namchuk, *Targets* **2002**, *1,* 125–129.

24 E. Martin, A. Wong, *J. Chem. Inf. Comput. Sci.* **2000**, *40,* 215–220.

25 R. E. Dolle, *J. Comb. Chem.* **2003**, *5,* 693–753.

26 C. K. Atterwill, M. G. Wing, *Altern. Lab. Anim.* **2000**, *28,* 857–867.

27 H. van de Waterbeemd, E. Gifford, *Nat. Rev. Drug Discov.* **2003**, *2,* 192–204.

28 P. W. Erhardt, *Pure Appl. Chem.* **2002**, *74,* 703–785.

29 K. H. Bleicher, H.-J. Böhm, K. Müller, A. I. Alanine, *Nat. Rev. Drug Discov.* **2003**, *2,* 369–378.

30 J. Drews, *Nature Biotech.* **1996**, *14,* 1516–1518.

31 J. Drews, S. Ryser, *Drug Inf. J.* **1996**, *30,* 97–108.

32 P. Tollman, P. Guy, J. Altshuler, A. Flanagan, M. Steiner, Boston Consulting Group Report: A Revolution in R&D. http://www.bcg.com/publications/files/eng_genomicsgenetics_rep_11_01.pdf, **2001**.

33 M. Mort, *Modern Drug Discov.* **2000**, *3,* 28–29.

34 Lehman Brothers and McKinsey & Co: The Fruits of Genomics (Lehman Brothers, New York, Jan. **2001**).

35 D. E. Szymkowski, *Drug Discov. Today* **2001**, *6,* 398–399.

36 Intelligent Drug Design, *Nature* **1996**, *384,* Suppl. to Issue No. 6604, 5.

37 S. P. Davies, H. Reddy, M. Caivano, P. Cohen, *Biochem. J.* **2000**, *351,* 95–105.

38 J. Bain, H. McLauchlan, M. Elliott, P. Cohen, *Biochem. J.* **2003**, *371,* 199–204.

39 M. Vieth, R. E. Higgs, D. H. Robertson, M. Shapiro, E. A. Gragg, H. Hemmerle, *Biochim. Biophys. Acta* **2004**, *1697,* 243–257.

40 C. A. Lipinski, F. Lombardo, B. W. Dolminy, P. J. Feeney, *Adv. Drug Deliv. Rev.* **1997**, *23,* 3–26.

41 B. E. Evans, K. E. Rittle, M. G. Bock, R. M. DiPardo, R. M. Freidinger, W. L. Whitter, G. F. Lundell, D. F. Veber, P. S. Anderson, R. S. L. Chang, V. J. Lotti, D. J. Cerino, T. B. Chen, P. J. Kling, K. A. Kunkel, J. P. Springer, J. Hirshfield, *J. Med. Chem.* **1988**, *31,* 2235–2246.

42 A. A. Patchett, R. P. Nargund, *Annu. Rep. Med. Chem.* **2000**, *35*, 289–298.
43 R. S. Chang, V. J. Lotti, T. B. Chen, M. E. Keegan, *Neurosci. Lett.* **1986**, *72*, 211–214.
44 D. Roemer, H. H. Buscher, R. C. Hill, R. Maurer, T. J. Petcher, H. Zeugner, W. Benson, E. Finner, W. Milkowski, P. W. Thies, *Nature* **1982**, *298*, 759–760.
45 I. C. Choong, J. A. Ellman, *Annu. Rep. Med. Chem.* **1996**, *31*, 309–318.
46 O. Roche, P. Schneider, J. Zuegge, W. Guba, M. Kansy, A. Alanine, K. Bleicher, F. Danel, E. M. Gutknecht, M. Rogers-Evans, W. Neidhart, H. Stalder, M. Dillon, E. Sjogren, N. Fotouhi, P. Gillespie, R. Goodnow, W. Harris, P. Jones, M. Taniguchi, S. Tsujii, W. von Der Saal, G. Zimmermann, G. Schneider, *J. Med. Chem.* **2002**, *45*, 137–142.
47 S. L. McGovern, E. Caselli, N. Grigorieff, B. K. Shoichet, *J. Med. Chem.* **2002**, *45*, 1712–1722.
48 S. L. McGovern, B. K. Shoichet, *J. Med. Chem.* **2003**, *46*, 1478–1483.
49 J. Seidler, S. L. McGovern, T. M. Doman, B. K. Shoichet, *J. Med. Chem.* **2003**, *46*, 4477–4486.
50 S. L. McGovern, B. T. Hefland, B. Feng, B. K. Shoichet, *J. Med. Chem.* **2003**, *46*, 4265–4272.
51 A. J. Ryan, N. M. Gray, P. N. Lowe, C.-W. Chung, *J. Med. Chem.* **2003**, *46*, 3448–3451.
52 G. W. Bemis, M. A. Murcko, *J. Med. Chem.* **1996**, *39*, 2887–2893.
53 G. W. Bemis, M. A. Murcko, *J. Med. Chem.* **1999**, *42*, 5095–5099.
54 X. Q. Lewell, D. B. Judd, S. P. Watson, M. M. Hann, *J. Chem. Inf. Comput. Sci.* **1998**, *38*, 511–522.
55 X. Q. Lewell, A. C. Jones, C. L. Bruce, G. Harper, M. J. Jones, I. M. Mclay, J. Bradshaw, *J. Med. Chem.* **2003**, *46*, 3257–3274.
56 R. P. Sheridan, *J. Chem. Inf. Comput. Sci.* **2003**, *43*, 1037–1050.
57 P. J. Hajduk, M. Bures, J. Praestgaard, S. W. Fesik, *J. Med. Chem.* **2000**, *43*, 3443–3447.
58 J. A. Ellman, *Acc. Chem. Res.* **1996**, *29*, 132–143.
59 S. H. DeWitt, A. W. Czarnik, *Acc. Chem. Res.* **1996**, *29*, 114–122.
60 R. A. Wiley, D. H. Rich, *Med. Res. Rev.* **1993**, *13*, 327–384.
61 C. J. Burns, R. D. Groneberg, J. M. Salvino, G. McGeehan, S. M. Condon, R. Morris, M. Morrissette, R. Mathew, S. Darnbrough, K. Neuenschwander, A. Scotese, S. W. Djuric, J. Ullrich, R. Labaudiniere, *Angew. Chem. Int. Ed. Engl.* **1998**, *37*, 2848–2850.
62 R. D. Groneberg, C. J. Burns, M. M. Morrissette, J. W. Ullrich, R. L. Morris, S. Darnbrough, S. W. Djuric, S. M. Condon, G. M. McGeehan, R. Labaudiniere, K. Neuenschwander, A. C. Scotese, J. A. Kline, *J. Med. Chem.* **1999**, *42*, 541–544.
63 D. Hoyer, M. M. A. Awad, J. M. Salvino, P. R. Seoane, R. E. Dolle, W. T. Houck, D. G. Sawutz, *Bioorg. Med. Chem. Lett.* **1995**, *5*, 367–370.
64 Y. Abe, H. Kayakiri, S. Satoh, T. Inoue, Y. Sawada, K. Imai, N. Inamura, M. Asano, C. Hatori, A. Katayama, T. Oku, H. Tanaka, *J. Med. Chem.* **1998**, *41*, 564–578.
65 Y. Abe, H. Kayakiri, S. Satoh, T. Inoue, Y. Sawada, N. Inamura, M. Asano, I. Aramori, C. Hatori, H. Sawai, T. Oku, H. Tanaka, *J. Med. Chem.* **1998**, *41*, 4062–4079.
66 Y. Abe, H. Kayakiri, S. Satoh, T. Inoue, Y. Sawada, N. Inamura, M. Asano, C. Hatori, H. Sawai, T. Oku, H. Tanaka, *J. Med. Chem.* **1998**, *41*, 4053–4061.
67 J. I. Armstrong, A. R. Portley, Y.-T. Chang, D. M. Nierengarten, B. N. Cook, K. G. Bowman, A. Bishop, N. S. Gray, K. M. Shokat, P. G. Schultz, C. R. Bertozzi, *Angew. Chem. Int. Ed. Engl.* **2000**, *39*, 1303–1306.
68 D. E. Verdugo, M. T. Cancilla, X. Ge, N. S. Gray, Y.-T. Chang, P. G. Schultz, M. Negishi, J. A. Leary, C. R. Bertozzi, *J. Med. Chem.* **2001**, *44*, 2683–2686.
69 Y. Kakuta, L. G. Pedersen, C. W. Carter, M. Negishi, L. C. Pedersen, *Nat. Struct. Biol.* **1997**, *4*, 904–908.
70 Y. Kakuta, T. Sueyoshi, M. Negishi, L. C. Pedersen, *J. Biol. Chem.* **1999**, *274*, 10673–10676.

71 C.-H. Lin, G.-J. Shen, E. Garcia-Junceda, C.-H. Wong, *J. Am. Chem. Soc.* **1995**, *117*, 8031–8032.

72 Y. T. Chang, N. S. Gray, G. R. Rosania, D. P. Sutherlin, S. Kwon, T. C. Norman, R. Sorahia, M. Leost, L. Meijer, P. G. Schultz, *Chem. Biol.* **1999**, *6*, 361–375.

73 A. J. Souers, A. A. Virgilio, S. A. Rosenquist, W. Fenuik, J. A. Ellman, *J. Am. Chem. Soc.* **1999**, *121*, 1817–1825.

74 A. J. Souers, A. A. Virgilio, S. S. Schurer, J. A. Ellman, T. P. Kogan, H. E. West, W. Ankener, P. Vanderslice, *Bioorg. Med. Chem. Lett.* **1998**, *8*, 2297–2302.

75 M. Gurrath, G. Müller, H. Kessler, M. Aumailley, R. Timpl, *Eur. J. Biochem.* **1992**, *210*, 911–921.

76 S. P. Rohrer, E. T. Birzin, R. T. Mosley, S. C. Berk, S. M. Hutchins, D. M. Shen, Y. Xiong, E. C. Hayes, R. M. Parmar, F. Foor, S. W. Mitra, S. J. Degrado, M. Shu, J. M. Klopp, S. J. Cai, A. Blake, W. W. S. Chan, A. Pasternak, L. Yang, A. A. Patchett, R. G. Smith, K. Champman, J. M. Schaeffer, *Science* **1998**, *282*, 737–740.

77 G. Müller, Topics in Current Chemistry **2000**, *211*, 17–59.

78 D. Leung, G. Abbenante, D. P. Fairlie, *J. Med. Chem.* **2000**, *43*, 305–341.

79 C. Southan, *Drug Discov. Today* **2001**, *6*, 681–688.

80 M. P. Weir, S. S. Bethell, A. Cleasby, C. J. Campbell, R. J. Dennis, C. J. Dix, H. Finch, H. Jhoti, C. J. Mooney, S. Patel, C. M. Tang, M. Ward, A. J. Wonacott, C. W. Wharton, *Biochem.* **1998**, *37*, 6645–6657.

81 M. J. O'Neill, J. A. Lewis, H. M. Noble, S. Holland, C. Mansat, J. E. Farthing, G. Foster, D. Nobel, S. J. Lane, P. J. Sidebottom, S. M. Lynn, M. V. Hayes, C. J. Dix, *J. Nat. Prod.* **1998**, *61*, 1328–1331.

82 S. J. F. MacDonald, M. D. Dowle, L. A. Harrison, G. D. Clarke, G. G. Inglis, M. R. Johnson, P. Shah, R. A. Smith, A. Amour, G. Fleetwood, D. C. Humphreys, C. R. Molloy, M. Dixon, R. E. Godward, A. J. Wonacott, O. M. Singh, S. T. Hodgson, G. W. Hardy, *J. Med. Chem.* **2001**, *45*, 3878–3890.

83 S. J. F. MacDonald, M. D. Dowle, L. A. Harrison, P. Shah, M. R. Johnson, G. G. Inglis, G. D. Clarke, R. A. Smith, D. Humphreys, C. R. Molloy, A. Amour, M. Dixon, G. Murkitt, R. E. Godward, T. Padfield, T. Skarzynski, O. M. Singh, K. A. Kumar, G. Fleetwood, S. T. Hodgson, G. W. Hardy, H. Finch, *Bioorg. Med. Chem. Lett.* **2001**, *11*, 895–898.

84 A. D. Borthwick, S. J. Angier, A. J. Crame, A. M. Exall, T. M. Haley, G. J. Hart, A. M. Mason, A. M. Pennell, G. G. Weingarten, *J. Med. Chem.* **2000**, *43*, 4452–4464.

85 A. D. Borthwick, A. J. Crame, P. F. Ertl, A. M. Exall, T. M. Haley, G. J. Hart, A. M. Mason, A. M. Pennell, O. M. Singh, G. G. Weingarten, J. M. Woolven, *J. Med. Chem.* **2002**, *45*, 1–18.

86 D. M. Andrews, S. J. Carey, H. Chaignot, B. A. Coomber, N. M. Gray, S. L. Hind, P. S. Jones, G. Mills, J. E. Robinson, M. J. Slater, *Organic Lett.* **2002**, *4*, 4475–4478.

87 D. M. Andrews, H. Chaignot, B. A. Coomber, A. C. Good, S. L. Hind, M. R. Johnson, P. S. Jones, G. Mills J. E. Robinson, T. Skarzynski, M. J. Slater, D. O. Somers, *Organic Lett.* **2002**, *4*, 4479–4482.

88 D. F. Veber, S. K. Thompson, *Curr. Opin. Drug Discov. Dev.* **2000**, *3*, 362–369.

89 H.-H. Otto, T. Schirmeister, *Chem. Rev.* **1997**, *97*, 133–171.

90 I. Schechter, A. Berger, *Biochem. Biophys. Res. Com.* **1967**, *27*, 157–162.

91 A. M. M. Mjalli, K. T. Chapman, M. MacCoss, N. A. Thornberry, E. P. Peterson, *Bioorg. Med. Chem.Lett.* **1994**, *4*, 1965–1968.

92 D. S. Yamashita, W. W. Smith, B. Zhao, C. A. Janson, T. A. Tomaszek, M. J. Bossard, M. A. Levy, H.-J. Oh, T. J. Carr, S. K. Thompson, C. F. Ijames, S. A. Carr, M. McQueney, K. J. D. Alessio, B. Y. Amagadzie, C. R. Hanning, S. Abdel-Meguid,

R. L. DesJarlais, J. G. Gleason, D. F. Veber, *J. Am. Chem. Soc.* **1997**, *119*, 11351–11352.

93 R. W. Marquis, D. S. Yamashita, Y. Ru, S. M. LoCastro, H.-J. Oh, K. F. Erhard, R. L. DesJarlais, M. S. Head, W. W. Smith, B. Zhao, C. A. Janson, S. S. Abdel-Meguid, T. A. Tomaszek, M. A. Levy, D. F. Veber, *J. Med. Chem.* **1998**, *41*, 3563–3567.

94 D. S. Yamashita, X. Dong, H.-J. Oh, C. S. Brook, T. A. Tomaszek, L. Szewczuk, D. G. Tew, D. F. Veber, *J. Comb. Chem.* **1999**, *1*, 207–215.

95 A. Lee, L. Huang, J. A. Ellman, *J. Am. Chem. Soc.* **1999**, *121*, 9907–9914.

96 P. Abato, J. L. Conroy, C. T. Seto, *J. Med. Chem.* **1999**, *42*, 4001–4009.

97 T. C. Sanders, C. T. Seto, *J. Med. Chem.* **1999**, *42*, 2969–2976.

98 J. L. Conroy, C. T. Seto, *J. Org. Chem.* **1998**, *63*, 2367–2370.

99 E. Vieira, A. Binggeli, V. Breu, D. Bur, W. Fischli, R. Güller, G. Hirth, H. P. Märki, M. Müller, C. Oefner, M. Scalone, H. Stadler, M. Wilhelm, W. Wostl, *Bioorg. Med. Chem. Lett.* **1999**, *9*, 1397–1402.

100 R. Güller, A. Binggeli, V. Breu, W. Fischli, G. Hirth, C. Jenny, M. Kansy, F. Montavon, M. Müller, C. Oefner, H. Stadler, E. Vieira, M. Wilhelm, W. Wostl, H. P. Märki, *Bioorg. Med. Chem. Lett.* **1999**, *9*, 1403–1408.

101 A. Schoop, C. Kallus, G. Müller, unpublished results.

2
Drug Discovery from Side Effects

Hugo Kubinyi

Many important therapeutic discoveries have resulted from serendipitous observations. Side effects of drugs or drug candidates in the clinics have paved the way to new applications of a drug or to the development of chemically modified analogs. Unexpected pharmacological effects against physiologically related or other, more diverse, targets have resulted in drug candidates with different modes of action. In the past decade, more systematic approaches have been followed: chemogenomics, the systematic investigation of the biological effects of certain classes of compounds in certain target families, and the selective optimization of drug side effects.

Such side effects may result from:

- A physiological reaction of the body to the action of the drug (e.g., the reflex tachycardia resulting from the antihypertensive activity of dihydropyridine calcium-channel blockers).
- Overdose of drugs with a narrow therapeutic range and/or unfavorable pharmacokinetics (e.g., phenprocoumon or warfarin, which exert their action in a delayed and indirect manner by inhibition of vitamin K biosynthesis).
- Action on different targets by the same mechanism (e.g., gastrointestinal bleeding after cyclooxygenase inhibition by acetylsalicylic acid, bradykinin-mediated cough as a side effect of angiotensin-converting enzyme inhibitors).
- Action on organs other than the target organ (e.g., peripheral tachycardic and hypertensive effects of dopamine after systemic application of the anti-Parkinson drug L-dopa, sedative side effects of lipophilic histamine H_1 antagonists).
- Lack of selectivity, i.e., inhibition, agonism, or antagonism at several different targets, a most common reason for drug side effects (e.g., respiratory depression by morphine, cardiotoxicity of certain drugs mediated by hERG channel inhibition).
- Inhibition of cytochrome P450 isoenzymes (e.g., nonlinear pharmacokinetics of propafenone, producing an exponential increase of plasma levels due to inhibition of its metabolism by CYP2D6, after application of higher doses).

Chemogenomics in Drug Discovery: A Medicinal Chemistry Perspective.
Edited by Hugo Kubinyi and Gerhard Müller

ISBN: 3-527-30987-X

- Drug–drug interactions resulting from cytochrome P450 inhibition or induction, a very common reason for adverse drug effects (e.g., terfenadine, which exerts fatal cardiotoxicity by hERG channel inhibition in the presence of a CYP3A4 inhibitor, whereas its active metabolite fexofenadine is not a hERG channel inhibitor).
- Genetic disposition, either by interaction of the drug with a mutant target or by the lack of certain (or mutant) metabolic enzyme (e.g., the inability of about 1–3% of the Caucasian population to metabolize S-warfarin, due to a CYP2C9 deficiency).

There must always be a significant advantage of the achievable therapeutic benefit, as compared to the risk of drug-related side effects. Severe side effects can only be tolerated in treatment of chronic degenerative or life-threatening diseases like arthritis, cancer, or AIDS. However, adverse drug side effects are frequently observed after medication; their high incidence, even as a relatively common cause of death, is only gradually being recognized.

However, a closer inspection of the history of drug discovery [1–3] shows that many new drug applications resulted from clinical observations of side effects or from the optimization of such unexpected side effects into new therapeutic areas. Only some prominent drugs that resulted from serendipitous observations of clinical side effects are discussed in the following sections. However, even these few examples show the importance of this source of new leads in drug research. Some more examples are discussed in special monographs on the history of drug research [1–3], as well as in some other medicinal chemistry and pharmacology textbooks [4–11] and reviews [12–14].

In addition to clinical observations of drug side effects, the optimization of side activities that are discovered by in vitro investigation plays an important role in drug research. Recently, Wermuth proposed using this approach as a general strategy for the "selective optimization of side activities" (the SOSA approach) [15]; in support of this concept he quotes Sir James Black, the 1988 Nobel Laureate in Physiology and Medicine, who stated "the most fruitful basis for the discovery of a new drug is to start with an old drug".

2.1
A Historical Perspective: The Great Time of Serendipitous Observations

The early history of drug discovery is characterized by many serendipitous drug discoveries [1–3, 9, 16]. After the preparation of nitrous oxide by Humphry Davy in the early 19th century, fun parties with this gas, and also with ether, became popular; people liked the euphoric effect after inhaling the chemicals. The anesthetic properties of nitrous oxide and ether were discovered in the 1840s just by chance, because participants of such events did not experience any pain after being hurt. The antianginal properties of organic nitrites were discovered at about the same time: inhalation of amyl nitrite vapor or oral uptake of a small amount of nitroglycerin resulted in severe headache but also in relief of angina pectoris symptoms. Today we complain about the long development time of new drugs. However, this has tradition: after the observation by the Italian chemist Ascanio Sobrero that even a small dose of nitroglycerin causes headache, published in 1847, the first application to a patient occurred only in 1878, 31 years later. People were obviously frightened of using an explosive as a remedy; in his late years, Alfred Nobel, who invented dynamite in 1867, had to take this drug himself [2, 3]. The incompatibility of disulfiram (Antabus®) with alcohol consumption was discovered when rubber workers, having contact with this antioxidant, complained about alcohol-induced periods of sickness; another version says that two pharmacologists who took this drug as an anthelmintic became ill at a cocktail party [1]. The hallucinogenic properties of lysergic acid diethylamide (LSD) were discovered after the accidental uptake of minor amounts of this compound by Albert Hofmann [9]. And, last not least, all artificial sweeteners of major importance, i.e., saccharin, cyclamate, and aspartame, were discovered by unexpected observations of their sweet taste [9].

A more or less systematic search for new drugs started in the last two decades of the 19th century. Although acetylsalicylic acid **1** (ASS, Aspirin®, Bayer; Figure 2.1) was originally designed as a 'better' derivative of salicylic acid, it is much more than just a prodrug. ASS is more active than its parent drug and is indeed better tolerated, but it causes gastrointestinal bleeding as a prominent side effect. In the 1970s it became clear that both its activity and its side effect are mediated by the same target. ASS inhibits cyclooxygenase, which converts arachidonic acid into prostacyclin, which is further converted into prostaglandins and thromboxane. Whereas inhibition of biosynthesis of the pain-mediating prostaglandins is responsible for the analgesic and antipyretic activities of ASS, inhibition of thromboxane biosynthesis is responsible for the increased bleeding tendency. Since thrombocytes have no nucleus and therefore no protein biosynthesis, platelet cyclooxygenase remains inhibited over the whole thrombocyte lifetime of about 120 days; correspondingly, thrombosis protection by aggregation inhibition can be achieved by application of only 100 mg or even less of ASS per day. Low-dose ASS treatment is now a standard therapy for the prevention of stroke, heart attack, and thrombosis.

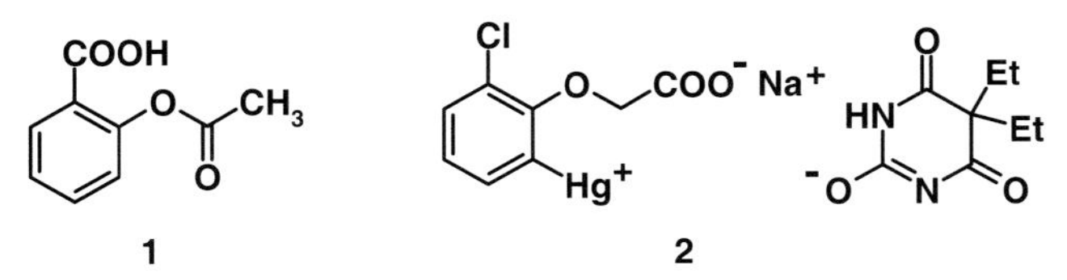

Figure 2.1 Acetylsalicylic acid **1** (Aspirin®, Bayer) is much more than a 'prodrug' of salicylic acid. Its major contribution to biological activity comes from a unique mechanism of action: the activated acetyl group is transferred to a serine hydroxyl group in the binding site of cyclooxygenase. Merbaphen **2** (Novasurol®, Bayer) was the first example of an organomercurial diuretic; some analogs with less severe side effects were the therapeutic standard from about 1920 to 1950.

The diuretic organomercurials are most probably the very first example of the discovery of a class of therapeutically useful drugs by a clinical side effect of one of their members. In 1888, mercury salicylate was introduced for the treatment of syphilis, followed by mersalyl in 1906 and arsphenamine (E 606), discovered by Paul Ehrlich in 1909. On October 7, 1919, a pale and weak 21-year-old female, Johanna M., was brought to the First Medical University Clinic in Vienna, in an insane status, with clear symptoms of severe neurosyphilis. Alfred Vogl, a 3rd-year medical student, was ordered to apply mercury salicylate, in a desperate attempt to help. Not knowing about the properties of this compound, he asked for a 10% aqueous solution for intramuscular injection. After a few days, when he had not received the solution, he was told that the compound was too insoluble. A colleague proposed trying a recently developed analog, merbaphen **2** (Novasurol®, Bayer; Figure 2.1), a water-soluble salt of an organomercurial compound with barbitone. After approval by his supervisors, he applied it to the suffering patient. To his great surprise, the daily urine production by the patient increased from 200–500 mL to 1200 mL, and after the third application of the drug even to 2000 mL. Application to other patients produced up to 10 L urine within 24 hours – a diuretic effect that had not been observed before! Merbaphen was too toxic for therapeutic application, but follow-on products held their place as diuretics till the 1950s, when another observation of a clinical side effect led to the discovery of the much safer sulfonamide diuretics (see below) [3, 4, 7].

A most fascinating area of rational drug research (indeed, rational drug research is not an invention of our time) is the systematic structural variation of morphine **3** (Figure 2.2). Morphine, discovered by Sertürner in 1806, has, in addition to its strong narcotic and analgesic activities, also antitussive, constipating, and respiratory depressant side effects. The creativity, intuition, and tenacity of generations of medicinal chemists have produced many morphine analogs, with much simpler chemical structures and different opiate receptor subtype selectivities [3, 4, 17]. However, the side effects could also be optimized. Whereas pethidine **4** (meperidine; originally designed as an anticholinergic atropine analog) and fentanyl **5** are prototypes of morphine-related strong analgesics (Figure 2.2), dextromethorphan **6** (the methyl ether of the D-enantiomer of the analgesic levorphanol; Figure 2.3) is completely devoid of analgesic activity; it retains only the antitussive properties of morphine. Loperamide **7** is also devoid of analgesic activity, despite its ability to pass the blood–brain barrier; after its uptake, the pGP transporter eliminates the drug by active transport; correspondingly, loperamide is used for the treatment of acute and chronic diarrhea. Further research in this area led to the highly potent neuroleptic haloperidol **8** (Figure 2.3). Whereas a pethidine analog with a butyrophenone side chain is still a strong analgesic, haloperidol **8**, bearing a hydroxyl group instead of the ester group, turned out to be a potent dopamine antagonist [1–3]. As such, it is used for the treatment of schizophrenia and other manic disorders. The morphine series is also one of several examples in which the chemical structures of agonists (e.g., morphine **3**) and antagonists (e.g., nalorphine **9**) are very closely related [6–8].

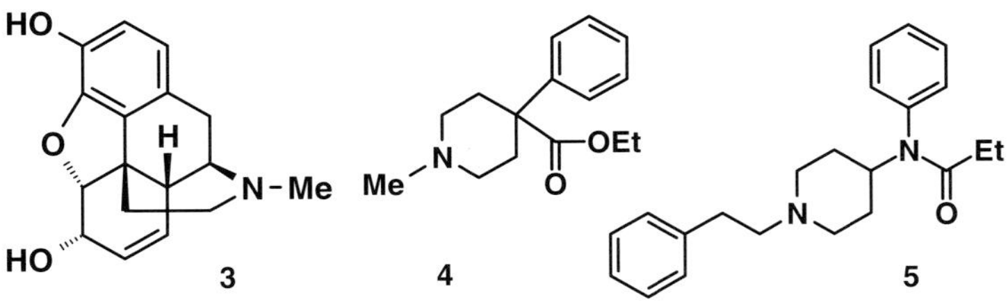

Figure 2.2 Morphine **3** was the lead structure for many structurally much simpler strong analgesics, e.g., pethidine **4** and fentanyl **5**.

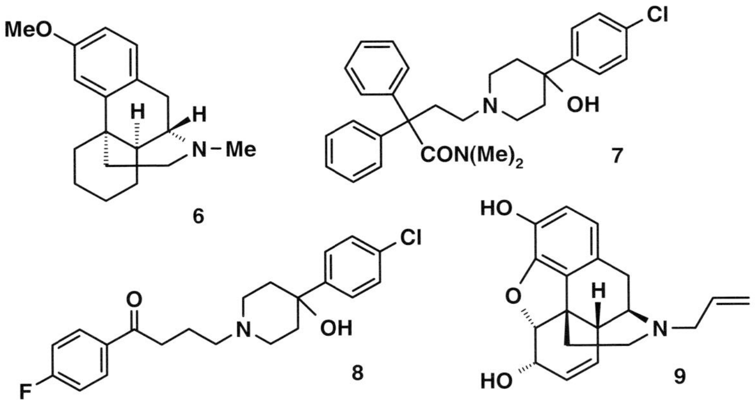

Figure 2.3 Dextromethorphan **6**, the unnatural enantiomer of a narcotic morphine analog, is an antitussive drug. The antidiarrhea drug loperamide **7** and the neuroleptic drug haloperidol **8** also resulted from structural modification of morphine. The morphine antagonist nalorphine **9** differs from the opioid agonist morphine **3** (Figure 2.2) only by having an *N*-allyl group instead of the *N*-methyl group.

2.2 Clinical Observations of Side Effects

As with the diuretic effect of organomercurials, many other new drugs have resulted from unexpected observations of clinical side effects. An especially rich source of new drugs resulted from the discovery of the antibacterial sulfonamide sulfamidochrysoidine **10** (Prontosil rubrum®, Bayer; Figure 2.4) by Gerhard Domagk in 1935. After Jacques and Therese Trefouel gave evidence that the metabolic cleavage product sulfanilamide **11** is the active agent, many groups started on the synthesis of analogs, to improve activity and pharmacokinetic properties [2, 3].

Massive doses of sulfanilamide, as well as of other sulfonamides, caused alkaline diuresis as a side effect. From 1940 onwards, the mechanism of this side effect was further investigated; it was confirmed that carbonic anhydrase inhibition was responsible for the diuresis. Presenting the whole story of the development of

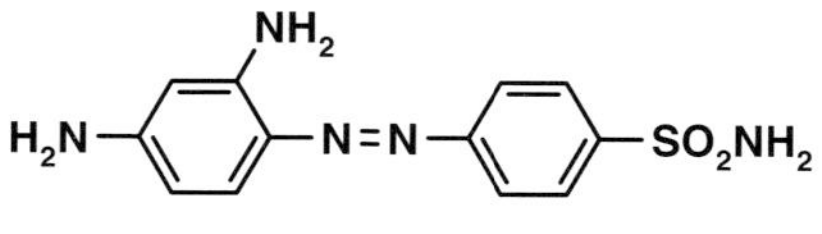

10 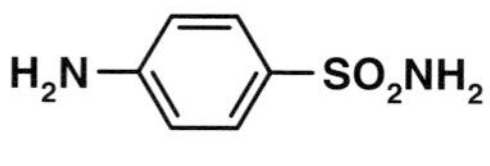 11

Figure 2.4 Sulfamidochrysoidine **10** (Prontosil rubrum®, Bayer) and related antibacterial sulfonamides act via the metabolite sulfanilamide **11**, which is an antimetabolite of *p*-aminobenzoic acid in the bacterial biosynthesis of dihydrofolic acid.

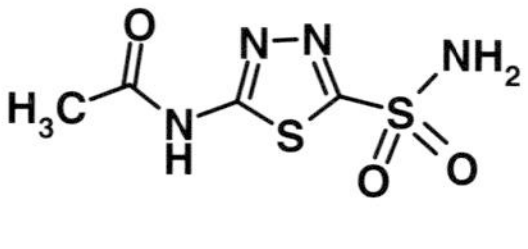

12 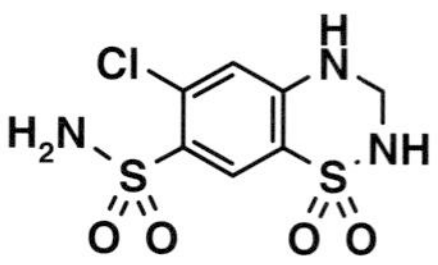 13 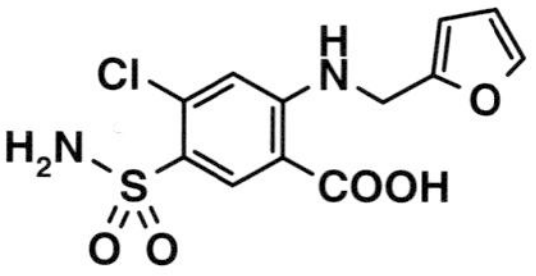 14 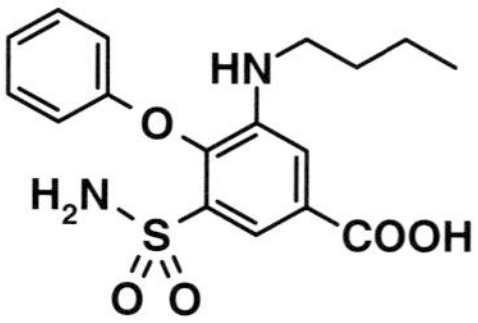 15

Figure 2.5 In addition to its antibacterial activity, sulfanilamide **11** (Figure 2.4) inhibits the enzyme carbonic anhydrase. Acetazolamide **12** is much more potent as a carbonic anhydrase inhibitor but its clinical use as diuretic was impaired by some serious side effects. Hydrochlorothiazide **13** is the prototype of orally active saluretic sulfonamide diuretics. Furosemide (frusemide) **14** and bumetanide **15** are so-called 'loop diuretics'.

diuretic sulfonamides [1–3] is beyond the scope of this chapter, but some prototypes **12–15** of this highly successful class of therapeutics, with different mechanisms of action, are presented in Figure 2.5. Today, these drugs are not only important diuretics but are also used in combination therapy of high blood pressure [4, 6, 7].

Due to the carbonic anhydrase inhibition of sulfonamides, it was also possible to design topically active analogs for the treatment of glaucoma, e.g., dorzolamide **16**, the result of a structure-based design (Figure 2.6) [18]. Sulfaguanidine **17** is a sulfonamide with only poor bioavailability; accordingly, it was tested against intestinal infections, but it turned out to be an inhibitor of thyroid hormone biosynthesis. This unexpected result paved the way to antithyroid drugs of the thiourea and thiouracil type [2, 3]. Dapsone **18** (Figure 2.6) may be considered a phenylog of sulfanilamide; still nowadays it is a standard drug for the treatment of leprosy [1].

In addition to the acceptable clinical side effects of many sulfonamides, some severe toxic effects were observed. In 1942, Marcel Janbon, the head of the medical faculty at Montpellier University, investigated an isopropylthiadiazole derivative of

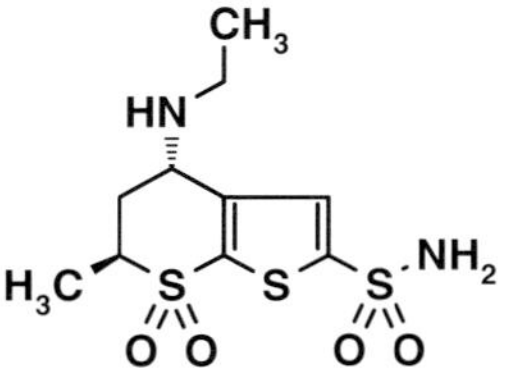

Figure 2.6 Dorzolamide **16**, a topically active carbonic anhydrase inhibitor, resulted from a structure-based ligand design; it is used for the treatment of glaucoma. Sulfaguanidine **17** inhibits thyroid hormone biosynthesis. A phenylog of sulfanilamide **11** (Figure 2.4), dapsone **18**, is used for the treatment of leprosy.

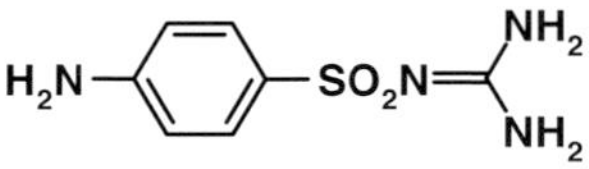

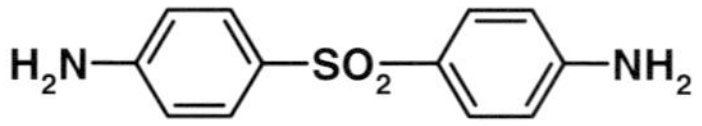

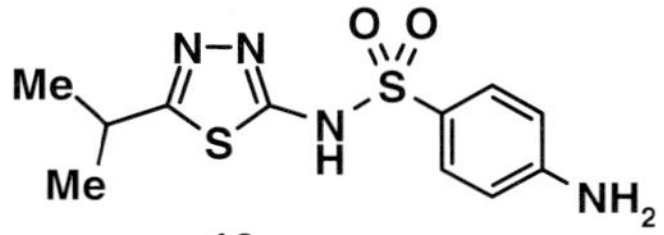

Figure 2.7 *N*-Isopropyl-thiadiazolyl sulfanilamide (IPTD) **19** was the first sulfonamide that showed antidiabetic properties in the clinics. Carbutamide **20** and tolbutamide **21** are also antidiabetics; tolbutamide **21** has a shorter biological half-life than carbutamide **20**, due to its methyl group instead of the chlorine atom; in addition, tolbutamide does not show antibacterial activity. Glibenclamide **22** is an antidiabetic drug with improved therapeutic properties.

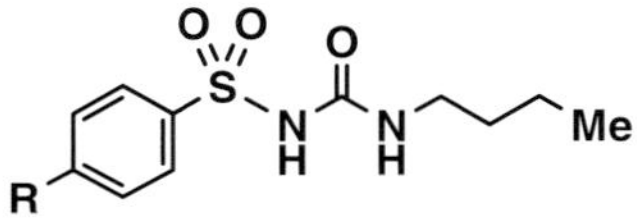

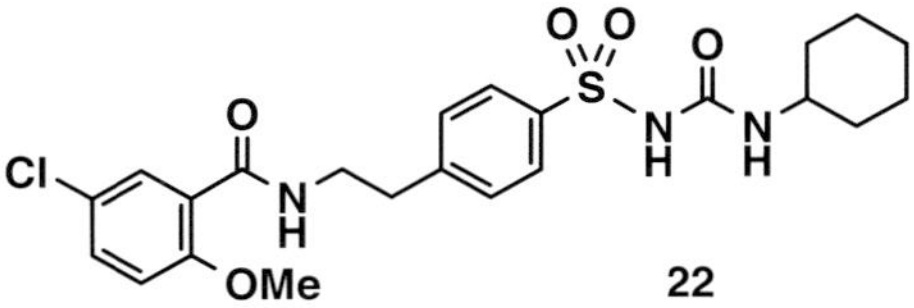

sulfanilamide, IPTD **19** (Figure 2.7), in typhoid patients. However, instead of being cured, some patients became very ill and a few of them even died. Quick recovery of the patients after intravenous glucose application led to the hypothesis that the compound produced severe hypoglycemia. In his PhD work, the medical student Auguste Loubatieres confirmed in animal experiments that the compound could indeed be used for the treatment of diabetes. However, due to bad experience with some other antidiabetic compounds and due to the general situation at the end of World War II, this proposal was not pursued [2, 3]. Only 12 years later, in February 1954, Klaus Fuchs at the Auguste Victoria Hospital in Berlin investigated a new sulfonamide for the treatment of severe infections, which was supplied by Boehringer Mannheim. After the high doses that were needed for treatment, his patients showed severe neurological symptoms as well as concentration and memory defects. After self-administration, he experienced all the signs of a hypoglycemic

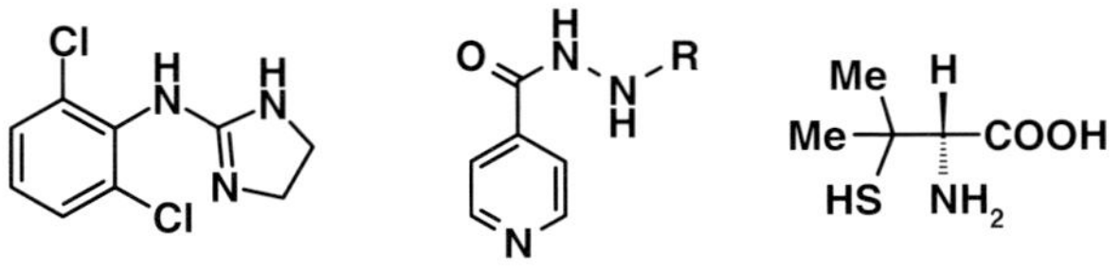

23

24 R = i-propyl
25 R = H

26

Figure 2.8 Clonidine **23** was designed as a nasal decongestant but it turned out to be a potent antihypertensive drug. Clinical tests revealed the antidepressant activity of iproniazid **24**, an isopropyl analog of the antituberculosis drug isoniazid **25**. D-Penicillamine **26** was originally used to treat Wilson's disease, to eliminate an excess of copper ions; later it was recognized to have beneficial effects in rheumatoid arthritis.

state, which disappeared after eating lunch. Fuchs and his supervisor Hans Franke investigated the compound in healthy and diabetic patients and were able to confirm their potential as relatively safe antidiabetic agents. The very first antidiabetic sulfonamides, carbutamide **20** and tolbutamide **21**, were later replaced by better-tolerated analogs like glibenclamide **22** (Figure 2.7) [2, 3, 6, 7, 10].

The anilino-imidazoline clonidine **23** (Catapresan®, Boehringer Ingelheim; Figure 2.8) was designed by the chemist Helmut Staehle as a nasal decongestant. When the secretary of a colleague caught a nasty cold, she was ready to test the new drug, telling them "I'll take anything if I can just get rid of these sniffles!" Shortly after taking the drug she became tired and fell asleep. After she was brought home, she continued sleeping for about 20 hours. A controlled self-experiment by her boss, the physician Martin Wolf, had the same outcome, with a heart rate reduction to about 40–48 beats s^{-1} and a blood pressure decrease to 90 vs. 60 mm Hg. Clearly, the compound was a potent antihypertensive drug, which was confirmed by further pharmacological and clinical investigations [3, 10, 13].

Iproniazid **24**, an alkyl analog of the antituberculous drug isoniazid **25** (Figure 2.8), surprisingly showed mood-improving activity in several depressed tuberculosis patients, which turned out to result from a monoamine oxidase (MAO) inhibitory activity. Since the compound was already registered as an antituberculosis drug and since it constituted the very first effective treatment of depression, more than 400 000 patients received it within only one year after the first announcement of its antidepressant activity [2, 33]. Later it was withdrawn from therapy, due to hepatotoxic side effects.

D-Penicillamine **26** (Figure 2.8) has for long time been used for the treatment of Wilson's disease, a metabolic disorder in which absorbed copper is deposited mainly in the liver and in the brain. Long-term application of this compound leads to suppression of rheumatoid arthritis, which now is its main therapeutic use [3].

Sildenafil (Viagra®, Pfizer), the first drug effective in male erectile dysfunction (MED), has a very interesting history. More than 30 years ago, the company May & Baker started research on antiallergic xanthine derivatives [19]. Their first leads **27**

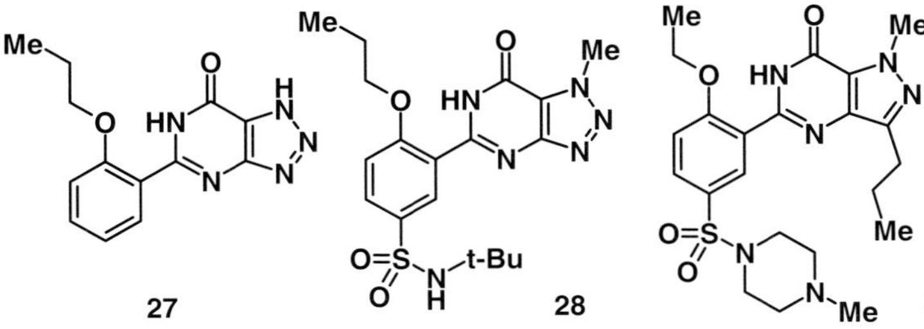

Figure 2.9 Already three decades ago, the clinical candidate zaprinast **27** (M&B 22, 948) and its sulfonamide analog **28** were synthesized. Whereas **27** is about 40 times more active as an antiallergic agent than the standard compound cromoglycate, **28** has a 1000-fold activity. Sildenafil **29** was originally investigated as an antianginal drug, but turned out to support and maintain penile erections.

and **28** (Figure 2.9), being between 40 times [19, 20] and 1000 times [21] more active than cromoglycate, the standard drug at this time, were structurally closely related to sildenafil. Zaprinast **27** (M&B 22, 948; Figure 2.9), was clinically tested as an orally active 'mast cell stabilizer' against histamine- and exercise-induced asthma. In addition to this activity, zaprinast has vasodilatory and antihypertensive side effects. In the mid 1980s, Nick Terrett and his team at Pfizer were searching for a new antihypertensive principle [22]. They followed the approach of enhancing the biological activity of the atrial natriuretic peptide (ANP) by prolonging the action of the second messenger of the corresponding receptor response. For this purpose, they were looking for a compound that would prevent the degradation of cyclic guanosine monophosphate (cGMP) by phosphodiesterase. As zaprinast **27** was one of the very few cGMP PDE inhibitors known in 1986, they started from this lead to improve its activity and selectivity. In 1989, the result of extensive structural modification was the PDE5-selective inhibitor sildenafil **29** (UK-92, 480; Figure 2.9), later clinically tested as an antianginal drug. The drug turned out to be safe and well tolerated but its clinical activity was disappointing. However, early in 1992, a 10-day toleration study in healthy volunteers led to the observation of a strange side effect. Among other effects, the patients reported some penile erections after the 4th or 5th day. Although it was not an obvious choice to test the new drug in male erectile dysfunction, its further clinical profiling went into this direction. After convincing clinical results, Viagra® was introduced into therapy in March 1998 [22].

2.3 Privileged Structures Bind to Many Different Targets

In 1988, Evans observed that organic compounds with certain structures "appear to contain common features which facilitate binding to various ... receptor surfaces, perhaps through binding elements different from those employed for binding of the natural ligands". From this observation he concluded that "certain privileged structures are capable of providing useful ligands for more than one receptor and that judicious modification of such structures could be a viable alternative in the search for new receptor agonists and antagonists" [23]. This was a generalization of observations that had been made before in studies of several classes of compounds, e.g., phenethylamines, tricyclic G protein-coupled receptor (GPCR) and transporter ligands, benzodiazepines, and steroids (e.g. [24]).

Phenethylamines include a wide variety of biologically active compounds. Depending on their lipophilicity, which correlates with their ability to penetrate the blood–brain barrier, they exert central nervous system activities (e.g., the lipophilic analogs amphetamine **30**, methamphetamine **31**, and MDMA **32**), peripheral activities (e.g., the polar analogs dopamine **33**, norepinephrine **34**, and epinephrine **35**) or both (e.g., ephedrine **36**), due to intermediate lipophilicity (Figure 2.10). The amino acid L-dopa **37** (Figure 2.10) is a special case: although it is even more polar than compounds **33–35**, it is absorbed and distributed into the brain by the amino acid transporter. In the brain, as well as in the periphery, it is then metabolically decarboxylated to dopamine; in combination with a polar dopa decarboxylase inhibitor, which acts only in the periphery, and a CNS-available monoamine oxidase inhibitor, it is used in the treatment of Parkinson's disease. The systematic chemical variation of dopamine and epinephrine has produced many highly selective, subtype-specific agonists, as well as antagonists (e.g. [3, 6, 7]).

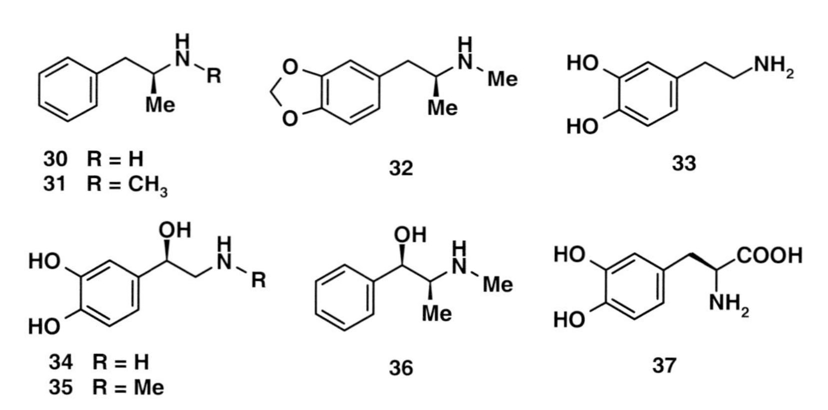

Figure 2.10 Amphetamine **30**, methamphetamine **31**, and methylenedioxymethamphetamine **32** (MDMA, ecstasy, XTC) are lipophilic compounds with good oral bioavailability; they easily cross the blood–brain barrier to exert central nervous system effects. Dopamine **33**, norepinephrine (noradrenalin) **34**, and epinephrine (adrenaline) **35** are polar phenethylamines; they have poor oral efficacy and do not pass the blood–brain barrier, producing only peripheral effects after intravenous application. Ephedrine **36** has intermediate lipophilicity; besides its peripheral effects it also acts as a central stimulant. Although L-dopa **37** is even more polar than dopamine **33**, it is orally active and crosses the blood–brain barrier by active transport mediated by the amino acid transporter.

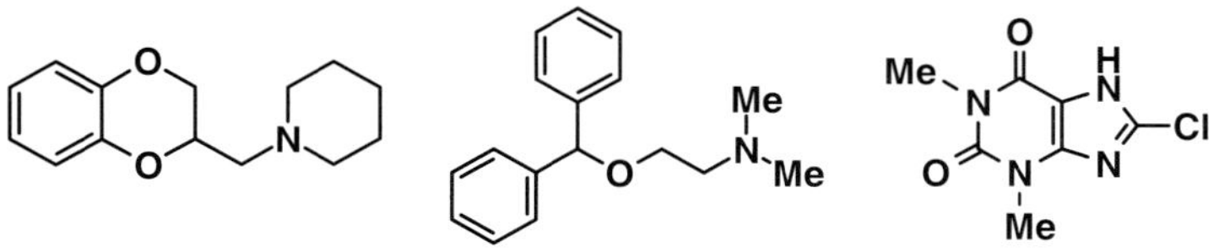

38 39 40

Figure 2.11 The antiadrenergic agent piperoxane **38** was the lead structure for the first antihistaminic drug diphenhydramine **39**. In an attempt to compensate for the sedative side effect of this compound, a complex with 8-chlorotheophylline **40**, dimenhydrinate, was investigated. In addition to its antihistaminic quality it is also effective against travel sickness.

The history of H_1 antihistaminics, neuroleptics, antidepressants, and some other drugs started from the observation that an antiadrenergic drug, piperoxan **38** (Figure 2.11), could also antagonize histamine in the guinea pig ileum. A breakthrough in therapy came after the synthesis of diphenhydramine **39** (Figure 2.11) in 1943 by George Rieveschl and Wilson Huber at the University of Cincinnati. Searle tried to get rid of the sedative side effect of this drug by combining it with the weak stimulant 8-chlorotheophylline **40** (Figure 2.11) in a complex, dimenhydrinate. Although the desired stimulating effect could not be observed, an interesting side effect appeared. A female patient, suffering from urticaria, realized that she could now travel in streetcars without becoming car sick, as before. This serendipitous observation led to the probably most curious 'clinical study' of all times: on November 27, 1947, the troop ship *General Ballou* sailed from New York to Bremerhaven. During the crossing, the sailors were treated with dimenhydrinate. Only 4% of those who received the drug became seasick, in contrast to about 25% of those who had received a placebo [2, 3].

Another antihistaminic, promethazine **41** (Figure 2.12), was the starting point for the development of potent neuroleptics. Henri Laborit, a French surgeon, was interested in preventing surgical shock by application of promethazine. Rhone-Poulenc supported his work by providing several analogs of this compound. One analog, chlorpromazine **42** (Figure 2.12), not only improved the condition of the patients due to its anti-shock action, but also seemed to make them more relaxed and less concerned about what was happening to them in the stressful preoperation period. On January 19, 1952, Joseph Hamon, Jean Paraire, and Jean Velluz treated, for the first time, a manic patient with chlorpromazine. After being injected with the compound he became calm and remained so for several hours; this day must be considered the start of successful drug treatment of psychotic diseases. Whereas promethazine preferentially antagonizes histamine H_1 receptors, chlorpromazine is a dopamine antagonist [2, 3]. In the mid-1950s, Roland Kuhn at the Cantonal Psychiatric Clinic in Münsterlingen, Switzerland, became interested in the tranquilizing properties of chlorpromazine; his work was supported by the synthesis of some new analogs by Geigy, Basle. Among these compounds, imipramine **43**

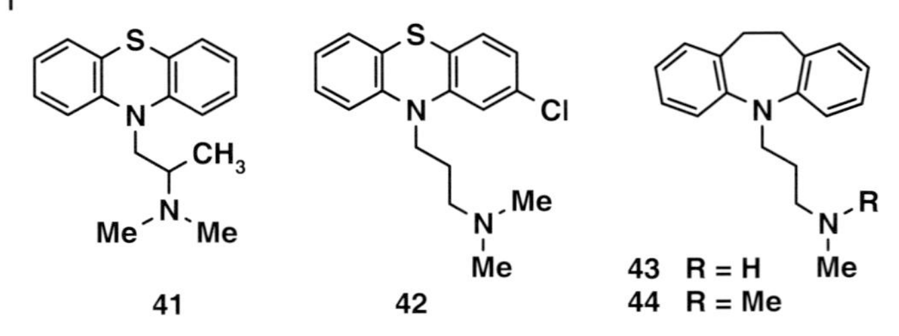

Figure 2.12 Despite a very close structural analogy, promethazine **41** is an antihistaminic drug, chlorpromazine **42** is a dopamine antagonist, and imipramine **43** and desipramine **44** are neurotransmitter uptake inhibitors. Correspondingly, **41** is used for the treatment of allergic inflammation, **42** for schizophrenia, and **43** and **44** for depression.

(Figure 2.12) turned out to have pronounced antidepressant activity. Imipramine, as well as its metabolite desipramine **44** (Figure 2.12), are neurotransmitter uptake blockers [2, 3].

The structure–activity relationships of the tricyclic compounds (and some other classes, e.g., the steroid hormones) prove that small structural variations may significantly alter the mode of action. Another example of even minor modifications being responsible for new mechanisms of action is that of some close analogs of the antihistaminic prototype diphenhydramine **39** (Figure 2.11), e.g., orphenadrine **45**, an atropine-type anticholinergic, and nefopam **46**, a non-opioid analgesic with largely unknown mechanism of action (Figure 2.13) [2, 3].

Benzodiazepines seem to be the most prominent class of privileged structures. Only some prototypes **47–50** (Figure 2.14) with GABA agonist, antagonist, inverse agonist, opiate agonist, and CCK antagonist activities are presented here [8, 23–27]. The CCK antagonist devazepide **51**, a structurally simplified analog of the microbial product asperlicin **52**, is about 4 orders of magnitude more active than its natural lead [7]. Other benzodiazepines are, e.g., muscle relaxants, antidepressants, neuroleptics, hypnotics, NK-1 receptor and vasopressin receptor antagonists, farnesyl transferase inhibitors, and potassium channel modulators.

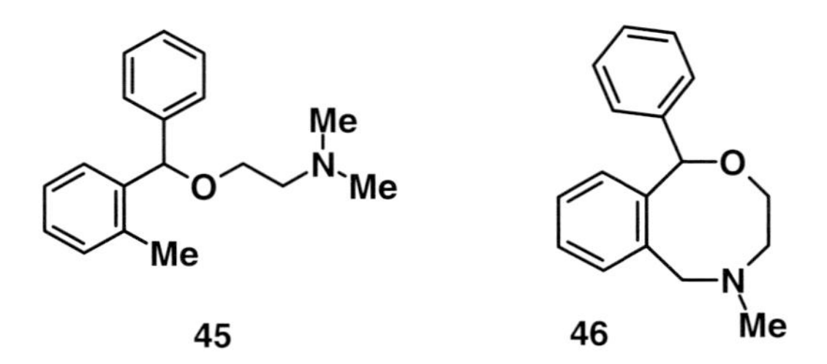

Figure 2.13 Minor structural modifications of the antihistaminic drug diphenhydramine **39** (Figure 2.11) produced the anticholinergic drug orphenadrine **45** and the non-opioid analgesic nefopam **46**.

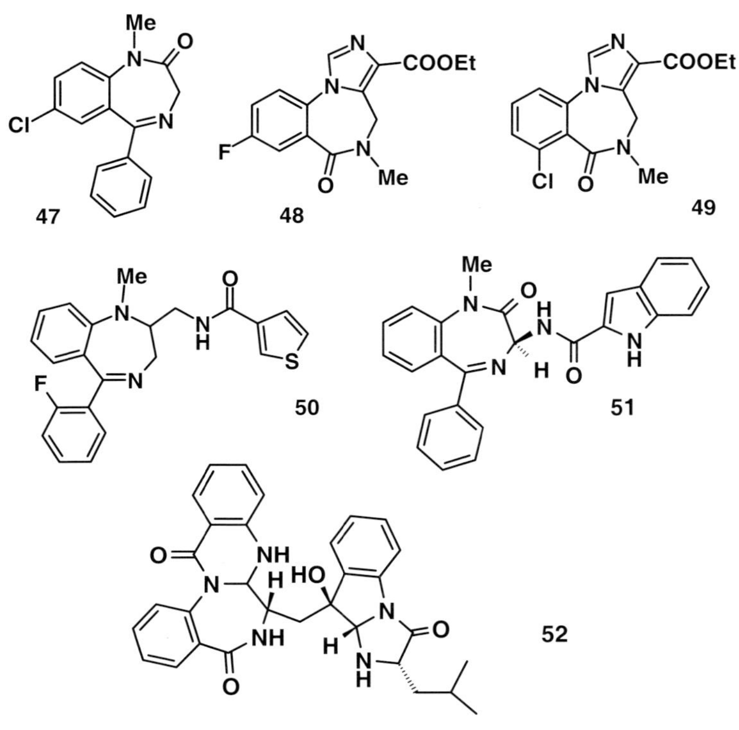

Figure 2.14 Benzodiazepines are a striking example of diverse biological activities of closely related structural analogs. The tranquilizer diazepam **47** is the prototype of a benzodiazepine agonist. The benzodiazepine antagonist flumazenil **48** is used as an antidote in treating benzodiazepine intoxication and after benzodiazepine use in surgery. Compound Ro-15-3505, **49**, is an inverse agonist, which acts as a proconvulsant. Tifluadom **50** is a strong opioid agonist, which selectively binds to the κ-opiate receptor, and a nanomolar cholecystokinin receptor antagonist. Devazepide **51** is an orally active cholecystokinin-B (CCK-B) antagonist, which is about four orders of magnitude more active than its structurally much more complex lead, the natural product asperlicin **52**.

2.4 Optimizing the Selectivity of Nonselective Lead Structures

Medicinal chemists always followed and still apply the principle of chemical and biological similarity. Whenever they discover an active lead, they modify its chemical structure more or less systematically, to find similar analogs with improved activities, selectivities, ADME (absorption, distribution, metabolism, elimination) properties; fewer side effects; and less toxic properties. However, as discussed above, structurally closely related analogs may have significantly different specificity or even a completely different mode of action.

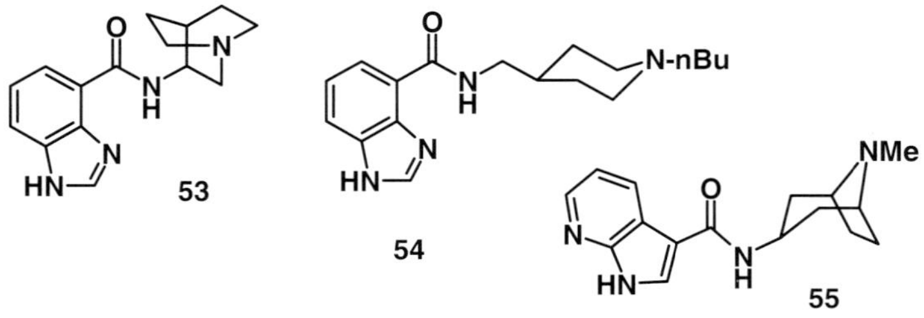

Figure 2.15 Compounds **53** and **54** differ only in their amine part, but **53** shows a greater than 300-fold selectivity for the 5-HT_3 receptor, as compared to the 5-HT_4 receptor, whereas **54** is at least three orders of magnitude more active at the 5-HT_4 receptor than at the 5-HT_3 receptor. The closely related compound **55** is an orally active antitussive drug.

Despite their close chemical relationship, the benzimidazole carboxamides **53** and **54** (Figure 2.15) show very different receptor subtype selectivities. Compound **53** has a more than 300-fold higher affinity for the 5-HT_3 ion channel than for the G-protein-coupled 5-HT_4 receptor (K_i 5-HT_3 = 3.7 nM vs. K_i 5-HT_4 > 1000 nM), whereas compound **54** binds almost exclusively to the 5-HT_4 receptor (K_i 5-HT_3 > 10 000 nM vs. K_i 5-HT_4 = 13.7 nM; selectivity > 700) [28, 29]. The chemically related compound DF-1012, **55** (Figure 2.15), is an orally active antitussive drug [30].

Integrins are cell-surface receptors; several of them recognize the RGD (arginine-glycine-aspartate) motif, obviously in different bioactive conformations. Thus, selective antagonists of these receptors have been developed: SB 214 857 (lotrafiban, SmithKline Beecham), **56**, and SB 223 245, **57** (Figure 2.16) [31, 32]. Both compounds are identical in their benzodiazepine part but differ in the amine residues attached to the carboxylic acid function. Compounds **53** and **54**, as well as compounds **56** and **57**, may be used as strong arguments for the potential of automated parallel syntheses. Although these compounds were optimized in a classical manner, they could also have resulted from combinatorial libraries, in which a single acid reacted with different amines.

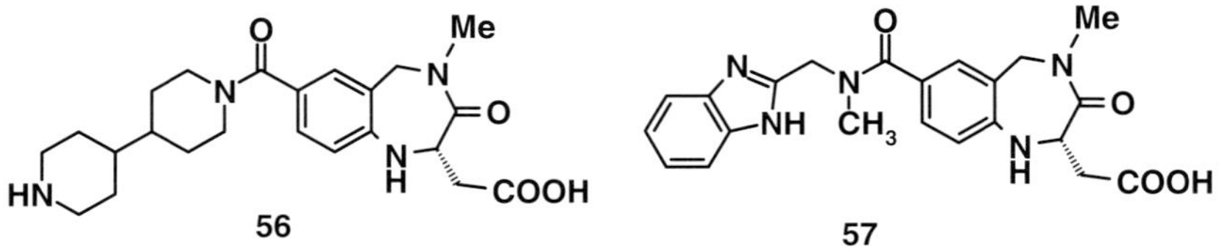

Figure 2.16 Like the 5-HT receptor ligands **53** and **54** (Figure 2.15), also lotrafiban **56** (failed in phase III clinical trials) and compound **57** differ only in their amine component but are highly selective for different integrins. Lotrafiban binds preferentially to the fibrinogen receptor (integrin $GP_{IIb/IIIa}$), with a selectivity factor of about 4000, whereas **57** binds preferentially to the vitronectin receptor (integrin $\alpha_v\beta_3$), with a selectivity factor of 15 000. Overall, this constitutes a selectivity ratio of more than seven orders of magnitude.

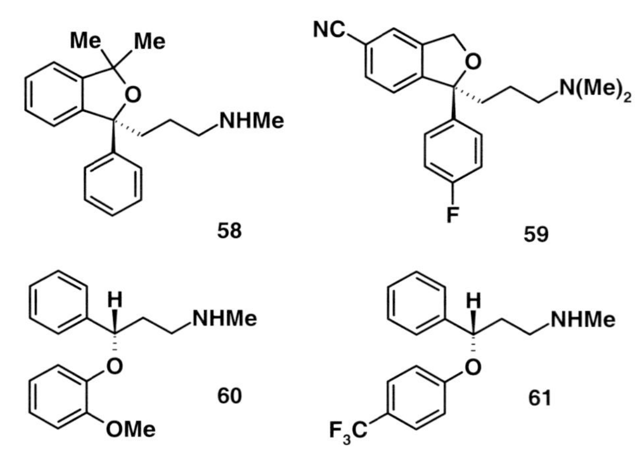

Figure 2.17 The structural analogs talopram **58** and citalopram **59** (upper compounds), as well as nisoxetine **60** and fluoxetine **61** (lower compounds), are chemically closely related. Whereas **58** and **60** (left compounds) are highly selective norepinephrine uptake inhibitors (selectivity factors of 550 and 180, respectively), the close analogs **59** and **61** (right compounds) are selective serotonin uptake inhibitors (selectivity factors of 3400 and 54, respectively).

Talopram **58** and citalopram **59** (Figure 2.17) are closely related in their chemical structures. Nevertheless, talopram is a norepinephrine uptake blocker with a selectivity factor of about 550 against serotonin uptake, whereas citalopram is a serotonin uptake blocker, with a selectivity of 3400 against norepinephrine uptake. A similar selectivity difference applies to the even more closely related pair nisoxetine **60**, with a norepinephrine uptake selectivity of about 180, and fluoxetine **61** (Figure 2.17), with a serotonin uptake selectivity of 54 [33].

Out of a large number of different peptidomimetic somatostatin analogs, compounds **62–66** (Figure 2.18) resulted from four different combinatorial libraries, with up to 350 000 members per library. Every compound has a more or less pronounced affinity to one of the five different somatostatin receptor subtypes sst1–sst5, with remarkable selectivity against the other subtypes (Table 2.1) [34].

Most cases of chronic myelogenous leukemia result from a cross-over between chromosomes 9 and 22, by which a longer chromosome 9+ and a shorter chromosome 22–, the so-called Philadelphia chromosome, are generated. The sequence at the fusion point of the two DNA strands in the 22– chromosome codes for a new protein, the so-called bcr-abl protein, with constitutionally enhanced tyrosine protein kinase activity [35]. At Novartis, research started from a general lead structure **67** (Figure 2.19), with protein kinase C (PKC)-inhibitory activity. When amide residues were introduced into an optimized PKC inhibitor **68**, bcr-abl inhibition was also observed for compound **69**. A methyl group in position R1 (compound **70**) then surprisingly abolished the undesired PKC activity. The result of the optimization was imatinib **71** (STI571, Gleevec®, Glivec®, Novartis; Figure 2.19), a highly selective bcr-abl kinase inhibitor [35].

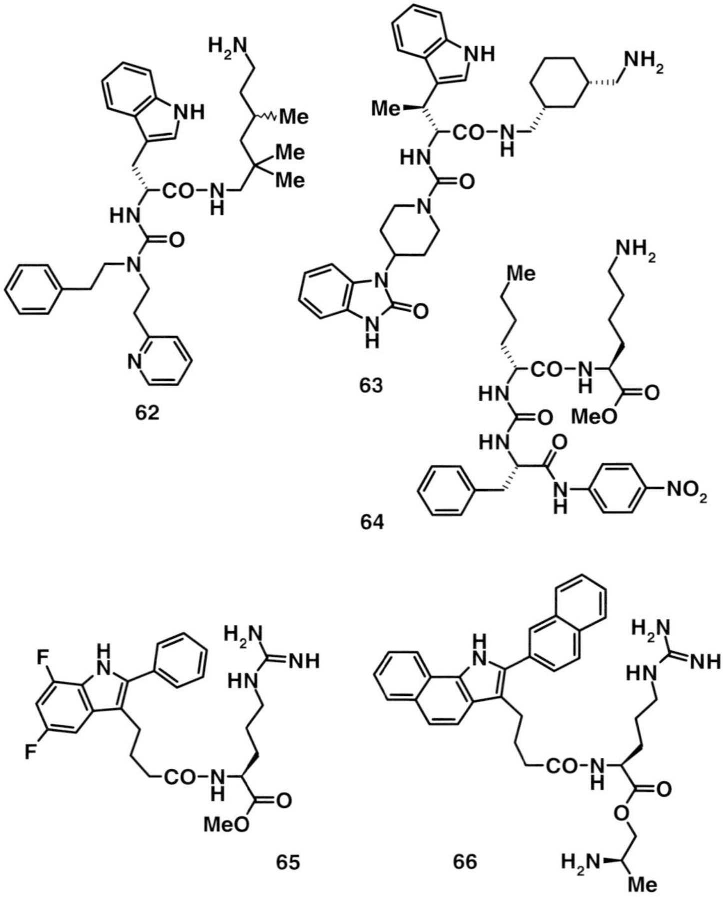

Figure 2.18 The tetradecapeptide somatostatin is a nanomolar-to-subnanomolar ligand of five different somatostatin receptor subtypes. Compounds **62**–**66** are structurally simplified analogs from four combinatorial libraries, with up to 350 000 members per library. Each compound shows a remarkable selectivity against the different sst1–sst5 receptor subtypes (Table 2.1). The orientation of the compounds follows a projection of their superposition with a Merck cyclopeptide.

Table 2.1 Somatostatin receptor subtype affinities of the tetradecapeptide somatostatin (K_i values in nM) and compounds **62**–**66** (Figure 2.18; K_d values in nM) [34].

Compound	*sst1*	*sst2*	*sst3*	*sst4*	*sst5*
Somatostatin	0.4	0.04	0.7	1.7	2.3
62	1.4	1875	2240	170	3600
63	2760	0.05	729	310	4260
64	1255	> 10000	24	8650	1200
65	199	4720	1280	0.7	3880
66	3.3	52	64	82	0.4

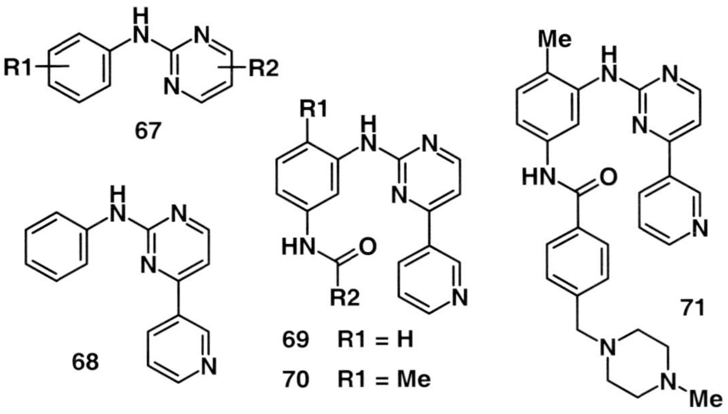

Figure 2.19 The lead structure **67** is a protein kinase C (PKC) inhibitor prototype. Whereas the optimized analog **68** is a strong PKC inhibitor, amides **69** also inhibit tyrosine kinases like the bcr-abl kinase. Introduction of a methyl group, to form **70**, abolishes the PKC activity. The optimized analog imatinib **71** (Gleevec®, Glivec®, Novartis) is a highly selective bcr-abl tyrosine kinase inhibitor.

2.5 Selective Optimization of Side Activities

Recently, Wermuth proposed an alternative and complementary strategy to high-throughput screening (HTS), the SOSA approach (SOSA = selective optimization of side activities) [15]. Instead of the laborious and expensive investigation of several hundred thousand compounds in HTS, he recommended screening new biological targets only with a small set of well-known drug molecules for which bioavailability and toxicity studies have already been performed. Hits from such a screening can then be used in a drug discovery program. Most drugs in human therapy do not interact with just one biological target; thus, if an interaction with some other target is unrelated to the primary therapeutic effect, affinities could be reversed in the optimization process, the former side effect now becoming the main effect and vice versa.

This approach can be illustrated, e.g., by the structural variation of β-antiadrenergic compounds. Several β-blockers have slight stimulating and hallucinogenic properties. If the side chain of the β-blocker prototype **72** (Figure 2.20) is cyclized, the antidepressant viloxazine **73** results [2, 3]. Another β-blocker-related compound is propafenone **74**, a class Ic antiarrhythmic [3, 6, 7]. Cyclization of a β-blocker structure to compound **75** produced, after optimization, the antihypertensive drug levocromakalim **76** (Figure 2.20). However, **76** is no longer a β-blocker, it is a potassium channel opener [15, 36].

A group of compounds for which a completely unrelated side effect was observed are 4-hydroxy-pyrones and 4-hydroxy-coumarons, which are chemically closely related to anticoagulant vitamin K antagonists. Screening at Parke-Davis showed

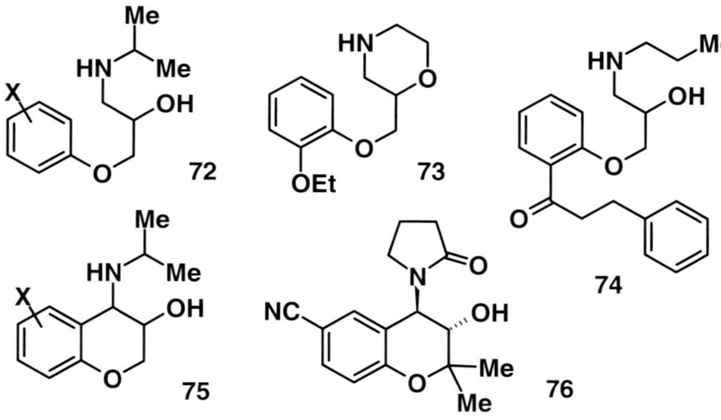

Figure 2.20 The general structure **72**, with different residues X, describes the prototype of a β-adrenergic antagonist. Cyclization of the side chain produced the antidepressant viloxazine **73**, whereas the *N*-*n*-propyl analog propafenone **74** turned out to be a class Ic antiarrhythmic with only weak β-antagonistic activity. An attempt to cyclize β-blockers to structures of the prototype **75** finally produced levocromakalim **76**; as expected, it had antihypertensive activity but its mode of action is different: instead of being a β-blocker, it is a potassium channel opener.

that the 4-hydroxy-pyrone **77** is a micromolar HIV protease inhibitor (K_i = 10 μM) [37]. Optimization of this prototype led to the phenethyl-substituted thio ether **78** (K_i = 35 nM) [38], which could be further optimized, by exchange of a phenyl group with an isopropyl residue and introduction of amino and hydroxyl groups, to CI-1029, **79** (K_i = 0.11 nM) [39, 40]. At Upjohn, phenprocoumon **80**, a therapeutically used anticoagulant, was independently discovered to be a moderately active HIV protease inhibitor (K_i = 1 μM) [41]. Optimization produced the bis-aralkyl-substituted 4-hydroxy-pyrone PNU-96 988, **81** (K_i = 38 nM) [41], and finally the picomolar inhibitor tipranavir **82** (*R*,*R* diastereomer: K_i = 8 pM). Surprisingly, the other diastereomers of tipranavir show a very low stereospecificity of drug action; they are also very potent HIV protease inhibitors (*R*,*S* diastereomer: K_i = 18 pM; *S*,*R* diastereomer: K_i = 32 pM; *S*,*S* diastereomer: K_i = 220 pM) [42].

The selective optimization of side activities was illustrated by Wermuth also with some examples from his own research [15, 43–46]. The antidepressant minaprine **83** (Figure 2.22) has some weak side activities. It is, e.g., a 17 μM muscarinic M_1 receptor ligand and a 600 μM acetylcholinesterase inhibitor (IC_{50}, electric eel AChE). Shift of the 4-methyl group of minaprine to the 5-position (compound **84**) increased its M_1 affinity to 550 nM. Further optimization led to the tropane analog **85** (IC_{50} musc M_1 = 50 nM) and its *o*-hydroxy derivative **86** (IC_{50} musc M_1 = 3 nM; Figure 2.22), an M_1 partial agonist [15, 43].

Elimination of the 4-methyl group of minaprine **83** (Figure 2.22) and exchange of the morpholine with a piperidine ring produced the AChE inhibitor **87** (Figure 2.23),

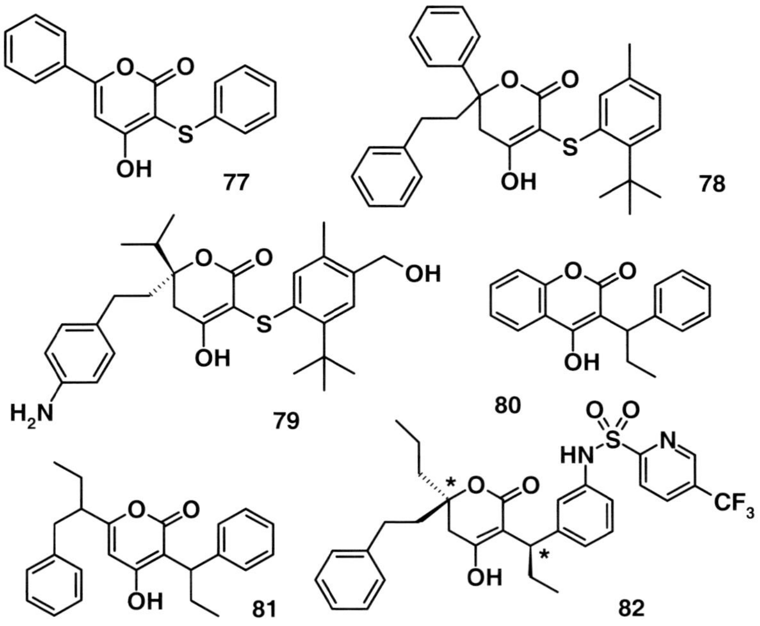

Figure 2.21 The achiral 4-hydroxypyrone **77** (K_i = 10 μM) is structurally related to anticoagulant vitamin K antagonists; it was discovered at Parke-Davis as a weakly active lead in a screening for HIV protease inhibitors. Optimization produced the thio ether **78** (K_i = 35 nM) and finally CI-1029, **79** (K_i = 0.11 nM). In an independent screening, Upjohn discovered that the therapeutically used anticoagulant phenprocoumon **80** (K_i = 1 μM) is a weak HIV protease inhibitor. Optimization at Pharmacia and Upjohn produced PNU-96 988, **81** (K_i = 38 nM), and the picomolar HIV protease inhibitor tipranavir **82** (*R*,*R* diastereomer: K_i = 8 pM).

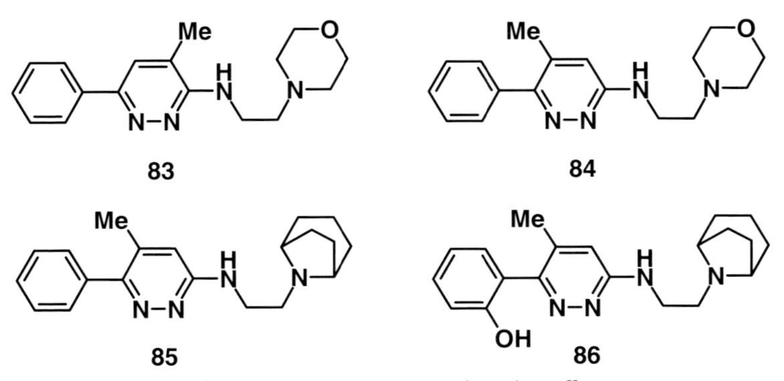

Figure 2.22 The antidepressant minaprine **83** is also a low-affinity ligand of the muscarinic M_1 receptor (K_i = 17 μM). Optimization of this side activity to the 5-methyl isomer **84** (K_i musc M_1 = 550 nM) and the tropane analog **85** (K_i musc M_1 = 50 nM) resulted in the *ortho*-hydroxy-substituted analog **86** (K_i musc M_1 = 3 nM).

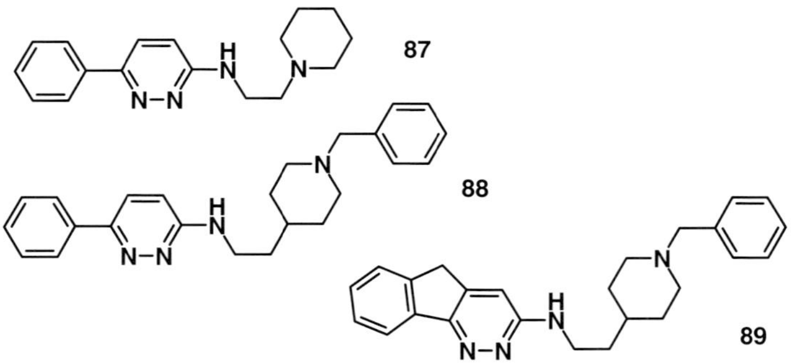

Figure 2.23 Minaprine **83** (Figure 2.22) is also a weak acetylcholinesterase (AChE) inhibitor (K_i AChE = 600 μM). Optimization of this side activity to deoxo,demethyl-minaprine **87** (K_i AChE = 13 μM) and an isomeric *N*-benzyl-piperazine **88** (K_i AChE = 120 nM) finally resulted in the potent AChE inhibitor **89** (K_i AChE = 10 nM).

with a K_i value of 13 μM. Variation of the side chain to an *N*-benzyl-piperidino derivative **88** increased the inhibitory activity to K_i AChE = 120 nM. Further optimization produced the cyclized analog **89** (K_i AChE = 10 nM: Figure 2.23) [15, 44, 45].

A close analog of minaprine **83** (Figure 2.22), compound **90**, is a low-affinity 5-HT_3 antagonist (IC_{50} 5-HT_3 = 425 nM). Whereas modifying the pyridazine to the phenyl-substituted phthalazine **91** did not significantly change the affinity (IC_{50} 5-HT_3 = 370 nM), shifting the phenyl substituent increased the affinity by about one order of magnitude (compound **92**, IC_{50} 5-HT_3 = 36 nM), most probably because of a better fit within the binding site. Correspondingly, the phthalazine **93** (IC_{50} 5-HT_3 = 10 nM), without a phenyl substituent, has a much higher affinity than the other analogs (Figure 2.24) [46].

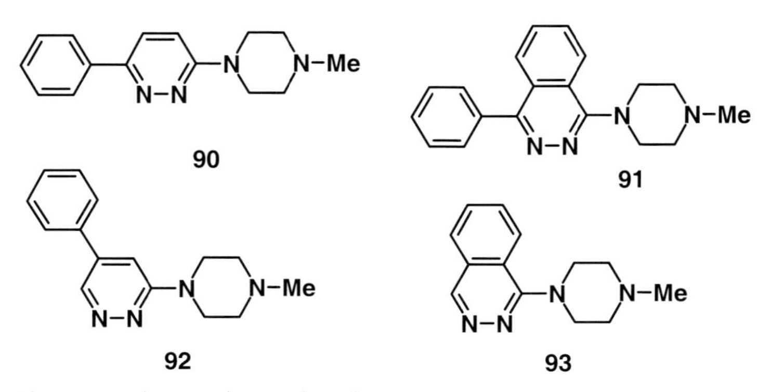

Figure 2.24 The 3,6-substituted pyridazine **90** (K_i 5-HT_3 = 425 nM), which is chemically closely related to minaprine **83** (Figure 2.22), is a low-affinity 5-HT_3 receptor ligand. Optimization of this side activity to the phenyl-substituted phthalazine **91** (K_i 5-HT_3 = 370 nM) and the 3,5-substituted pyridazine **92** (K_i 5-HT_3 = 36 nM) resulted in the nanomolar 5-HT_3 ligand **93** (K_i 5-HT_3 = 10 nM).

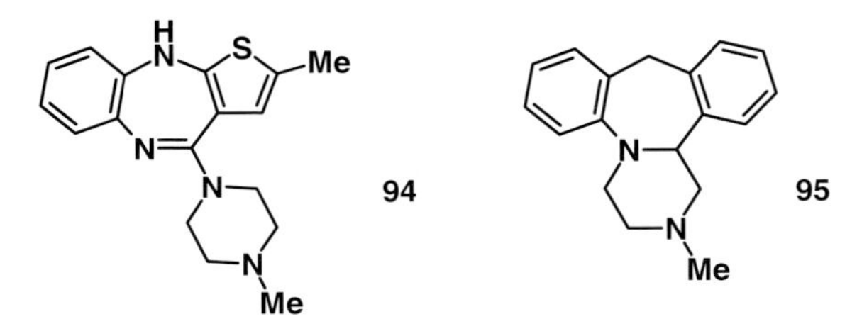

Figure 2.25 The atypical neuroleptic olanzapine **94** is a highly promiscuous ligand of many different G protein-coupled receptors (Table 2.1). The antidepressant mianserin **95** also is a promiscuous ligand; in addition to its α_2-blocking activity, it blocks serotonin uptake and is a histamine, 5-HT_2, and 5-HT_3 antagonist.

The atypical neuroleptic olanzapine **94** (Figure 2.25), one of the most successful drugs of recent years, is a highly promiscuous, nanomolar ligand of more than a dozen different GPCRs (Table 2.2) [47–49]. Thus, its real mechanism of action remains unclear, as is also true for the antiadrenergic, antihistaminic, and antiserotonergic antidepressant mianserin **95** (Figure 2.25), the so-called 'good humor pill' of the 1970s. Thus, despite the fact that many structurally modified analogs have already been synthesized and tested, such drugs could be the starting point for different follow-on drugs with modified selectivities [2, 3].

Drug candidates have also been derived from herbicides and fungicides. By screening the BASF library for endothelin receptor ligands, compound **96** was

Table 2.2 GPCR and 5-HT_3 binding affinities of olanzapine **94** in different in-vitro models; K_i values from two different sources [47–49].

Receptor	*K_i values [nM]*
5-HT_{2A}	2.5; 4
5-HT_{2B}	12
5-HT_{2C}	11; 29
5-HT_3	57
dop D_1	31; 119
dop D_2	11
dop D_4	27
musc M_1	1.9; 2.5
musc M_2	18
musc M_3	13; 25
musc M_4	10; 13
musc M_5	6
adr α_1	19
adr α_2	230
hist H_1	7

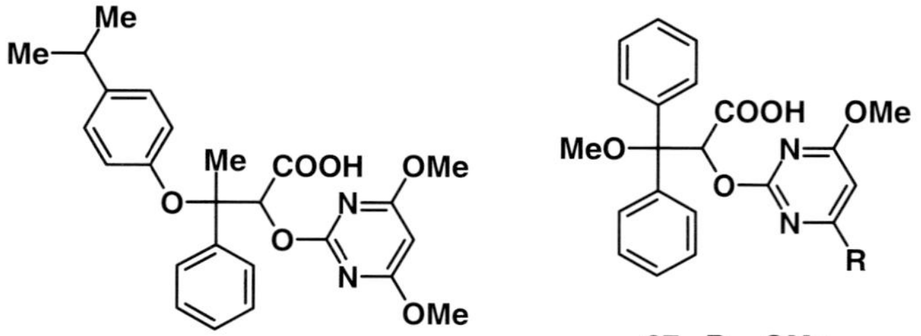

Figure 2.26 The potential herbicide **96** has been discovered to be a moderately active, selective ET_A receptor antagonist. Optimization produced the nanomolar to subnanomolar ET_A antagonists **97** and **98**.

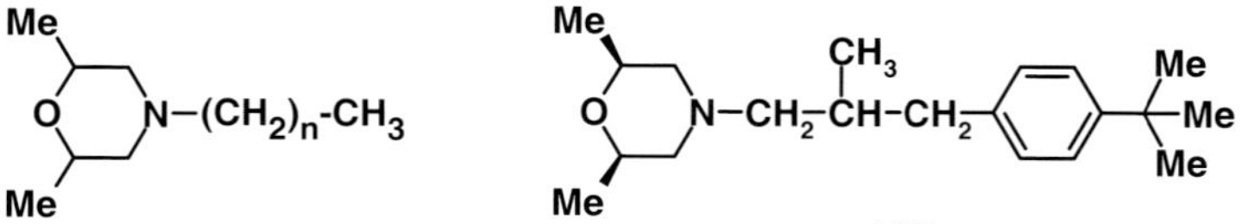

Figure 2.27 The fungicides tridemorph **99** (n = 12 is the major component, but n = 10, 11, and 13 are present in minor amounts) and fenpropimorph **100** are picomolar σ_1 receptor ligands which still await their optimization to drug candidates.

discovered; although this structure was originally designed as a potential herbicide, it turned out to be a submicromolar, ET_A-specific ligand (K_i ET_A = 0.25 μM, K_i ET_B = 3 μM; Figure 2.26). Optimization by elimination of one steric center yielded the nanomolar antagonists **97** (K_i ET_A = 6 nM, K_i ET_B = 1000 nM) and **98** (K_i ET_A = 0.12 nM, K_i ET_B = 29 nM) [50].

Sigma receptors were originally considered to be a subtype of opiate receptors. Only recently has it become clear that, despite the fact that they bind some opiates, they are neither G protein-coupled receptors nor do they have any other homology to mammalian proteins. The closest related protein is yeast sterol C_8–C_7 isomerase (ERG2 protein). According to this relationship, the BASF fungicides tridemorph **99** (K_i σ_1 = 39 pM, guinea pig liver; K_i σ_1 = 23 pM, guinea pig brain; replacement of (+)-^{3}H-pentazocine) and fenpropimorph **100** (K_i σ_1 = 11 pM, guinea pig liver; K_i σ_1 = 5 pM, guinea pig brain) are picomolar ligands of σ_1 receptors (Figure 2.27) [51]. So far, they have not been converted into drugs, despite the fact that σ receptors play a functional role in many important physiological processes.

2.6 Summary and Conclusions

The discussed examples provide convincing evidence that, in addition to many drugs that were serendipitous discoveries, many others have resulted from the observation of side effects, in the laboratory, in clinics, or during their therapeutic application. Today, we possibly focus too much on single targets that are investigated in vitro. Hidden treasures may be discovered by testing 'old chemistry' against new targets, by systematically optimizing some side effects of known drugs, and by rescuing drugs that failed because of problems in their metabolism or hERG channel inhibition. Thus, it might well be that known drugs are a much better source of lead structures for new projects than we anticipated so far. As a consequence, we will experience a successful comeback of traditional medicinal chemistry [8, 15].

References

1 A. Burger, *A Guide to the Chemical Basis of Drug Design*, John Wiley & Sons, New York **1983**.

2 W. Sneader, *Drug Discovery: The Evolution of Modern Medicines*, John Wiley & Sons, Chichester **1985**.

3 W. Sneader, *Drug Prototypes and Their Exploitation*, John Wiley & Sons, Chichester **1996**.

4 G. Ehrhart, H. Ruschig, *Arzneimittel. Entwicklung, Wirkung, Darstellung*, Verlag Chemie, Weinheim **1972**.

5 F. H. Clarke, *How Modern Medicines are Discovered*, Futura Publishing, Mount Kisko, NY **1973**.

6 E. Mutschler, H. Derendorf (Eds.), *Drug Actions: Basic Principles and Therapeutic Aspects*, CRC Press, Stuttgart, Boca Raton **1995**.

7 D. J. Abraham (Ed.), *Burger's Medicinal Chemistry & Drug Discovery*, John Wiley & Sons, Hoboken, NJ **2003**.

8 C. G. Wermuth (Ed.), *The Practice of Medicinal Chemistry*, Academic Press, London **1996**; 2nd edit., Academic Press, London **2003**.

9 R. M. Roberts, *Serendipity: Accidental Discoveries in Science*, John Wiley & Sons, New York **1989**.

10 E. Bäumler, *Die großen Medikamente. Forscher und ihre Entdeckungen schenken uns Leben*, Gustav Lübbe Verlag, Bergisch Gladbach **1992**.

11 C. R. Ganellin, S. M. Roberts (Eds.), *Medicinal Chemistry. The Role of Organic Chemistry in Drug Research*, 2nd edit., Academic Press, London **1993**.

12 W. Sneader, in General Principles, P. D. Kennewell (Ed.), *Volume 1 of Comprehensive Medicinal Chemistry*, C. Hansch, P. G. Sammes, J. B. Taylor (Eds.), Pergamon, Oxford **1990**, pp. 7–80.

13 G. de Stevens, *Progr. Drug Res.* **1986**, *30*, 189–203.

14 J. F. Ryan (Ed.), *The Pharmaceutical Century. Ten Decades of Drug Discovery*, Supplement to ACS Publications, American Chemical Society, Washington, DC **2000**.

15 (a) C. G. Wermuth, *Med. Chem. Res.* **2001**, *10*, 431–439; (b) C. G. Wermuth, *J. Med. Chem.* **2004**, *47*, 1303–1314.

16 H. Kubinyi, *J. Receptor & Signal Transduction Research* **1999**, *19*, 15–39.

17 H. Buschmann, T. Christoph, E. Friderichs, C. Maul, B. Sundermann (Eds.), *Analgesics: FromChemistry and Pharmacology to Clinical Application*, Wiley-VCH, Weinheim **2002**.

18 J. J. Baldwin, G. S. Ponticello, P. S. Anderson, M. E. Christy, M. A. Murcko, W. C. Randall, H. Schwam, M. F. Sugrue, J. P. Springer, P. Gautheron, J. Grove, P. Mallorga, M.-P. Viader,

B. M. McKeever, M. A. Navia, *J. Med. Chem.* **1989**, *32*, 2510–2513.

19 B. J. Broughton, P. Chaplen, P. Knowles, E. Lunt, S. M. Marshall, D. L. Pain, K. R. H. Wooldridge, *J. Med. Chem.* **1975**, *18*, 1117–1122.

20 T. Fujita, in *Quantitative Drug Design*, C. A. Ramsden (Ed.), *Vol. 4 of Comprehensive Medicinal Chemistry. The Rational Design, Mechanistic Study & Therapeutic Application of Chemical Compounds*, C. Hansch, P. G. Sammes, J. B. Taylor (Eds.), Pergamon, Oxford **1990**, pp. 497–560.

21 K. R. H. Wooldridge, personal communication **1976**.

22 J. Kling, Modern *Drug Discov.*, Nov./Dec. **1998**, pp. 31–38.

23 B. E. Evans, K. E. Rittle, M. G. Bock, R. M. DiPardo, R. M. Freidinger, W. L. Whitter, G. F. Lundell, D. F. Veber, P. S. Anderson, R. S. L. Chang, V. J. Lotti, D. J. Cerino, T. B. Chen, P. J. Kling, K. A. Kunkel, J. P. Springer, J. Hirshfield, *J. Med. Chem.* **1988**, *31*, 2235–2246.

24 A. A. Patchett, R. P. Nargund, *Ann. Rep. Med. Chem.* **2000**, *35*, 289–298.

25 D. Römer, H. H. Büscher, R. C. Hill, R. Maurer, T. J. Petcher, H. Zeugner, W. Benson, E. Finner, W. Milkowski, P. W. Thies, *Nature* **1982**, *298*, 759–760.

26 R. S. Chang, V. J. Lotti, T. B. Chen, M. E. Keegan, *Neurosci. Lett.* **1986**, *72*, 211–214.

27 B. E. Evans, M. G. Bock, K. E. Rittle, R. M. DiPardo, W. L. Whitter, D. F. Veber, P. S. Anderson, R. M. Freidinger, *Proc. Natl. Acad. Sci. USA* **1986**, *83*, 4918–4922.

28 A. Ursini, A. M. Capelli, R. A. Carr, P. Cassara, M. Corsi, O. Curcuruto, G. Curotto, M. Dal Cin, S. Davalli, D. Donati, A. Feriani, H. Finch, G. Finizia, G. Gaviraghi, M. Marien, G. Pentassuglia, S. Polinelli, E. Ratti, A. M. Reggiani, G. Tarzia, G. Tedesco, M. E. Tranquillini, D. G. Trist, F. T. Van Amsterdam, A. Reggiani, *J. Med. Chem.* **2000**, *43*, 3596–3613; erratum in *J. Med. Chem.* **2000**, *43*, 5057.

29 M. L. Lopez-Rodriguez, M. J. Morcillo, B. Benhamu, M. L. Rosado, *J. Comput.-Aided Mol. Design* **1997**, *11*, 589–599.

30 R. Aslanian, J. A. Hey, N.-Y. Shih, *Ann. Rep. Med. Chem.* **2001**, *36*, 31–40.

31 J. M. Samanen, F. E. Ali, L. S. Barton, W. E. Bondinell, J. L. Burgess, J. F. Callahan, R. R. Calvo, W. Chen, L. Chen, K. Erhard, G. Feuerstein, R. Heys, S.-M. Hwang, D. R. Jakas, R. M. Keenan, T. W. Ku, C. Kwon, C.-P. Lee, W. H. Miller, K. A. Newlander, A. Nichols, M. Parker, C. E. Peishoff, G. Rhodes, S. Ross, A. Shu, R. Simpson, D. Takata, T. O. Yellin, I. Uzsinskas, J. W. Venslavsky, C. K. Yuan, W. F. Huffman, *J. Med. Chem.* **1996**, *39*, 4867–4870.

32 R. M. Keenan, W. H. Miller, C. Kwon, F. E. Ali, J. F. Callahan, R. R. Calvo, S. M. Hwang, K. D. Kopple, C. E. Peishoff, J. M. Samanen, A. S. Wong, C. K. Yuan, W. F. Huffman, *J. Med. Chem.* **1997**, *40*, 2289–2292.

33 K. Gundertofte, K. P. Bogeso, T. Liljefors, in *Computer-Assisted Lead Finding and Optimization* (Proceedings of the 11th European Symposium on Quantitative Structure–Activity Relationships, Lausanne 1996), H. van de Waterbeemd, B. Testa, G. Folkers (Eds.), Verlag Helvetica Chimica Acta and VCH, Basel, Weinheim **1997**, pp. 445–459.

34 S. P. Rohrer, E. T. Birzin, R. T. Mosley, S. C. Berk, S. M. Hutchins, D. M. Shen, Y. Xiong, E. C. Hayes, R. M. Parmar, F. Foor, S. W. Mitra, S. J. Degrado, M. Shu, J. M. Klopp, S. J. Cai, A. Blake, W. W. Chan, A. Pasternak, L. Yang, A. A. Patchett, R. G. Smith, K. T. Chapman, J. M. Schaeffer, *Science* **1998**, *282*, 737–740; erratum *Science* **1998**, *282*, 1646.

35 R. Capdeville, E. Buchdunger, J. Zimmermann, A. Matter, *Nature Rev. Drug Discov.* **2002**, *1*, 493–502.

36 G. Stemp, J. M. Evans, in Me*dicinal Chemistry. The Role of Organic Chemistry in Drug Research*, 2nd edit., C. R. Ganellin, S. M. Roberts (Eds.), Academic Press, London **1993**, pp. 141–162.

37 J. V. N. Vara Prasad, K. S. Para, E. A. Lunney, D. F. Ortwine, J. B. Dunbar, D. Ferguson,

P. J. Tummino, D. Hupe, B. D. Tait, J. M. Domagala, C. Humble, T. N. Bhat, B. Liu, D. M. A. Guerin, E. T. Baldwin, J. W. Erickson, T. K. Sawyer, *J. Am. Chem. Soc.* **1994**, *116*, 6989–6990.

38 B. D. Tait, S. Hagen, J. Domagala, E. L. Ellsworth, C. Gajda, H. W. Hamilton, J. V. N. Vara Prasad, D. Ferguson, N. Graham, D. Hupe, C. Nouhan, P. J. Tummino, C. Humblet, E. A. Lunney, E. A. Pavlovsky, J. Rubin, S. J. Gracheck, E. T. Baldwin, T. N. Bhat, J. W. Erickson, S. V. Gulnik, B. Liu, *J. Med. Chem.* **1997**, *40*, 3781–3792.

39 S. E. Hagen, J. V. N. Vara Prasad, F. E. Boyer, J. N. Domagala, E. L. Ellsworth, C. Gajda, H. W. Hamilton, L. J. Markoski, B. A. Steinbaugh, B. D. Tait, E. A. Lunney, P. J. Tummino, D. Ferguson, D. Hupe, C. Nouhan, S. J. Gracheck, J. M. Saunders, S. VanderRoest, *J. Med. Chem.* **1997**, *40*, 3707–3711.

40 F. E. Boyer, J. V. N. Vara Prasad, J. M. Domagala, E. L. Elsworth, C. Gajda, S. E. Hagen, L. J. Markoski, B. D. Tait, E. A. Lunney, A. Palovsky, D. Ferguson, N. Graham, T. Holler, D. Hupe, C. Nouhan, P. J. Tummino, A. Urumov, E. Zeikus, G. Zeikus, S. J. Gracheck, J. M. Sanders, S. VanderRoest, J. Brodfuehrer, K. Iyer, M. Sinz, S. V. Gulnik, J. W. Erickson, *J. Med. Chem.* **2000**, *43*, 843–858.

41 S. Thaisrivongs, P. K. Tomich, K. D. Watenpaugh, K.-T. Chong, W. J. Howe, C.-P. Yang, J. W. Strohbach, S. R. Turner, J. P. McGrath, M. J. Bohanon, J. C. Lynn, A. M. Mulichak, P. A. Spinelli, R. R. Hinshaw, P. J. Pagano, J. B. Moon, M. J. Ruwart, K. F. Wilkinson, B. D. Rush, G. L. Zipp, R. J. Dalga, F. J. Schwende, G. M. Howard, G. E. Padbury, L. N. Toth, Z. Zhao, K. A. Koeplinger, T. J. Kakuk, S. L. Cole, R. M. Zaya, R. C. Piper, P. Jeffrey, *J. Med. Chem.* **1994**, *37*, 3200–3204.

42 S. R. Turner, J. W. Strohbach, R. A. Tommasi, P. A. Aristoff, P. D. Johnson, H. I. Skulnick, L. A. Dolak, E. P. Seest, P. K. Tomich, M. J. Bohanon, M.-M. Horng, J. C. Lynn, K.-T. Chong, R. R. Hinshaw, K. D. Watenpaugh, M. N. Janakiraman, S. Thaisrivongs, *J. Med. Chem.* **1998**, *41*, 3467–3476.

43 C. G. Wermuth, J. J. Bourguignon, R. Hoffmann, R. Boigegrain, R. Brodin, J. P. Kan, P. Soubrie, *Bioorg. Med. Chem. Lett.* **1992**, *2*, 833–838.

44 J.-M. Contreras, Y. M. Rival, S. Chayer, J.-J. Bourguignon, C. G. Wermuth, *J. Med. Chem.* **1999**, *42*, 730–741.

45 J.-M. Contreras, I. Parrot, W. Sippl, Y. M. Rival, C. G. Wermuth, *J. Med. Chem.* **2001**, *44*, 2707–2718.

46 Y. Rival, R. Hoffmann, B. Didier, V. Rybaltchenko, J.-J. Bourguignon, C. G. Wermuth, *J. Med. Chem.* **1998**, *41*, 311–317.

47 F. P. Bymaster, D. O. Calligaro, J. F. Falcone, R. D. Marsh, N. A. Moore, N. C. Tye, P. Seeman, D. T. Wong, *Neuropsychopharmacology* **1996**, *14*, 87–96.

48 F. P. Bymaster, D. L. Nelson, N. W. DeLapp, J. F. Falcone, K. Eckols, L. L. Truex, M. M. Foreman, V. L. Lucaites, D. O. Calligaro, *Schizophr. Res.* **1999**, *37*, 107–22.

49 J. M. Schaus, F. P. Bymaster, *Ann. Rep. Med. Chem.* **1998**, *33*, 1–10.

50 H. Riechers, H.-P. Albrecht, W. Amberg, E. Baumann, H. Bernard, H.-J. Böhm, D. Klinge, A. Kling, S. Müller, M. Raschack, L. Unger, N. Walker, W. Wernet, *J. Med. Chem.* **1996**, *39*, 2123–2128.

51 F. F. Moebius, R. J. Reiter, M. Hanner, H. Glossmann, *Brit. J. Pharmac.* **1997**, *121*, 1–6.

3
The Value of Chemical Genetics in Drug Discovery

Keith Russell and William F. Michne

3.1
Introduction

To understand what chemical genetics is and how it can add value to the drug discovery process, we must first consider some of the challenges and needs of the pharmaceutical industry. The process of discovering new drugs is a highly complex multidisciplinary activity requiring very large investments of time, intellectual capital, and money. Today the average cost of bringing an NCE to market is on the order of $ 900 million [1]. For every 5000 compounds synthesized, only one makes it to the market. Only three of ten drugs generate revenue that meets or exceeds average R&D costs, and 70% of total returns are generated by only 20% of the products [2]. Given this gloomy backdrop it is even more disturbing to learn that, despite the proliferation of many new technologies of great potential (and great cost), pharmaceutical productivity levels have not increased in the past ten years (as shown graphically in Figure 3.1).

Pharmaceutical R&D costs continue to grow exponentially, driven in part by investments in new technologies, but the return on this investment remains elusive. There are many reasons for these disturbing trends. If we consider the pharmaceutical industry as primarily a generator of knowledge (defining knowledge as compiled and interpreted information that can be acted upon) and focus on the knowledge creation process, we can shed some light on how the current situation, a productivity gap, emerged. Working harder is not likely to overcome this productivity gap to deliver more drugs. Working smarter, doing things differently, and focusing on what we actually need to deliver, i.e., *knowledge*, may be a new way to approach the problem. Ultimately, spanning the 'knowledge gap' will lead us to the efficient exploitation of the human genome to discover new drugs to meet major medical needs.

Chemogenomics in Drug Discovery: A Medicinal Chemistry Perspective.
Edited by Hugo Kubinyi and Gerhard Müller

ISBN: 3-527-30987-X

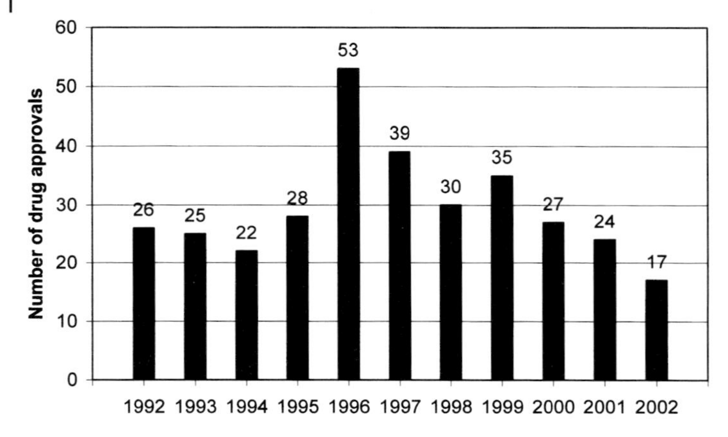

Figure 3.1 US drug approvals during the past ten years.

3.2 Knowledge Management in Drug Discovery

Pharmaceutical companies create and sell knowledge, e.g., knowledge that a drug product will rid patients of the symptoms of their disease while not causing serious side effects. The resources that go into the production of the drug pale alongside the resources needed to discover the knowledge of what the drug will do when administered to a patient. In the early years of drug discovery it was often true that the literature provided a significant knowledge base for our efforts. Two approaches were taken: (1) function-based screening, where one did not know what the target was but could easily screen for small molecules that possessed the right biology [3]; and (2) 'rational drug discovery', where one has knowledge of the target and its function [4]. What was needed were small molecules that would interact with the target in the right way before being optimized for in vivo activity and safety.

The existing and evolving chemistry and biology literature fueled these efforts. It is probably also true to say that the medical problems addressed in these early days of drug discovery represented the more accessible opportunities. Often the biology was not only reasonably well understood, but it was reasonably easy to study and measure. Examples of biological effects that were tackled include blood pressure, acid secretion, and cytotoxicity. The situation today is very different. We now face many new targets we know little about and biology that is complex to study and understand. In addition to these issues, advances in our knowledge of distribution, metabolism, and pharmacokinetics, as well as toxicology and pharmacogenetics, have led to the introduction of discovery processes that front-load measurement of such small-molecule properties. This also raises the bar for passage of compounds through the process – making the process more difficult and slower. While this may lead to lower output of development candidates, it should also lead to lower failure rates later in development, i.e., improvements in quality.

3.3
Knowledge Gaps, Their Importance, and How to Address Them

The human genome has been solved and optimistic promises have been made. It is clear that the human genome did not deliver knowledge (i.e., something immediately useful); rather, it delivered a massive amount of data. Significant advances have also been made in cell biology and systems biology. The relationship between genes/proteins derived from the human genome and their function as a part of a biological system constitutes the knowledge gap, and our appreciation of the extent of this void is still emerging. The human genome is thought to consist of ca. 30 000 genes. Each gene can potentially produce several proteins via alternative splicing and post-translational modification, and every protein can potentially combine with other proteins to form many different protein complexes. Clearly, the number of different proteins and protein complexes is much larger than 30 000. To add further complexity, small molecules (that we hope will become drugs) can interact with different sites on a protein or via different mechanisms to further expand the diversity of possible outcomes from the interaction of small molecules with a protein target. We do not know what many gene products (proteins) do, either physiologically or pathologically, and we do not really know how many of these proteins can interact with small-molecule ligands [5]. There are many genes about which we know nothing at all. In summary, there is clearly a vast knowledge gap between knowing a gene and knowing the function (physiology and pathology) of its protein product (Figure 3.2). The enormity of this knowledge gap has been underestimated by the pharmaceutical industry.

To illustrate the size of the knowledge gap, consider the following (admittedly approximate) analysis from the area of substance P. Substance P antagonists have emerged in recent years as potential new treatments for depression, although none

Figure 3.2 The knowledge gap represents the large gap in understanding that exists between genetic information from the human genome project and information regarding biological function from cell and systems biology.

have yet been approved for this use. Substance P has been known since 1937, and since that time (67 years!) there have been over 6500 papers published providing significant new information on substance P. Thousands of scientists have worked on generating this information during this period. It is sobering that our understanding of Substance P's role in depression is still in its infancy. No one pharmaceutical company can generate this volume of information. New faster and more efficient methods must be developed to fill these knowledge gaps. Partnership with the academic community will become increasingly important as the number of druggable targets expands.

3.4 Target Validation: The Foundation of Drug Discovery

One critical piece of knowledge to the pharmaceutical industry relates to knowledge of a drug target and its link to a disease process. In the context of small-molecule drug discovery, we define target validation in a broader sense as including knowledge of the protein target and its specific interaction with small molecules, and the consequences of this interaction in terms of modifying a disease process. In fact, drug discovery is *primarily* focused on the biology of a target in the presence of a drug, i.e., drug-induced biology. It begins with a chemical effect – the interaction of a ligand with a protein at a specific site in a specific manner – and ends in patients' gaining benefit from taking a drug derived from the application and exploitation of this knowledge. Target validation that simply links a specific protein and its function to a disease state does not include reference to whether a small molecule can modulate the function of the protein. The protein may not therefore constitute a true target since it is not a target for a small-molecule ligand and efforts to do target validation on such a protein will ultimately lead to a negative outcome. We can (and do) proceed to work on drug discovery before we have all the knowledge we need. The absence of this knowledge constitutes the major risk of drug discovery. One way to proceed is to focus on obtaining the most critical knowledge first. This is the knowledge that modulation of a protein target by a small molecule can ultimately lead to a clinical benefit in patients.

3.5 Chemical Genetics – How Chemistry Can Contribute to Target Identification and Validation

Target validation (TV) is the foundation of drug discovery and requires greater attention if we are to reduce the risk of failure after significant investment. Traditionally, target validation has been thought of as a biology problem. Thinking in terms of what knowledge we need makes it clear that the problem does not neatly fall into any particular discipline and is better characterized as an integrated biology and chemistry problem. A schematic target validation roadmap is shown in Figure 3.3, where the entire validation path from a chemical effect through various

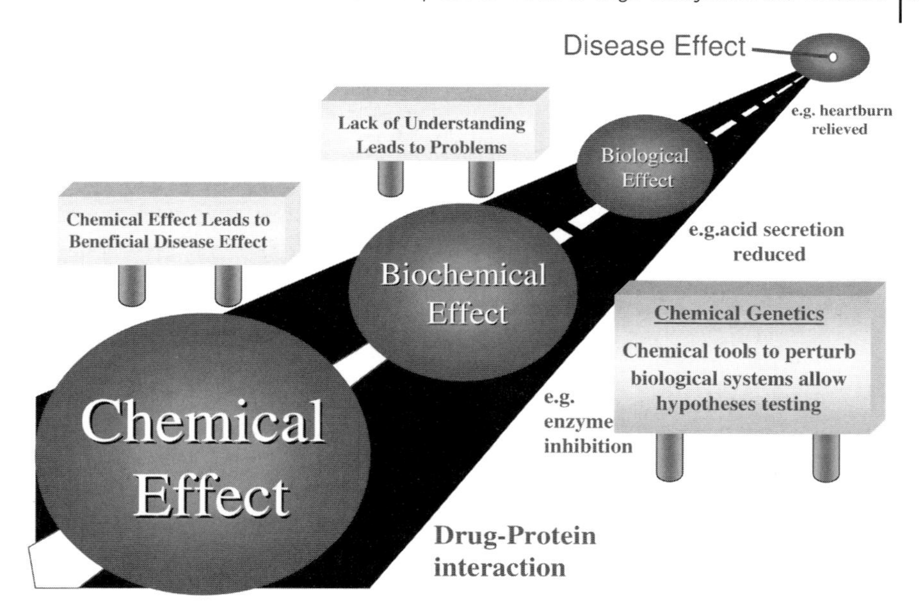

Figure 3.3 The knowledge roadmap for target validation, beginning with a chemical effect between a small molecule and a protein target and ending with a beneficial clinical effect on a person with a disease. Chemical genetics approaches provide some assistance in pursuing this path.

levels of biological effects to a clinical effect is outlined. To begin with, an understanding of the function of a particular gene product can often be achieved through the methods of classical genetics. However, the process can be slow and tedious. For example, developing a mouse carrying the mutation of interest could take months or years. Indeed, if the gene product is essential, the organism may not survive long enough to be studied. On the other hand, the situation wherein a molecule is available that alters the function of the gene product has a number of advantages. However, we should recognize that significant chemical effort is often required. The phenotype of interest is conditional, in that it is present only when the molecule is present, allowing the study of essential gene products. It is also tunable, i.e., the intensity of the phenotype can be adjusted by controlling the concentration of the molecule.

Chemical genetics is the purposeful modulation of protein function through its interaction with a small molecule. The principles of chemical genetics were established in the rich history of using small molecules to explore biological function and, in this sense, chemical genetics is not new. What is new is the development of a systematic approach to studying biological function with small molecules – this is the emerging field of chemical genetics. Just as genetic changes can alter protein function, so can small molecule–protein interactions [6]. It is important to appreciate that, by interaction of a ligand with a protein, we mean interaction of a small molecule at a specific site on a protein causing a specific protein change, conformational or otherwise, ultimately leading to a specific biological effect. Small molecules can often interact with multiple sites on proteins and cause a multitude of consequences such as agonism, antagonism, partial agonism, modulation, competitive and noncompetitive inhibition, etc. They can also interact at junctions between protein subunits. The sophistication of small molecule–protein interactions and their biological consequences cannot be easily reproduced by techniques such as gene knockin/out or the use of siRNA, by which genes/proteins are simply removed or increased in concentration in a biological system. Having said that, knockout models have certainly contributed significantly to drug discovery and will continue to do so [7]. The power of chemical genetics resides in this sophistication of the small molecule–protein interaction and the precise way we can (in principle) modulate the function of a protein. As a precursor to drug discovery it serves the purpose of focusing us on where small molecule drug discovery really begins – with the chemical interaction between a small molecule and a protein.

The knowledge gap outlined above can be thought of as a cycle linking the target (a protein or protein complex) with a function ultimately linked to an effect important in a disease process (Figure 3.4). Going from target to function represents the knowledge path of target validation. Going from a function to a target represents the knowledge path of target identification (TI) or deconvolution. Chemical genetics approaches can be applied to both knowledge paths. Application to the target validation path is called *reverse chemical genetics*. Application to the target identification/deconvolution path is referred to as *forward chemical genetics*. At the heart of this approach to knowledge generation in TI/TV is the simple concept that small molecules are used to perturb biological systems. Manipulation of a biological system in a controlled manner by small molecules allows us to study these systems more systematically.

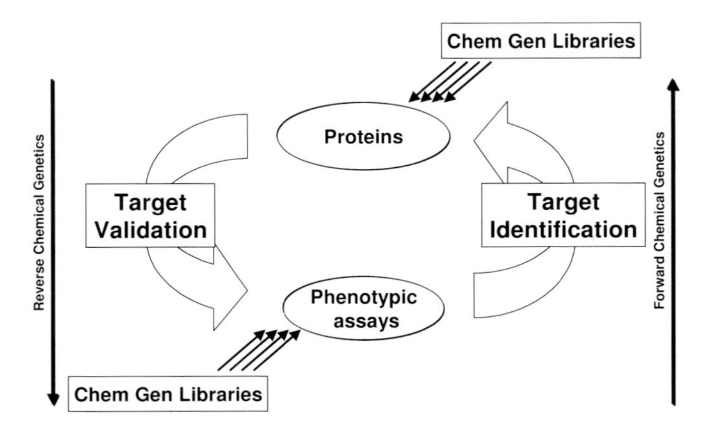

Figure 3.4 Chemical genetics tools (libraries) can help uncover the function of proteins (target validation) and the protein target responsible for biological function (target identification) in a phenotype assay.

3.6
Integration of Chemistry and Biology: Importance and Issues

Given that the foundation of target validation is a ligand–protein interaction (a chemical effect) and its consequence (a biochemical/biological effect), we can expect that advances in this area will come from a close integration of chemistry and biology. Some key questions at the interface of chemistry and biology that are fundamental to chemical genetics include – why are some molecules biologically active while others are not? What is the biological profile of a small molecule's structure and how do we dissect this into what each part (fragment) of the small molecule is doing to each protein target? Is there a protein 'code' for recognition of small molecules that is used by every protein in the proteome? The following sections begin to address these questions.

3.7
Finding New Chemical Tools and Leads

A *chemical tool* is small molecule that is sufficiently potent and selective for a protein target to be used in the identification and validation of that target. It could, although it need not, meet the rigorous absorption, distribution, metabolism, excretion, and toxicology criteria required of a lead to start an optimization project. How do we find such tools? The total number of 'reasonable' drug-like molecules has been estimated [8] as approximately 10^{63} discrete molecules, a number so large that synthesizing all of them is simply impossible. Natural products were designed by nature to bind to proteins and other macromolecular targets and represent powerful chemical tools for use in chemical genetics. Numerous examples exist in which natural products have been identified that modulate biological function. The natural products are then used to identify proteins that they interact with and so to begin deconvolution (forward chemical genetics) of the target responsible for the biological effect. For example, fumagillin inhibits new blood vessel growth (angiogenesis), and analogs of this compound are now in Phase 3 trials. Using fumagillin as a starting point, chemical tools (e.g., biotinylated analogs) were constructed to bind and tag cellular proteins. One of these proteins, methionine aminopeptidase, has been identified as the likely target for this class of molecules [9]. Some other examples of natural products used in forward chemical genetics are shown in Table 3.1. Cases in which these natural products were then used to deconvolute the target protein are noted. Interestingly, some of the top-selling drug classes originated from a forward chemical genetics approach, e.g., the gastric acid secretion inhibitors omeprazole and esomeprazole were discovered by a process that began with screening for antisecretory agents that lowered stomach acid [3].

Interestingly, given the discussion of the importance of understanding small molecule–target protein interactions early in drug discovery, there is renewed interest in reexamining many older drugs to more fully understand how they work [10].

Table 3.1 Natural products used to identify targets.

Biological effect	***Protein target***	***Natural product***	***Reference***
Angiogenesis	Methionine aminopeptidase	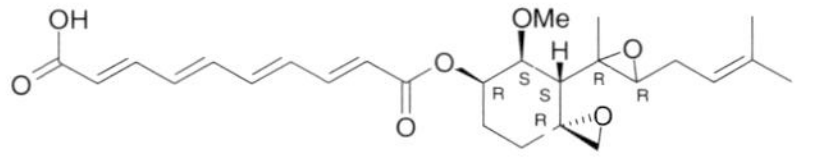 fumagillin	9
Immuno-suppressive, anticancer	Microtubule binder/ stabilizer	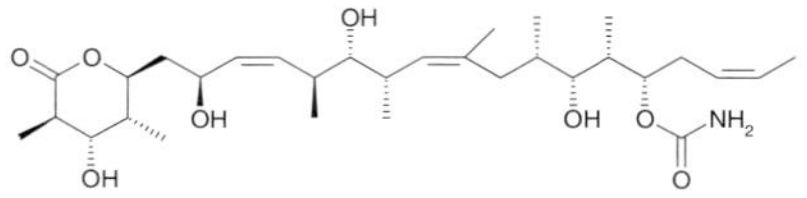 discodermolide	59
Immuno-suppresion, IL-2 production inhibition	Calcineurin (a protein phosphatase) inhibition	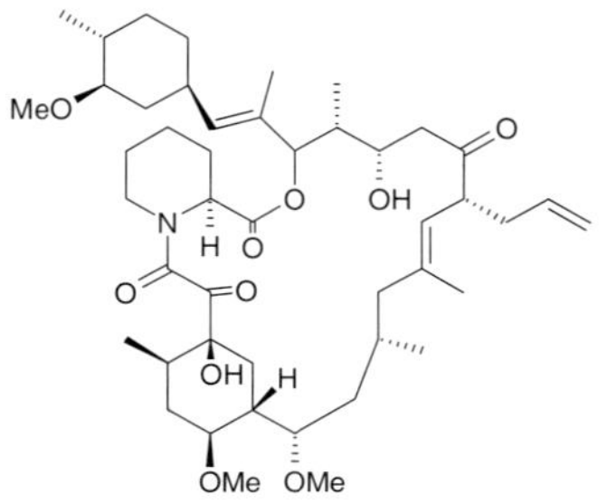 FK506	17
Histone deacetylase inhibitor		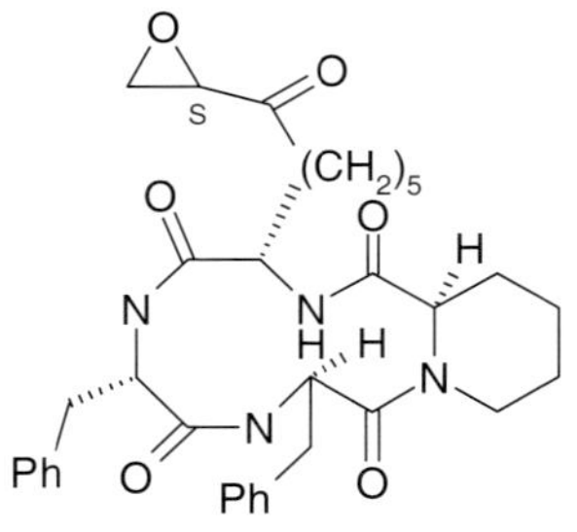 trapoxin	60
Proteasome inhibition		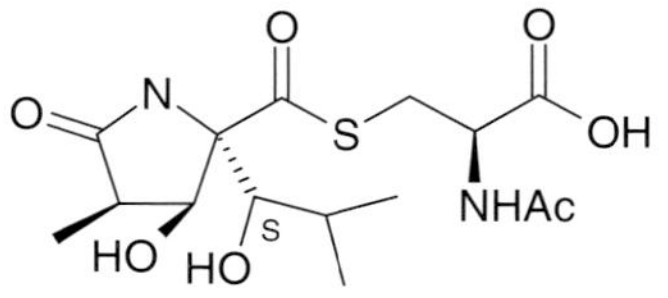 lactacystin	61

An advantage of the chemical genetics approach is that the small molecules identified in biological screens can act both as conditional switches for inducing phenotypic changes *and* as probes/chemical tools for identifying protein targets implicated in those phenotypic changes. However, identifying the molecular target and mechanisms by which the small molecules affect biological systems (target deconvolution) can sometimes be difficult. Classical deconvolution approaches, such as affinity chromatography and biochemical fractionation using photoactivatable and other affinity ligands to pull out the target protein, often work well [11]. More recently, genomics-based techniques have been added to the deconvolution toolset [12].

Beyond natural products, finding chemical tools to modulate biological systems is a difficult step and shares many of the risks associated with finding leads in a drug discovery program [13]. Strategies for finding small-molecule tools representing two poles on a continuum of approaches are illustrated by structure-based design and the high-throughput screening approach. Given our focus on knowledge generation, it is interesting to note that molecules at either end of this spectrum also reflect different levels of information content. Individual molecules used in high-throughput screening teach us (if we are fortunate) about a simple IC_{50} or EC_{50}. Molecules that additionally teach us how they bind to their molecular target provide us with much more useful information, especially when we consider what to do next to improve or change the biology of the molecule (Figure 3.5).

Schreiber has been a pioneer in this rapidly developing area of chemical biology. He has constructed several structurally complex screening libraries using a diversity-oriented synthesis approach and has used these libraries to uncover chemical tools to begin to unravel complex biology. Using this approach, Schreiber discovered a small-molecule chemical tool that he named uretupamine, which interacts with the protein Ure2p. Ure2p represses the transcription factors Gln3p and Nil1p.

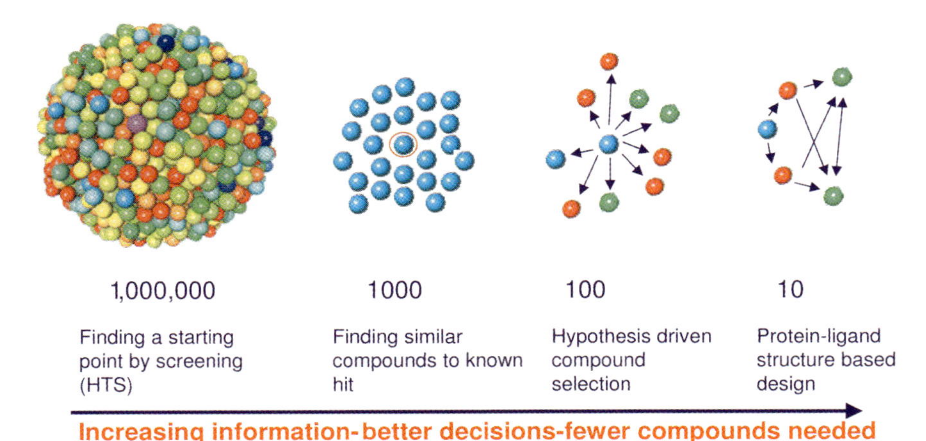

Figure 3.5 The spectrum of approaches to finding chemical tools or leads, illustrating the inverse relationship between information content and number of compounds needed.

Uretupamine was found to specifically modulate a subset of glucose-sensitive genes downstream of *Ure2p*. As noted earlier, this type of behavior, modulating a subset of the function of Ure2p, cannot be replicated by gene knockouts (e.g., knockout of the *URE2* gene) or siRNA approaches and represents a real challenge for proteomics to identify and control all inputs and outputs of a protein [14]. He used the natural product FK506 to uncover its target FKBP12 [15] and then went on to design specific molecular probes derived from FK506, guided by crystal structures of FKBP, FK506, and calcineurin to uncover its mechanism of action as a 'small-molecule dimerizer' of FKBP12 and calcineurin [16]. The formation of this ternary complex led to inhibition of the protein phosphatase activity of calcineurin [17]. This discovery, together with the discovery by Gerald Crabtree of NFAT proteins, helped define the calcium–calcineurin–NFAT signaling pathway, now known to be essential for immune function, heart development, and the acquisition of memory in the hippocampus [18].

Peter Schultz's team used a combinatorial library of purines to identify agents that could disassemble multinucleated myotubes into mononucleated fragments (a morphological differentiation screen). A new microtubule-binding molecule, mysoseverin, was identified in this way [19].

Structure-based design has been employed in some powerful examples of chemical genetics by teams led by Kevan Shokat and by John Koh. Shokat's team has studied the function of kinases by engineering designed modifications into both the kinase and kinase inhibitor ligand to create highly selective chemical tools that can then be used to probe the function of individual kinases in complex kinase cascades (for an explanation of the basic concept see Figure 3.6) [20].

John Koh's team have focused their efforts on nuclear hormone receptors, including the vitamin D receptor, in an effort to target specific clinical problems. Koh studied a mutant version of the vitamin D receptor (an arginine located in the binding pocket is mutated to a leucine) that binds vitamin D with only one thousandth the affinity of the normal receptor. Analogs of vitamin D were synthesized, based on computer modeling of their interaction in the mutant vitamin D receptor. Some of these compounds were found to bind 500 times better than vitamin D to the mutant receptor. This work may ultimately lead to drugs to treat a disease known as vitamin D resistant rickets [21]. Koh previously demonstrated the feasibility of this approach with other nuclear hormone members, including thyroid hormone receptor [22].

David Corey's team has also employed this approach, termed 'engineered orthogonal ligand–receptor pairs', in studies of retinoid x receptor to find 'near drugs'. These near drugs are chemical tools used to discern the biology of the retinoid x receptors [23].

To date, the results of efforts to find new biologically active molecules through preparation of large libraries based solely on diversity considerations have been disappointing. On the other hand, the collective experience of the global bioorganic and medicinal chemistry community indicates that biological activity is not uniformly distributed in chemistry space; rather, it is found within discrete regions. Since we cannot know the locations of these regions a priori, we might look to

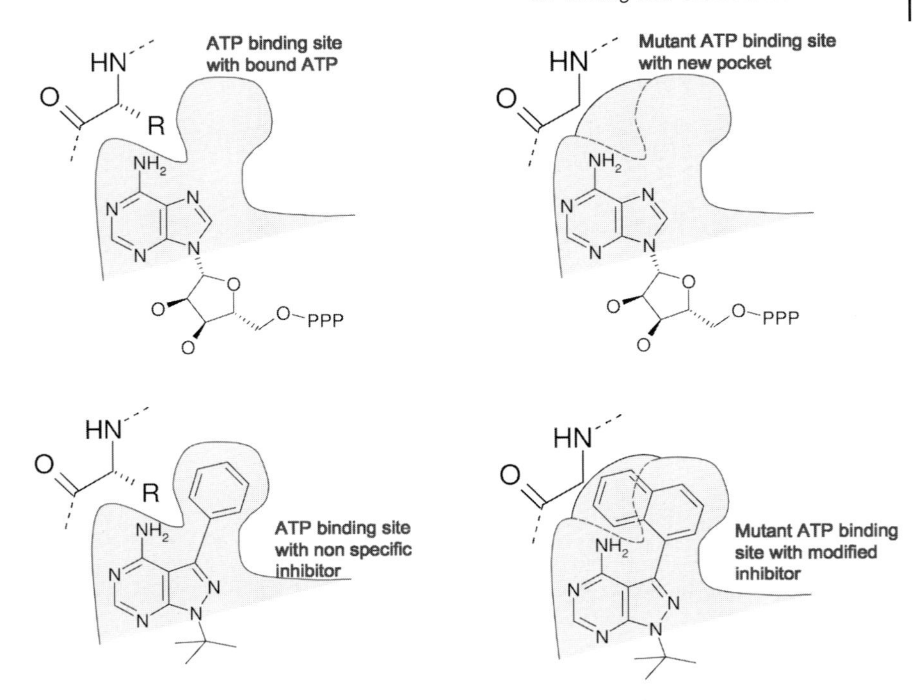

Figure 3.6 Replacing a bulky amino acid with glycine in the ATP-binding site of a kinase enlarges the site. ATP binding and catalytic activity are unaffected. The nonselective kinase inhibitor can now be modified to create a molecule that selectively blocks the mutant enzyme.

known biologically active molecules to guide our search. There have been several approaches to doing this. Many natural products derived from plants and animals have evolved over time to have specific biological effects on either the parent organism or an unrelated one. The pool of natural products is extremely large with respect to both numbers and structural diversity. Not surprisingly, a number of methods to produce natural product libraries have emerged [24]. Some companies provide prefractionated extracts of unknown structures for screening. Structures are determined after a hit is found. Many companies have established libraries of single pure natural products. Yet another approach is the assembly of libraries of derivatized natural products. Finally, one can develop syntheses of natural product core structures and, using combinatorial techniques, decorate the cores with diverse elements. In this way it is possible to prepare large libraries of peripherally diverse compounds related to natural products for general screening. The following library (Fig. 3.7) is illustrative [25]. It contains over 2 million compounds that are both sterically and functionally complex. Little biological activity was observed; for the purposes of the pharmaceutical industry this result might be viewed as somewhat disappointing, given the size of the library and the effort invested in preparing it. Why were more active compounds not found?

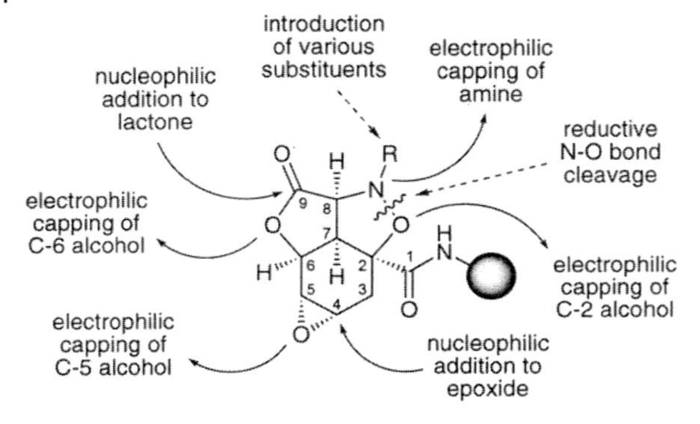

Figure 3.7 Potential coupling sites on a natural product-related core-based diversity library. (Reprinted from [25] with permission, copyright 1999, American Chemical Society).

One reason might be related to the high degree of overall molecular complexity of the library. Hann and coworkers [26] reported an in-depth analysis of the relationship between molecular complexity and the probability of finding leads. They derived a model system in which ligand complexity and ability to bind to a protein target could be studied statistically. They found that, as systems became more complex, the chance of observing a useful interaction for a randomly chosen ligand fell dramatically. Thus, there may be an optimal complexity for molecules in a screening library. Smaller libraries of less-complex molecules are likely to be more productive in terms of finding relevant chemistry space, with enhancements in potency and selectivity resulting from iterative rounds of synthesis and testing to increase complexity. Although the compounds were not derived from a library, a comparison of glutamic acid to LY354740 and MGS0028 is illustrative (Figure 3.8).

Glutamic acid is a relatively simple molecule with several degrees of rotational freedom, and obviously interacts with all glutamate receptors, both ionotropic and metabotropic. LY354740 is arguably more complex with respect to stereochemistry and rigidity, is much more potent than glutamate at Group 2 mGluR's, and has no activity at iGluR's [27]. MGS0028 is even more complex with respect to functionality and heteroatoms and, although no more selective than LY354740, it is about 20 times more potent [28]. Most chemists would no doubt agree that molecular complexity increases from glutamic acid to LY354740 to MGS0028, but there have been few attempts to quantify molecular complexity. Bertz [29] developed a general

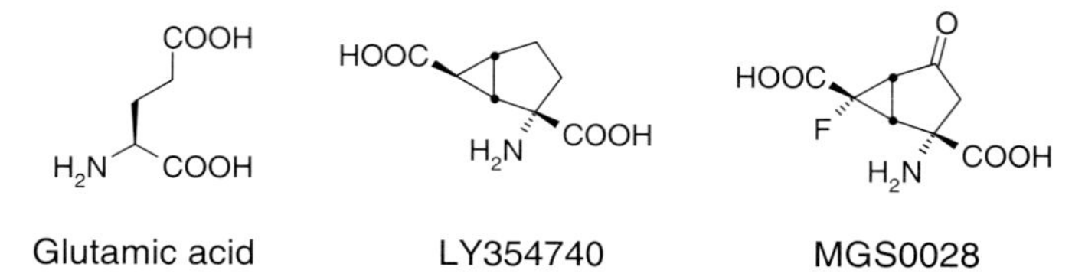

Figure 3.8 Increasing structural complexity of glutamate analogs.

index of molecular complexity based on concepts of graph theory and information theory and included features such as branching, rings, multiple bonds, heteroatoms, and symmetry. In the work reported by Hann, the number of bits set in the Daylight 2D structure representation was taken as an indication of the internal bond complexity, but the method does not capture notions of stereochemistry and rigidity.

A rather different approach to natural product-based libraries is being promoted by Waldmann and coworkers [30]. Recent results in structural biology and bioinformatics indicate that the number of distinct protein families and folds is fairly limited. Often, many proteins use the same structural domain in a more or less modified form created by divergent evolution. Protein families can have similar folds, even though they at first seem to have completely different sequences and/or catalyze quite different chemical reactions with a different arrangement of active-site residues. However, proteins in these families evolved from the same ancestors and can still bind similar ligands [31]. If ligand types or frameworks for certain domain families are already known from the investigation of evolutionarily related proteins, the underlying structure of this ligand may be employed as the guiding principle for library development. Such ligands would provide targeted, biologically validated starting points in structural space for the development of relatively small compound libraries, which should yield significantly higher hit rates than much larger libraries designed exclusively on the basis of available and proven chemical transformations.

Accordingly, they synthesized a library of nakijiquinone analogs (Figure 3.9) [32], the only natural products known to be inhibitors of the Her-2/Neu receptor tyrosine kinase, and investigated them as possible inhibitors of the receptor tyrosine kinases involved in angiogenesis. This led to the identification of inhibitors of IGF1R, Tie-2, and VEGFR-3, with IC50's in the range of 0.5–18 μM.

The growing awareness that biological activity is not uniformly distributed throughout chemistry space has led to a number of efforts to determine those molecular attributes that are drivers of that activity. At an elementary level, Ghose and coworkers [33] carried out quantitative and qualitative characterization of known drug databases with respect to computed physicochemical property profiles, such

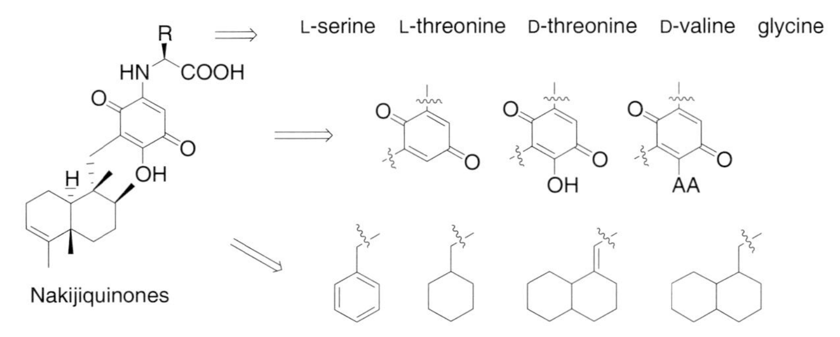

Figure 3.9 Molecular composition of the nakijiquinone library. (Reprinted from [32] with permission, copyright 2003, American Chemical Society).

as log P, molar refractivity, molecular weight, and number of atoms, as well as characterization based on the occurrence of functional groups and important substructures. For many parameters, they defined a *qualifying range* covering ≥ 80% of the compounds. They also found that the benzene ring is the most abundant substructure, slightly more abundant than all heterocyclic rings combined, and that nonaromatic heterocycles were twice as abundant as aromatic heterocycles. The most abundant functional groups were tertiary aliphatic amines, alcohols, and carboxamides.

Bemis and Murcko [34] carried out an extensive structure-based analysis using shape description methods to analyze a database of commercially available drugs and prepare a list of common drug shapes. A useful way of organizing this structural data is to group the atoms of each drug molecule into ring, linker, framework, and side-chain atoms. On the basis of the 2D molecular structures (without regard to atom type, hybridization, or bond order), there were 1179 different frameworks among the 5120 compounds analyzed. However, the shapes of half of the drugs in the database were described by the 32 most frequently occurring frameworks. This suggests that the diversity of shapes in the set of known drugs is extremely low. Within the set of 32 frameworks, 23 contained at least two six-membered rings linked or fused together, and only three had more than five rotatable bonds. In a second method of analysis, in which atom type, hybridization, and bond order were considered, more diversity was seen: there were 2506 different frameworks among the 5120 compounds in the database, and the most frequently occurring 42 frameworks accounted for only one-fourth of the drugs. Subsequently, the same workers analyzed the side chains of the same set of drugs [35]. On the basis of the atom pair shape descriptor (taking into account atom type, hybridization, and bond order), there were 1246 different side chains among the 5090 compounds analyzed. The average number of side chains per molecule was 4, and the average number of heavy atoms per side chain was 2. Ignoring the carbonyl side chain, there were approximately 15 000 occurrences of side chains. Of these 15 000, approximately 11 000 were from the 'top-20' group of side chains. This suggests that the diversity that side chains provide to drug molecules is also quite low. The authors have combined this information to generate new structures that are likely to be drug-like and synthetically accessible. They used this approach to generate a set of molecules optimized for blood–brain barrier penetration [36].

Ajay and coworkers [37] used a Bayesian neural network to distinguish between drugs and nondrugs. They evaluated commercial databases of drug (Comprehensive Medicinal Chemistry, CMC) and nondrug (Available Chemicals Directory, ACD) molecules with respect to 1D and 2D parameters. The former contain information about the entire molecule, like molecular weight, and the latter contain information about specific functional groups. Their results correctly predicted over 90% of the compounds in the drug database while classifying about 10% of the molecules in the nondrug database as drug-like. The neighborhoods defined by their model are not similar to those generated by standard Tanimoto similarity calculations, and thus new and different information is being generated by these models, as shown in Figure 3.10.

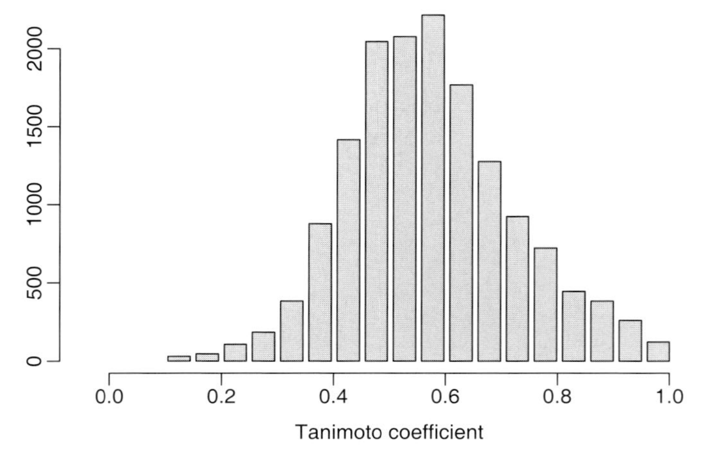

Figure 3.10 Histogram of Tanimoto coefficients based on topological torsions of the most similar CMC molecule for each of the drug-like molecules from the ACD. (Reprinted from [37] with permission, copyright 1998, American Chemical Society).

Further efforts have been made to distinguish between drugs and nondrugs. Sadowski and Kubinyi [38] developed a scoring scheme for rapid and automatic classification of molecules into drugs and nondrugs. The method was set up by using atom type descriptors for encoding the molecular structures and by training a feed-forward neural network for classifying the molecules. It was parameterized and validated by using large databases of drugs (World Drug Index, WDI) and nondrugs (ACD). The method revealed features in the molecular descriptors that either qualify or disqualify a molecule for being a drug and classified 83% of the ACD and 77% of the WDI appropriately.

Clark and coworkers [39] investigated techniques for distinguishing between drugs and nondrugs using a set of molecular descriptors derived from semiempirical molecular orbital (AM1) calculations. These descriptors had been used successfully to build absorption, distribution, metabolism, and excretion-related QSPR models. A principal-components analysis was carried out for the descriptors in property space. The third-most significant principal component of this set of descriptors served as a useful numerical index of drug-likeness, but no others were able to distinguish between drugs and nondrugs. The set of descriptors was extended, and ultimately three descriptors were used to train a Kohonen artificial neural net for the entire Maybridge dataset. Projecting the drug database onto the map so obtained resulted in clear distinction between drugs and nondrugs.

Figure 3.11 demonstrates that there is no simple relationship between drug-likeness and standard 2D similarity measures of molecules. Martin and coworkers [40] addressed this question in a study using Daylight fingerprints. They showed that, for IC50 values determined as a follow-up to 115 high-throughput screening assays, there is only a 30% chance that a compound that is ≥0.85 Tanimoto similar to an active is itself active.

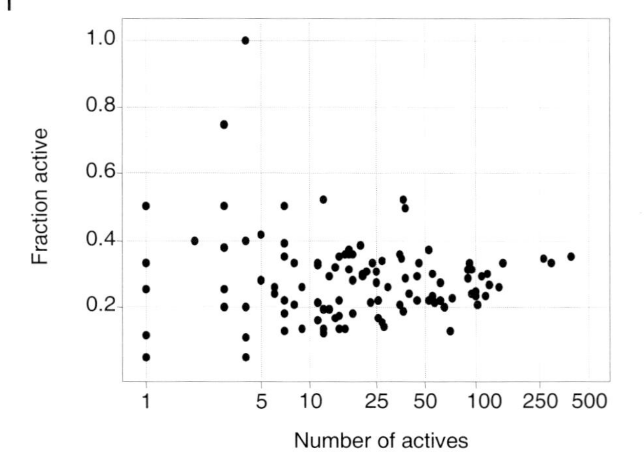

Figure 3.11 The fraction of molecules that are similar to any active that are themselves active, as a function of the number of actives with similars. (Reprinted from [40] with permission, copyright 2002, American Chemical Society).

These workers also asked whether biologically similar compounds have similar chemical structures. Considering such classic example pairs as the nicotinic agonists acetylcholine and nicotine or the dopaminergic agonists dopamine and pergolide (Figure 3.12), the expected answer is no. In fact, the highest Tanimoto similarity within this group of four compounds is between nicotine and pergolide, and the second-highest is between nicotine and dopamine. Nevertheless, in general, the Daylight and Unity fingerprints are more similar for compounds with the same biological properties than for compounds with different biological activities. What might at first be perceived as a disappointing level of similarity-predicted actives might be the result of compounds binding in subtly different ways to the same receptor or to different but related populations of receptors.

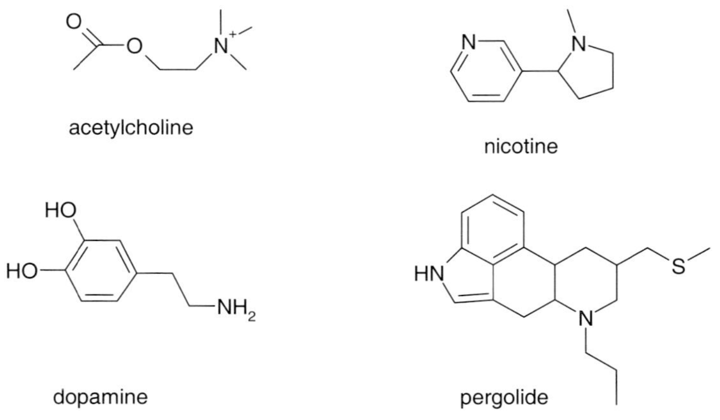

Figure 3.12 Pairs of cholinergic and dopaminergic agonists.

Pearlman and Smith [41] indicated that such distance-based algorithms are quite satisfactory for simple subset selection, but are considerably less useful for all other diversity-related tasks. In their view, traditional descriptors make rather poor chemistry space metrics for three reasons: many of the traditional descriptors are highly correlated, some traditional descriptors are strongly related to pharmacokinetics but only weakly related to receptor affinity, and traditional descriptors convey very little information about substructural differences that are the basis of structural diversity. They defined BCUT metrics in a manner that incorporates both connectivity information and atomic properties relevant to intermolecular interaction, i.e., atomic charge, polarizability, and H-bond donor and acceptor abilities. Given a set of active compounds that all bind to a given receptor in the same way, it is certainly reasonable to expect that these active compounds should be positioned near each other in a small region of chemistry space if the chemistry space metrics are valid. They developed the Activity-Seeded Structure-Based clustering algorithm, which provides a method for directly testing that expectation in the typical case in which the chemistry space dimensionality is greater than three and, thus, simple visual inspection of the distribution of active compounds is difficult or impossible. Given a number of compounds for which a particular receptor has significant affinity, they can then identify the receptor-relevant subspace for that receptor by identifying the axes along which compounds are tightly clustered. The algorithm also accounts for the possibility of multiple receptor binding modes by allowing more than one cluster of actives per relevant axis. In addition to their own application to ACE inhibitors as an illustration of the method, Stanton [42] independently applied this method to a QSAR study of dihydrofolate reductase inhibitors. The resulting model was highly predictive, as shown in Figure 3.13. It is apparent that the BCUT metrics are

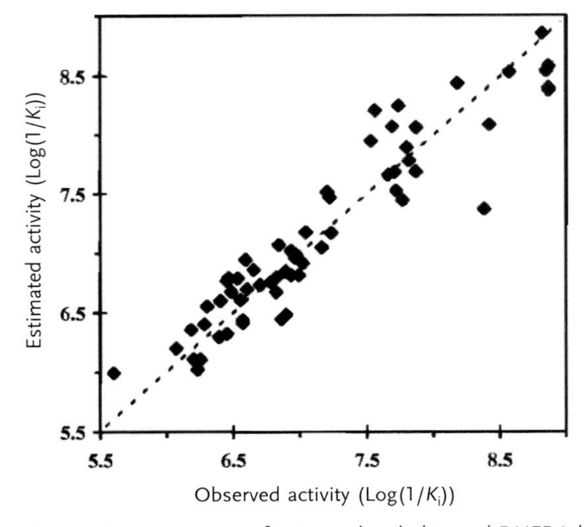

Figure 3.13 Comparison of estimated and observed DHFR inhibitor activity values using a BCUT-based model (reprinted from [42] with permission, copyright 1999, American Chemical Society).

measuring particular structural features that can be related to the observed properties of a variety of molecules. They appear to perform quite well in capturing structural information important for understanding polar intermolecular interactions.

BCUT metrics are being used increasingly in QSAR studies and library design. A particularly interesting study was done by Pirard and Pickett [43], who presented studies with BCUTs for the classification of ATP site-directed kinase inhibitors active against five different protein kinases, three from the serine/threonine family and two from the tyrosine kinase family. In combination with a chemometric method, the BCUTs were able to correctly classify the ligands according to their target. The authors concluded that BCUTs are indeed a useful set of descriptors for design tasks, extracting information in a manner relevant to describing ligand–receptor interactions. They are particularly suited to the design of targeted libraries and virtual screening of compound collections, as they are quick to calculate while containing more information than a standard 2D fingerprint type descriptor.

3.8 Is Biological Selectivity an Illusion?

We have illustrated the enormity of chemistry space and the focus on biologically relevant chemistry space, but what about biology space itself? How many biologically relevant targets are there? Although this number has been estimated to be around 3000 [5], it may well be much larger than this if we extrapolate from what we know about particular target classes, e.g., GPCRs, where there are many potential druggable targets and many potential pharmacologies, from agonists to antagonists to modulators to inverse agonists. In a typical drug discovery program, selectivity of potential development candidates is often assessed against a panel of 50–100 biologies. Clearly, this does not cover a very large fraction of available biology space. In fact, many compounds originally thought to be very selective were later found to have effects against many other targets. For example, cholesterol-lowering HMG-CoA reductase Inhibitors (statins) are among the world's top-selling drugs. It was recognized recently that statins possess additional biology. e.g., anti-inflammatory activity, that is not explained by their interaction with this enzyme. High-throughput screening of large chemical libraries has identified lovastatin (a statin) as an extracellular inhibitor of LFA-1. Lovastatin was shown to decrease LFA-1-mediated leukocyte adhesion to ICAM-1 and T-cell co-stimulation. Unexpectedly, lovastatin was found to bind to a hitherto unknown site in the LFA-1 I (inserted) domain, as documented by nuclear magnetic resonance spectroscopy and crystallography [44].

Some structural classes, e.g., benzodiazepines, are well known to exhibit diverse biology depending on the precise substituent pattern and conformation. Selective ligands with common cores have been obtained against many protein targets (Figure 3.14). The existence of such privileged structures suggests that some common structural binding motifs on proteins are reused across many different protein families [31]. It is widely accepted that few if any of the known biologically active molecules are exclusively selective for a single biological target. This forms

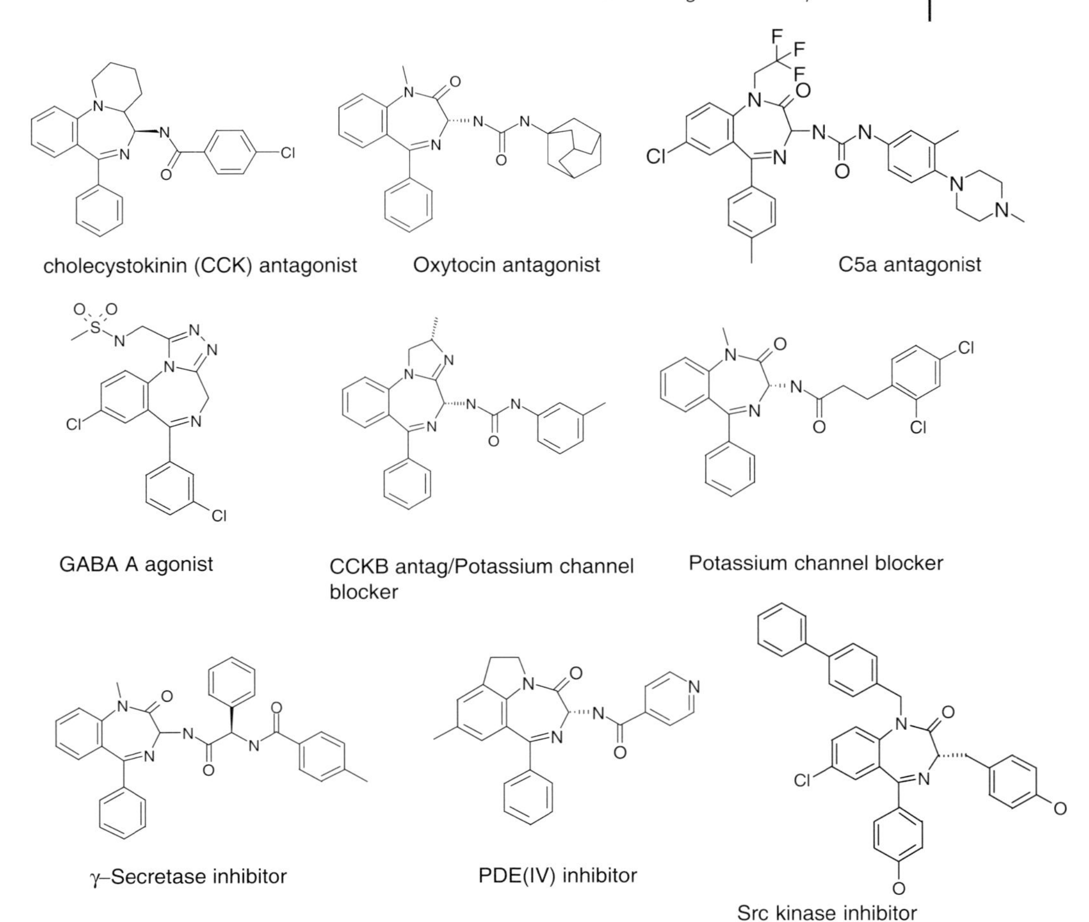

Figure 3.14 The classic privileged structure – the benzodiazepine nucleus with small structural modifications – is capable of many different biologies.

the basis for the discovery of new uses for existing drugs and the explanation of side effects observed for all drugs. Indeed, in a commentary on the molecular basis for the binding promiscuity of antagonist drugs, LaBella [45] stated that it is unlikely that binding-site dimensions, geometry, charge environments, hydrophobic surfaces, and other features will ever be known to the extent that drug design technology will yield a compound with absolute specificity for one species of functional protein. On a molecular level this may well be a consequence of there being a relatively small number of protein families and folding motifs (see above). These considerations are being applied in interesting ways to quickly find new biologically active compounds. For example, Kauvar [46] and Dixon [46] have developed a method called affinity fingerprinting, for predicting ligand binding to proteins. In this method, the binding potency of a small molecule is measured against a panel of reference proteins, in which the panel members have been empirically selected to provide binding sites that are well diversified with regard to interactions with small molecules. The resulting set of pIC50's constitutes the molecule's molecular fingerprint. Libraries of compounds can be evaluated and the collection of corresponding fingerprints entered into a database. From this large set, a subset is then chosen to represent the diversity of the set. The subset is then screened against a new target protein. Those compounds with the best pIC50's against the new protein are used to query the database to find other compounds with the same or similar fingerprint. Repetition of the cycle quickly finds the best-binding compounds in the collection. These can then serve as seeds for combinatorial expansion, presumably accelerating the lead discovery process.

We have used a related strategy to analyze the performance of our corporate collection in high-throughput screening over the past several years [47]. Our panel of proteins consists of drug targets of interest and spans several target classes, including GPCRs, several classes of enzymes, ion channels, etc. Our thesis is that a compound that exhibits biological activity in any target class is more likely to exhibit activity in another unrelated class than is a compound that has never exhibited biological activity of any kind. We initially used a relatively small set of assays and screened compounds and identified about 3500 compounds that were biologically active in at least one assay and met our internal criteria with respect to molecular weight, cLogP, polar surface area, and other chemistry-based filters. About 10% of these compounds were found to exhibit activity in other assays. The number of active compounds was then to expanded about 10 000, and the number of assays to 40 [48]. The hit rate of the general corporate collection was normalized to a frequency of 1 and compared to the hit rate of the 10 000 known biologically active set. The results are shown in Figure 3.15.

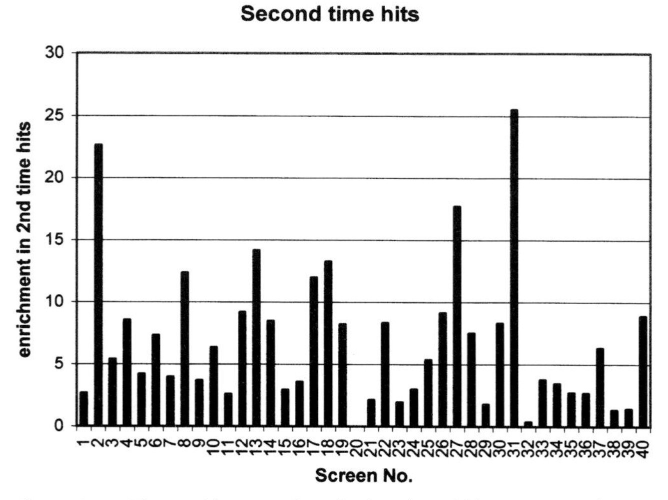

Figure 3.15 Observed hit rates for a biology-based library on a scale in which the hit rate of the general collection was normalized to 1.

Clearly, the hit rate exceeds that of the general collection in the majority of screens. However, recent publications have sounded a cautionary note. Roche and coworkers [49] reported the development of a virtual screening method for the identification of 'frequent hitters'. These compounds appear as hits in many different biological assays covering a wide range of targets for two main reasons: (1) the activity of the compound is not specific for the target; and (2) the compound perturbs the assay or the detection method. They found that, with an increasing drug-likeness of the database, a decreasing fraction of frequent hitters is predicted. Sheridan [50] reported finding multiactivity substructures by mining databases of drug-like compounds. Shoichet and coworkers [51] described a common mechanism underlying this phenomenon. In their study they observed that several nonspecific inhibitors formed aggregates 30–400 nm in diameter and that these aggregates were likely responsible for the inhibition. With these two reports in mind, we returned to our corporate database and identified, again after suitable filtering, a set of 72 000 biologically active compounds. We then selected a subset of about 25 000 compounds based on the following criteria: (1) compounds with confirmed activity in at least two assays, (2) compounds with confirmed activity in no more than five assays, (3) compounds tested in at least ten assays. We felt that this simple approach would give us a set of information-rich compounds largely free of frequent hitters. Using Daylight 2D fingerprints and a Tanimoto distance of 0.3, the set consists of 9200 clusters, of which there are almost 5100 singletons. We propose that this richly diverse subset is an ideal starting platform for the design of screening libraries and for the discovery of new privileged structures. Interestingly, with respect to physical properties, the subset is slightly more lipophilic and has slightly larger polar surface area than the general collection, but the distribution of molecular weights and the numbers of hydrogen-bond donors and acceptors is the same. We conclude that the currently accepted drug-like physical properties boundary conditions are necessary but not sufficient to define biological activity and that other, poorly understood, factors are the true drivers of such activity. We continue to explore just what those factors might be.

3.9 Synthesis of Chemical Genetics Libraries: New Organic Synthesis Approaches to the Discovery of Biological Activity

The recognition that the intersection of biology space is limited within chemistry space has encouraged the development of new strategies in organic synthesis for the discovery of biological activity. For example, Ellman and coworkers [52] have developed combinatorial target-guided ligand assembly. In this method, a set of potential binding elements is prepared in which each molecule incorporates a common chemical linkage group. The set of potential binding elements is screened to identify all binding elements that interact even weakly with the biological target. A combinatorial library of linked binding elements is prepared in which the binding elements are connected through a set of flexible linkers. The library is then screened to identify the tightest-binding ligands. Using this approach they identified a potent (IC_{50} = 64 nM) inhibitor of the nonreceptor tyrosine kinase c-Src. An extension of this strategy has been developed by Lehn and others [53]. So-called dynamic combinatorial chemistry uses self-assembly processes to generate libraries. In contrast to the stepwise assembly of molecules in the library, this method allows for the generation of libraries based on continuous interconversion among the library constituents. Addition of the target ligand or receptor creates a driving force that favors formation of the best-binding constituent. Sharpless and coworkers [54] have investigated a slightly different approach. Rather than using a set of interconverting constituents, they allow the target to select building blocks and synthesize its own inhibitor. Dubbed 'click chemistry,' it depends on the simultaneous binding of two ligands, decorated with complementary reactive groups, to adjacent sites on the protein. Their colocalization is then likely to accelerate the reaction that connects them. The reaction of course must be selected so as to not take place in undesired ways within biochemical systems. One such reaction is the cycloaddition of azides to acetylenes to yield 1,2,3-triazoles. As a proof of principle, AChE was used to select and synthesize a triazole-linked bivalent inhibitor by using known site-specific ligands as building blocks. This resulted in the discovery of an inhibitor with a K_d in the range of 77–410 fM (femtomolar), depending on the species. This is the most potent noncovalent AChE inhibitor known to date, by approximately two orders of magnitude.

The standard approach to parallel synthesis of libraries is to start with a polyfunctional common core and elaborate those functions with diversity elements. With just a few diversity locations and the large number of commercially available diversity reactants, this can result in libraries consisting of tens or hundreds of thousands, or even more, members. Nevertheless, such libraries retain the common core for all members, which necessarily limits the total diversity of the library. Far more challenging, and arguably more valuable to the efficient exploration of chemistry space, would be the synthesis of libraries whose members are based on disparate cores. Schreiber [55] is addressing the problem of skeletal diversity by using a synthesis strategy that involves transforming substrates with different appendages that pre-encode skeletal information into products that have different skeletons, with the use of common reaction conditions.

Our own interest in this problem was the result of our work on the biology-based collections discussed above. We found that roughly only half the compounds were available as solid samples for further study, and the remainder were dropped from consideration for that reason. The efficient resynthesis of hundreds or thousands of disparate compounds was simply not practical. Or was it? Perhaps there was an easy way to sort multiple syntheses into common starting materials and reactions and to carry them out in parallel. To that end, we used LeadScope software [56] as our management tool. Normally, LeadScope links chemical and biological data, allowing chemists to explore large sets of compounds by a systematic substructural analysis using a predefined set of 27 000 structural features. More importantly for our purposes, two sets can be compared with respect to these features. We chose the ACD database as our second set. We could then easily select those starting materials that would give rise to many products via different routes. We then ran as many reactions as possible using parallel synthesis methods. We have used this method for syntheses of up to four steps and have been able to maintain a productivity level of one compound per chemist per day, 25 mg scale, purified ≥85%, and characterized by LC/MS and NMR.

We are developing an approach to true simultaneous synthesis of disparate core compounds. Most molecules of the size and complexity we are interested in would likely be prepared in no more than five steps. The actual transformations are usually limited to the chemistry background and experience of the chemist(s) involved in the project. However, the routes need not be so limited. Indeed, consider the generation of tens or hundreds of routes to each compound of interest. The problem then becomes one of how to prepare the maximum number of compounds using the minimum set of common chemistries, staging the routes as necessary so as to maximize the overlap of reagents and conditions. The generation of syntheses is software based. Two or three decades ago there was a lot of effort to develop software to predict the most efficient syntheses of complex organic molecules; most have been abandoned. We chose to use the SynGen program [57] for the very reason that it usually produces several routes to a molecule, each of which begins with a commercially available starting material and whose transformations usually have a literature precedent. Common chemistries can be grouped at three levels: (1) reaction type, e.g., acylation of amines; (2) reagent type, e.g., acylation of secondary amines; and (3) specific reagents, e.g., acylation of diethyl amine. Each level is specifically encoded by the program, making searching, sorting, and matching fairly easy. We will not necessarily choose the shortest route to each molecule, since it is entirely possible that some longer routes would give rise to additional commonalities, thereby allowing the preparation of a larger total number of compounds. We are in the process of testing this concept using a set of 100 very different structures and will report the results in due course.

3.10 Information and Knowledge Management Issues

The integration of chemistry and biology that constitutes the engine for chemical genetics presents a major challenge for existing models of information and knowledge management. The management of information and knowledge is so critical as to deserve a place as one of the three critical components necessary to truly enable chemical genetics (Figure 3.16). Linking chemical structures with biology in a systematic way has challenged pharmaceutical companies and software vendors for many years, and several proprietary and off-the-shelf solutions now exist. Typically, these products are not scaleable or flexible enough to deal with the problems exposed by chemical genetics.

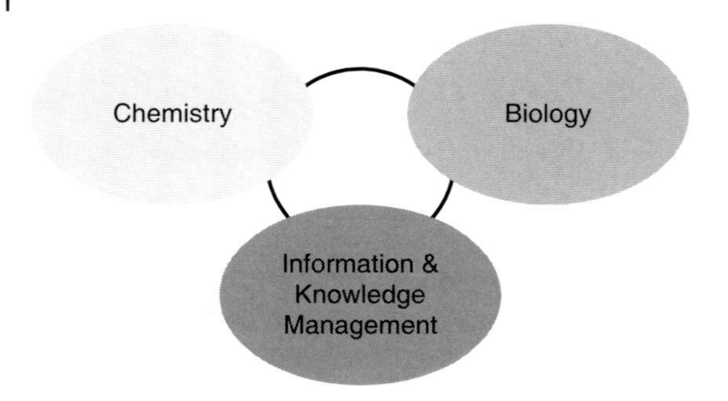

Figure 3.16 Chemical genetics requires the integration of the three critical elements of chemistry, biology, and information/knowledge management.

3.11 Annotation of Small Molecules

Several groups have realized the information management challenges posed by chemical genetics. The US National Cancer Institute is developing a powerful open-access database called ChemBank that will link small-molecule structure and associated effects on proteins, cell pathways, and tissue formation [58]. Additionally the effect of small molecules on an organism's phenotype will also be captured. ChemBank is a chemical genetics database, which has been described as a chemical version of GenBank, the online repository of genetic data [55]. The NCI plans to synthesize and screen thousands of molecules for their biological activity. Annotation of small molecules should allow for much closer integration of chemical structure and biological activity. Use of such annotated compounds (sometimes referred to as information-rich compounds) as chemical tools for probing biological systems promises to be a fruitful area of future research.

The central informatics issue in chemical genetics is annotation of chemical structures in the same way as annotation of genes, i.e., annotation of the biology and other properties of a chemical structure. In a typical single-drug discovery project, it is common for many structures to be profiled by a single biological screen generating a simple vertical data format (Figure 3.17). In chemical genetics we focus on single compounds annotated with many biologies – a horizontal data format (Figure 3.17).

NCI is asking scientists from all over the world to deposit information on the effects of small molecules on cells on the micro (gene expression) [13] and macro levels in ChemBank. One of the hopes here is to link phenotypic changes with structures and to use this information in predicting the mechanism of action of drugs.

Structure	*Biology* (IC_{50}, μM)
mol 1	1.00
mol 2	2.00
mol 3	1.50
mol 4	0.10
mol 5	0.00
mol 6	1.20
mol 7	12.10
mol 8	0.20
mol 9	0.80
mol 10	2.20
mol 11	0.00
mol 12	0.04
mol 13	5.10
mol 14	0.40
mol 15	0.90
mol 16	0.02
mol 17	4.10
mol 18	6.30
mol 19	0.00
mol 20	0.20
mol 21	0.70
mol 22	2.90
mol 23	5.90
mol 24	13.20
mol 25	1.60
mol 26	0.20
mol 27	0.01
mol 28	7.30
mol 29	8.50

Vertical to horizontal data sets

- Small molecule annotation *(cf. gene annotation)*
- Structure Activity Relationships *(vs. many biology's)*

Structure	*Biology 1* (IC_{50}, μM)	*Biology 2* (IC_{50}, μM)	*Biology 3* (IC_{50}, μM)	*Biology 4* (IC_{50}, μM)
mol 1	1.00	0.01	2.00	1.50
mol 2	2.00	2.40	30.00	30.00

Structure	*Biology 5* (IC_{50}, μM)	*Biology 6* (IC_{50}, μM)	*Biology 7* (IC_{50}, μM)	*Biology 8* (IC_{50}, μM)
mol 1	0.10	30.00	1.20	12.10
mol 2	30.00	0.02	30.00	12.10

Figure 3.17 Chemical genetics databases require the annotation of individual compounds with many biologies, in contrast to the more traditional way of capturing the assay results of many compounds against a single biology.

3.12 Summary

Bridging the knowledge gap between the data provided by the human genome project and our knowledge of biological processes and systems is a requirement for the efficient and effective exploitation of this knowledge in drug discovery. We see this knowledge gap as being best bridged by a truly interdisciplinary approach and a close integration of chemistry and biology – in both thinking and experiment. Chemical genetics provides a framework for the systematic study of small molecules to perturb and thus understand biological systems. The adoption of chemical genetics thinking is already growing in influence among chemists and biologists, and the fruits of this integrated approach to drug discovery promises to be an exciting and rewarding area of research for the next decade.

References

1 K. I. Kaitin (Ed.), Post-approval R&D raises total drug development costs to $ 897 million, *Tufts Center for the Study of Drug Development Impact Report* **2003**, 5(3), can be found at http://csdd.tufts.edu/InfoServices/Publications.asp#ResearchBibliography.

2 Pharmaceutical innovation: an analysis of leading companies and strategies, *Reuters Business Insight* **2002**, can be found at http://www.delphipharma.com/innovation.htm.

3 L. Olbe, E. Carlsson, P. Lindberg, *Nature Rev. Drug Discov.* **2003**, *2*, 132–139.

4 J. W. Black, *Science* **1989**, 245, 486.

5 A. L. Hopkins, C. R. Groom, *Nature Rev. Drug Discov.* **2002**, *1*, 727–736.

6 (a) S. L. Schreiber, *C&EN* **2003**, *81*, 51–61; (b) S. L. Schreiber, *Bioorg. Med. Chem.* **1998**, *6*, 1127–1152; (c) B. R. Stockwell, *Trends Biotechnol.* **2000**, *18*, 449–455.

7 B. P. Zambrowicz, A. T. Sands, *Nature Reviews* **2003**, *2*, 38–51.

8 (a) R. S. Bohacek, C. McMartin, W. C. Guida, *Med. Res. Rev.* **1996**, *16*, 3–50; (b) H. C. Kolb, M. G. Finn, K. B. Sharpless, *Angew. Chem. Int. Ed.* **2001**, *40*, 2004–2021.

9 H. J. Kwon, *Curr. Med. Chem.* **2003**, *10*, 717–726.

10 E. Shorter, *Nature Rev. Drug Discov.* **2002**, *1*, 1003–1006.

11 (a) J. Taunton, C. A. Hassig, S. L. Schreiber, *Science* **1996**, *272*, 408–411; (b) A. Karin, M. Winnik, *Proc. Natl. Acad. Sci. USA* **1968**, *60*, 668–674; (c) M. A. Raftery, M. W. Hunkapiller, C. D. Strader, L. E. Hood, *Science* **1980**, *208*, 1454–1457; (d) M. Noda, H. Takahashi, T. Tanabe, M. Toyosato, Y. Furutani, T. Hiroshe, M. Asai, S. Inayama, T. Miyata, S. Numa, *Nature* **1982**, *299*, 793–797; (e) L. Kornel, *Acta Endocrinol. Suppl. (Copenh)* **1973**, *178*, 1–45; (f) R. W. King, *Science* **1997**, *277*, 973–974; (g) P. P. Sche, K. M. McKenzie, J. D. White, D. J. Austin, *Chem. Biol.* **1999**, *6*, 707–716.

12 (a) B. R. Stockwell, J. S. Hardwick, J. K. Tong, S. L. Schreiber, *J. Am. Chem. Soc.* **1999**, *121*, 10662–10663; (b) M. J. Marton, J. L. DeRisi, H. A. Bennett, V. R. Iyer, M. R. Meyer, C. J. Roberts, R. Stoughton, J. Burchard, D. Slade, H. Dai, D. E. Bassett, L. H. Hartwell, P. O. Brown, S. H. Friend, *Nature Med.* **1998**, *4*, 1293–1301.

13 T. Gura, *Nature* **2000**, *407*, 282–284.

14 F. G. Kuruvilla, A. F. Shamji, S. M. Sternson, P. J. Hergenrother, S. L. Schreiber, *Nature* **2002**, *416*, 653–656.

15 P. J. Belshaw, S. D. Meyer, D. D. Johnson, D. Romo, Y. Ikeda,

M. Andrus, D. G. Alberg, L. W. Schultz, J. Clardy, S. L. Schreiber, *Synlett* **1994**, 381–392.

16 G. D. Van Duyne, R. F. Standaert, P. A. Karplus, S. L. Schreiber, J. Clardy, *J. Mol. Biol.* **1993**, *229*, 105–124.

17 J. Liu, J. D. Farmer, W. S. Lane, J. Friedman, I. Weisman, S. L. Schreiber, *Cell* **1991**, *66*, 807–815.

18 S. L. Schreiber, G. Crabtree, *Harvey Lect.* **1997**, *89*, 373–380.

19 G. R. Rosania, Y.-T. Chang, D. Sutherlin, H. Dong, D. J. Lockhart, P. G. Schultz, *Nature Biotechnol.* **2000**, *18*, 304–308.

20 (a) A. C. Bishop, *Curr. Biol.* **1998**, *8*, R257-R266; (b) T. Liu, K. Shah, F. Wang, L. Witucki, K. M. Shokat, *Chem. Biol.* **2000**, *5*, 91–101; (c) A. C. Bishop, C. Kung, K. Shah, L. Witucki, K. M. Shokat, Y. Liu, *J. Am. Chem. Soc.* **1999**, *121*, 627–631; (d) A. C. Bishop, J. A. Ubersax, D. T. Petsch, D. P. Matheos, N. S. Gray, J. Biethrow, E. Shimizu, J. Z. Tsein, P. G. Schultz, M. D. Rose, J. L. Wood, D. O. Morgan, K. M. Shokat, *Nature* **2000**, *407*, 395–399.

21 (a) S. L. Swann, J. Bergh, M. C. Farach-Carson, C. A. Ocasio, J. T. Koh, *J. Am. Chem. Soc.* **2002**, *124*, 13795–13805; (b) S. L. Swann, J. J. Bergh, M. C. Farach-Carson, J. T. Koh, *Org. Lett.* **2002**, *4*, 3863–3866.

22 (a) H. Fen Ye, K. E. O'Reilly, J. Koh, *J. Am Chem. Soc.* **2001**, *123*, 1521–1522; (b) Y. Shi, J. T. Koh, *Chem. Biol.* **2001**, *8*, 501–510; (c) J. T. Koh, M. Putnam, M. Tomic-Canic, C. M. McDaniel, *J. Am. Chem. Soc.* **1998**, *121*, 1984–1985.

23 D. F. Doyle, D. A. Braasch, L. K. Jackson, H. E. Weiss, M. F. Boehm, D. J. Mangelsdorf, D. R. Corey, *J. Am Chem. Soc.* **2001**, *123*, 11367–11371.

24 U. Abel, C. Koch, M. Speitling, F. G. Hansske, *Curr. Opin. Chem. Biol.* **2002**, *6*, 453–458.

25 D. S. Tan, M. A. Foley, B. R. Stockwell, M. D. Shair, S. L. Schreiber, *J. Am. Chem. Soc.* **1999**, *121*, 9073–9087.

26 M. M. Hann, A. R. Leach, G. Harper, *J. Chem. Inf. Comput. Sci.* **2001**, *41*, 856–864.

27 J. A. Monn, M. J. Valli, S. M. Massey, R. A. Wright, C. R. Salhoff, B. G. Johnson, T. Howe, C. A. Alt, G. A. Rhodes, R. L. Robey, K. R. Griffey, J. P. Tizzano, M. J. Kallman, D. R. Helton, D. D. Schoepp, *J. Med. Chem.* **1997**, *40*, 528–537.

28 A. Nakazato, T. Kumagai, K. Sakagami, R. Yoshikawa, Y. Suzuki, S. Chaki, H. Ito, T. Taguchi, S. Nakanishi, S. Okuyama, *J. Med. Chem.* **2000**, *43*, 4893–4909.

29 S. H. Bertz, *J. Am. Chem. Soc.* **1981**, *103*, 3599–3601.

30 R. Breinbauer, I. R. Vetter, H. Waldmann, *Angew. Chem. Int. Ed.* **2002**, *41*, 2879–2890.

31 L. Holm, *Curr. Opin. Chem. Biol.* **1998**, *8*, 372–379.

32 L. Kissau, P. Stahl, R. Mazitschek, A. Giannis, H. Waldmann, *J. Med. Chem.* **2003**, *46*, 2917–2931.

33 A. K. Ghose, V. N. Viswanadhan, J. J. Wendoloski, *J. Comb. Chem.* **1999**, *1*, 55–68.

34 G. W. Bemis, M. A. Murcko, *J. Med. Chem.* **1996**, *39*, 2887–2893.

35 G. W. Bemis, M. A. Murcko, *J. Med. Chem.* **1999**, *42*, 5095–5099.

36 Ajay, G. W. Bemis, M. A. Murcko, *J. Med. Chem.* **1999**, *42*, 4942–4951.

37 Ajay, W. P. Walters, M. A. Murcko, *J. Med. Chem.* **1998**, *41*, 3314–3324.

38 J. Sadowski, H. Kubinyi, *J. Med. Chem.* **1998**, *41*, 3325–3329.

39 M. Brüstle, B. Beck, T. Schindler, W. King, T. Mitchell, T. Clark, *J. Med. Chem.* **2002**, *45*, 3345–3355.

40 Y. C. Martin, J. L. Kofron, L. M. Traphagen, *J. Med. Chem.* **2002**, *45*, 4350–4358.

41 R. S. Pearlman, K. M. Smith, *J. Chem. Inf. Comput. Sci.* **1999**, *39*, 28–35.

42 D. T. Stanton, *J. Chem. Inf. Comput. Sci.* **1999**, *39*, 11–20.

43 B. Pirard, S. D. Pickett, *J. Chem. Inf. Comput. Sci.* **2000**, *40*, 1431–1440.

44 G. Weitz-Schmidt, *Trends Pharmacol. Sci.* **2002**, *23*, 482–486.

45 F. S. LaBella, *Biochem. Pharmacol.* **1991**, *42 Suppl.*, S1-S8.

46 (a) L. M. Kauvar, D. L. Higgins, H. O. Villar, J. R. Sportsman,

Å. Engqvist-Goldstein, R. Bukar, K. E. Bauer, H. Dilley, D. M. Rocke, *Chem. Biol.* **1995**, *2*, 107–118; (b) S. L. Dixon, H. O. Villar, *J. Chem. Inf. Comput. Sci.* **1998**, *38*, 1192–1203.

47 W. Michne, unpublished results.

48 A. Tinker, unpublished results.

49 O. Roche, P. Schneider, J. Zuegge, W. Guba, M. Kansy, A. Alanine, K. Bleicher, F. Danel, E. Gutknecht, M. Rogers-Evans, W. Neidhart, H. Stalder, M. Dillon, E. Sjögren, N. Fotouhi, P. Giillespie, R. Goodnow, W. Harris, P. Jones, M. Taniguchi, S. Tsujii, W. von der Saal, G. Zimmerman, G. Schneider, *J. Med. Chem.* **2002**, *45*, 137–142.

50 R. P. Sheridan, *J. Chem. Inf. Comput. Sci.* **2003**, *43*, 1037–1050.

51 S. L. McGovern, E. Caselli, N. Grigorieff, B. K. Shoichet, *J. Med. Chem.* **2002**, *45*, 1712–1722.

52 D. J. Maly, I. C. Choong, J. A. Ellman, *Proc. Nat. Acad. Sci. USA* **2000**, *97*, 2419–2424.

53 O. Ramström, J. Lehn, *Nature Rev. Drug Discov.* **2002**, *1*, 26–36.

54 W. G. Lewis, L. G. Green, F. Grynszpan, Z. Radiæ, P. R. Carlier, P. Taylor, M. G. Finn, K. B. Sharpless, *Angew. Chem. Int. Ed.* **2002**, *41*, 1053–1057.

55 M. D. Burke, E. M. Berger, S. L. Schreiber, *Science* **2003**, *302*, 613–618.

56 G. Roberts, G. J. Myatt, W. P. Johnson, K. P. Cross, P. E. Blower, *J. Chem. Inf. Comput. Sci.* **2000**, *40*, 1302–1314.

57 (a) J. B. Hendrickson, *Knowl. Eng. Rev.* **1997**, *12*, 369–386; (b) Further information and a demonstration of the program can be found at http://syngen2.chem.brandeis.edu/syngen.html.

58 D. Adam, *Nature* **2001**, *411*, 873.

59 (a) D. T. Hung, J. B. Nerenberg, S. L. Schreiber, *J. Am. Chem. Soc.* **1996**, *118*, 11054–11080; (b) R. E. Longley, D. Caddigan, D. Harmody, M. Gunasekera, S. P. Gunasekera, *Transplantation* **1991**, *52*, 656.

60 J. Taunton, C. A. Hassig, S. L. Schreiber, *Science* **1996**, *272*, 408–411.

61 G. Fenteany, R. F. Standaert, W. S. Lane, S. Choi, E. J. Corey, S. L. Schreiber, *Science* **1995**, *268*, 726–731.

4
Structural Aspects of Binding Site Similarity: A 3D Upgrade for Chemogenomics

Andreas Bergner and Judith Günther

4.1
Introduction

4.1.1
Binding Sites: The Missing Link

The idea of chemogenomics is just starting to take shape. One approach, which leaves room for many definitions as to what techniques and applications the concept of chemogenomics comprises, considers it to be the effort of creating links between chemistry space and the genome space. This notion may appear rather vague; however, it points directly to the interface between biology and chemistry where chemogenomics is expected to assume its definite form. Undeservedly, the wealth of protein structural data is often disregarded in this area. The aim of this article is to review this perception, and to demonstrate that using the perspective and methods of structural biology can enhance the way in which chemogenomics is integrated into pharmaceutical research and development.

The idea that a small molecule, active with a particular target protein, is very likely to be active also with a sequence-related protein is by no means new. Over several decades medicinal chemists have acquired valuable experience as to how to systematically explore chemistry space around a given lead structure, how to establish structure–activity relationships, and how to use such knowledge for refining the selectivity profile of the drug candidate. Often, selectivity with a related protein can be achieved by relatively small modifications to the original small molecule's structure.

With the decoding of the human genome it has become apparent that the development of pharmaceutically active substances has so far targeted only a very small fraction of the human proteome. It is commonly assumed that many more druggable targets are available that offer new perspectives for drug development [1]. Whether due to convergent or divergent evolution, the genome space contains clusters of gene (and accordingly target) families whose mutual similarity is conventionally described by the sequence homology of the target proteins. Given

Chemogenomics in Drug Discovery: A Medicinal Chemistry Perspective.
Edited by Hugo Kubinyi and Gerhard Müller

ISBN: 3-527-30987-X

these advances in the genomics area, the experience previously acquired for a particular target and compound class can now be exploited in an unprecedented way, reusing relevant information and know-how. A small molecule well-profiled for a particular target provides an excellent starting point for the exploration of its neighbors in genome space and possibly even for the subsequent development of drug candidates for related proteins.

Although the term chemogenomics obviously neglects protein structural aspects, and structural genomics [2, 3] has become an established field in its own right, ultimately, the similarity of two proteins on the level of their native 3D structure provides the basis for the binding of structurally related small molecules. In particular, the protein cavity accommodating the small molecule largely determines the recognition features of the target protein. Thus, the characteristics of a binding site, often illustrated as the *lock* into which a drug molecule fits *like* a key [4], provide the missing link needed for a thorough understanding of the correlation between chemistry space and genome space that chemogenomics aims to achieve.

Although relationships between small-molecule structures and protein families can be established on a purely empirical basis, and 3D protein-structure information is not a necessary precondition, it would be foolish not to consider such information whenever it is available. With the rapidly growing number of protein structures collated in the PDB [5], the chance of finding either the experimentally determined 3D structure of the target protein or at least one of a closely related protein that allows a sufficiently reliable homology model to be built [6] are constantly increasing.

This section approaches chemogenomics from the viewpoint of structural biology, focusing on the relevant aspects of binding site characteristics and similarities. The authors believe that the implementation of this perspective can be advantageous in virtually all stages of the drug development process. Clearly, the impact of structural biology is extremely beneficial for lead finding and lead optimization; nevertheless, it can also be used to facilitate drug discovery projects in the early stages.

4.1.2
Target Assessment

The number of biologically validated targets known to date is large, forcing pharmaceutical companies to carefully select the targets to be pursued in a lead-finding project. With an increasing number of feasible targets being discovered through DNA chip technologies, the need for prioritizing target candidates for biological validation and then selecting the most promising targets is gaining importance. One approach to target selection, along the lines of chemogenomics, is to take advantage of all the knowledge collected in projects that failed at a very late stage of development, leaving active compound(s) with well-tailored ADMET profile(s). Such data can then be used as a starting point for searching the available genome data for related proteins. The identified proteins can then be critically assessed with respect to their potential for representing biologically valid drug targets. If one of the initially identified proteins indeed turns out to be a valid target, a new lead-finding project can take advantage of all the knowledge collected on the target family in general. The 'fallen angel' provides an excellent starting point for structural modifications, and one can benefit from the fact that the chemistry for this compound class is already well established.

Predicting the success of a target project before it has begun remains visionary, but structural biology can certainly assist in identifying those targets that are likely to pose particular difficulties, thus rendering them less promising. This can be particularly helpful if the biological validation data for the target candidates are all equally sound. Comparison of the binding sites for different proteins belonging to one family allows an assessment of whether or not selectivity between two proteins can be feasibly achieved. This is particularly valuable in the situations mentioned above, where a 'fallen angel' inspires the initiation of a new project focusing on a related target protein. In the course of a long-pursued project a protein crystal structure often becomes available, enabling an estimation of whether selectivity towards the old, unsuccessful target can be achieved.

As a first approximation, the sequence of the protein of interest can be mapped onto the known 3D structure of a homologous protein (e.g., a previously investigated target protein of the same class). Tools for mapping sequential features onto protein structures, including intuitive visualization features, were recently developed by Lion Bioscience [7] and are publicly available through a web service [8]. Thus, the 3D structure of the target protein does not necessarily have to be solved.

The same holds true for assessing the druggability of a target by analyzing the shape and physicochemical properties of a binding site. Large, shallow binding sites with unbalanced proportions of polar and hydrophobic atoms exposed to the binding site surface appear less promising than deep crevices, which can bury large portions of a ligand and bind it via both H-bonding and hydrophobic interactions [9].

4.1.3 Lead Finding

High-throughput screening (HTS) of enormous compound libraries has been pursued in almost all pharmaceutical companies for more than a decade and has not resulted in the initially expected number of hits suitable for further lead optimization. Furthermore, virtual high-throughput screening (vHTS) methods have increasingly been used as a complementary means of finding small molecules that are active with a particular target [10, 11]. *In silico* methods allow for the screening of millions of molecules within a few days. Although pre- or post-filtering techniques for focusing on drug-like molecules, often based on filters such as Lipinski's rule of five [12], have been developed, other requirements for an initial hit to be promising, such as synthetic accessibility of the compound class, cannot be considered well with vHTS methods. For both HTS and vHTS, the sheer number of compounds does not improve the chance of finding the right molecule, and due to the size of chemical space, a complete sampling is nearly impossible (the number of possible molecules with a molecular weight less than 500 Da has been estimated to be 10^{200}, 10^{60} of which might possess drug-like properties [13]). Therefore, increasing attention is being paid to finding so-called privileged structures. Such compounds may not exhibit the desired potency or selectivity profile, but they provide a promising starting point for further exploration of the surrounding chemical space in order to find related compounds that are suitable as lead candidates. Along these lines, efforts in combinatorial chemistry have focused on the synthesis of target family based libraries, which are preferentially screened whenever a lead structure for a member of the respective target family is to be found. To provide enhanced hit rates for such targets, target family based libraries feature a scaffold that qualifies the designed structures to bind to various members of the target family. To this end, the ligands have to form interactions to binding site residues that are well conserved within this family (and, if possible, which also contribute well to binding affinity). At the same time, the substituents attached to the scaffold should be designed for exploring regions of high structural variability within the protein pockets, thus raising the chances of finding fairly selective compounds in the screened library.

Fragment screening techniques are increasingly being utilized for identifying suitable scaffolds. Both X-ray crystallography [14] and NMR [15] have proved to be useful methods for extracting small molecules with moderate affinity from a mixture of compounds. The compounds in such cocktails are synthetically easily accessible and small enough to leave room for chemical modification by attaching further functional groups. Some groups have used computational screening techniques for prioritizing fragments and picking out hits that appear to fit well into the protein pocket of interest for subsequent experimental screening [16–18]. A fragment that forms well-conserved interactions within the binding pockets of the particular protein family can provide a privileged structure. Knowledge of the binding mode of the fragment can guide further synthesis, for example, by pointing to further attachment sites for new substituents to be added to the core and by estimating the spatial and physicochemical requirements for the substituent.

4.1.4
Lead Optimization

Once lead finding has been accomplished and the stage of lead optimization has begun, detailed knowledge about the binding site of the target protein becomes even more important, especially if the selectivity profile of the lead compound is suboptimal. If the binding mode of a small molecule in the pocket of the target protein has been determined, the detected structural differences between two binding sites can be systematically exploited to guide further synthetic efforts. If crystallization of the target protein turns out to be difficult, a drug candidate can alternatively be cocrystallized with a closely related protein (e.g., an anti-target); this approach is usually referred to as the surrogate approach. Apart from sequential insertions and deletions that have a major effect on the binding site's shape, substitutions of corresponding amino acid residues are the most obvious differences one can take advantage of if selectivity between two closely related proteins is to be achieved. If the side chains of the residues are sufficiently different in size and/or physicochemical properties, a single interaction can be sufficient to obtain reasonably large differences in affinity (see Section 4.5.5, *Selectivity Issues*). In this situation, mapping of sequential differences onto the structure of one of the proteins or, alternatively, construction of a homology model, is often sufficient for guiding chemical modification of the lead structure. If the amino acids expose fairly similar recognition features to the ligand or interact with it only via backbone atoms, smaller differences affecting the overall shape of the binding sites could be targeted by suitably tailored compounds. Here, experimental determination of both protein structures complexed with the current lead candidate is highly advisable. Even more subtle differences, such as different extents of protein flexibility in the binding pocket, are fairly difficult to exploit, because the underlying effects are poorly understood and X-ray crystallography can give only a very limited picture of these phenomena.

A tool for analyzing and comparing the binding sites of sequence-related proteins is available within the receptor–ligand database Relibase [19, 20]. Superposition and visualization of any combination of such similar proteins from the PDB can be done using a free web service [21]. The enhanced version of Relibase, Relibase+, provides an automatic analysis of their structural similarities and differences, including backbone and side chain movements, conserved solvation sites, and volume overlap of bound ligands. A related tool utilizing a database of prealigned binding sites is Ligbase [22].

These scenarios highlight the importance of 3D structural information in different steps of drug design by means of chemogenomics. Clearly, a thorough understanding of the nature of protein binding pockets, alongside the means for evaluating common features and differences of such cavities, is of great relevance for the success of such efforts.

This chapter is organized into five sections. The next section provides an introduction to the structural biology of binding sites and sheds some light on why nature usually uses pockets for intermolecular recognition processes. Sections 4.3 and 4.4 review computational methods for detecting binding sites, given a 3D protein structure, as well as different approaches for describing binding site similarities among a set of protein structures. Section 4.5 looks at applications of binding site comparisons, focusing on some of the popular target classes and highlighting how the consideration of binding site similarities can inspire and promote drug discovery projects at different stages. The review concludes with a future vision outlining the implementation of methods for analyzing and comparing protein binding sites within the framework of chemogenomics efforts.

4.2
Structural Biology of Binding Sites

The biological function of most proteins depends on specific interactions with other molecules binding to particular surface areas, the binding sites. Binding sites can be defined as clusters of amino acids whose structural, dynamic, and physicochemical properties directly affect the interaction and transformation of the binding molecules. These molecules can be, for example, other proteins, nucleic acids, or organic ligands, or, classifying them by their function, effectors, substrates, inhibitors, cofactors, agonists, or antagonists. Binding sites constitute the arena in which the function(s) of a protein are turned into action. In spite of the dazzling array of protein functions, researchers have tried to identify structural determinants capable of distinguishing binding sites from other surface areas or, in other words, to understand what makes a binding site a binding site [23]. To act as a functional unit, a binding site has to possess several characteristics that are also reflected in its structure. The following section discusses the energetic, functional, specificity-related, and evolutionary aspects that restrain, and thus characterize, the constitution of binding sites, from a 3D-structural perspective in the light of recent research. The bound ligands, the actual focus of attention in medicinal chemistry, are sidelined in this section. The question of how ligand similarity and binding site similarity are related is discussed in more detail following this introduction.

Early attempts to understand the nature of binding sites focused on the chemical composition, i.e., the amino acid distribution in protein binding sites. A study by Villar and Kauvar [24] revealed an accumulation of some residues, in particular Arg, His, Trp, and Tyr. Young et al. [25] found that protein–protein interface areas often correspond to the strongest hydrophobic clusters on the protein surface. Also, about 10% of protein structures (total dataset size: 419) appear to exhibit at least one large cluster of charged amino acid residues [26, 27]. Typically, negatively charged clusters are involved in the formation of metal binding sites, whereas mixed-charge clusters occur in stable protein–protein interactions. However, these observations on their own are not really suitable for reliably detecting binding sites.

4.2.1 Energetic, Thermodynamic, and Electrostatic Aspects

A binding site has to be assembled in such a way that the binding of an interacting molecule is energetically feasible. This may seem trivial; however, recent studies provide several different perspectives linking some surface properties with the thermodynamics of binding. Generally, binding depends on formation of contacts between chemical groups, including van der Waals contacts and H bonds. The larger the number of contacts, the tighter the binding will be. Obviously, an increased contact surface area corresponds to an increased number of potential contacts, thus facilitating stronger binding. A structural means for increasing the contact surface area, which is particularly relevant for the binding of small molecules, is the formation of a cleft or cavity on the protein surface. The implications of active-site clefts have been broadly analyzed and discussed by Laskowski et al. [28]. This study also stresses the importance of the burial of enzyme substrates in clefts. The shielding of the reaction center from surrounding water molecules is essential for many biochemical reactions, in particular those involving electron transfer processes. Moreover, because of its burial in a pocket, a substrate molecule encounters an environment with a significantly decreased local dielectric constant.

This allows the enzyme to generate the strong electrostatic forces required for enzyme catalysis.

A study by Pettit and Bowie analyzing the roughness of protein surfaces, qualified as fractal dimension [29], showed that functional sites are generally much rougher than other areas on the surface of a protein [30]. It has been postulated that the roughness of a protein surface is also related to binding [29]. The fractal dimension correlates with the surface area squeezed into a fixed volume. In a rougher surface patch, the effective surface area (per volume) is larger than in a smooth patch, thus allowing for more energetically favorable van der Waals contacts facilitating tight binding. The study showed that particularly small binding sites exhibit surface roughness values significantly above average; for larger interaction sites there is apparently no need for squeezing more contact (area) into a small volume. The study concluded that, although surface roughness alone does not guarantee binding, smoothness effectively precludes binding if the binding interface area is small.

In recent years, it has become evident that the energy of stabilization of a protein structure is not evenly distributed throughout the molecule. A series of site-directed mutagenesis studies have revealed that functionally important residues energetically destabilize the protein; often the mutation of such residues yields more stable proteins (see Sancluz-Ruiz and Mahatadze for a review [31]). An interesting theoretical approach exploiting this observation, based on continuum electrostatic methods (see Honig and Nicholls for a review [32]), has been reported by Elcock [33]. For six selected proteins, Elcock showed that residue-based calculations of the electrostatic free energy enable the identification of amino acid residues found to be energetically destabilizing in experiments. By implication, these are supposed to be of functional relevance. The study showed that the residues identified by the method cluster on the protein surface, representing the functional binding site; the discussion also mentions that false positives can be easily detected. A large-scale study based on a 216 protein dataset supports this idea and suggests that residues estimated to be destabilizing are also more likely to be conserved.

Related methods for the prediction of hotspots on protein–protein interfaces using virtual mutagenesis and virtual alanine scanning have recently been reviewed by DeLano [34].

Another study on structural stability, by Luque and Freire [35], revealed more dual characteristics of binding sites. According to their study, binding sites appear to comprise areas of both high and low stability. Interestingly, low-stability areas in the regulatory binding sites of allosteric enzymes appear to be essential for propagation of the signal to the catalytic site, as exemplified by glycerol kinase. The method they used is based on the COREX algorithm. COREX calculates the stability constant for each amino acid residue based on the generation of a large ensemble of partially folded local conformations used for estimating the probability, and thus the stability, of these states. It is worth adding that an increase in structural stability by point mutations in areas that undergo conformational changes upon ligand binding can have a major effect on the binding affinity, even if the respective amino acid residue is distally located from the binding site. The study demonstrates this nicely for HIV-1 protease.

A noteworthy diagnostic tool for the identification of enzyme active sites, THEMATICS (theoretical microscopic titration curves), has been described by Ondrechen et al. [36]. The approach is related to the hypothesis that ionizable residues in the active site of enzymes require a complex perturbed electrostatic field to regulate their acid or base strengths, so as to achieve the protonation state needed for proper enzymatic activity. This cannot be accomplished by a simple decoupled acid-dissociation reaction. The method employs theoretical titration curves plotting the net charge (which depends on the pK_a) of each ionizable residue against the pH. Calculation of the pK_a values is carried out with finite-difference Poisson–Boltzmann (PB) methods. The study impressively demonstrates that perturbed titration curves exhibit a distinctive shape that is different from standard curves and that they mostly represent amino acids of the active site. THEMATICS has been tested with triosephosphate isomerase, aldose reductase, and phosphomannose isomerase showing that most residues belonging to titration curves with a perturbed shape are part of the active site or are situated very close to it.

4.2.2 Functional Aspects

If the functionally relevant process is to take place at a given binding site, the molecular machinery itself must be implemented as part of the binding site. (In other types of sites, such as the regulatory sites of allosteric enzymes, the binding site has to transmit the signal given by an interacting molecule to a distant functional site, triggering highly specific responses.). For example, the catalytic ability of enzymes rests on a specific spatial arrangement of chemical groups building up the molecular machinery through which the chemical and structural steps of the biotransformation are orchestrated. This also means that the interaction partner has to be bound and anchored in a particular conformation, enabling the catalytic machinery of the binding site to carry out the biotransformation. Deep clefts are particularly well suited for facilitating anchoring; this, in addition to the energetic advantages described above, is probably the reason why, in most enzymes, the largest cleft on the protein surface represents the functional active site [28]. In fact, the study conducted by Laskowski et al. [28] showed that, in 83.6% of the structures contained in an enzyme dataset (size: 67), the functional active site corresponds to the largest cleft; in another 9%, to the second largest cleft. Also, the largest cleft tends to be much larger than all other clefts present on the protein surface. In contrast to the active sites of enzymes, functional sites involved in protein–protein interactions are characterized by shallower, flat surfaces [37]. Methods for identifying active sites of enzymes are therefore often based on purely geometrical considerations, detecting clefts and depressions on the protein surface. These include programs such as APROPOS (automated protein pocket search) [38] and CAST [39] (see the two references for details about the underlying geometrical methods). CASTp (computed atlas of surface topography) provides a free online resource for cavities in proteins. It should be noted that, for multichain proteins, the success of such methods relies on the biologically relevant multimer used as input. A web resource [40] providing such information for the PDB is the PQS (protein quaternary structure) database [41].

A heuristic approach to analyzing the properties of residues directly involved in enzyme catalysis, using secondary structure, solvent accessibility, flexibility, conservation of quaternary structure, and function, was recently published by Bartlett et al. [42]. Such studies will help to provide a more general picture of the environment of enzyme active sites.

4.2.3 Specificity versus Function

The interaction between a ligand and its target protein has to be specific, i.e., the binding site can be expected to feature chemical properties complementary to those of the interaction partner, facilitating molecular recognition. The discrimination between function and specificity is important since it determines which attributes and structural features of a binding site have to be conserved among a series of proteins for maintaining the function, and which properties can be allowed greater variation. For example, all proteases include a specific motif responsible for accomplishing the hydrolysis of polypeptides. There are a limited number of these motifs, representing different mechanisms for the same catalytic reaction. For example, serine and cysteine proteases feature catalytic triads with Ser or Cys nucleophiles; other specific motifs are found in metalloproteinases and aspartic proteases. Serine proteases, regardless of their sequence homology, also contain a pocket substructure referred to as the oxyanion hole [43], which facilitates stabilization of the tetrahedral transition state via formation of H bonds between the substrate and the enzyme. Thus, from a functional point of view, there is a limited set of motifs representing the catalytic machinery. In contrast, there is a wealth of very different protease binding sites featuring a huge variety of diverse chemical and electrostatic properties, governing the selectivity and specificity of proteolytic enzymes.

4.2.4 Evolutionary Aspects

Generally, fewer mutations, and thus a higher degree of conservation, are observed in functionally relevant residues than in other parts of a protein, since a loss of functionality leads to the dismissal of a protein mutant in evolution. This was shown by Ma et al. [44] for protein–protein interfaces. Their study shows that binding hotspots tend to be conserved, thus differentiating between binding sites and the remainder of the molecular surface. The authors of the study propose that the most conserved polar residues make the interface rigid, thus minimizing binding entropy, due to the decrease in conformational flexibility.

A method for detecting conserved residues, called evolutionary tracing (ET), was developed by Lichtarge et al. [45]. Several groups have embarked on developing methods for detecting functional sites based on ET and related methods (see

Lichtarge and Sowa [46] for a recent review). Generally, these methods are based on mapping evolutionary data onto the 3D surface of the protein, so as to identify clusters of conserved residues representing the binding site. Recent approaches include ConSurf, by Armon and coworkers [47], and an enhanced method, Rate4Site, by Pupko et al. [48, 49]. Some applications of ConSurf have been described by Glaser et al. [50]. A related method for assessing functional inheritance within protein superfamilies was reported by Aloy et al. [51].

Another advanced evolutionary method for identifying functionally relevant clusters was reported by Landgraf et al. [52]. Their approach uses multiple sequence alignment data for both the overall (global) structure and residue-specific alignments (local). It has been shown that the use of regional conservation scores overcomes some of the disadvantages of using ET only, particularly for transient interfaces, as exemplified for MAP kinase ERK2.

These examples show that binding site formation is governed by physical, chemical, and evolutionary constraints and that these principles can be used for uncovering functional binding sites.

4.3 Methods for Identifying Binding Sites

4.3.1 Integrated Methods for the Prediction of Binding Sites

The *conditio sine qua non* for structure-based drug design is the identification and functional annotation of the relevant binding site(s) in a target protein. A number of methods, closely related to the characteristics of binding sites and the restraints imposed on the formation of functional structural units, are discussed in Section 4.2. The most commonly used methods can be classified into geometry-based methods for cavity detection, methods for identifying specific patterns, and evolutionary methods.

Recently, some more advanced methods have been reported that integrate the disparate features used for the characterization of functional binding sites. It can be expected that the cooperative effect of using all the information available will greatly enhance the reliability of binding site prediction and detection tools.

One approach employing neural networks for the prediction of active sites in enzymes was recently reported by Gutteridge et al. [53]. In this approach, a neural network is used to estimate the likelihood of a residue being catalytically active, utilizing both evolutionary and structural information. The neural network is trained on experimentally confirmed active sites. A network score is calculated for each residue, based on the weights derived during training. A clustering algorithm, equipped with a significance test, identifies accumulations of highly scored residues at the protein surface. High weights are assigned to network parameters such as conservation, diversity of position, relative solvent accessibility, and charged residues, whereas secondary structure and uncharged residues contribute less to the network scoring. A success rate of 69% is reported for correctly detecting the active sites (spatial overlap of predicted and real site > 50%) and another 25% for partially correct detection (spatial overlap < 50%). Successful examples for correctly identifying the (known or proposed) active site of proteins include, e.g., the SET domain containing histone lysine methyltransferase, intron endonuclease I-TevI, and α-L-arabinanase. Putative active sites have been suggested for FemA (factor essential for methicillin resistance). However, the main problem of the method remains the generation of a high number of false positives. Nevertheless, the study nicely demonstrates the benefit of integrating structural and sequence-related (evolutionary) information in binding site prediction methods.

A related approach integrating sequence information (conservation), geometric information (cleft detection), and data on local stability calculated by Poisson–Boltzmann methods was reported by Ota et al. [54]. The method was used for predicting catalytic residues (polar atoms only) in enzymes. A number of putative active sites for a series of hypothetical proteins were found and are discussed in the study.

4.3.2 Sampling the Protein Surface

A different concept for predicting binding sites is exemplified by docking-related methods specifically designed for probing a protein surface for energetically favorable interactions (see the recent review by Sotriffer and Klebe [55] and the method described by Silberstein et al. [56]). A multi-scale approach has been reported by Glick et al. [57] and Davies et al. [58]. Their method aims to locate binding sites for specific ligand–protein pairs, using simple feature points for describing the characteristics of the ligand. Sampling of the protein surface is an iterative procedure; the number of feature points is increased in each step. Representation of the probe on different scales allows for initially finding general clefts and surface depressions, followed by a refined scanning for preferred ligand positions.

4.4 Methods for Detecting Binding Site Similarity

The development of methods for comparing 3D protein structures and for searching for similarities, so as to understand evolutionary and functional relationships, is one of the most challenging and thriving areas of structural bioinformatics. Similarity between protein structures can be searched for on different levels of structural hierarchy. These include methods for determining similarity of primary structures (amino acid sequences), for comparing secondary structures or small spatial motifs [59, 60], and for investigating the similarities of tertiary and quaternary structures dedicated to the analysis and comparison of protein folds [61, 62]. In contrast to the enormous variability in sequence coding for functionally relevant proteins, current estimates suggest that there are only 1000–5000 distinct, stable, polypeptide chain folds in nature [3]. Methods for detecting binding site similarity, the focus of this chapter, represent an intermediate area crossing this structural hierarchy. They offer a highly complementary approach to fold and sequence comparison methods. The methods available to date can be roughly classified into two groups, although there are no definite boundaries. First, there are approaches for finding specific structural motifs, defined as topological arrangements of functionally important atoms or amino acid residues. Such methods appear to be most promising for finding functional motifs in, for example, enzyme active sites, where a distinct structural arrangement of some key components is essential for the protein's function, so that major structural variations are not possible. The second group of methods comprises approaches that try to encapsulate the general flavor of binding sites in terms of their chemical or electrostatic nature, by using descriptors that are independent of specific tertiary patterns, making them more tolerant in terms of finding structural matches.

4.4.1
Searches for Specific Structural Motifs

Related to the set of methods belonging to the first group (above), it is worth mentioning studies specifically designed for investigating particular binding site 3D motifs. Fetrow and Skolnick [63] used fuzzy functional forms (FFF) to describe protein active sites in terms of conformation and geometry. FFFs were constructed and successfully used to detect glutaredoxins/thioredoxins and T_1 ribonuclease active sites within datasets comprising high-resolution structures and threading models.

Zhao et al. [64] developed a grid-based method for deriving recognition templates for adenylate binding sites. Previous studies revealed some fuzziness in adenylate binding pockets, which lack universally conserved residues [65]. This hampers easy construction of a recognition template that incorporates all the relevant structural and energetic features of the binding motif. The approach by Zhao et al. [64] is based on grid-based affinity potentials and aims to produce a comprehensive description of all conserved active site features. It is related to methods that estimate the likelihood of intermolecular interactions in a binding site, such as GRID [66] and SuperStar [65], but employs combined maps derived from superposed structures, referred to as consensus affinity maps. These consensus maps are used for generating recognition templates, which are given by the expected interaction energies assigned to each atom position in the purine ring. The predictive power of the method was demonstrated by identifying adenylate binding sites in a series of dinucleotide binding proteins. The method can discriminate adenine- from guanine-specific pockets when the respective recognition templates are used.

4.4.2
General Methods for Searching Similar Structural Motifs

One approach, belonging to the first group of methods mentioned above, for performing searches using 3D templates is TESS (template search and superimposition) [67]. TESS is based on a geometrical hashing algorithm and allows searching a database of protein structures using user-defined query templates consisting of any arbitrary geometrical arrangement of atoms and amino acid residues. TESS also enables specification of more generalized templates. For example, the catalytic triad Ser-His-Asp, present in trypsin-like proteases, can be generalized as Nuc-His-El, where Nuc stands for a nucleophilic group and El denotes an electrostatic group stabilizing the His residue of the triad. The consensus template derived in this way also includes the catalytic triads Ser-His-Glu of lipases and Asp-His-Asp of haloalkane dehalogenases. A database consisting of 3D enzyme active site templates derived using TESS, PROCAT [67, 68], is available on the web [69].

A related method for searching triad-type sidechain patterns using a multi-dimensional index tree was reported by Hamelryck [70]. With this approach, mirror images of patterns are detected, which appear to be very common among metal binding sites.

Another method, FEATURE, searches microenvironmental patterns, which are represented as a statistical model of a given set of functional sites. FEATURE is based on a supervised learning algorithm that estimates the significance of physicochemical properties present in each functional site. A study utilizing FEATURE revealed previously unknown features conserved among the active sites of non-homologous serine proteases [71]. These include an abundant number of amino acid residues with a high number of freely rotatable bonds in the region near the active site entrance. The authors speculate that this flexibility supports the accommodation of the substrate molecule in the binding site. Also, an increased polarity between the catalytic serine and the oxyanion hole, accompanied by a fairly well-conserved amide opposite the oxyanion hole, is reflected in the property descriptors used in the learning algorithm, indicating additional electronic stabilization of the transition state. The relationship between trypsin-like and carboxypeptidase active sites as approximate enantiomers is also discussed. Recently, a web-based service [72], WebFEATURE, has been established [73], which currently includes statistical models for magnesium [74], calcium, chloride, and ATP binding site motifs (see also section 4.5), and these motifs can be searched in a single protein structure.

Methods such as TESS and FEATURE require specification of a protein-based query template. An approach to the detection of 3D side chain patterns, without predefinition of a query motif or prior knowledge of the active site or binding site, has been devised by Russell [75]. The method is based on a string-matching procedure originally developed for fold recognition. To reduce the initial search space, a number of amino acids are excluded from the search, mostly unreactive amino acids having only carbon atoms in their side chains (Ala, Gly, Ile, Leu, Phe, Pro, Val) and all amino acid positions that are not well conserved. The conservation analysis is carried out by multiple sequence alignments. The search procedure detects amino acid side chains that are present in two structures in approximately the same orientation. A weighted rmsd for the pair of side chains is calculated, and a statistical significance test estimating the probability of actually observing a given rmsd is carried out. The probability is derived from analyzing the distribution of random structural patterns. An all-against-all (to our knowledge the first one ever carried out) comparison of SCOP (structural classification of proteins [76], see web service [77]) representatives, in addition to confirming already known functional motifs (such as catalytic triads and tetrads, metal binding centers, and Mg-ATP binding motifs), revealed new examples of evolutionary converged motifs. These include, for example, a di-zinc binding pattern present in phosphatases and aminopeptidases, a motif common to chitobiase and neuraminidase, and a motif shared by DNAse I and endocellulase. Recent developments have involved assessing the statistical significance of local structural similarities [78], and a web service, PINTS (pattern in nonhomologous tertiary structures) [79, 80], utilizing an amended search method and an improved significance check, has been set up. PINTS enables similar patterns to be uncovered in new structures and assesses their significance, allowing for the prediction of functional relationships among structurally different proteins. PINTS is continuously being updated. At the time of writing, it was possible to carry out searches for protein vs. pattern and pattern vs. protein and also to do pairwise comparison of protein structures.

A related approach for detecting recurring side chain patterns (DRESPAT) was recently developed by Wangikar et al. [81]. Picking up on some ideas developed by Russell [75] (considering one functional atom per side chain only, ignoring hydrophobic residues), the method treats structural patterns as complete subgraphs comprising three to six nodes that represent non-carbon side chain atoms. All possible structural patterns are generated for all proteins to be investigated, and the patterns recurring most frequently are selected based on geometrical considerations (rmsd) and on evaluating a statistical significance value based on the number of proteins in the dataset, the recurrence frequency, and the number of atoms in the pattern. In total, 128 datasets were generated, representing groups of non-redundant representatives of SCOP superfamilies, 17 of which were investigated in more detail in the study. These include, for example, catalytic triads and tetrads present in serine, aspartyl, and cysteine proteases and lipases, EF-hand proteins, a series of metal binding proteins, SH3 domains, and restriction endonucleases. Depending on the rmsd thresholds chosen for the pattern selection, most of the biologically relevant patterns known to be present in the structures can be found (with a high rmsd cutoff value). Unfortunately, this is accompanied by finding a huge number of false positives. In contrast, decreasing the rmsd cutoff results in a high number of false negatives. Generally, the method appears to perform best for finding patterns comprising four, five, or six atoms; however, it appears to function fairly poorly for finding three-atom patterns.

Common features between any two protein structures with different folds can be detected by using GENFIT, which was developed by Lehtonen et al. [82]. GENFIT locates similar local structures in a protein, using an algorithm for finding equivalent C_α atoms contained in unique equivalent protein fragments. By restricting the search to a limited subset of atoms that represent cofactor binding sites, binding site similarity among proteins with different folds could be identified. This was demonstrated for selected binding sites for pyridoxal phosphate (PLP) [83] and ATP [84].

Three-dimensional side chain patterns of amino acids can also be found with ASSAM, as reported by Spriggs et al. [85]. ASSAM employs a common subgraph isomorphism technique. Each amino acid is represented as a vector connecting the main chain position with a functionally relevant position in the side chain. All vectors are specified with three points: start (S), middle (M), and end (E). In the graph-theoretical approach, the vectors represent the nodes of the graph. The edges are given as the distances between vectors (nodes) and comprise six components each: SS, SM, SE, MM, ME, and EE. Auxiliary programs can be used for generating the appropriate input from a set of coordinates representing the 3D query motif. Recent developments allow for the specification of more generic queries and also of patterns including main chain, secondary structure, and solvent accessibility information, as well as disulfide bridges. The method has been tested for several 3D query motifs, including phosphate binding proteins and the catalytic triads of α-chymotrypsin and papain. A discussion of the α-chymotrypsin example demonstrated the ability of the method to reasonably detect such patterns.

Another program suite for finding templates and particular motifs in a huge preprocessed database containing common amino acid configurations was reported by Oldfield [86].

The methods summarized so far are independent of the order of the binding site residues in the primary sequence. By including the order dependence of sequence patterns, protein surface sequence patterns can be utilized for binding site comparisons. In an approach reported by Binkowski et al. [87], all residues constituting a particular binding site are extracted from the primary sequence and concatenated in the same order, forming a short sequence motif (the approach uses precalculated binding sites stored in the CASTp database [88]). These motifs can then be used for initial surface patch similarity searches, which are followed by methods for investigating the spatial match of the patterns found.

4.4.3 Similar Shape and Property Searches

The second group of approaches, seeking similarities in the shapes and chemical surface properties of binding sites, include recently developed tools such as CavBase [89, 90], eF-Site [91], and SuMo [92]. An earlier technique using surface shape only for binding site comparisons was reported by Rosen et al. [93]. The reliability of this geometric surface-matching approach has been shown for the catalytic triad of serine proteases and chorismate mutase.

CavBase was developed by Schmitt et al. [89, 90] and is fully integrated into the protein–ligand data mining system Relibase+ [19, 20]. With CavBase, cavities are detected on the basis of a purely geometrical grid-based approach, Ligsite [94]. Ligsite effectively rasters the protein structure and evaluates the local degree of burial for each grid point. Areas above a certain threshold are considered to represent cavities in the protein surface. After cavity detection, the amino acid residues lining the cavity are transformed into simplified 3D property descriptors, referred to as pseudocenters. The current implementation of CavBase features five types of pseudo-centers, namely donor, acceptor, donor–acceptor, π-aromatic, and aliphatic. Geometrical considerations concerning the directionality of possible surface interactions are used to analyze which pseudo-centers project their chemical properties to the surface. Pseudo-centers that match the criteria governing the surface property are assigned to corresponding surface patches (defined on a grid); all other pseudo-centers are omitted. All information regarding pseudo-centers, surface patches, and corresponding amino acid residues is stored in the CavBase database. Similarity searches with CavBase are based on a clique detection algorithm and can be performed by using either all pseudo-centers representing a query cavity or a selected subset of pseudo-centers representing, e.g., a particular subpocket. For clique detection, the pseudo-centers of a cavity represent the nodes of a graph, and the distances between them are the edges of the graph. The algorithm detects the largest common subgraph of two given graphs. A scoring function based on calculating the overlap of surface patches belonging to matching pairs of pseudo-centers is used to rank the solutions found.

The study of Schmitt et al. nicely showed that CavBase can detect binding site similarity for a number of examples, regardless of sequence or fold similarity. These include trypanothione reductase and a subpocket of HIV protease, which were found to share some similarity with the adenine binding pocket of cAMP-dependent kinase. Recently, Weber et al. [95] discovered unexpected cross-reactivity between the COX-2 specific sulfonamide inhibitor celecoxib and members of the structurally unrelated carbonic anhydrase family. Using CavBase, a database containing 9433 cavities was searched for similarities with subpockets of the celecoxib binding site in COX-2. The subpockets lining the sulfonamide moiety (25 pseudo-centers) and the trifluoromethyl group (7 pseudocenters) of celecoxib were used as query subpockets, and corresponding subpockets were detected in carbonic anhydrase. Recent developments of CavBase include an improved clique detection method based on clique hashing [96], which has enhanced the performance of the cavity comparison algorithm. This, in conjunction with improved similarity scoring functions, will enable more unexpected binding site similarities to be found within large structure databases in a high-throughput fashion.

Another similarity search method, also based on clique detection, is eF-Site [91, 97], an improved version of an older approach [98]. In eF-Site, the physicochemical properties of the surface are described by the electrostatic potential on the surface, calculated by numerically solving the Poisson–Boltzmann equation. Currently, the eF-Site database comprises more than 7000 entries from the PDB, including molecular surface and electrostatic potential data. For the graph-theoretical search approach, the nodes are vertices of triangles representing the molecular surface. The electrostatic potential and the local surface curvature are assigned to each node. The suitability of the method has been demonstrated by comparing proteins exhibiting completely different folds but sharing similar functions. A database search using the entire surface of phosphoenolpyruvate carboxykinase (PDB entry 1ayl) as the query was carried out, finding a number of proteins containing mono-nucleotide binding sites. Furthermore, an eF-Site search with a ligand-free structure of a 'hypothetical' protein as query, which was later shown to bind ATP, revealed a series of ATP binding proteins as hits. The study did not, however, discuss the issue of significance and scoring of hits when applying the method to a large database to discover functional similarities.

A recent approach for detecting common sites in proteins is SuMo (surfing the molecules) [92, 99]. This approach is related to that of CavBase, but in this work individual amino acid residues are transformed into descriptors that represent different chemical groups. These can be, for example, hydroxyl or aromatic and, according to their chemical nature, are assigned to one or more amino acid. The positions of the chemical groups (represented by points in space) are then used to build up triangles. A graph of adjacent triangles representing the query surface area is then subjected to a graph-theoretical approach for actually performing the similarity search. Potentially similar patches initially found by this approach are further refined based on a geometrical approach, taking into account the local atom density as a descriptor of the degree of burial of atoms and groups. The method has been successfully applied to the detection of similarities among serine proteases comprising the Asp-His-Ser catalytic triad (γ-chymotrypsin, subtilisin) and between legume lectins. For the lectins, SuMo was able to reasonably distinguish between functionally active (i.e., carbohydrate binding) and inactive representatives among the 106 legume lectin structures in the test dataset. However, we think that, in disagreement with Jambon et al., pattern-based methods, such as TESS, or the approach described by Russell are better suited to the detection of specific 3D motifs such as catalytic triads. Also, the investigation of lectins was restricted to a selected set of lectins and thus lacks any indication of how the approach would perform in terms of producing false hits in database searches.

A concept in the spirit of CavBase was reported by Stahl et al. [100]. Their approach employs a cavity detection algorithm related to Ligsite, based on calculating access values for positions representing the solvent space, followed by extracting the cavities as contiguous clusters of points with high access values. The solvent-accessible surfaces are calculated for protein residues forming cavities, using the Connolly algorithm [101]. A descriptor for the possible interaction types (aliphatic, H-bond donor or acceptor, aromatic face or edge) is assigned to each surface point, based on geometrical considerations taking the orientation of functional groups into account. Similarity searches are performed using Kohonen self-organizing neural networks [102]. Kohonen networks are a commonly used means for (nonlinearly) projecting high-dimensional dependencies into low-dimensional (here, 2D) descriptions. In this case, the neural network was trained by using cross-correlation vectors representing the distances between points on the solvent-accessible surface, in this way encoding the spatial distribution of the properties associated with the surface points. The training set contained 175 structures from different structural families. The results showed a clustering into different groups of enzymes in the 2D Kohonen map, including carbonic anhydrase, alkaline phosphatase, and metalloproteinases. The latter were split into three independent clusters. Interestingly, some of the outliers could be easily explained: the method failed for structures containing shallow pockets (superoxide dismutase) and structures containing a large variable loop region in their binding site (β-lactamase). The predictive power of the method was tested for 18 zinc enzymes not present in the training set. With only one exception, the method was able to distinguish the active site out of the five largest cavities in the protein being considered. Furthermore, these pockets were nicely assigned to the clusters belonging to the correct enzyme type.

Initial results based on ideas that are related to the CavBase approach were reported by Pickering et al. [103]. Their method encodes the characteristics of a binding site by assigning a shape index, a curvedness value, and chemical features (based on their parent amino acid residue) to each of the vertices on a Connolly surface. Likewise, as in the other approaches described above, the surfaces are represented as graphs, and the best match between two surfaces is detected with a clique detection algorithm. Initial results include calculations on the NAD binding sites of alcohol dehydrogenases (ADH) from different species. A comparison of various ADH binding sites with the NAD binding site of the more distantly related glyceraldehyde-3-phosphate dehydrogenase revealed that typically 30%–35% of the features match in both cavities.

4.5 Applications of Binding Site Analyses and Comparisons in Drug Design

4.5.1 Protein Kinases and Protein Phosphatases as Drug Targets

Chemogenomics efforts have so far focused on protein families encompassing a large number of drug targets or target candidates. Both protein kinases and protein phosphatases represent such families and are the focus of many ongoing research projects in pharmaceutical companies. In this chapter, we embark on a tour of sequence space by taking a closer look at these two protein families.

Phosphorylation and dephosphorylation of proteins play a fundamental role in the regulation of protein activity. The enzymes responsible for these transformations, protein kinases and phosphatases, act as mutual opponents in the up- and down-regulation of individual protein functional activity. The addition or removal of a phosphoryl group, usually attached to a Ser, Thr, or Tyr residue sidechain, initiates a conformational change triggering the activation or deactivation of the substrate protein. Protein kinases and phosphatases are involved in many crucial cellular events such as signal transduction processes, the modulation of which are of major importance for a variety of pathological conditions. An important therapeutic area in which protein kinases in particular have attracted much attention is oncology. Apart from these, both kinases and phosphatases are validated targets for the treatment of, for example, diabetes, cardiovascular and inflammatory diseases, and autoimmune disorders.

4.5.2 Relationships of Fold, Function, and Sequence Similarities

In spite of the closely related functional roles of protein kinases and phosphatases in many biochemical pathways, a comparison of both enzyme classes exemplifies the variability of relationships among function, sequence, and fold.

In principle, nature has decoupled protein function and protein fold. The most commonly known example for a fold conveying a broad variety of functions is the TIM barrel. First found in triosephosphate isomerase, the TIM barrel also occurs in proteins as diverse as aldose reductase, enolase, and adenosine deaminase (see, e.g., the review by Nagano et al. [104]). To date, the TIM barrel fold, as a generic scaffold, is associated with 15 different types of enzymatic functions.

On the other hand, a particular protein function can be realized with different protein folds, and an example of this are protein phosphatases. Protein phosphatases feature two distinctively different catalytic mechanisms for hydrolytically cleaving phosphorylated amino acid residues. The active sites of serine/threonine protein phosphatases (PPs) contain two metal centers that directly activate a water molecule for nucleophilic attack of the phosphate ester bond. In contrast, protein tyrosine phosphatases (PTPs) [105] possess a Cys residue present in the active site loop containing the conserved PTP signature motif HCXXXXXRS. The Cys sidechain acts as the attacking nucleophile in the formation of a phosphocysteine intermediate, which is eventually hydrolyzed by a water molecule [106]. The same catalytic mechanism is also shared by dual-specificity phosphatases (see below).

For both classes of protein phosphatases representing the two different dephosphorylation mechanisms, different folds of the catalytic domain are known. The PP class can be subdivided into the PPM family (e.g., PP2C) and the PPP family (e.g., PP1) which differ in fold. Different architectures found for the PTP domains include classical pTyr-specific PTPs, low molecular weight PTPs, dual-specificity phosphatases, and CDC25 phosphatases. Apart from the active site loop PTP signature motif, these subfamilies share little or no sequence similarity. However, a significant 3D structural similarity between their binding sites can be established (Figure 4.1).

An entirely different picture emerges for protein kinases. In spite of the evolutionary differentiation of serine/threonine and tyrosine kinases, which is apparent on the sequence level, the catalytic mechanism is conserved and always involves transfer of the γ-phosphate group of the substrate cofactor ATP. (Only the individual mechanisms of the preceding kinase activation are very different.) Since the substrate binding pocket of protein kinases appears to be difficult to target by small molecules, usually the cofactor binding pocket is the focus of interest in current kinase inhibitor development. A conserved Lys residue, present in the N-terminal subdomain, along with amino acid residues of the glycine-rich loop (GXGXXGXV) interact with the phosphate groups of ATP. The primary Mg^{2+} ion is coordinated by a conserved Asp residue present in the DFG motif. The Asp and Asn residues present in the conserved DXXXN motif play a role in catalysis and in coordinating a secondary Mg^{2+} ion, respectively. Whether the phosphoryl transfer

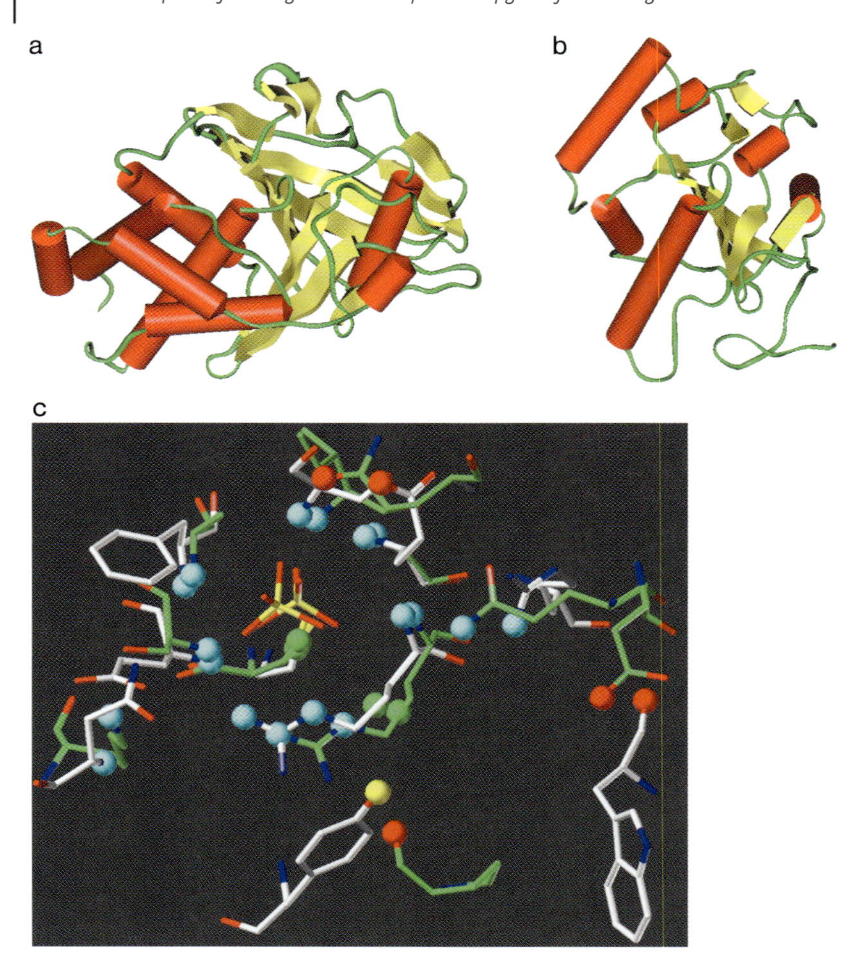

Figure 4.1 Protein tyrosine phosphatase 1B (2hnq (a)) and CDC25B (1cwt (b)) employ the same catalytic mechanism for hydrolysis of phosphorylated substrates but share no sequence homology and exhibit very different folds. The similarity of their binding sites can, however, be detected on the level of the interaction properties exposed to a ligand. A CavBase calculation found 14 pairs of matching pseudo-centers, resulting in the superposition of the two binding sites shown in (c). The positions of the sulfate ions in the two structures match remarkably well. PTP-1B is shown with carbon atoms in green and CDC25 with white carbon atoms. The pseudocenters are shown as spheres, and interaction types are indicated by colors (blue = donor, red = acceptor, yellow = donor/acceptor, green = aliphatic). (a) and (b) were prepared with Insight II [144], and (c) was prepared with SYBYL [145].

involves an associative (S_N2-like) or dissociative (S_N1-like) transition state [107, 108] has not yet been clarified.

All catalytic domains of protein kinases exhibit the same characteristic fold, with the ATP-binding niche being located in the cleft between the N-terminal and C-terminal subdomains (for kinases operating on non-protein substrates, however, other folds are also found [109]). The uniqueness of this picture is even more surprising, given that the human genome is estimated to include approximately 520 protein kinases [110], compared to a mere 150 protein phosphatases [111].

4.5.3 Druggability

The abundance of kinases, along with the conserved nature of many of their ATP-binding-site residues, has cast the suitability of protein kinases as drug targets into doubt. There has been a long debate as to whether protein kinases can be considered promising drug targets at all and whether selectivity between closely related kinase structures can possibly be achieved. Since the launch of Gleevec® [112], this discussion has tapered off. Interestingly, the X-ray crystal structure of Abl kinase in complex with Gleevec® revealed an enormous movement of the activation loop. Thus, the conformational flexibility of protein kinases [113, 114] might extend their range of structural differentiation and thereby improve the chances of finding selective inhibitors. At the same time, it must be stressed that most of the known kinase inhibitors [115–117] have been developed for cancer treatment, where selectivity against all other related targets is often not critical or even desirable. Yet, for certain conditions protein kinases appear to be valid, druggable targets.

The druggability of protein phosphatase binding pockets has some problematic aspects as well, though they are different from those encountered with protein kinases. To accomplish phosphorylation, protein kinases have to bind ATP as a cofactor. Thus, the need to recognize the ATP molecule constitutes the fundamental similarity of the binding niches for all protein kinases. Conversely, phosphatases must be able to bind phosphate groups. By comparison with ATP, phosphate is a small structural fragment, which can be accommodated in protein cavities in many different ways, leaving room for varying shape and constitution of phosphate-binding pockets [118]. This is in accordance with the diversity of folds among the phosphatase family [119]. Most notably, anchoring a phosphate group can already be achieved with a very small pocket. In fact, some protein phosphatases feature just a small protein surface depression for phosphate binding and achieve specific intermolecular recognition with the partner protein via flat, extended interaction interfaces, as is typical of protein–protein interactions [120]. Dual-specificity phosphatases (DSPs) employ such flat-shaped binding sites as a direct consequence of the need to bind both Ser/Thr and Tyr residues [121]. If DSP binding pockets were deep enough to accommodate entire phospho-tyrosine moieties, as given for PTPs, neither Ser nor Thr residues could reach the bottom of these cavities. In other words, protein architecture here complies with the principle 'form follows function'.

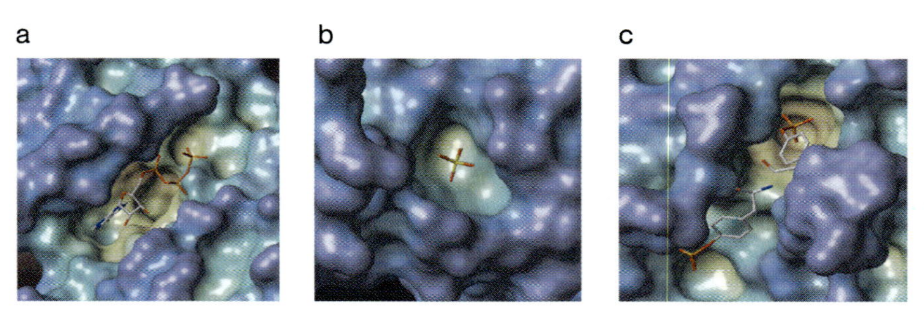

Figure 4.2 The global shapes of binding pockets are an important aspect of assessing the druggability of a target family. A distinctly different picture emerges for protein phosphatases and protein kinases. While the ATP binding pocket of all protein kinases emerges as a deep narrow cleft between the N-terminal and the C-terminal subdomains (CDK2 structure 1fin (a)), the binding pockets of phosphatases are, in general, more open and shallow. However, there are significant differences among various phosphatases. A binding site should be capable of burying large portions of a small molecule ligand, enabling tight, specific binding. For shallow binding sites as in KAP (1fpz (b)) this can be difficult (the Ligsite [94] algorithm does not even detect a cavity here). The presence of subpockets adjacent to the catalytic pocket offers alternative interaction areas for ligands, enhancing the chance of finding a suitable drug candidate. For example, in PTP-1B, a validated drug target in the treatment of diabetes, a second aryl binding site was detected by Puius et al. [146] (1pty (c)); in this structure two phosphor-tyrosine molecules are bound to the active site. The figure was prepared with SYBYL [145].

Unfortunately, the flat shape of such binding pockets can impose severe restrictions in terms of druggability. It appears to be difficult to construct drug-like molecules that bind tightly and specifically to such shallow surface areas (Figure 4.2). Unless the accommodation of a ligand induces local conformational changes in the protein in such a way that a real pocket is formed upon binding, lead finding projects for this type of target are more likely to get stuck with a compound series exhibiting moderate binding but lacking a concise SAR. Sadly, the prediction of such induced-fit effects remains an unsolved problem.

4.5.4
Relationship between Ligand Similarity and Binding Site Similarity

The classification of binding pockets based on their recognition patterns brings us closer to shedding some light on links between genome (or sequence) space and chemical space. Establishing such links is by no means simple, and only few studies have attempted to do so directly. From the numerous studies undertaken by independent research groups, which resulted in completely different drug candidate molecules for the same target protein, it is obvious that one pocket can bind small molecules that differ strikingly in structure. Thus, a given starting point in sequence space cannot be linked to a particular area in chemistry space in a one-to-one relationship unless the underlying structural description captures the features that determine molecular recognition.

A noteworthy chemometrical method (CHEMDOCK) for establishing complementary relationships between receptors and ligands has been devised by Oloff et al. [122]. Their approach employs atomic descriptors derived by quantum chemical methods (TAE/RECON descriptors [123]), which represent both the ligand structures and the protein binding pockets in global a descriptor space. The software was trained by using a dataset comprising 99 PDB protein–ligand structures (SMoG dataset [124]). The correct ligand for any of the receptors included in the test set could be identified within the ten best hits, with the average rank order of the native ligand being the third on the hit list. The approach is widely applicable, since links can be established in either direction. Knowledge of a receptor's active site structure facilitates straightforward identification of complementary ligands from large databases, and starting from a given ligand structure may equally well identify possible complementary receptor cavities.

In the context of chemogenomics, a related question is also worth addressing: can two proteins binding to the same or very similar ligands be expected to share similarity in their binding sites, and if so, to what extent? Although we are not aware of any systematic studies related to this question, several studies on similarity aspects of adenine binding sites can be referred to, to provide a guideline. Adenine is part of the ATP cofactor of kinases, but is also a common substructure of other enzyme cofactors such as AMP, ADP, NAD, and FAD, making it one of the most widespread chemical groups present in a large number of different protein structures. Since many adenine binding proteins represent targets of major pharmaceutical interest, unraveling the determinants of adenine binding has attracted a number of researchers in recent years. Various studies have been reported which aimed to identify structurally invariant patterns in adenine binding pockets [65, 125–129]. A study by Moodie et al. [65] revealed that complementary shape and electrostatic properties between the adenylate group and the protein can be achieved via a number of alternative amino acid residue arrangements, without these residues being conserved. These findings are supported in related studies by Denessiouk et al. [126–128]. A recent study by Cappello et al. [129] suggests that the structural diversity of adenine binding pockets appears to be even larger than previously described. Not only can different amino acids form the same kind of interaction, recognition of the ligand can even be accomplished by different interaction patterns. Generally, the adenine moiety is sandwiched between mostly nonpolar areas above and below the ring plane. However, there are a broad variety of in-plane interactions, and a number of different H-bonding patterns were identified around the rim of the purine ring system. Notably, the number of H-bonds formed is generally smaller than the number of theoretically feasible H-bonds, and, in agreement with the findings of Moodie et al. [65], water molecules appear to be important H-bonding partners for adenine. The H-bonding patterns were used to establish a simple pattern recognition classification scheme, based on encoding the actual involvement of polar adenine atoms in H-bonding as bit strings. The study showed that these recognition motifs appear to be conserved only among very closely related proteins. Even within protein families, significant differences in the adenine binding site composition were observed; whereas, in contrast, very similar binding sites belonging to different folds could also be found.

These studies indicate that protein cavities binding the same ligand do not necessarily exhibit a high degree of structural similarity. Thus, if one is looking only for related targets that could bind the same molecule, considering genome space, then taking a given ligand structure as a starting point is a reasonable approach. Trying to identify all target proteins binding to a given ligand in this way is certainly inadequate. However, searching the neighborhood of a given target can facilitate the identification of targets for which selectivity problems might be encountered, thus assisting project planning, such as early setup of assays, etc. Nevertheless, unexpected selectivity problems can always emerge with other proteins that are unrelated on the sequence and even functional level.

4.5.5 **Selectivity Issues**

Achieving an appropriate selectivity profile is one of the major challenges in drug design. Selectivity problems arise if, unintentionally, an active compound interacts with proteins other than the target protein, modulating their functional activity. On a microscopic scale, this implies that the active compound binds to protein cavities present in one or more antitargets.

As discussed above, two binding sites (in different proteins) exhibiting some degree of structural similarity may or may not bind the same ligand. The selectivity of a ligand towards such binding sites depends on the structural elements of the protein pockets involved in ligand binding. If only conserved structural features present in both pockets are used to facilitate ligand binding, no selectivity can be expected. Utilizing recognition features unique to one of the structures will, in contrast, enable selectivity. Although drug design efforts usually aim to develop selective inhibitors, the identification of a scaffold representing a nonspecific ligand can also be extremely valuable, provided that the scaffold offers high optimization potential. Selectivity toward members of the same protein family can be introduced by attaching appropriate substituents to the core structure. In targeting protein tyrosine phosphatases, the Novo Nordisk group succeeded in finding a general, competitive, efficient, and lead-like inhibitor, 2-oxalylamino-benzoic acid (OBA) [130]. The X-ray structure of OBA in complex with PTP-1B (PDB code 1c85) was determined. The binding mode of the OBA ligand largely resembles that of tyrosine phosphate, as found in the natural substrate, and includes H-bond formation with the PTP signature motif (Figure 4.3). Analogous to the phosphate group, the carboxylate anchor forms four H bonds with the protein (two H-bonds with the guanidinium group of Arg221 and one H-bond with each backbone nitrogen atom of Ser216 and Ala217). Other H-bonds bridge the carboxy group of the oxalylamino moiety with the backbone nitrogen atom of Gly220 and the *o*-carboxylate group with Asp181. Asp181 acts as a general acid and is therefore protonated. Also, a weak salt bridge (contact distance: 3.41 Å) between the *o*-carboxylate group and Lys120 is formed. Although the phenyl ring in OBA is shifted up relative to the phenyl

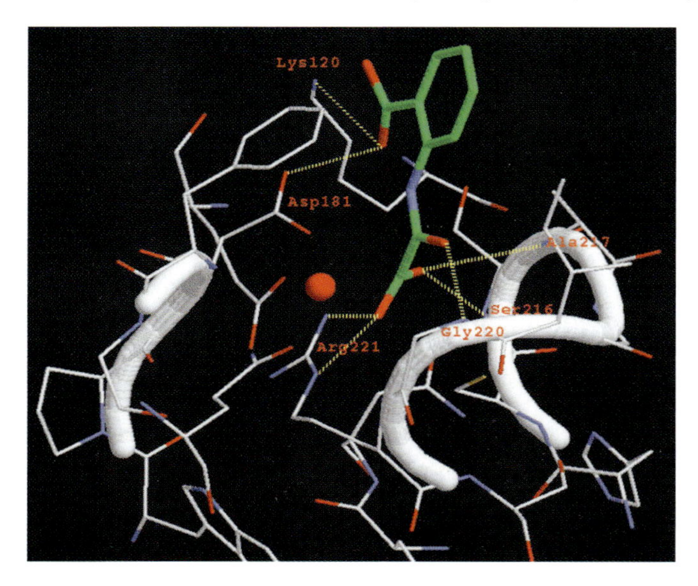

Figure 4.3 Binding mode of 2-oxalylaminobenzoic acid (OBA) in the catalytic pocket of PTP-1B (1c85). The inhibitor was developed as a phenyl phosphate mimetic and provided the starting point for the development of inhibitors selective toward individual PTPs. OBA binding largely resembles the interactions between PTP and its natural substrate. The figure was prepared with RasMol [147].

ring in tyrosine phosphate, it occupies largely the same hydrophobic pocket. OBA appears to be the most potent 'minimal unit' phenylphosphate mimetic obtained so far.

The OBA scaffold has been further developed into potent and selective PTP-1B inhibitors. A sequence alignment of the catalytic domains from 106 known vertebrate PTPs was carried out. This information, in combination with the crystal structure of PTP-1B, was used to identify residues unique to PTP-1B reasonably close to its active site. In this way, a set of four residues was found, namely, Arg47, Asp48, Met258, and Gly259. Initial attempts at optimizing the ligand focused on interactions with Asp48, since many PTPs have an uncharged Asn in this position. Introducing a charged nitrogen atom into the core structure enabled formation of a new salt bridge with Asp48 in PTB-1B, while, presumably, the presence of repulsive forces between the positive charge and the Asn sidechains present in other PTPs resulted in the desired gain in selectivity [131].

Further optimization steps focused on enhancing the selectivity for PTPs possessing the conserved Asp48 residue, but differing at position 259. In PTP-1B, position 259 is a Gly residue, whereas, for example, PTPα has a Gln residue here. The bulkier sidechain of the Gln residue changes the substrate recognition properties of the phosphatase. Hence, the guiding principle for enhancing the selectivity at this optimization step was that of exploiting steric hindrance, rather than differences in the electrostatic properties of a given pocket area [132]. With PTPα as a model for the anti-targets, an analog of the compound lacking the basic nitrogen atom responsible for PTP-1B selectivity was used as a starting point for chemical modification. Thus, any improvement in selectivity could unambiguously be attributed to residue 259. By extending the inhibitor towards the so-called 258/259 gateway region, activity against PTPα was completely lost, supporting the rationale of the design. Moreover, the affinity of the inhibitor to PTP-1B was improved by a factor of 100. The crystal structure of PTP-1B in complex with the newly designed inhibitor revealed a feasible reason for the improved binding: Asp48 was found in a different rotameric state compared to the structures belonging to the first compound series, where it formed a salt bridge to the basic nitrogen atom. A water molecule, not observed in the other structures, facilitates ligand binding via a water-mediated protein–ligand contact bridging to Asp48. This additional contact also provides an explanation for the seemingly contradictory finding of reasonably high selectivity against the tyrosine phosphatase SHP-1, which has a Gly residue in position 259.

In summary, the manipulation of both the attractive and repulsive forces for residue 48 and exploiting steric differences for residue 259 independently lead to an increase in selectivity. The authors of this work [132] conclude with the convincing hypothesis that combination of these efforts could yield inhibitors with significantly improved selectivity against the majority of related PTPs.

The situation described above, where a crystal structure is available only for the target protein but not for the relevant anti-targets, is very common. Nevertheless, if crystal structures were available for all (or a sufficient number of) the anti-targets, advanced possibilities could be opened up for analyzing the structural determinants of selectivity. An elegant approach for systematically exploiting 3D protein structural data with the aim of identifying selectivity-related differences in binding sites was presented by Kastenholz et al. [133]. Its application to kinases was recently reported by Naumann and Matter [134]. In their study, the molecular recognition properties of the ATP pockets of 26 different kinase structures were investigated. All structures contained in the dataset were superimposed, and GRID [66] fields were calculated using the N1, O, and DRY probes (representing H-bond donor, H-bond acceptor, and hydrophobic interaction properties) for each protein. The calculated interaction energies of all grid points and all probes were concatenated in a vector. These 26 vectors represent the rows in an X-matrix used for a subsequent chemometric analysis by PCA and CPCA (consensus principal component analysis) methods [135]. The PCA and CPCA score plots, termed *target family landscapes* by the authors, allow a classification of the binding sites according to the similarities of their binding patterns. The first principal component (PC1) separates CDK and MAP/receptor kinases from the family of PKA kinases. The CDK structures fall into two separate clusters, representing the cyclin-bound (activated) and the inactivated state of a kinase. The two states differ significantly in conformation. The second PC separates MAP and other receptor kinases from the CDK family.

For an interpretation of the GRID/PCA model in structural terms, contour plots for individual probes were derived from the loadings of the first and second PC, respectively. These plots highlight regions of observed differences with respect to interaction with a particular probe. Thus they indicate positions where a particular ligand functional group should optimally be placed, to initially achieve or enhance selectivity. In fact, the results were in good agreement with the experimental selectivity profile of a series of 2,6,9-substituted purine compounds used as a scaffold for CDK inhibitors. The fact that hydrophobic groups attached to the purine N9 atom improve the selectivity towards CDK (and against PKA) could be nicely explained by the PC1 contour plot for the DRY probe. The compounds roscovitine and purvalanol A and B, all selective toward CDK, contain an isopropyl group in this position, which appears to be more favorable than a smaller methyl group (as present in olomoucine). In CDK2, the subpocket accommodating these alkyl groups is lined with Val64, Phe80, and Ala144, while the PKA subpocket exhibits more polar features, with a Thr residue replacing the Ala residue and a Met residue replacing the Phe residue. The PC2 contour plot demonstrates that the same substitution also improves selectivity against MAP kinases, in which the corresponding residues Ile82, Gln103, and Ala144 form a more polar environment. Furthermore, roscovitine, purvalanol A, and purvalanol B carry an ethyl or isopropyl group at the C2′ position of the hydroxyethyl group. These moieties bind to another hydrophobic subpocket (lined with Ile10 and Val18). The impact of this subpocket on specificity can also be identified in the PC1 contour plot. The favorable hydroxyethyl moiety attached to N1′ in roscovitine, olomoucine and both purvalanol compounds points to a contour patch in the N1 probe-derived PC2 plot.

In addition to providing an explanation for the experimentally determined selectivity profiles of known CDK inhibitors, the study also pointed out further opportunities for forming selective interactions that have not been exploited by any of the compounds in the series. With an increasing number of kinase structures being solved, the range of applications for this approach [133, 134] will increase, providing a valuable tool for designing selective inhibitors.

4.5.6 Caveats

Switching from the sequence level to the 3D structure level allows one to focus directly on the molecular determinants of ligand binding, i.e., the physicochemical properties featured in the ligand binding site. There are, however, several caveats to remember. So far, it has been assumed that the level of sequence similarity in a pair of proteins correlates with the structural similarity of their binding sites (in fact, the clustering of the different kinase families in the *target family landscapes* supports this concept). However, this is a simplified picture that cannot be universally applied. Neither the physicochemical properties that binding site residues project onto the cavity surface nor the 3D arrangement of such interaction centers are unambiguous and unalterable.

4.5.7
Protein Flexibility

One of the most common problems that binding site comparisons involve is that of protein flexibility [136]. The dynamic nature of proteins can appear on different levels of the structural hierarchy, ranging from the thermal motion of individual side chains, over shifts of backbone segments, to movements of entire domains [20]. As previously mentioned, protein conformational changes are often triggered by the binding of a ligand. By means of standard X-ray crystallography, only single snapshots of 3D structures can be generated, and assertions regarding flexibility can only be made based on B-factor distributions and interpretation of the electron density. Thus, structural changes in protein structures triggered by, for example, variations in the crystallization conditions, or, more importantly, binding of different ligands, cannot be captured by a single crystal structure. However, a series of crystal structures representing different relevant structural states can facilitate an understanding of the conformational flexibility of proteins and their binding sites [137]. Since all existing methods for comparing binding sites refer to rigid coordinates, they are bound to fail if two distinctly different conformations are compared. Thus, existing similarities that can be easily detected on the sequence level might not be revealed.

The problem of protein flexibility can be exemplified with both the families of protein kinases and protein phosphatases.

Apart from major loop rearrangements, as found with Gleevec, the ATP pocket of kinases can also undergo more subtle structural adaptations upon binding of a ligand. Staurosporine is an unspecific kinase inhibitor that, due to its large aromatic ring system, distends the respective binding cavity. Taking the structure of CDK2 in complex with staurosporine (1aq1) as a reference and searching for similar pockets with CavBase, one finds that the ranking of the derived pockets does not reflect the sequence relationship of the different kinases. Interestingly, the pocket of the Src kinase 1byg, which binds staurosporine as well but shares a sequence identity of only 26% with CDK2, appears to be more similar to 1aq1 (35 matching pseudo-centers, similarity score 5051) than the pocket of the sequence-identical CDK2 structure 1fvv, which represents the cyclin-bound state of the kinase with an oxindole inhibitor (28 matching pseudo-centers with 1aq1, similarity score 4691). This example illustrates two effects: on the one hand, binding of the same ligand can increase the structural similarity between distantly related pockets; on the other hand, binding of different ligands can decrease the structural similarity between closely related pockets.

Marked induced-fit effects are also found for protein phosphatases. The PTP-1B active site is surrounded by several surface loops, which are important for catalysis and substrate recognition. Binding of pTyr, tyrosine-phosphorylated substrates, or inhibitors like OBA (see above) induces an 8 Å movement of the so-called WPD loop, which brings Asp181 (the general acid) into the catalytic site. This movement closes the active site pocket and in turn traps the substrate. A comparison of the ligand-free structure 2hnp to the ligand-bound structure 1bzj shows how the

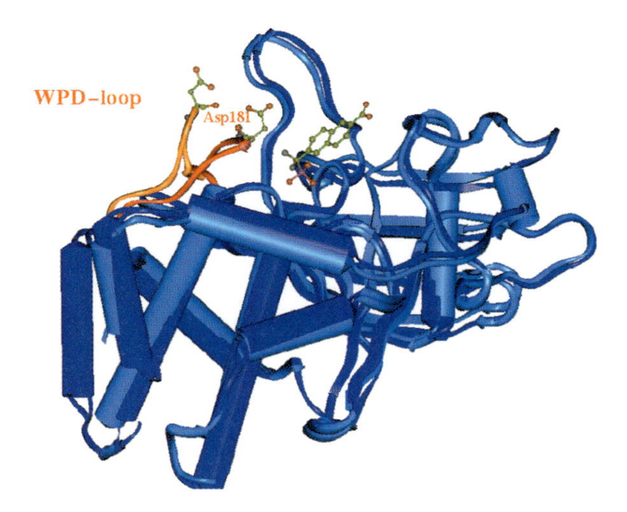

Figure 4.4 Comparison of a ligand-free (2hnp) with a complexed (1bzj) structure of PTP-1B reveals a pronounced induced-fit effect. Upon inhibitor binding, the WPD loop closes over the catalytic pocket. The same structural rearrangement is also produced by substrate binding and involves shifting the general acid Asp181 required for catalysis in the active site. The figure was prepared with Insight II [144].

transition of the WPD loop from the open to the closed conformation affects the overall binding site shape (Figure 4.4). Effectively, one of the cavity walls is built up by the loop closure.

Statistics including large and diverse sets of protein structures have revealed that the 20 amino acid residue types exhibit distinctly different levels of flexibility [138]. Gln residues appear to be amazingly flexible (given their medium size) and are used in the following section for exemplifying some of the caveats that complicate binding site comparisons.

4.5.8 Ambiguities in Atom Type Assignment

In X-ray crystallography, the decision as to which of the terminal atoms of Gln (and Asn) sidechains is oxygen or nitrogen is prone to error. The X-ray scattering power of oxygen and nitrogen is very similar. As a result, the electron density of Gln and Asn sidechains usually appears symmetric, thus hampering unambiguous assignment of the correct atom type. (A similar situation is encountered with the C_γ and O_γ atoms in Thr side chains, as well as the $N_{\varepsilon2}$ and $N_{\delta1}$ atoms in His residues). Unless the resolution is very high, which would allow for identification of the hydrogen atoms bound to the amide nitrogen atom, only indirect methods making use of chemical knowledge can be applied to overcome the ambiguities. By taking the H-bond characteristics of the oxygen atom (acceptor) and nitrogen atom (donor) into consideration and analyzing the local environment of these atoms for potential H-bond partners, for 75% of the residues the atoms can be unambiguously assigned [139]. Nevertheless, an investigation of the stability of potential H-bond networks provided evidence that there are still a substantial number of misassignments in the PDB. In many cases, calculations revealed more stable networks when the positions of the oxygen and nitrogen atoms in Gln/Asn side chains were switched [140]. In addition, some structures are genuinely difficult, resulting in arbitrary assignments.

Clearly, donors and acceptors have to be considered as non-matching properties in binding site comparisons (this of course does not hold for bifunctional groups such as hydroxyl groups, which can act as either hydrogen donors or acceptors). Wrong atom type assignments, leading to erroneous property descriptions, can hamper the success of binding site comparison methods by producing false or missing hits. The ostensibly promising workaround of unifying atom assignments among corresponding Gln residues in a set of pockets could, however, conceal real flexibility effects. An example of this situation is given by the elastase structures 1ela and 1elc (Figure 4.5). In 1ela, Gln200 acts as an H-bond donor via its $N_{\varepsilon 2}$ atom, bridging to a carbonyl group in the ligand. In 1elc, Gln200 is flipped with respect to 1ela, and its $O_{\varepsilon 2}$ atom interacts with an amide nitrogen atom in the ligand.

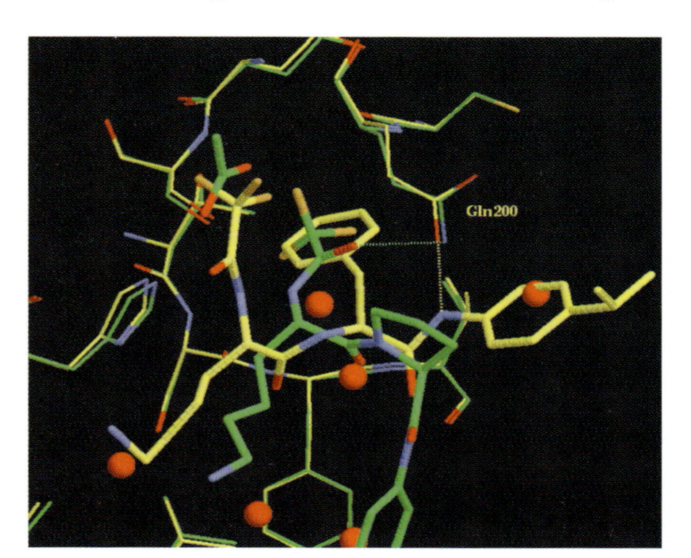

Figure 4.5 Assignment of atom types for the N and O atoms in Gln sidechains can be difficult and often relies on interpreting the H-bond partners in the local environment. For the two elastase structures 1ela (green carbon atoms) and 1elc (yellow carbon atoms), a donor and an acceptor group in the ligand form close contacts with the Gln amide group, respectively. This allows unambiguous assignment of the O and N atom types. The position and orientation of the amide plane are almost identical in the two structures; only the O and N positions are switched. The figure was prepared using RasMol [147].

4.5.9
Versatility of Interaction Types

Apart from the caveats discussed so far, H-bonds are not the only type of intermolecular interaction in which Gln side chains can be involved. As for peptide backbone bonds, the π-face of the amide moiety can interact with ligands via π–π interactions. Thus, even if the side chain orientation can be unambiguously determined from the crystal structure, the type of interaction occurring between a ligand and a Gln side chain is not predetermined. Since π–π interactions usually occur with aromatic ring systems of a ligand, within a series of ligands the fragments that interact with a particular Gln residue can even be chemically very different.

In summary, the general flexibility of Gln side chains, the variability of their donor and acceptor properties enabled by amide flipping, and their ability to present different characteristics (polar group or π-face) to a ligand, all reveal Gln residues to be a kind of chemical chameleon. Gln residues contribute greatly to the ability of some protein binding pockets to adapt to a large variety of diverse ligand structures.

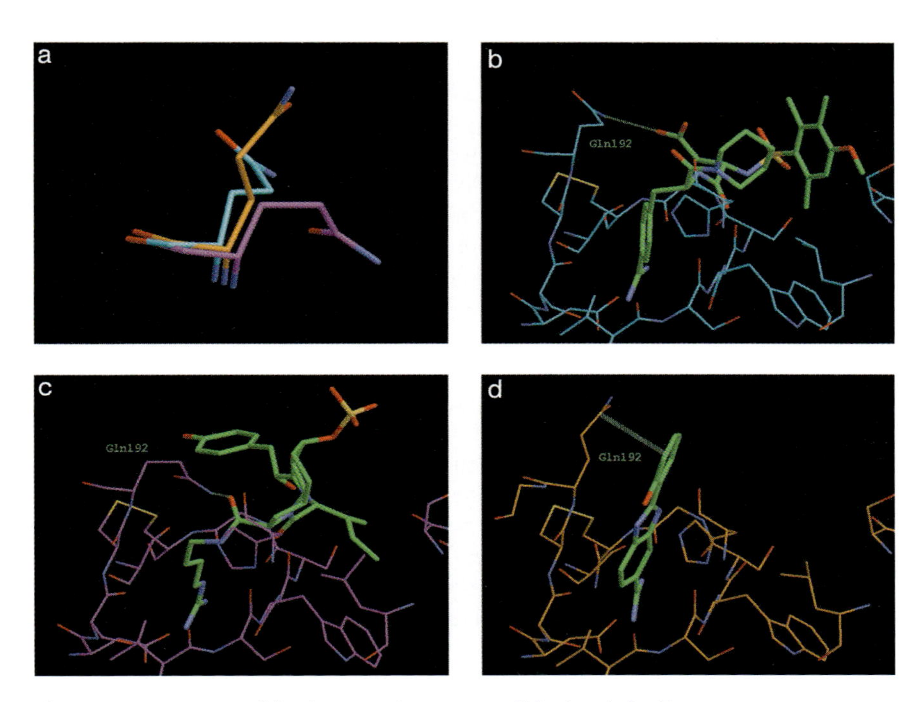

Figure 4.6 Comparison of the three trypsin structures 1k1n (cyan), 1aq7 (magenta), and 1gi0 (orange) illustrates the side chain flexibility of Gln192 and its ability to take part in different interaction types with chemically diverse ligand structures in the binding site. Different rotamers are found among the structures (a). While in 1k1n (b) and 1aq7 (c), H bonds are formed with ligand acceptor groups, the ligand in 1gi0 (d) interacts with the Gln192 side chain via π–π stacking. The figures were prepared with RasMol [147].

An example of this is found in the serine protease trypsin (Figure 4.6). The conserved Gln192 residue shows substantial variability in terms of its spatial orientation and interactions with various ligands. A comparison of the three trypsin structures 1aq7, 1k1n, and 1gi0 revealed pronounced side chain reorientations depending on the ligand bound in the active site. In 1aq7 and 1k1n, Gln192 forms H-bonds to the ligands. In 1gi0, the amide group covers one side of a phenol ring in a coplanar arrangement. This π–π stacking interaction reduces the solvent accessibility of the hydrophobic face of the ligand. The variations in the interaction patterns are accompanied by significant rearrangements of the Gln192 side chain.

Changes in the interaction type of Gln side chains can also occur without major spatial reorientations, as shown by a comparison of the blood coagulation factor Xa structures 1fax and 1xkb. In 1fax, the $N_{\varepsilon 2}$ atom of Gln192 forms an H-bond with the carboxylate group of the ligand. Conversely, in 1xkb the face of the amide plane forms a π-stacking interaction with one aromatic ring of the biphenyl moiety.

4.5.10
Crystallographic Packing Effects

It is commonly accepted that protein X-ray crystal structures represent the 'real' structure and thus reflect the molecular situation of the relevant biological system. However, in a crystal, protein surface areas are involved in interactions with neighboring molecules, forming the contacts that hold the crystal together. In certain circumstances, such interactions can also severely affect the binding sites for small ligands, leading to binding modes that are unlikely to represent the physiological situation. Such problems arising from crystal contacts not only demand careful examination when applying structural knowledge to rational drug design projects [141], but they also complicate binding site comparisons in chemogenomics approaches. This can be exemplified further by structures of factor Xa.

In 1fax, the ligand (3-letter code DX9) fills the entire unprimed active site, forming favorable interactions in the S1 (salt bridge between naphthylamidine group and Asp189) and S4 (H bond between terminal imido group and Glu97.O) specificity pockets (Figure 4.7). As in 1fax, the benzamidine moiety of the ligand in 1xkb (3-letter code 4PP) also forms a salt bridge with Asp189 in the S1 pocket. However, the ligand extends less far into the S4 pocket, only partially filling it. Notably, a lysine side chain (Lys79 in the EGF-like domain, B chain) found in the crystallographic packing environment invades this empty area, forming an H bond with Glu97.O. The pyridyl ring of 4PP, not centered in the S4 pocket, forms an H bond with Thr98.O. Another H bond between Gln56 (B-chain) in the crystal packing and the carboxylate group of 4PP suggests a pulling force shifting the ligand out of the S4 pocket [142].

Although the problems discussed above can impose some restrictions on the use of binding site comparisons in chemogenomics programs, the broad array of methods available still offers significant support for the majority of cases. Careful assessment of the target structure and the relevant structural data can certainly guide the choice of appropriate methods. For example, the application of methods

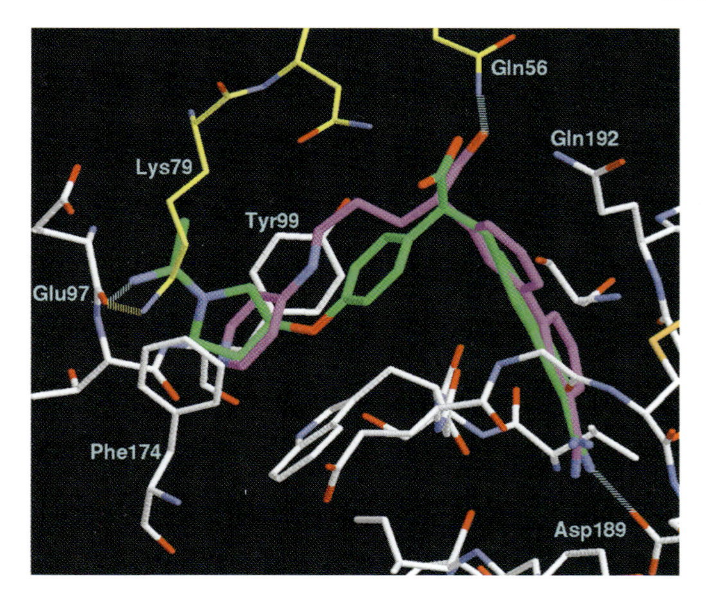

Figure 4.7 Crystallographic packing effects in factor Xa binding sites, shown for PDB structures 1fax and 1xkb. 1fax is shown with white carbon atoms, the superimposed 1xkb structure (rms deviation of binding site C_α atoms: 0.38 Å) is omitted. The ligand of 1fax is shown in green and the ligand of 1xkb in magenta. Residues present in the crystal packing environment of 1xkb are shown in yellow. The figure was prepared with RasMol [147].

allowing for the detection of partial matches (common subpockets) appears to be advantageous for binding sites that show a significant degree of flexibility. Once a structural relationship between two binding sites can be established, an analysis of differences is just as interesting as the analysis of similarities, especially if selectivity issues are to be addressed.

Clearly, the quantitative methods for describing binding site similarities, required for large-scale database searches in the spirit of 'omics' efforts, still need to be improved. The development of powerful scoring functions is an area of ongoing research, and hopefully some of the shortcomings will be resolved in the near future. Nevertheless, the examples discussed in this section demonstrate, within the realms of possibility, the usefulness of binding site comparison methods.

4.6 Summary and Outlook

On a molecular level, the binding sites of proteins represent the locations where the functions of a protein take place. In this article, we have reviewed the characteristics and constitutional principles of binding sites from the 3D perspective of structural biology and have demonstrated the application of such knowledge for detecting functional binding sites. Furthermore, methods for establishing structural relationships and for investigating binding site similarities have been described. In the context of chemogenomics, such methods can be utilized for revealing local structural similarities among proteins and thus for finding cavities or subcavities likely to bind a ligand or functional group already known from previous projects. Since fold is even more conserved than sequence, protein targets that are related on the genome level can be expected to share a certain degree of structural similarity in their binding sites. This is all the more true because binding site residues usually exceed other protein residues in their degree of conservation during their evolution. A practical limitation is given for binding sites exhibiting pronounced flexibility. In such situations, representation of the accessible conformational space can be incomplete unless many crystal structures of the target protein complexed with different ligands are available. Thus, for two crystallographically characterized, closely related proteins the detection of similarity can occasionally be difficult on the structural level, even though recognizing the similarity based on their sequences would be trivial.

On the other hand, the analysis of binding sites in terms of shape and physicochemical properties rather than on the level of sequence relationships provides an outstanding advantage: as discussed above, two binding sites can exhibit a high degree of structural similarity without the respective proteins sharing any sequence homology. Implementation of this idea in the drug discovery pipeline opens up entirely new possibilities, extending beyond what is genuinely encompassed by the concept of chemogenomics. Approaches such as the transfer of an established active compound class from one target to another appear promising as long as the binding sites are similar. A relationship of these targets in genome space is not a necessary precondition. Although most often, two proteins with similar binding-site features are also sequence-related, comparisons of binding sites throughout the entire pool of PDB structures can sometimes reveal unexpected binding-site similarities. Such findings can provide innovative ideas for drug design and even allow for jumps between genomic target classes. Likewise, even the similarity of two subpockets belonging to binding sites that differ in other parts can stimulate structure-based drug design. A molecular fragment bound to a reference structure could be equally well embedded in the matching subpocket of a second protein. Attaching such a chemical moiety to an existing lead structure for the latter target could therefore be a promising approach for optimizing the ligand structure. De-novo design programs could increasingly resort to such knowledge.

Considering chemogenomics in the context of other 'omics' approaches, it appears that some possible synergies have not been exploited so far. Driven by advances in the automation of protein crystallization and in X-ray data collection and analysis, the current efforts in structural genomics can be expected to lead to an enormous increase in the number of known protein structures. Naturally, this will provide a wealth of previously unknown information on binding sites and their 3D arrangements. Nevertheless, one has to remember that the major aim of structural genomics is not to produce data ideally suited to binding site related analyses and studies as described above. Structural genomics is focused on acquiring protein structures in a high-throughput fashion. Its goal is to make at least one 3D protein structure per fold representative available, so as to completely cover the 'fold space'. Currently, only 700 out of 1000–5000 estimated protein folds [3] are available as experimental structures.

A more practical perspective considers the goal of structural genomics as the experimental determination of the minimum number of structures that are required for building all other structures by homology modeling techniques. This usually requires a sequence identity of at least 30% between the template and the modeled structure. In addition, it is obvious that structural genomics does not aim to solve the structures of a large number of representatives of the same protein family [2]. However, comprehensive knowledge of all structures belonging to a particular protein family could provide an optimal basis for chemogenomical approaches utilizing binding site information. Moreover, in structural genomics programs, the proteins to be investigated are often chosen according to the availability of suitable crystals and not necessarily by their pharmaceutical relevance. Jhoti [143] compared this situation with the development of combinatorial chemistry: an impressive number of synthetically easily accessible molecules were generated as the first generation of compounds, but these were then found to be of very limited value. Finally, it should be mentioned that the emphasis on revealing the overall architecture of proteins is accompanied by minor interest in structural detail. Thus, optimization of crystallization conditions for obtaining high-resolution crystal structures is rarely pursued in structural genomics programs. However, information with this level of detail is highly desirable for addressing, for example, selectivity issues from a chemogenomics point of view.

Hopefully, the focus of structural genomics will shift in the future, so that more attention can be paid to the potential applications of structural data in drug design. Undoubtedly, the technical advances that promoted the development of structural genomics can also be applied for the in-depth study of particular target families. Often, a successful crystallization protocol for a particular member of a protein family can facilitate the crystallization of another protein in the same family. The information thus derived will certainly stimulate chemogenomics efforts and encourage researchers to look more often at proteins through the eyes of a structural biologist.

Acknowledgements

We gratefully acknowledge the invaluable help of Karen Lipscomb with the English and with proofreading. We also thank Dr. Ursula Egner for very stimulating discussions and hints, as well as Drs. Robin Taylor, Hans Briem, and Roman Hillig for useful comments on the manuscript.

References

1 A. L. Hopkins, C. R. Groom, *Nat. Rev. Drug Discov.* **2002**, 727–730.
2 S. Goldsmith-Fischman, B. Honig, *Protein Sci.* **2003**, 1813–1821.
3 S. K. Burley, J. B. Bonanno, *Methods Biochem. Anal.* **2003**, 591–612.
4 E. Fischer, *Ber. Dtsch. Chem. Ges.* **1894**, 2985–2993.
5 J. Westbrook, Z. Feng, L. Chen, H. Yang, H. M. Berman, *Nucleic Acids Res.* **2003**, 489–91.
6 M. R. Chance, A. R. Bresnick, S. K. Burley, J. S. Jiang, C. D. Lima, A. Sali, S. C. Almo, J. B. Bonanno, J. A. Buglino, S. Boulton, H. Chen, N. Eswar, G. He, R. Huang, V. Ilyin, L. McMahan, U. Pieper, S. Ray, M. Vidal, L. K. Wang, *Protein Sci.* **2002**, 723–738.
7 A. Schafferhans, J. E. Meyer, S. I. O'Donoghue, *Nucleic Acids Res.* **2003**, 494–498.
8 http://srs3d.ebi.ac.uk/.
9 A. L. Hopkins, C. R. Groom, *Ernst Schering Res. Found. Workshop* **2003**, 11–17.
10 J. Bajorath, *Nat. Rev. Drug Discov.* **2002**, 882–894.
11 P. D. Lyne, *Drug Discov. Today* **2002**, 1047–1055.
12 C. A. Lipinski, F. Lombardo, B. W. Dominy, P. J. Feeney, *Adv. Drug Deliv. Rev.* **1997**, 3–25.
13 P. M. Dean, E. D. Zanders, D. S. Bailey, *Trends Biotechnol.* **2001**, 288–292.
14 V. L. Nienaber, P. L. Richardson, V. Klighofer, J. J. Bouska, V. L. Giranda, J. Greer, *Nat. Biotechnol.* **2000**, 1105–1108.
15 J. Fejzo, C. Lepre, X. Xie, *Curr. Top. Med. Chem.* **2003**, 81–97.
16 R. Carr, H. Jhoti, *Drug Discov. Today* **2002**, 522–527.
17 H. J. Boehm, M. Boehringer, D. Bur, H. Gmuender, W. Huber, W. Klaus, D. Kostrewa, H. Kuehne, T. Luebbers, N. Meunier-Keller, F. Mueller, *J. Med. Chem.* **2000**, 2664–2674.
18 T. Niimi, M. Orita, M. Okazawa-Igarashi, H. Sakashita, K. Kikuchi, E. Ball, A. Ichikawa, Y. Yamagiwa, S. Sakamoto, A. Tanaka, S. Tsukamoto, S. Fujita, K. Tatsuta, Y. Maeda, K. Chikauchi, *J. Med. Chem.* **2001**, 4737–4740.
19 M. Hendlich, A. Bergner, J. Günther, G. Klebe, *J. Mol. Biol.* **2003**, 607–620.
20 J. Günther, A. Bergner, M. Hendlich, G. Klebe, *J. Mol. Biol.* **2003**, 621–636.
21 http://relibase.ccdc.cam.ac.uk/.
22 A. C. Stuart, V. A. Ilyin, A. Sali, *Bioinformatics* **2002**, 200–201.
23 D. Ringe, *Curr. Opin. Struct. Biol.* **1995**, 825–829.
24 H. O. Villar, L. M. Kauvar, *FEBS Lett.* **1994**, 125–130.
25 L. Young, R. L. Jernigan, D. G. Covell, *Protein Sci.* **1994**, 717–729.
26 Z. Y. Zhu, S. Karlin, *Proc. Natl. Acad. Sci. USA* **1996**, 8350–8355.
27 S. Karlin, Z. Y. Zhu, *Proc. Natl. Acad. Sci. USA* **1996**, 8344–8349.
28 R. A. Laskowski, N. M. Luscombe, M. B. Swindells, J. M. Thornton, *Protein Sci.* **1996**, 2438–2452.
29 M. Lewis, D. C. Rees, *Science* **1985**, 1163–1165.
30 F. K. Pettit, J. U. Bowie, *J. Mol. Biol.* **1999**, 1377–1382.
31 J. M. Sanchez-Ruiz, G. I. Makhatadze, *Trends Biotechnol.* **2001**, 132–135.
32 B. Honig, A. Nicholls, *Science* **1995**, 1144–1149.
33 A. H. Elcock, *J. Mol. Biol.* **2001**, 885–896.
34 W. L. DeLano, *Curr. Opin. Struct. Biol.* **2002**, 14–20.
35 I. Luque, E. Freire, *Proteins* **2000**, 63–71.
36 M. J. Ondrechen, J. G. Clifton, D. Ringe, *Proc. Natl. Acad. Sci. USA* **2001**, 12473–12478.
37 S. Jones, J. M. Thornton, *Proc. Natl. Acad. Sci. USA* **1996**, 13–20.
38 K. P. Peters, J. Fauck, C. Frömmel, *J. Mol. Biol.* **1996**, 201–213.
39 J. Liang, H. Edelsbrunner, C. Woodward, *Protein Sci.* **1998**, 1884–1897.
40 http://pqs.ebi.ac.uk/.
41 K. Henrick, J. M. Thornton, *Trends Biochem. Sci.* **1998**, 358–361.
42 G. J. Bartlett, C. T. Porter, N. Borkakoti, J. M. Thornton, *J. Mol. Biol.* **2002**, 105–121.
43 A. K. Whiting, W. L. Peticolas, *Biochemistry* **1994**, 552–561.

44 B. Ma, T. Elkayam, H. Wolfson, R. Nussinov, *Proc. Natl. Acad. Sci. USA* **2003**, 5772–5777.

45 O. Lichtarge, H. R. Bourne, F. E. Cohen, *J. Mol. Biol.* **1996**, 342–358.

46 O. Lichtarge, M. E. Sowa, *Curr. Opin. Struct. Biol.* **2002**, 21–27.

47 A. Armon, D. Graur, N. Ben-Tal, *J. Mol. Biol.* **2001**, 447–463.

48 T. Pupko, R. E. Bell, I. Mayrose, F. Glaser, N. Ben-Tal, *Bioinformatics* **2002**, S71–77.

49 http://consurf.tau.ac.il/.

50 F. Glaser, T. Pupko, I. Paz, R. E. Bell, D. Bechor-Shental, E. Martz, N. Ben-Tal, *Bioinformatics* **2003**, 163–164.

51 P. Aloy, E. Querol, F. X. Aviles, M. J. Sternberg, *J. Mol. Biol.* **2001**, 395–408.

52 R. Landgraf, I. Xenarios, D. Eisenberg, *J. Mol. Biol.* **2001**, 1487–1502.

53 A. Gutteridge, G. J. Bartlett, J. M. Thornton, *J. Mol. Biol.* **2003**, 719–734.

54 M. Ota, K. Kinoshita, K. Nishikawa, *J. Mol. Biol.* **2003**, 1053–1064.

55 C. Sotriffer, G. Klebe, *Farmaco* **2002**, 243–251.

56 M. Silberstein, S. Dennis, L. Brown, T. Kortvelyesi, K. Clodfelter, S. Vajda, *J. Mol. Biol.* **2003**, 1095–1113.

57 M. Glick, D. D. Robinson, G. H. Grant, W. G. Richards, *J. Am. Chem. Soc.* **2002**, 2337–2344.

58 E. K. Davies, M. Glick, K. N. Harrison, W. G. Richards, *J. Comput. Chem.* **2002**, 1544–1550.

59 X. Pennec, N. Ayache, *Bioinformatics* **1998**, 516–22.

60 G. J. Kleywegt, *J. Mol. Biol.* **1999**, 1887–1897.

61 J. E. Bray, A. E. Todd, F. M. Pearl, J. M. Thornton, C. A. Orengo, *Protein Eng.* **2000**, 153–165.

62 A. Harrison, F. Pearl, R. Mott, J. Thornton, C. Orengo, *J. Mol. Biol.* **2002**, 909–926.

63 J. S. Fetrow, J. Skolnick, *J. Mol. Biol.* **1998**, 949–968.

64 S. Zhao, G. M. Morris, A. J. Olson, D. S. Goodsell, *J. Mol. Biol.* **2001**, 1245–1255.

65 S. L. Moodie, J. B. Mitchell, J. M. Thornton, *J. Mol. Biol.* **1996**, 486–500.

66 P. J. Goodford, *J. Med. Chem.* **1985**, 849–857.

67 A. C. Wallace, N. Borkakoti, J. M. Thornton, *Protein Sci.* **1997**, 2308–2323.

68 A. C. Wallace, R. A. Laskowski, J. M. Thornton, *Protein Sci.* **1996**, 1001–1013.

69 http://www.biochem.ucl.ac.uk/PROCAT/PROCAT.html.

70 T. Hamelryck, *Proteins* **2003**, 96–108.

71 S. C. Bagley, R. B. Altman, *Fold Des.* **1996**, 371–379.

72 http://feature.stanford.edu/webfeature.

73 M. P. Liang, D. R. Banatao, T. E. Klein, D. L. Brutlag, R. B. Altman, *Nucleic Acids Res.* **2003**, 3324–3327.

74 D. R. Banatao, R. B. Altman, T. E. Klein, *Nucleic Acids Res.* **2003**, 4450–4460.

75 R. B. Russell, *J. Mol. Biol.* **1998**, 1211–1227.

76 A. G. Murzin, S. E. Brenner, T. Hubbard, C. Chothia, *J. Mol. Biol.* **1995**, 536–540.

77 http://scop.mrc-lmb.cam.ac.uk/scop.

78 A. Stark, S. Sunyaev, R. B. Russell, *J. Mol. Biol.* **2003**, 1307–1316.

79 http://www.russell.embl.de/pints.

80 A. Stark, R. B. Russell, *Nucleic Acids Res.* **2003**, 3341–3344.

81 P. P. Wangikar, A. V. Tendulkar, S. Ramya, D. N. Mali, S. Sarawagi, *J. Mol. Biol.* **2003**, 955–978.

82 J. V. Lehtonen, K. Denessiouk, A. C. May, M. S. Johnson, *Proteins* **1999**, 341–355.

83 K. A. Denessiouk, A. I. Denesyuk, J. V. Lehtonen, T. Korpela, M. S. Johnson, *Proteins* **1999**, 250–261.

84 K. A. Denessiouk, J. V. Lehtonen, M. S. Johnson, *Protein Sci.* **1998**, 1768–1771.

85 R. V. Spriggs, P. J. Artymiuk, P. Willett, *J. Chem. Inf. Comput. Sci.* **2003**, 412–421.

86 T. J. Oldfield, *Proteins* **2002**, 510–528.

87 T. A. Binkowski, L. Adamian, J. Liang, *J. Mol. Biol.* **2003**, 505–526.

88 T. A. Binkowski, S. Naghibzadeh, J. Liang, *Nucleic Acids Res.* **2003**, 3352–3355.

89 S. Schmitt, D. Kuhn, G. Klebe, *J. Mol. Biol.* **2002**, 387–406.

90 S. Schmitt, M. Hendlich, G. Klebe, *Angew. Chem. Int. Ed. Engl.* **2001**, 3141–3144.

91 K. Kinoshita, H. Nakamura, *Protein Sci.* **2003**, 1589–1595.

92 M. Jambon, A. Imberty, G. Deleage, C. Geourjon, *Proteins* **2003**, 137–145.

93 M. Rosen, S. L. Lin, H. Wolfson, R. Nussinov, *Protein Eng.* **1998**, 263–277.

94 M. Hendlich, F. Rippmann, G. Barnickel, *J. Mol. Graph Model* **1997**, 359–363, 389.

95 A. Weber, A. Casini, A. Heine, D. Kuhn, C. T. Supuran, A. Scozzafava, G. Klebe, *J. Med. Chem.* **2004**, *47*, 550–557.

96 N. Weskamp, D. Kuhn, E. Hüllermeier, G. Klebe, *German Conference on Bioinformatics, Proceedings Volume I* **2003** (Ed. H. W. Mewes), Belleville Verlag.

97 http://ef-site.protein.osaka-u.ac.jp/eF-site.

98 K. Kinoshita, J. Furui, H. Nakamura, *J. Struct. Funct. Genomics* **2002**, 9–22.

99 http://sumo-pbil.ibcp.fr/cgi-bin/sumo-welcome.

100 M. Stahl, C. Taroni, G. Schneider, *Protein Eng.* **2000**, 83–88.

101 M. L. Connolly, *Science* **1983**, 709–413.

102 T. Kohonen, *Biol. Cybern.* **1982**, 59–69.

103 S. J. Pickering, A. J. Bulpitt, N. Efford, N. D. Gold, D. R. Westhead, *Comput. Chem.* **2001**, 79–84.

104 N. Nagano, C. A. Orengo, J. M. Thornton, *J. Mol. Biol.* **2002**, 741–765.

105 http://www.science.novonordisk.com/PTP/ (This outstanding website by Novo Nordisk provides a comprehensive resource for classification and amino acid sequence analysis of protein tyrosine phosphatases.).

106 Z. Y. Zhang, *Acc. Chem. Res.* **2003**, 385–392.

107 K. Parang, J. H. Till, A. J. Ablooglu, R. A. Kohanski, S. R. Hubbard, P. A. Cole, *Nat. Struct. Biol.* **2001**, 37–41.

108 A. Matte, L. W. Tari, L. T. Delbaere, *Structure* **1998**, 413–419.

109 S. Cheek, H. Zhang, N. V. Grishin, *J. Mol. Biol.* **2002**, 855–881.

110 G. Manning, D. B. Whyte, R. Martinez, T. Hunter, S. Sudarsanam, *Science* **2002**, 1912–1934.

111 M. D. Jackson, J. M. Denu, *Chem. Rev.* **2001**, 2313–2340.

112 R. Capdeville, E. Buchdunger, J. Zimmermann, A. Matter, *Nat. Rev. Drug Discov.* **2002**, 493–502.

113 R. A. Engh, D. Bossemeyer, *Pharmacol. Ther.* **2002**, 99–111.

114 M. Huse, J. Kuriyan, *Cell* **2002**, 275–282.

115 P. Cohen, *Nat. Rev. Drug Discov.* **2002**, 309–315.

116 P. Traxler, G. Bold, E. Buchdunger, G. Caravatti, P. Furet, P. Manley, T. O'Reilly, J. Wood, J. Zimmermann, *Med. Res. Rev.* **2001**, 499–512.

117 G. Scapin, *Drug Discov. Today* **2002**, 601–611.

118 T. S. Widlanski, J. K. Myers, B. Stec, K. M. Holtz, E. R. Kantrowitz, *Chem. Biol.* **1997**, 489–492.

119 It is worth mentioning recent investigations on the phosphate binding sites of protein–mononucleotide complexes (Kinoshita et al., *Protein Eng.* **1999**, 11–14) and pyridoxal phosphate binding proteins (Denesyuk et al., *J. Mol. Biol.* **2002**, 155–172). Both studies highlight the presence of distinctively different phosphate recognition motifs, which are conserved within the protein families they belong to.

120 H. Song, N. Hanlon, N. R. Brown, M. E. Noble, L. N. Johnson, D. Barford, *Mol. Cell* **2001**, 615–626.

121 W. Q. Wang, J. P. Sun, Z. Y. Zhang, *Curr. Top. Med. Chem.* **2003**, 739–748.

122 S. Oloff, Y. Xiao, J. Feng, W. Deng, C. Breneman, A. Tropsha, *ACS Meeting* (New York) **2003**.

123 C. M. Breneman, T. R. Thompson, M. Rhem, M. Dung, *Comput. Chem.* **1995**, 161–169.

124 A. V. Ishchenko, E. I. Shakhnovich, *J. Med. Chem.* **2002**, 2770–2780.

125 Y. Y. Kuttner, V. Sobolev, A. Raskind, M. Edelman, *Proteins* **2003**, 400–411.

126 K. A. Denessiouk, J. V. Lehtonen, T. Korpela, M. S. Johnson, *Protein Sci.* **1998**, 1136–1146.

127 K. A. Denessiouk, M. S. Johnson, *Proteins* **2000**, 310–326.

128 K. A. Denessiouk, V. V. Rantanen, M. S. Johnson, *Proteins* **2001**, 282–291.

129 V. Cappello, A. Tramontano, U. Koch, *Proteins* **2002**, 106–115.

130 R. B. Andersen, J. Neuhard, *J. Biol. Chem.* **2001**, 5518–5524.

131 L. F. Iversen, H. S. Andersen, S. Branner, S. B. Mortensen, G. H. Peters, K. Norris, O. H. Olsen, C. B. Jeppesen, B. F. Lundt, W. Ripka, K. B. Moller, N. P. Moller, *J. Biol. Chem.* **2000**, 10300–10307.

132 L. F. Iversen, H. S. Andersen, K. B. Moller, O. H. Olsen, G. H. Peters, S. Branner, S. B. Mortensen, T. K. Hansen, J. Lau, Y. Ge, D. D. Holsworth, M. J. Newman, N. P. Hundahl Moller, *Biochemistry* **2001**, 14812–14820.

133 M. A. Kastenholz, M. Pastor, G. Cruciani, E. E. Haaksma, T. Fox, *J. Med. Chem.* **2000**, 3033–3044.

134 T. Naumann, H. Matter, *J. Med. Chem.* **2002**, 2366–2378.

135 J. A. Westerhuis, T. Kourti, J. F. McGregor, *J. Chemom.* **1998**, 301–321.

136 S. J. Teague, *Nat. Rev. Drug Discov.* **2003**, 527–541.

137 It should be mentioned that the significance of structural differences can be assessed by means of X-ray crystallography by using local density correlation maps (G. J. Kleywegt, *Acta Crystallogr. D Biol. Crystallogr.* **1999**, 1878–1884). This can help to avoid problems linked to structure elucidation methods, prior to drawing conclusions regarding the biological significance of structural variations.

138 R. Najmanovich, J. Kuttner, V. Sobolev, M. Edelman, *Proteins* **2000**, 261–268.

139 R. W. Hooft, C. Sander, G. Vriend, *Proteins* **1996**, 363–376.

140 R. W. Hooft, G. Vriend, C. Sander, E. E. Abola, *Nature* **1996**, 272.

141 A. Bergner, J. Günther, M. Hendlich, G. Klebe, M. L. Verdonk, *Biopolymers (Nucleic Acid Sciences)* **2002**, 99–110.

142 Docking calculations using GOLD (G. Jones, P. Willett, R. C. Glen, A. R. Leach, R. Taylor, *J. Mol. Biol.* **1997**, 727–48) suggest that the 4PP ligand could fit into the binding site in a more extended conformation, filling up the S4 pocket and forming an H-bond with the Glu97.O oxygen atom; (B. Wegscheid, A. Bergner, unpublished results).

143 H. Jhoti, *Drug Discov. Today* **2001**, 1261–1262.

144 Insight II (version 2000.1), Accelrys Inc., San Diego, CA, USA.

145 SYBYL (version 6.9), Tripos, Inc., St. Louis, MO, USA.

146 Y. A. Puius, Y. Zhao, M. Sullivan, D. S. Lawrence, S. C. Almo, Z. Y. Zhang, *Proc. Natl. Acad. Sci. USA* **1997**, 13420–13425.

147 R. A. Sayle, E. J. Milner-White, *Trends Biochem. Sci.* **1995**, 374.

II
Target Families

Chemogenomics in Drug Discovery: A Medicinal Chemistry Perspective.
Edited by Hugo Kubinyi and Gerhard Müller

ISBN: 3-527-30987-X

5
The Contribution of Molecular Informatics to Chemogenomics. Knowledge-based Discovery of Biological Targets and Chemical Lead Compounds

Edgar Jacoby, Ansgar Schuffenhauer, and Pierre Acklin

Summary

A central aspect of chemogenomics refers to the systematic exploration of target families and aims at identification of all possible ligands of all target families. The elucidation of the human genome in 2001 stimulated such chemogenomics approaches by the fact that now almost all members of a target family are visible at the DNA sequence level. Targets are no longer viewed as singular objects having no interrelationship, and the systematic exploration of selected target families appears to be a promising way to speed up and further industrialize target-based drug discovery, especially in the target-identification and lead-finding processes. Here, we summarize chemogenomics knowledge-based strategies that are currently being investigated for target identification and lead finding. The underlying principle of chemogenomics knowledge-based strategies is that similar ligands bind to similar targets. We will show how previously generated knowledge in both the biological and chemical knowledge spaces is useful in simultaneously identifying both new targets and their potential ligands. Since the entire process is knowledge-driven, we emphasize the integration of cheminformatics and bioinformatics into a molecular informatics platform for drug discovery. This chapter has four parts: (1) molecular information systems for targets and ligands; (2) bioinformatics discovery of targets within subfamilies with conserved molecular recognition; (3) cheminformatics discovery of potential ligands of target subfamilies with conserved molecular recognition; and (4) knowledge-based combinatorial library design strategies within homogenous target subfamilies. Special focus is given to the chemogenomics of G-protein coupled receptors, which constitutes one of our main fields of interest.

Chemogenomics in Drug Discovery: A Medicinal Chemistry Perspective.
Edited by Hugo Kubinyi and Gerhard Müller

ISBN: 3-527-30987-X

5.1 Introduction

To organize drug discovery around target families is not new, and indeed in the past a number of target families, e.g., the monoamine GPCRs (G-protein coupled receptors) were explored systematically, so that today, selective ligands are known for a large number of the receptors in these families [1]. Discoveries of new binding sites for known hormones or drugs in the early days of molecular pharmacology were only later followed by identification of the corresponding molecular receptors, which were often subtypes or subsubtypes of previously investigated receptors. The elucidation of the human genome further stimulated the approach, because now almost all members (sequences) of a target family are visible and accessible. The promotion of the approach to an industrial discovery technology was probably first emphasized by researchers at Glaxo Wellcome, who, in a supplement to *Nature* magazine in 1996 on redesigning drug discovery, discussed the concept of systematizing drug discovery within target families, based on analysis of the gene families that had been successfully explored for drug discovery at that time [2]. The Glaxo Welcome scientists highlighted obvious advantages of system-based approaches, for example, combining advances in gene cloning and expression, automation, combinatorial chemistry, and bioinformatics. Since then, numerous pharmaceutical and biotechnology companies have built their business models on these principles. Most notably, the pioneering company Vertex Pharmaceuticals designed its entire drug discovery process around target families, in contrast to the traditional disease-area-oriented organization of pharmaceutical research [3–5].

The promise that chemogenomics, which aims to identify all possible chemical ligands and drugs of all target families, will continue to affect the drug discovery process is very high. Indeed, because of the commonalities existing inside a target family or a homogenous subgroup of it, especially for aspects of molecular recognition, it is only logical to expect that, through additional focus within target families, it will be possible to discover ligands and drugs for new targets at an increased rate and to improve the 'innovation deficit' in the pharmaceutical industry [4, 6–8]. Chemogenomics approaches are expected to be especially effective within previously well-explored target families, for which, in addition to protein sequence and structure information, considerable knowledge on pharmacologically active chemical classes and SAR (structure–activity relationship) data exists. A recent retrospective analysis of the pharmacologically explored targets to date for which high affinity drug-like compounds have been discovered, shows that, first, their number (399) represents a rather small set, and that, second, nearly half of these targets fall into just six gene families: GPCRs, serine/threonine and tyrosine protein kinases, zinc metallopeptidases, serine proteases, nuclear hormone receptors, and phosphodiesterases [6, 9]. Since the genome is rich with additional members of these families, the authors of the study [6] concluded that these proven druggable gene families are still underexploited and will continue to be a source of cost-effective medical innovation. This chapter summarizes chemogenomics knowledge-based strategies for target-identification and lead-finding that are currently being investigated by us and other groups.

The fact that similar ligands bind to similar targets is the underlying principle of chemogenomics knowledge-based strategies for drug discovery. This principle was first summarized and generalized by Stephen Frye at Glaxo Welcome as the SARAH (structure–activity relationship homology) concept, which aims to group drug discovery targets into families based on the relatedness of the SAR data of their ligands [10]. The evolutionary conservation of binding-site architecture within a target family or subfamily translates into conservation of the architectures of ligands that bind to these targets. Since the approach is knowledge-driven, we emphasize the integration of cheminformatics and bioinformatics into a molecular informatics platform for chemogenomics.

5.2 Molecular Information Systems for Targets and Ligands

Based on the SARAH concept [10], it is logical to expect that pharmacological investigations of new members of the main known druggable gene families should benefit from knowledge-based compound selection and design strategies that try to extract relevant characteristics from the established knowledge.

However, given that the cheminformatics and bioinformatics worlds have evolved more or less independently, it is first necessary to establish classification and annotation schemes that link the chemical and biological knowledge spaces.

Annotation and classification efforts in bioinformatics have focused mainly on gene sequences and protein structures. Comprehensive gene ontologies like GO (Gene Ontology) [11] – annotating the biological process, molecular function, and cellular component of gene products – are the ultimate goal of this research. More specifically, several nomenclature and classification committees have established comprehensive class-specific molecular information systems for proteins in general or for proteins associated with drug therapeutic effects (e.g., enzymes [12], GPCRs [1, 13], NRs (nuclear receptors) [13], and LGICs (ligand-gated ion channels) [14]), which are accessible through the Internet. Compared to this, only limited effort has been made on annotation schemes for ligands. Ligand molecular information systems have mainly evolved from the need to track literature, patent, and clinical status information. Catalogues like the MDDR (MDL Drug Data Report) [15], WDI (World Drug Index) [16], CMC (Comprehensive Medicinal Chemistry) [17], IDdb (Investigational Drugs database) [18], or PharmaProjects [19] are typical database systems that provide structural information on ligands together with molecular target or therapeutic class information.

Because the molecular target information provided within the ligand database systems contains only the target name, if anything, and does not provide any further relationships among the targets, the potential of these systems for lead-finding applications remains limited. Ligands of close homologous receptors are, for instance, generally accepted as putative starting points in lead-finding programs for receptors for which no specific ligands are yet known [6, 20]. Therefore, ligand classification schemes that reflect phylogenetic or other relationships of conserved molecular recognition should be useful for lead finding. The main purpose of such a ligand ontology concept should be that ligands of specified levels can easily be collated to serve as comprehensive reference sets for cheminformatics-based similarity searches and for library design or compound selection in purchasing campaigns of target-class-focused collections.

Within this context, while we were in the Combinatorial Chemistry group at Novartis in 2000, we initiated annotation schemes for ligands of four major target families of interest [21]. The MDDR01.1 [15] database, which includes target information for a large number of its ligands, constituted the underlying ligand dataset. Our ligand–target ontology is based on the classification references established by the EC [12], GPCRDB [13], NuclearDB [13], and LIGCDB [14]. By linking MDDR activity keys to the targets of the classification schemes, we were able to group the MDDR ligands within their macromolecular target classes. In total, 309 of the 799 activity keys used in MDDR01.1 could be linked to a target, which allowed annotating 53 211 of the total 113 821 compounds (Figure 5.1).

Linking the leaf nodes of the ligand–target classification tree to the sequence accession codes (e.g., SWISS-PROT AC) of the precise molecular targets allows BLAST-type sequence similarity-based identification of the ligands of the next homologous receptors; for this purpose relating potential ligands to sequences. The molecular information system can also be used for analysis of corporate HTS (high-throughput screening) and profiling data, which are a very rich source of structure–activity data, including large amounts of proprietary data. If one links each assay to a target node in the classification scheme, it is possible to select all

Figure 5.1 Molecular information system highlighting the GPCR ontology. The ligand–target classification of each of the four target families considered is based on the references established by the EC [12], GPCRDB [13], NuclearDB [13], and LIGCDB [14]. These reference systems use different classification criteria and allow different numbers of classification levels. Most of the classifiable compounds (53 211 out of 113 821) are active on enzymes (28 418) and GPCRs (20 961); substantially fewer LGIC (2941) and NR (1443) ligands were classifiable. Within the enzymes, hydrolases – especially peptidases – were previously the most intensively investigated, followed by oxidoreductases and transferases (the latter class includes the kinases). The EC naming system, which is based on considerations of chemical catalysis, is restricted to exactly four levels of hierarchy, resulting in limitations; for example, it does not discriminate between the different types of protein tyrosine kinases. The second important group of targets are the GPCRs for which most structures are active against the peptide-binding or amine-binding class A GPCRs. The GPCRDB uses an unlimited number of hierarchy levels, and the scheme distinguishes GPCR subtypes and subsubtypes. The same is true for the scheme of NuclearDB, which was built by the same researchers following the same guidelines, based on consideration of the pharmacological nature of the ligands and the results of sequence analyses. The LGICs are classified in three different superfamilies without evolutionary relationship. For further details see [21].

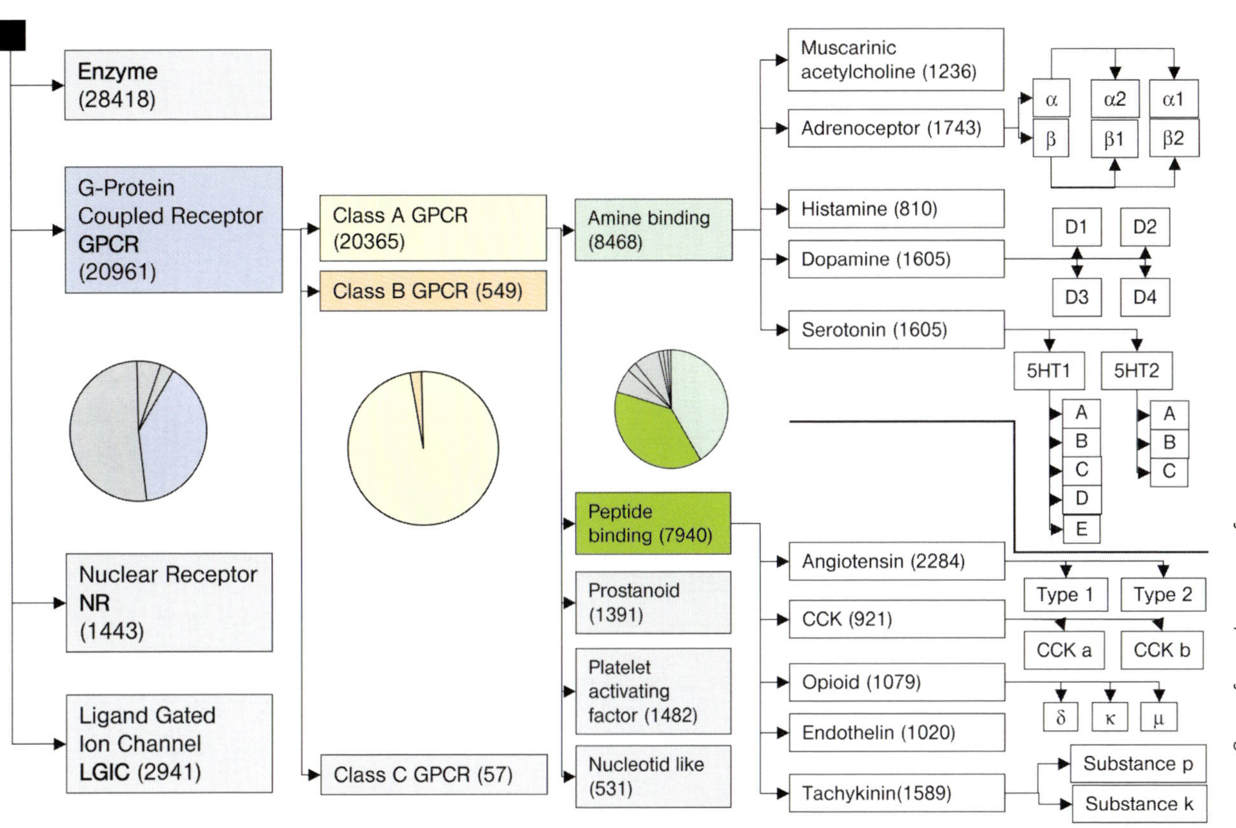

Figure 5.1 Legend see page 142.

assays related to a target family. In a second step, all compounds that showed activity in at least one assay can be used to collect the compounds active against a target family. These compounds can then be used in assays on related targets or can serve as reference structures for further in-silico screening or design of target-class-focused libraries. Both disciplines rely on the possibility of retrieving comprehensive sets of ligands that are likely to share a conserved molecular recognition mode. Alternatively, such a system enables previously known active sets to be subtracted from HTS hit lists, thus providing a strategy for enhancing the potential for discovery of truly novel chemotypes from HTS campaigns. An additional value of the molecular information system is that it is a basis for rational navigation systems for the fast-growing number of datasets generated within the HTS and profiling factories.

The Novartis molecular information system addresses four major target classes and will have to be extended to include novel target families in the genome. The EC classification system should be extended to differentiate subfamilies like the protein kinases [22–24]. The models of classification for GPCRs and NRs, which are based on sequence analyses, are possibly best suited for the purpose of identifying ligands that share commonalities in molecular recognition. The standardized gene grouping and family resources of the HUGO (Human Genome Organization) Genome Nomenclature Committee can be a reference for this [25]. The usefulness of a classification is always closely linked to the nature of the desired application, and several challenges were encountered when the Novartis ligand–target ontology was first implemented. Regarding the EC system, we and others recognized that EC families often do not correlate well with structural families, due to two general problems: first, the EC system may cluster structurally dissimilar proteins as functionally similar; and second, the EC system may cluster structurally similar proteins as functionally dissimilar [26, 27]. Furthermore, many multidomain enzymes are multifunctional and thus belong in multiple classes. Developments of new classification systems for describing enzyme function at the sequence/structure and chemistry levels of granularity were thus suggested for use in inferring functional properties from sequence and structural similarities [27].

The nonstatic nature of annotations and classifications also has to be taken into account. The 880 human GPCRs, including 342 unique functional nonolfactory human GPCR sequences, were recently reclassified on the basis of phylogenetic analysis, resulting in the GRAFS (glutamate, rhodopsin, adhesion, frizzled/taste2, and secretin families) classification system, which differs in detail from the GPCRDB classification [28, 29]. Ways of addressing multiple classifications in parallel thus need to be implemented. InterPro [30], an integrated documentation resource of protein families, domains, and functional sites promises to become an invaluable database, in view of the further generalization of protein family data. InterPro was created in 1999 as a means of amalgamating the major protein signature databases (PROSITE [31], Pfam [32], PRINTS [33], ProDom [34], SMART [35], and TIGRFAMs [36]) into a single comprehensive resource that provides results in a single format, rationalizing the results that would be obtained by searching the member databases individually. Ongoing InterPro developments, which include integration of PIR superfamilies [37] and integration of the structural classifications from SCOP (structural classification of proteins) [38] and CATH (class, architecture, topology, homology) [39], will enhance the utility of the database in the field of protein classification, facilitating retrieval of protein family information, identification of domain and family relationships, and classification of multidomain proteins. *i*ProClass is an other integrated database of protein family, function, and structure information which is comparable to InterPro [40].

The field of ligand–target classification turns out to be quite complex and is related to the bioinformatics projects on ontologies for proteomics that aim at a systematic definition of the structure and function of proteins that scales to the genome level [41, 42]. Because the current ontologies in the fields of protein structure (primary, secondary, tertiary, and quaternary) and function have been developed separately and remain largely isolated, a key point remains their integration [41]. The description of active sites and binding sites in protein structures is here recognized as one potential connection point that describes the protein function. Classifications based on molecular interactions [41, 43], in which each protein is associated with a row vector that consists of the probability of its binding to various ligands – the central chemogenomics idea – may thus become prominent in the future [41, 44].

It is noteworthy that chemogenomics knowledge-based companies such as Aureus Pharma [45], Inpharmatica [46], and Jubilant [47] are developing comprehensive molecular information systems for a variety of target classes, including GPCRs, kinases, ion channels, and proteases. Their main contribution is to comprehensively integrate data from patents and selected literature, including 2D structures of the ligands, target sequence and classification, mechanism of action, structure–activity data, assay and bibliographic information, together with chemical and biological search engines. Additional ongoing academic and commercial developments in the area of ligand–target molecular information systems address targets of adverse reactions [48, 49], targets implied in ADME mechanisms, and targets that define metabolic and signaling pathways [47, 50–52]. The Cerep BioPrint [53, 54] and NIMH Psychoactive Drug Screening Program [55] database are two recent pharmacoinformatics systems with strong focus on GPCR pharmacology and profile structure–activity data.

5.3
Bioinformatics Discovery of Target Subfamilies with Conserved Molecular Recognition

The sequence and functional similarities within a gene family usually indicate a general conserved binding-site architecture and molecular recognition of ligands [6, 20]. This suggests that, if one member of gene family were able to bind a ligand, other members with conserved molecular recognition should also be able to bind compounds with similar chemical structure. Based on this principle, the investigation of sequence similarities through phylogenetic or fingerprint analyses is a commonly used strategy to classify new orphan members of gene families and to facilitate identification of the endogenous ligands.

Within the GPCR family, for which several bioinformatics studies have now comprehensively mined the human genome for potential members [28, 29], phylogenetic analyses were useful for predicting that sphingosine-1-phosphate, the endogenous ligand of the EDG_1 (endothelial differentiation gene) GPCR, is also the ligand of the EDG_3, EDG_5, EDG_6, and EDG_8 GPCRs [56]. Also, the ligand for and the pharmacology of the human histamine HHR_4 GPCR was predicted through phylogeny [56]. On the other hand, examples are known for which sequence homology can be misleading; for example, a receptor originally known as P_2Y_7 (BLT1) was thought to be a nucleotide receptor based on its similarity to P_2Y purinoceptors, but it was shown to be activated by an unrelated ligand, leukotriene B_4 [57, 58]. This indicates that bioinformatics deorphanization strategies based on the overall sequence have limitations and that, for ligand pairing of the remaining 140 orphan GPCRs, additional wet experiments are needed [59].

Because, in general, a broad diversity of chemical structures are recognized endogenously and selectively as substrates, inhibitors, or agonists (e.g., peptide and protein hormones, nucleosides and nucleotides, and lipids) by individual members of the same target family, it is expected that conserved molecular recognition will only exist within subfamilies of each gene family. For successful lead-finding strategies, it is thus necessary to classify the orphan members of each target family based on the relatedness of their molecular recognition. Understanding the principles of molecular recognition in combination with residue- and motif-based 1D and 3D bioinformatics datamining are becoming essential elements for implementing successful chemogenomics knowledge-based strategies. Our recent analysis of monoamine-related GPCRs illustrates such a datamining approach to searching for orphan GPCRs deposited in the 2001 versions of the SwissProt and SPTREMBL databases [60]. The conserved aspartate residue D3.32 [61] in TM3 (transmembrane helix three) was demonstrated by 2D mutation experiments (in which both the ligand and receptor are mutated according to the underlying interaction hypothesis) to be responsible for the recognition of the charged amino group of monoamine ligands by their GPCRs [62]. Focusing on the central importance of the D3.32 residue and using the $D3.32X_{16}$(DE)R(YFH) motif in TM3 as a sequence signature defining relatedness to the monoamine GPCR subfamily, we identified 50 human GPCRs for which the sequence comparisons, both for the 7TM (seven transmembrane) domain and for the ligand-binding sites, are presented in the form of a dendrogram Figures 5.2 and 5.7. The 50 receptors include 7 orphan GPCRs (two of which are now known to correspond to pseudogenes) and, somewhat surprisingly, 9 peptide class A GPCRs – somatostatin and opiate receptors – which, by our definition, are related to the monoamine GPCRs. The dendrogram analysis of the 7TM domain shows that the peptide and monoamine GPCRs are separated into two distinct, non-intermixing groups, and that the identified orphan receptors fall into three groups. GPR14, recently identified as the urotensin receptor [63], and GPR24, recently identified as the MCH (melanin-concentrating hormone) receptor [64], are peptide receptor singletons connected to the somatostatin and opiate node. GPR7 and GPR8 fall into a separate cluster directly connected to the somatostatin receptors; in 2002, subsequent to our study, both receptors were ligand-

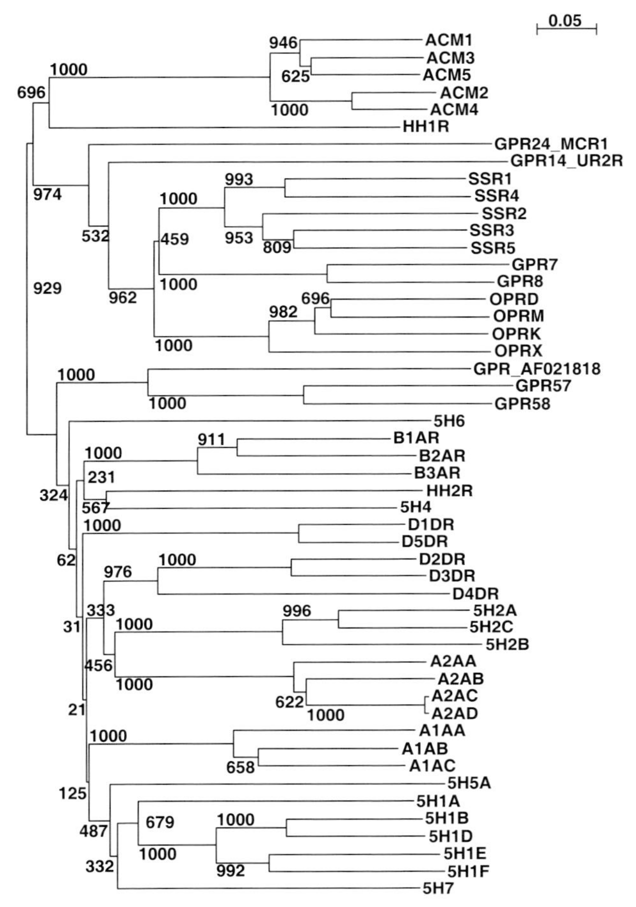

Figure 5.2 Neighbor-joining tree of sequence similarity in the 7TM domains of human monoamine-related GPCRs. The receptors are coded according to the SwissProt nomenclature scheme; orphan receptors are coded with the prefix 'GPR' followed by an index number. Distance corresponds to percent sequence identity, scale is indicated by a 5% bar. The tree is rooted by outgrouping the node of the H_1 and muscarinic receptors. The numbers on the branches are the result of bootstrap analysis (1000 replicates). For further details see [60].

paired and are, respectively, the NPW (neuropeptide W) and NPB (neuropeptide B) receptors [59]. GPR_AF021818 (PNR), and the pseudogenes GPR57 and GPR58 form an orphan cluster connected to the monoamine receptors. These three receptors belong to the trace amine family: after our study, 22 members were identified in humans and rodents which all conserve the D/E3.32X_{16}(DE)R(YFH) signature motif [65]. Trace amine receptors have a potential role in psychiatric and neurological disorders [66].

The challenging hypothesis resulting from this analysis is that all the identified receptors should be able to recognize ligands of the tertiary amine chemotype known to be active as synthetic agonists or antagonists to many monoamine GPCRs. Indications that this hypothesis is valid can be found in the facts that the tertiary amine chemotype binds to opiate GPCRs (e.g., morphines) and that ergoline [67] and benzoquinoline [68] ligands bind as antagonists to somatostatin GPCRs, as was recently discovered at Novartis. Interestingly, the overall architecture of these somatostatin GPCR ligands is similar to those of previously well explored monoamine GPCR ligands, thus indicating that the identified peptide- and monoamine-binding GPCRs do indeed contain conserved elements for molecular recognition. This hypothesis is supported by mutagenesis data, demonstrating the role of D3.32 for ionic interaction with the endogenous ligands of the opioid [69], somatostatin [70, 71], and MCH peptide receptors [72]. The tertiary amine chemotype should thus also be a valid starting point for lead-finding programs investigating the identified orphan monoamine-related GPCRs. This can easily be tested by focusing screening on the ligands of the monoamine-related receptors that were previously successfully investigated and listed in our molecular information system, together with similar ligands for them, which were identified in cheminformatics similarity searches (see Section 5.4).

It is worth emphasizing that sequence similarity searches based on overall sequence identity can lead to different conclusions than motif-based analyses: for instance, the two orphan receptors GPR61 and GPR62 were reported to have overall sequence identities of 30% with the human $5HT_6$ receptor and were thus classified as monoamine-like receptors [73]. Strikingly, both of them show mutations of the D3.32 residue and should therefore belong to a different subfamily. The art thus lies in identifying those characteristic motifs that define a conserved mode of molecular recognition. Such motifs can in principle be rationalized only after detailed knowledge of ligand–receptor interaction is available based on the 3D structure of the ligand–receptor complex or on mutagenesis experiments.

Noteworthy in this perspective is the recent work done at Pfizer and Biofocus (see Section 5.5), where, based on analysis of sequence data, mutation data, and physicochemical properties of the ligands, approaches were outlined for discovering sequence patterns characteristic of specific ligand classes. Pfizer applied the approach successfully to the construction of a sequence motif characteristic of monoamine GPCRs [74].

The sequence motifs identified in this way may be different from the motifs/fingerprints listed in the Prosite [31], Blocks [75], and especially PRINTS [33, 76] databases, which use single or multiple conserved regions as consensus signatures to describe families or subfamilies and which are used as automated diagnostic tools for inserting new members into the framework of the already classified members.

A further element of proof of our transposition hypothesis can be recognized in the recently described characterization of the binding site of CCR_{2b} chemokine receptor spiropiperidine antagonists (see Figure 5.5), binding to a common chemokine GPCR motif within the 7TM bundle [77]. Again, these ligands are of the tertiary amine chemotype and also bind with high affinity to some monoamine GPCRs. Surprisingly, the CCR_{2b} receptor lacks the critical aspartate D3.32 in TM3. Instead, the residue glutamate E7.39 (E291 in the CCR_{2b}) in TM7 that is conserved among chemokine receptors is responsible for recognition of the tertiary amine chemotype; residue positions [61] 7.39 and 3.32 are close together and are positioned on opposite sides of the central binding cavity within the rhodopsin-based models of class A GPCRs. Chemokine GPCRs can thus, for certain aspects of molecular recognition, be classified as monoamine-related GPCRs.

Such examples reveal the challenges for designing ligand selectivity during lead optimization, which should ideally include profiling against all targets with expected conserved molecular recognition. Mining the human genome for new targets homologous to previously well-characterized targets, for which the principles of ligand recognition are well rationalized and for which ligand and SAR information are abundant, appears highly promising. This of course does not eliminate the necessary target-validation step.

5.4
Cheminformatics Discovery of Potential Ligands of Target Subfamilies with Conserved Molecular Recognition

The principle of chemogenomics knowledge-based strategies for ligand identification is that similar ligands bind to similar targets. The similarity principle is prominent in medicinal chemistry, although it is well known as the similarity paradox, i.e., that very minor changes in chemical structure, such as the introduction of an additional methyl group, can result in total loss of activity [78–82]. The type of molecular similarity relevant to chemogenomics knowledge-based approaches is illustrated in Figure 5.3, which shows several monoamine-GPCR ligands and drugs, which, for a given molecular architecture type, look structurally similar to the medicinal chemist's eye and are biologically active on multiple members of this target family.

Computer-based similarity searching is one core discipline of cheminformatics, and in the 1990s many molecular descriptors, similarity metrics, and similarity-ranking methods were developed and are now available in commercial software packages [82, 84–88]. In the view of chemogenomics, cheminformatics similarity-searching methods that are able to identify, not only ligands binding to the same target as the reference ligand(s), but also potential ligands of other homologous targets for which no ligands are yet known, are essential tools for further exploration of previously successful target families. Until now, very few studies have investigated the power of similarity-searching methods on targets different from the reference target. Similarity-searching methods are now mostly used to identify lead compounds for a target for which a large number of reference compounds are already known – allowing competitors to find catch-up lead molecules. A very important concept was introduced in 1999 by researchers at Roche, aiming at similarity methods able to enable 'scaffold hopping', which allow to escape from a patented

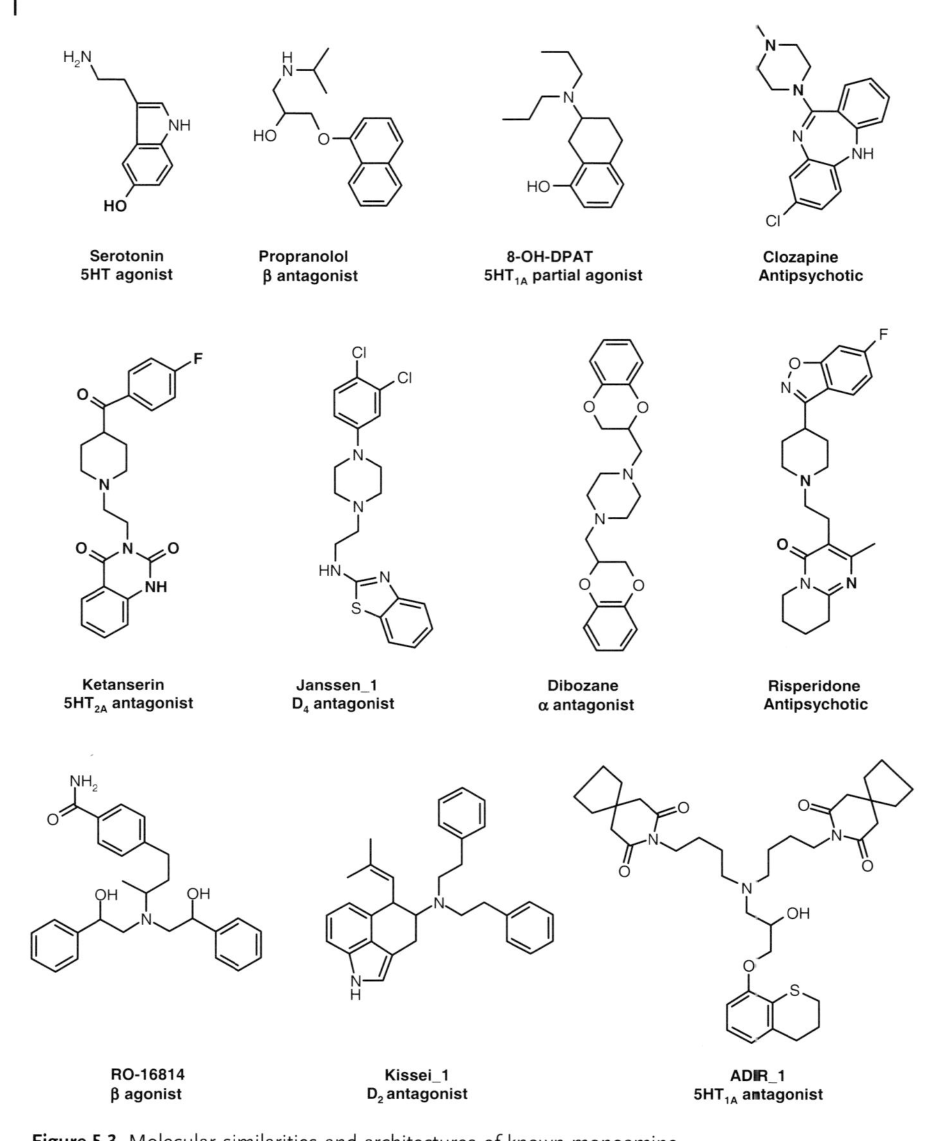

Figure 5.3 Molecular similarities and architectures of known monoamine GPCR ligands included in the Novartis molecular information system [60, 83]. Ligands that are the same size as the endogenous ligands are called 'simple one-site filling' ligands here (top row). In addition to this natural architecture, ligands exist in which two or three such 'simple' ligand fragments are joined around a basic positively charged group: these ligands are called 'double' and 'triple' ligands (middle and bottom rows). Exploration of the architectures of known monoamine GPCR ligands constitutes a direct way of providing evidence for the existence of three binding sites. The 'triple' ligands especially demonstrate the existence of three binding sites (see Figure 5.6).

chemical class [89]. Today, several such methods, using pharmacophores [89], feature trees [90], or reduced graphs [91], that abstract descriptions of the reference and candidate molecules are available.

Chemogenomics homology-based similarity searching requires molecular representations that reflect the conserved aspects of molecular recognition between ligands and their targets. The Similog keys, which we described recently [92], were designed to meet this and the 'scaffold-hopping' requirements. The Similog keys are counts of atom triplets in which each triplet is characterized by the interatom graph distances and the types of its atoms. The atom-typing scheme classifies each atom by its function as an H-bond donor or acceptor and by its electronegativity and bulkiness. In combination with various distance-averaging methods based on the Tanimoto coefficient, we showed in retrospective in-silico screening experiments, which included ligand sets of several target families (GPCRs, NRs, and proteases) derived from our molecular information system, that the Similog keys perform better than other conformation-independent molecular representations like the Unity 2D fingerprints [86], 2D topological descriptors [87], ISIS public key count [93], or the E-state descriptors (Figure 5.4) [94].

Our analyses showed that similarity searching based on Unity 2D fingerprints or Similog keys are equally effective in the identification of molecules binding to the same target as the reference set – the classical application scenario of similarity searching. However, use of the Similog keys is more effective in identification of ligands binding to targets homologous to the reference target – the chemogenomics scenario. We attribute this superiority to the fact that the Similog keys provide a generalization of the chemical elements and that the keys are counted instead of merely noting their presence or absence in a binary form. The Similog keys thus capture the potential points of interactions between the ligands and the target proteins. The difference in the performance of the distance-averaging methods is attributed to the fact that the centroid method in particular is able to enhance commonalities displayed in the pharmacophore representations of the reference compounds, crystallizing in this manner the main repeated pharmacophore features. The results obtained suggest that ligands for a new target within a previously well-explored family with supposed conserved molecular recognition can be identified by the following three-step procedure: (1) select at least one target with known ligands that is homologous to the new target – both must belong to the same subfamily with conserved molecular recognition; (2) combine the known ligands of the selected target(s) into a reference set; and (3) search candidate ligands for the new targets by their similarity to the reference set using the Similog method. This clearly enlarges the scope of similarity searching from the classical application to a single target to the identification of candidate ligands for whole target families. Applying homology-based similarity searching, one expects that ligands binding to the reference target are accumulated preferentially and that, in consequence, unselective ligand candidates are identified. The method is thus less suitable for the lead-optimization process, in which target selectivity is one of the main objectives, but it is most suitably applied early in the lead-finding process. To a first approximation, the only knowledge required to identify homologous targets with

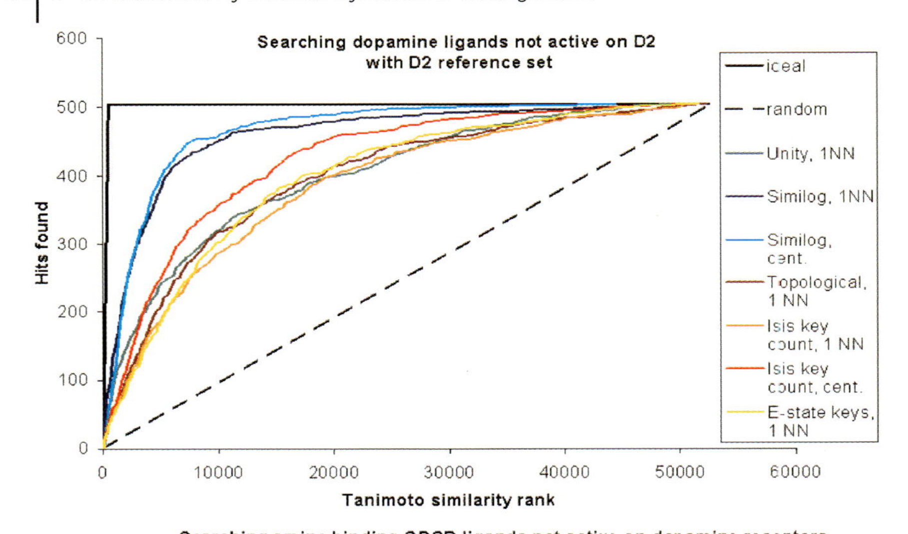

Figure 5.4 Retrospective in-silico screening experiments comparing the retrieval performance of Similog keys with that of other molecular descriptors in chemogenomics homology-based similarity-searching applications. The MDDR01.1 dataset was split randomly into two halves; the first was used to obtain reference sets and the second as a test set. Enrichment curves showing the number of recovered hits as a function of the similarity score (rank) of the candidate test set compared with the reference query set are shown. D_2 GPCR ligands were used to search for: (1) ligands of the other dopamine GPCRs excluding D_2 hits (upper panel); and (2) ligands of the other monoamine GPCRs excluding all dopamine GPCR hits (lower panel). The two searches illustrate the approximate differences in retrieval performance for increasing phylogenetic distance. The Similog keys were compared with other descriptors, including the Unity-2D fingerprints, 2D topological descriptors, ISIS public key count, and the E-state descriptors. These descriptors were used in combination with two Tanimoto distance-averaging methods: 1NN (nearest neighbor) and cent. (centroid). For further details see [92].

known ligands is the sequence of the target protein; rigorous searches should be based on a detailed analysis and comparison of the binding-site residues, requiring knowledge of a 3D structure model.

As a recent successful application of the method within Novartis, we can cite a search for antagonists to $5HT_7$ GPCR, which has the $5HT_{1A}$ receptors as its next neighbor (see Figures 5.2 and 5.7) for which reference ligands are abundantly known. Searching with $5HT_{1A}$ reference compounds by using the Similog centroid method within a designed tertiary amine combinatorial library (see Section 5.5), we were able to identify a 10% hit rate ($pK_B < 5$ μM) when only a biological assay with limited capacity (88 compounds) was available. The hits were arylpiperazines (see Figure 5.8), which in follow-up studies were also active on other monoamine GPCRs.

Homology-based similarity searching is also invaluable for design of target-family-focused combinatorial libraries, for compilation of compound-acquisition selections, and for intelligent structuring of the HTS collection; it is currently being used for these applications at Novartis. Whether the performance of our method might be further increased by using data-fusion models based on scoring and rank methods and a combination of several descriptors simultaneously [95] or by using profile-scaling methods [96] remains to be evaluated; such methods were, in the classical scenario, recently shown to increase the similarity-search performance of molecular fingerprints.

Directly related to the similarity-searching methods are cheminformatics methods that try to align the chemical and biological spaces based on mapping procedures. The goal here is to identify which parts (islands) of the chemical space correspond to a specific target family or therapeutic activity and vice versa. Testing such approaches is motivated by the observation that, within the different therapeutic classes of drugs, molecular properties like AlogP, molecular weight, or calculated molar refractivity show distinct statistical distributions. The average computed AlogP for antipsychotic drugs is, for instance, one unit higher than that of antihypertensive drugs [97]. Low-dimensional projections using principal component analysis based on pharmacophore fingerprints [98, 99] and singular-value decomposition projections based on MACCS fingerprints, pharmacophore holograms, and topological descriptors [100, 101] have showed that such mapping is in some cases feasible. The studies included multiple chemical classes active on different targets not belonging to the same target family, and revealed, in particular, that points that are close together on the 2D maps have similar chemical structures. By leaving out the dimension-reduction aspect, which is necessary for graphical representation and for eliminating correlations among descriptors, it has been shown that accurate partitioning of compounds belonging to diverse activity classes is possible in principle. Further investigation is required to address the question of how well these methods are suited to differentiation within subfamilies of a given target class like the class A GPCRs or protein kinases. The recent study of Pirard and Pickett is encouraging in this respect, showing that BCUT descriptors can be used for classifying ATP site-directed kinase inhibitors active against five different protein kinases: three from the serine/threonine family and two from the tyrosine kinase family [102, 103].

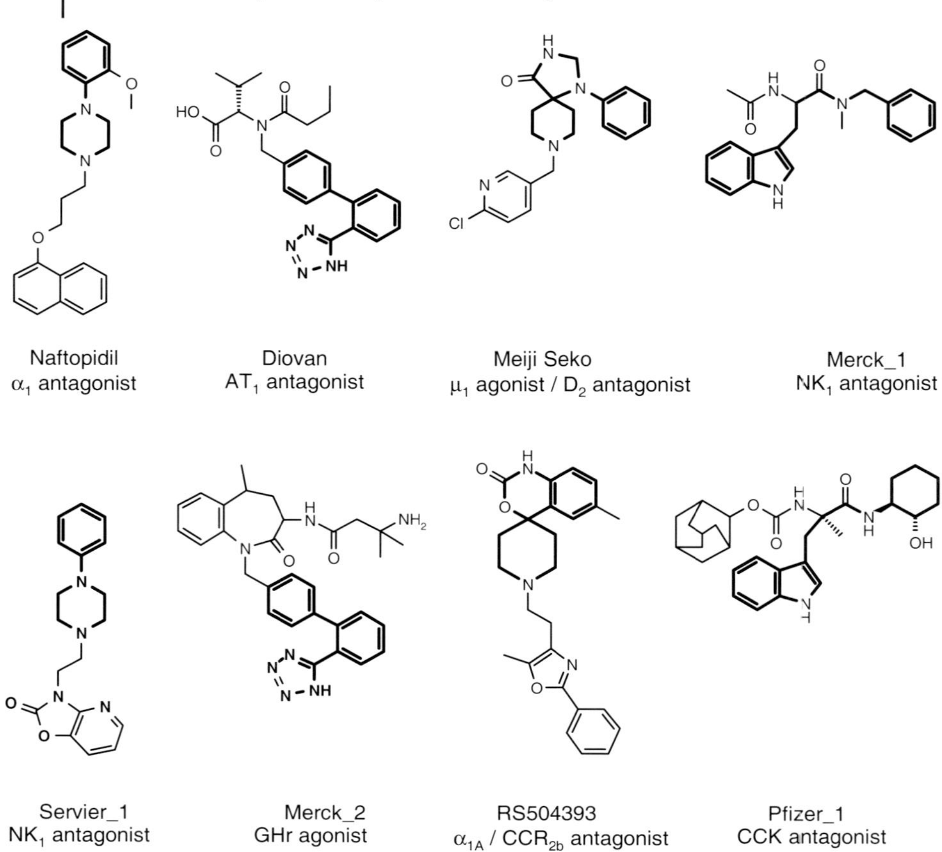

Figure 5.5 Examples of empirically identified privileged structures of GPCR ligands. The privileged structures are highlighted. A ligand pair is shown for each privileged motif. A 'privileged' structure does not always correspond to an exact substructure, as shown for the capped amino acid and spiropiperidine types [8, 44, 104–106].

The use of privileged substructures or molecular master keys, whether target-class specific or not, is an accepted concept in medicinal chemistry [8, 44, 104, 105]. The approach emphasizes scaffolds that have been shown to be able to provide ligands for diverse receptors within the known target classes and was, for instance, successfully used in the design of GPCR antagonists and protease inhibitors (Figure 5.5).

The design of combinatorial libraries around these master keys has proven to be a very successful strategy, allowing multiple members of a target family to be addressed [8, 44, 105, 106]. The development of cheminformatics methods and procedures enabling the automatic identification and extraction of such privileged structures is a recently developed discipline. Most of the known privileged structures were identified empirically (e.g., GPCR antagonists) or were designed with structure-based and mechanism-based approaches (e.g., protease inhibitors). Automatic privileged structure identification methods are especially needed in the context of differentiating frequent HTS hitters and for generating knowledge from HTS data [107, 108]. The development of fast maximum common subgraph isomorphism algorithms [109], as implemented for example in the Bioreason HTS data analysis software, addresses this need and allows the analysis of multiple assays in parallel [110]. The Leadscope chemical classification system, which uses a hierarchy of over 27 000 chemical features, based on the building blocks of medicinal chemistry, can also be used for this purpose [111].

Noteworthy in this perspective is the work by Bemis and Murcko, who developed a method for decomposing molecules into frameworks, sidechains, and linkers and analyzed the statistical occurrence of the frameworks within a subset of drugs listed in the CMC catalog [112]. This analysis revealed that only 32 frameworks describe the shapes of half the drugs in the CMC set (~5000 compounds) and resulted in the design of the SHAPES NMR screening library – a limited but diverse library of small molecules derived from the shapes most commonly found – which Vertex uses within the lead-finding process [113].

For both similarity-searching and privileged-structure strategies, it was correctly pointed out that intellectual property considerations can become a capital issue [8]. This stresses again the above-mentioned need for scaffold hopping and bioisostere-identification methods and also the value of proprietary chemical and biological data [114–117].

5.5 Knowledge-based Combinatorial Library Design Strategies within Homogenous Target Subfamilies

Based on the commonly accepted strategy that ligands of closely homologous receptors are generally accepted as putative starting points in lead-finding programs for receptors for which no specific ligands are yet known, and based on the premise that the existence of multiple binding sites and modular ligand architectures – cross-linking the individual sites – are fundamental to knowledge-based strategies aiming to transition toward orphan receptor systems, we recently proposed a chemogenomics knowledge-based ligand-design strategy for lead finding and combinatorial library design [60, 83]. The strategy is based on integration of both the deconvolution of known ligands of homologous receptors into their component fragments and a structural bioinformatics comparison of the binding sites for the individual ligand fragments. In essence, by analysis of both the ligand architectures and the structures of the component 'one-site filling' fragments of known ligands, it should be possible, by referring to the locally most directly related and characterized receptors in the ligand space, to identify those component ligand fragments, which, based on binding-site similarities, are potentially best-suited for designing ligands tailored to the new target receptor. This strategy was presented with respect to monoamine-related GPCRs, for which positioning analyses in the sequence space of the 7TM domains of the receptors and in the sequence spaces of the three previously identified distinct ligand-fragment binding regions of the monoamine GPCRs were carried out, with the objective of characterizing orphan receptors and monoamine receptors for which no specific ligands are yet known (see Figure 5.6).

Applied to the $5HT_{1E}$ and $5HT_5$ receptors (for which no selective ligands are yet known – see IUPHAR receptor compendium [1]) and the 5 orphan receptors GPR7, GPR8, GPR14, GRP24, and GPR_AF021818, which are depicted in Figures 5.2 and 5.7, the two strategies result in different conclusions for ligand design proposals.

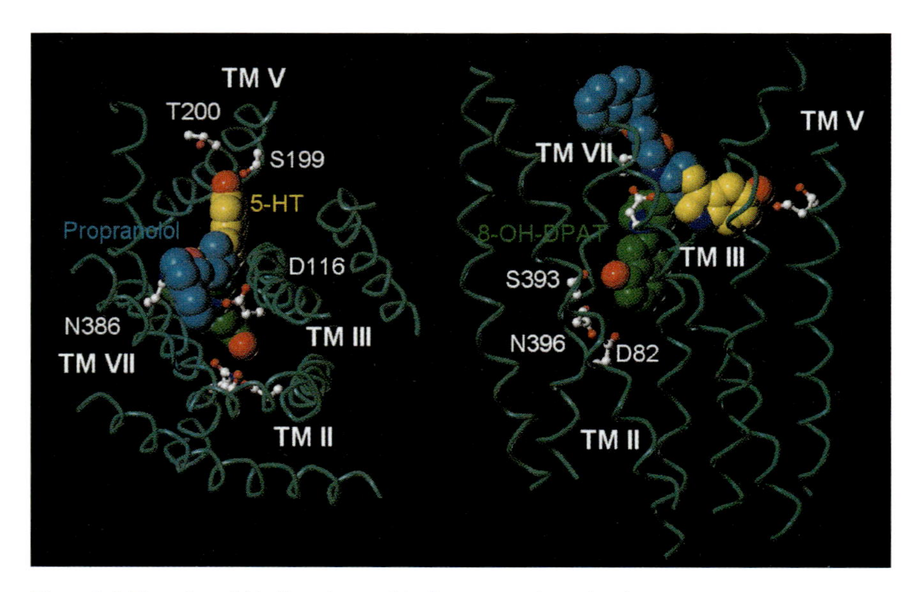

Figure 5.6 Three-ligand binding-site models for monoamine-related GPCRs illustrated by a rhodopsin-based 3D model of the $5HT_{1A}$ receptor (left: extracellular view; right: side view). We recently proposed a three-binding-sites hypothesis for the molecular recognition of ligands by monoamine GPCRs [60, 83] by combining (1) analyses of the architectures of known monoamine GPCR ligands (see Figure 5.3); (2) analyses of molecular models of the ligand–receptor interactions; and (3) structural bioinformatics analyses of the sequence similarities of the three distinct binding regions of 'one-site filling' ligand fragments within the monoamine GPCR family. For the $5HT_{1A}$ receptor, which provided a frame for the discussion of other, related, ligand–GPCR interactions, mutagenesis studies map three spatially distinct binding regions, which correspond to the binding sites of the 'small, one-site filling' ligands 5-HT (serotonin: yellow), propranolol (blue), and 8-OH-DPAT (8-hydroyxy-*N*,*N*-dipropylaminotetralin: green), respectively. All three binding sites are located within the highly conserved 7TM domain of the receptor and overlap at residue Asp3.32 (D 116) in TM3, which constitutes the key anchor site for basic monoamine ligands. The three distinct binding sites are also reflected by the architectures of known high-affinity ligands that crosslink two or three 'one-site filling' fragments around a basic amino group. See Figure 5.3 for chemical structures; for further details see [60, 83].

Based on the overall sequence similarity of the 7TM domain of the receptors (Figure 5.2), known $5HT_{1F}$ ligands are suggested as potential $5HT_{1E}$ ligands; known $5HT_1$ and $5HT_7$ ligands are suggested as potential $5HT_5$ ligands; serotonin, dopamine, and adrenoceptor ligands are suggested as potential GPR_AF021818 ligands; somatostatin and opiate receptor ligands are suggested as potential GPR14 and GPR24 ligands; and somatostatin receptor ligands are suggested as potential GPR7 and GPR8 ligands. Compared to this, the 'three binding sites' hypothesis localizes the homology to the different binding sites. The following hints are provided according to inspection of the neighbor-joining trees of the individual fragment-binding sites depicted in Figure 5.7. For potential $5HT_{1E}$ ligands, the $5HT_{1D}$ and $5HT_{1B}$ ligand fragments are the nearest neighbors in the '5HT' site, and the $5HT_{1F}$ ligand fragments are the nearest neighbors for the 'propranolol' and '8-OH-DPAT' sites. For potential $5HT_5$ ligands, the $5HT_{1A}$ ligand fragments are the nearest neighbors in the '5HT' site, and $5HT_7$ ligand fragments are the nearest neighbors in the 'propranolol' and '8-OH-DPAT' sites. For potential GPR7, GPR8, and GPR24 ligands, the dendrogram analyses indicate no preference for opiate over somatostatin receptor ligand fragments. Similarly, because of the low bootstrap values, no further conclusions can be reached regarding preferable component fragments of potential GPR_AF021818 ligands. For potential GPR14 ligands, monoamine ligand fragments, in general, are suggested at the '5HT' site, whereas opiate and somatostatin peptide GPCR ligand fragments are the nearest neighbors in the 'propranolol' and '8-OH-DPAT' sites, as observed from the overall sequence similarity analysis. GPR14 represents a quite interesting local binding-site similarity pattern. Compared to the predictions based on the strategy of analyzing the overall sequence similarity of the 7TM domain, the number of starting points is hence increased and, more importantly, the sequence homology is localized to the different binding sites.

Within a target subfamily with a presumed conserved ligand-recognition type, like the conserved Asp3.32 anchor site for the recognition of amine ligands in the monoamine-related GPCRs analyzed here, the proposed approach is to identify the next neighbors for each binding site and to accordingly identify the previously used ligand fragments, which should be recombined on an appropriate scaffold to yield ligands for a newly investigated target. Databases of site-specific ligand fragments are the keystones of such a knowledge-based system. Their generation is, in principle, possible through deconvolution of the known ligands guided by SAR and by molecular similarity consideration. Given the promiscuity of some fragments (e.g., symmetric ligands – see Dibozane in Figure 5.3), one has to be cautious before drawing definitive conclusions about the actual positioning of the fragments [60, 83]. In the particular example of the design of monoamine-related GPCR ligands, the general approach might be hampered by the occasional difficulty of attributing individual ligand fragments to a specific receptor site. Pragmatically, these limitations to the generation of site-specific ligand fragment databases can however be approached by pooling fragments into multiple pools and by designing generic combinatorial libraries of known privileged active fragments around appropriate scaffolds. This strategy was used at Novartis for designing the TAM

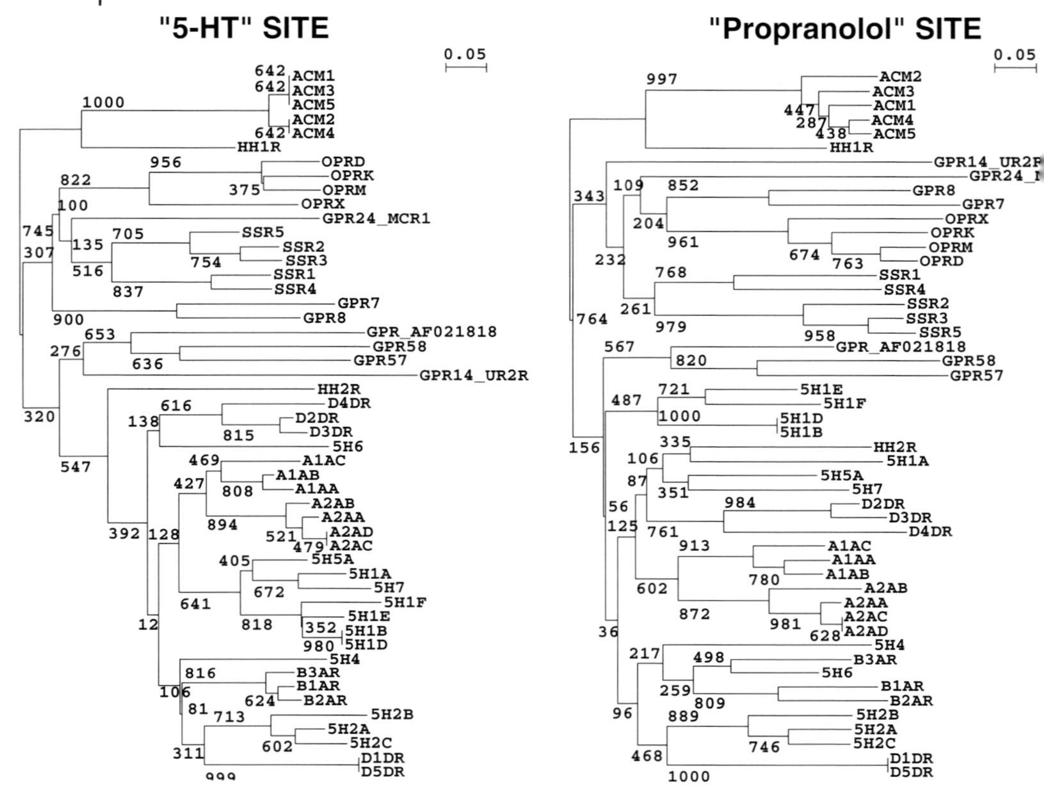

Figure 5.7

(tertiary amine) combinatorial libraries which, using reductive amination of selected aldehydes and secondary amines, contributed ~20 000 tertiary amines to the general HTS collection. Prototype structures of TAM libraries are shown in Figure 5.8.

Researchers at Biofocus have developed the thematic analysis method for designing focused class A GPCR libraries [118, 119]. Their method, which, like our approach, is based on the SARAH concept, is more general and systematic. SARs were analyzed in detail across the whole class A GPCR family, and family–activity relationships were used to develop a new classification process based on pairing sequence themes and ligand structural motifs. A sequence theme is a consensus collection of amino acids within the central binding cavity, and a motif is a specific structural element binding to such a particular microenvironment of the binding site. The Biofocus analysis resulted in a compilation of themes and motifs, which to date are used at Biofocus to generate focused discovery libraries for class A GPCRs and to increase the lead optimization efficiency for these targets. The individual Biofocus libraries target subsets of GPCRs, including orphans, that share a predefined combination of themes consisting of a central dominant theme and peripheral ancillary themes. The library scaffold is designed so that it complements

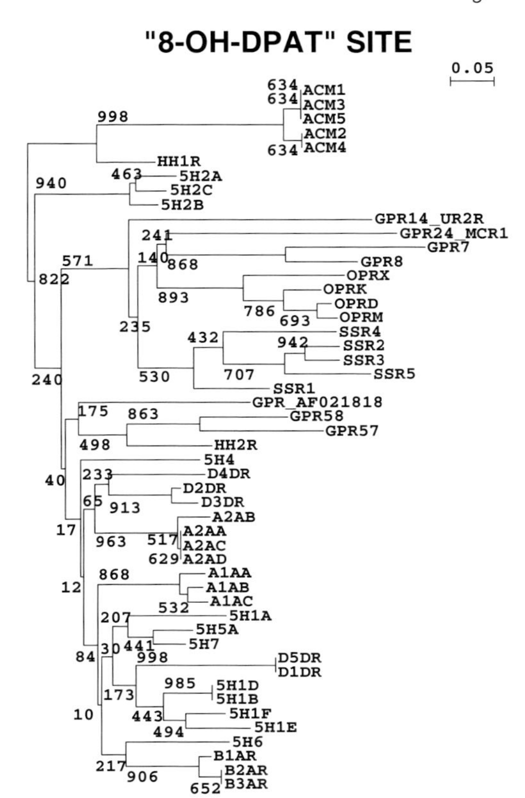

Figure 5.7 Neighbor-joining trees of the sequence homology of each binding site for monoamine-related GPCRs. The underlying sequence blocks correspond to transmembrane residues identified within the 6-Å contact spheres of 5-HT (panel A), propranolol (panel B) and 8-OH-DPAT (panel C), respectively, in the rhodopsin-based models of the 5-HT_{1A} receptor–ligand complexes. Details as in the legend for Figure 5.2.

the central theme and is amenable to the incorporation of a variety of structural motifs addressing the individual sequence themes. Each library, consisting of approximately 1000 compounds, can thus be thought of as representing a number of predefined themes, which are either present or absent in any given receptor – allowing an appropriateness score to be computed for each receptor through this kind of fingerprinting. Thematic analysis is also used to aid lead optimization by analysis of those themes that are involved or not involved in binding a particular hit molecule and by the exploitation of new combinations of used and unused themes to increase affinity and selectivity [118, 119].

In both the Novartis and Biofocus approaches, the identification of building blocks is essential. Several computational approaches to deconvoluting known drugs into their component fragments exist and are discussed here. Two different approaches were developed at Glaxo. Lewell et al. [120] developed a method for drug motif-based diverse monomer selection using clustering and similarity analysis. The candidate building blocks were first clustered with Jarvis–Patrick clustering, and then the overall similarity and the substructure similarity of each centroid to 30 000 compounds from the WDI catalogue was calculated; this drug motif knowledge

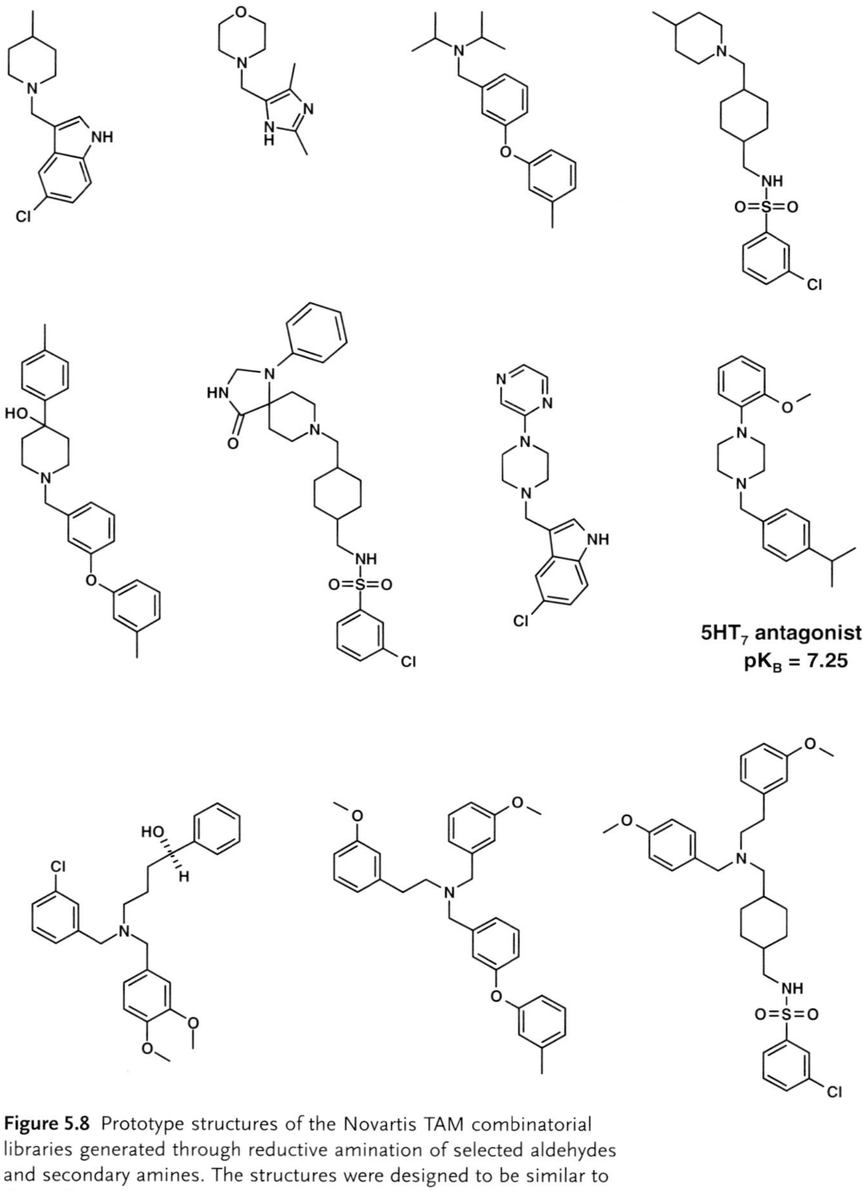

5HT$_7$ antagonist
pK_B = 7.25

Figure 5.8 Prototype structures of the Novartis TAM combinatorial libraries generated through reductive amination of selected aldehydes and secondary amines. The structures were designed to be similar to those shown in Figure 5.3, and all three architectures of known monoamine GPCR ligands are included. The TAM libraries are being screened in current GPCR campaigns, and high hit rates are observed especially for the monoamine-related GPCRs. The library includes many new combinations of known active fragments and privileged GPCR motifs. In addition to addressing new receptors, this should allow the discovery of interesting multireceptor profiles of potential pharmacological interest.

was then used for the selection of building blocks. As a follow-up development, the same group described the RECAP (retrosynthetic combinatorial analysis procedure) method, a computational technique that fragments molecules based on chemical knowledge of 11 retrosynthetic steps [121]. Combined with statistical frequency analysis and applied to databases of biologically active molecules, RECAP allows the identification of building block fragments compatible with common synthesis methods and rich in biologically recognized elements. When applied to the WDI catalogue, the procedure identified privileged motifs for specific therapeutic classes and resulted in the creation of a WDI fragment knowledge base that stores drug motifs together with general and class-specific frequency information. The application to ligand design by recombining RECAP fragments was then later illustrated by Schneider et al. [122] with the evolutionary de-novo construction of compounds for several specific targets, but not yet for entire target classes, applying a dynamic programming algorithm and feature tree representations.

In an approach different from the molecular fragment or structural motif-based approaches, several groups have developed knowledge-based library design strategies using neural networks based on molecular descriptors. These approaches are based on the pioneering work of Sadowski and Kubinyi at BASF and Ajai et al. at Vertex, who introduced neural network methods for discriminating drug and nondrug molecules and for designing CNS (central nervous system) active molecules [123–125]. Since then, several groups have applied neural networks for the design of broad gene family focused libraries, especially targeting GPCRs and proteases [126, 127], and more recently, specifically to distinguish family subgroups like class A monoamine GPCRs and class A peptide-binding GPCRs [128].

5.6 Conclusions

Chemogenomics knowledge-based approaches draw a logical roadmap for the discovery of targets and lead compounds within the framework of previously well explored target families for which an abundance of both chemical and biological knowledge exists. To establish standardized molecular informatics platforms and real drug discovery ontologies at the genome level, which integrate the relevant chemical and biological knowledge, is thus being pursued within academic and industrial drug discovery organizations and informatics-based discovery technology companies [47, 129, 130]. Fundamental to the success of such knowledge-based strategies is the question of how broad or narrow the similarity of the binding sites is locally and globally. Emerging structural bioinformatics methods that aim to compare the binding sites of the entire protein-structure world will be very important to this goal and will potentially contribute to the design of selective compounds [131–134].

One immediately favorable drug-discovery strategy in the post genomic age might be to identify, by bioinformatics analysis, new targets that are directly related – having conserved molecular recognition – to previously successfully explored targets and to screen the ligand sets that are similar to the ligands of these targets. These ligands can be identified by cheminformatics methods and should be abundantly available within corporate compound archives. Given that, based on a recent analysis of the target portfolio of the pharmaceutical industry, only 120 proteins are the targets of today's successfully marketed drugs [6] and that there are in total about 400–500 protein targets, which have been pharmacologically explored up to now [6, 9], this appears to be a focused effort.

A now established in-silico technology, which we did not highlight in this chapter, is HTD (high-throughput docking) for protein structure-based in-silico screening. Recent developments within the chemogenomics area include the application of HTD to the evaluation of multiple libraries against multiple targets and the docking of single compounds against the comprehensive protein structure database [135, 136]. In the context of the GPCR family, a number of HTD studies based on homology models have been published recently [137, 138].

An important question is related to the more distant targets and to the members of novel target families, which have not been explored in the past and for which, in consequence, little or no knowledge exists. This situation is obviously more challenging, and drug discovery chemistry will probably have to first generate new structural classes to make the application of chemogenomics knowledge-based strategies possible also here; HTD, fragment-based de-novo design, and NMR/X-ray diffraction studies, as well as needle screening approaches, are potential catalysts to this aim [139–143].

Acknowledgments

Drs K. Azzaoui, P. Floersheim, D. Hoyer, A. Marzinzik, H. Mattes, P. Meier, M. Popov, H.-J. Roth, K. Seuwen, P. Schoeffter, U. Schopfer, T. Troxler, E. Vangrevelinghe, and J. Zimmermann (Novartis) and Profs A. Bairoch (Swiss Institute of Bioinformatics, Geneva), R. Stoop (ETH Zürich), and P. Willett (University of Sheffield, Sheffield) are acknowledged for various support and discussions. Parts of the work reported here were done within the context of the Swiss KTI research project "Information-based Approaches in Drug Design", which we also gratefully acknowledge. The text of this chapter is partly based on our earlier publications [21, 60, 83, 92, 140, 143].

References

1 The IUPHAR Committee on Receptor Nomenclature and Drug Classification. *The IUPHAR Compendium of Receptor Characterization and Classification*, 2nd ed., IUPHAR Media **2000**.

2 J. Lehmann et al., *Nature* **1996**, 384, Supp, 1–5.

3 P. R. Caron, M. D. Mullican, R. D. Mashal, K. P. Wilson, M. S. Su, M. A. Murcko, *Curr. Opin. Chem. Biol.* **2001**, 5, 464–470.

4 M. Murcko, P. Caron, *Drug Discov. Today* **2002**, 7, 583–584.

5 M. N. Namchuk, *Targets* **2002**, 1, 125–129.

6 A. L. Hopkins, C. R. Groom, *Nat. Rev. Drug Discov.* **2002**, 1, 727–730.
7 G. Wess, *Drug Discov. Today* **2002**, 4, 533–535.
8 K. H. Bleicher, *Curr. Med. Chem.* **2002**, 9, 2077–2084.
9 J. Drews, *Nat. Biotechnol.* **1996**, 14, 1516–1518.
10 S. V. Frye, *Chem. Biol.* **2001**, 6, R3–R7.
11 The Gene Ontology Consortium, *Nat. Genet.* **2000**, 25, 25–29.
12 http://www.chem.qmw.ac.uk/iubmb/enzyme (**2003**).
13 F. Horn, G. Vriend, F. E. Cohen, *Nucl. Acids Res.* **2001**, 29, 346–349.
14 N. Le Novère, J.-P. Changeux, *Nucl. Acids Res.* **2001**, 29, 294–295.
15 MDL Drug Data Report, MDL ISIS/HOST software, MDL Information Systems, Inc. http://www.mdl.com/ (**2003**).
16 Derwent World Drug Index, MDL ISIS/HOST software, Derwent Information Ltd. http://thomsonderwent.com/chem (**2003**).
17 Comprehensive Medicinal Chemistry (CMC-3D) Database, MDL ISIS/HOST software. CMC is an updated electronic version of *Comprehensive Medicinal Chemistry*, Pergamon Press **1990**, Vol. 6, http://www.mdl.com/ (**2003**).
18 Iddb: Investigational Drugs Database, Current Drugs Ltd., http://www.current-drugs.com/products/iddb (**2003**).
19 PharmaProjects, PJB Publications Ltd. http://www.pjbpubs.com/ (**2003**).
20 T. Nishioka, K. Sumi, J. Oda, in *Probing Bioactive Mechanisms*, P. S. Magee, D. R. Henry, J. H. Block (Eds.), American Chemical Society, New York **1989**, pp. 105–122.
21 A. Schuffenhauer, J. Zimmermann, R. Stoop, J. J. van der Vyver, S. Lecchini, E. Jacoby, *J. Chem. Inf. Comput. Sci.* **2002**, 42, 947–955.
22 S. Cheek, H. Zang, N. V. Grishin, *J. Mol. Biol.* **2002**, 320, 855–881.
23 D. R. Robinson, Y.-M. Wu, S. F. Lin, *Oncogene* **2000**, 19, 5548–5557.
24 G. Manning, D. B. Whyte, R. Martinez, T. Hunter, S. Sudarsanam, *Science* **2002**, 298, 1912–1934.
25 http://www.gene.ucl.ac.uk/nomenclature/genefamily.shtml (**2003**).
26 V. Anantharaman, L. Aravind, E. Koonin, *Curr. Opin. Chem. Biol.* **2003**, 7, 12–20.
27 P. C. Babbitt, *Curr. Opin. Chem. Biol.* **2003**, 7, 230–237.
28 R. Fredriksson, M. C. Lagerström, L. G. Lundin, H. B. Schiöth, *Mol. Pharmacol.* **2003**, 63, 1256–1272.
29 D. K. Vassilatis, J. G. Hohmann, H. Zeng, F. Li, J. E. Ranchalis, M. T. Mortrud, A. Brown, S. S. Rodriguez, J. R. Weller, A. C. Wright, J. E. Bergmann, G. A. Gaitanaris, *Proc. Natl. Acad. Sci. USA* **2003**, 100, 4903–4908.
30 N. J. et al. Mulder, *Nucl. Acids Res.* **2003**, 31, 315–318.
31 C. J. Sigrist, L. Cerutti, N. Hulo, A. Gattiker, L. Falquet, M. Pagni, A. Bairoch, P. Bucher, *Brief Bioinform.* **2002**, 3, 265–274.
32 A. Bateman, E. Birney, L. Cerruti, R. Durbin, L. Etwiller, S. R. Eddy, S. Griffiths-Jones, K. L. Howe, M. Marshall, E. L. Sonnhammer, *Nucl. Acids Res.* **2002**, 30, 276–280.
33 T. K. Attwood, M. D. Croning, A. Gaulton, *Protein Eng.* **2002**, 15, 7–12.
34 F. Servant, C. Bru, S. Carrère, E. Courcelle, J. Gouzy, D. Peyruc, D. Kahn, *Brief Bioinform.* **2002**, 3, 246–251.
35 I. Letunic, L. Goodstadt, N. J. Dickens, T. Doerks, J. Schultz, R. Mott, F. Ciccarelli, R. R. Copley, C. P. Ponting, P. Bork, *Nucl. Acids Res.* **2002**, 30, 242–244.
36 D. H. Haft, J. D. Selengut, O. White, *Nucl. Acids Res.* **2003**, 31, 371–373.
37 C. H. Wu, H. Huang, L. Arminski, J. Castro-Alvear, C. Chen, Z. Z. Hu, R. S. Ledley, K. C. Lewis, H. W. Mewes, B. C. Orcutt, B. E. Suzek, A. Tsugita, C. R. Vinayaka, L. S. Yeh, J. Zhang, W. C. Barker, *Nucl. Acids Res.* **2002**, 30, 35–37.
38 L. Lo Conte, S. E. Brenner, T. J. P. Hubbard, C. Chothia, A. Murzin, *Nucl. Acid Res.* **2002**, 30, 264–267.
39 F. M. G. Pearl, D. Lee, J. E. Bray, I. Sillitoe, A. E. Todd, A. P. Harrison, J. M. Thornton, C. A. Orengo, *Nucl. Acids Res.* **2000**, 28, 277–282.
40 H. Huang, W. C. Barker, C. Chen, C. H. Wu, *Nucl. Acids Res.* **2003**, 31, 390–392.

41 N. Lan, G. T. Montelione, M. Gerstein, *Curr. Opin. Chem. Biol.* **2003**, 7, 44–54.
42 R. Stevens, C. A. Goble, S. Bechhofer, *Brief Bioinform.* **2000**, 4, 398–414.
43 P. D. Karp, *Bioinformatics* **2000**, 16, 269–285.
44 T. Klabunde, G. Hessler, *ChemBioChem* **2002**, 3, 928–944.
45 http://www.aureus-pharma.com/ (**2003**).
46 http://www.jubilantbiosys.com/ (**2003**).
47 http://www.inpharmatica.com/ (**2003**).
48 Z. L. Ji, L. Y. Han, C. W. Yap, L. Z. Sun, X. Chen, Y. Z. Chen, *Drug Safety* **2003**, 26, 685–690.
49 L. Z. Sun, Z. L. Ji, X. Chen, J. F. Wang, Y. Z. Chen., *Bioinformatics* **2002**, 18, 1699–1700.
50 http://www.genego.com/ (**2003**).
51 Z. L. Ji, L. Z. Sun, X. Chen, J. C. Zheng, L. X. Yao, L. Y. Han, Z. W. Cao, J. F. Wang, W. K. Yeo, C. Z. Cai, Y. Z. Chen, *Drug Discov. Today* **2003**, 8, 526–529.
52 X. Chen, X. L. Ji, Y. Z. Chen, *Nucl. Acids Res.* **2002**, 30, 412–415.
53 C. M. Krejsa, D. Horvath, S. L. Rogalski, J. E. Penzotti, B. Mao, F. Barbosa, J. C. Migeon, *Curr. Opin. Drug Discov. Devel.* **2003**, 6, 470–480.
54 http://www.cerep.com/ (**2003**).
55 http://pdsp.cwru.edu/pdsp.htm (**2003**).
56 P. Joost, A. Methner, *Genome Biol.* **2002**, 3, 1–16.
57 A. Wise, K. Gearing, S. Rees, *Drug Discov. Today* **2002**, 7, 235–246.
58 T. Yokomizo, T. Izumi, K. Chang, Y. Takuwa, T. Shimizu, *Nature* **1997**, 387, 620–624.
59 N. Robas, M. O'Reilly, S. Katugampola, M. Fidock, *Curr. Opin. Pharmacol.* **2003**, 3, 121–126.
60 E. Jacoby, *Quant. Struct.–Act. Relat.* **2001**, 20, 115–123.
61 A. M. Van Rhee, K. A. Jacobson, *Drug Develop. Res.* **1996**, 37, 1–38.
62 C. D. Strader, I. S. Sigal, R. B. Register, M. R. Candelore, E. Rands, R. A. Dixon, *Proc. Natl. Acad. Sci. USA* **1987**, 84, 4384–4388.
63 R. S. Ames et al., *Nature* **1999**, 40, 282–286.
64 Y. Shimomura, M. Mori, T. Sugo et al., *Biochem. Biophys. Res. Commun.* **1999**, 261, 622–626.
65 B. Borowsky, N. Adham, K. A. Jones, R. Raddatz, R. Artymyshyn, K. L. Ogozalek, M. M. Durkin, P. P. Lakhlani, J. A. Bonini, S. Pathirana, N. Boyle, X. Pu, E. Kouranova, H. Lichtblau, F. Y. Ochoa, T. A. Branchek, C. Gerald, *Proc. Natl. Acad. Sci. USA* **2001**, 98, 8966–8971.
66 T. A. Brancheck, T. P. Blackburn, *Curr. Opin. Pharmacol.* **2003**, 3, 90–97.
67 P. Pfäffli, P. Neumann, R. Swoboda, P. Stutz, *PCT Int. Appl.* **1998**, WO 9854183.
68 P. Neumann, P. Pfäffli, M. P. Seiler, R. Swoboda, *PCT Int. Appl.* **1997**, WO9703054.
69 C. K. Surratt, P. S. Johnson, A. Moriwaki, B. K. Seidleck, C. J. Blaschak, J. B. Wang, G. R. Uhl, *J. Biol. Chem.* **1994**, 269, 20548–20553.
70 R. B. Nehring, W. Meyerhof, D. Richter, *DNA Cell Biol.* **1995**, 14, 939–944.
71 J. Strnad, J. R. Hadcock, *Biochem. Biophys. Res. Commun.* **1995**, 216, 913–921.
72 D. MacDonald, N. Murgolo, R. Zhang, J. P. Durkin, X. Yao, C. D. Strader, M. P. Graziano, *Mol. Pharmacol.* **2000**, 58, 217–225.
73 A. Marchese, S. R. George, L. F. Kolakowski, K. R. Lynch, B. F. O'Dowd, *Trends Pharmacol. Sci.* **1999**, 9, 370–375.
74 E. S. Huang, *Protein Sci.* **2003**, 12, 1360–1367.
75 J. G. Henikoff, E. A. Greene, S. Pietrokovski, S. Henikoff, *Nucl. Acids Res.* **2000**, 28, 228–230.
76 A. Gaulton, T. K. Attwood, *Curr. Opin. Pharmacol.* **2003**, 3, 114–120.
77 T. Mirzadegan, F. Diehl, B. Ebi et al., *J. Biol. Chem.* **2000**, 275, 25562–25571.
78 Y. C. Martin, J. L. Kofron, L. M. Traphagen, *J. Med. Chem.* **2002**, 45, 4350–4358.
79 D. Horvath, C. Jeandenans, *J. Chem. Inf. Comput. Sci.* **2003**, 43, 680–690.
80 D. Horvath, C. Jeandenans, *J. Chem. Inf. Comput. Sci.* **2003**, 43, 691–698.
81 H. Kubinyi, in *Perspectives in Drug Discovery and Design*, vols. 9–11 (*3D QSAR in Drug Design: Ligand/Protein*

Interactions and Molecular Similarity), H. Kubinyi, G. Folkers, Y. C. Martin (Eds.), Kluwer, Dordrecht **1998**, pp. 225–252.

82 J. Bajorath, *Nat. Rev. Drug Discov.* **2002**, 1, 882–894.

83 E. Jacoby, J. L. Fauchère, E. Raimbaud, S. Ollivier, A. Michel, M. Spedding, *Quant. Struct.–Act. Relat.* **1999**, 18, 561–572.

84 R. P. Sheridan, S. K. Kearsley, *Drug Discov. Today* **2002**, 4, 903–911.

85 D. Wilton, P. Willett, K. Lawson, G. Mullier, *J. Chem. Inf. Comput. Sci.* **2003**, 43, 469–474.

86 http://www.tripos.com/ (**2003**).

87 http://www.accelrys.com/ (**2003**).

88 http://www.scitegic.com/ (**2003**).

89 G. Schneider, W. Neidhart, T. Giller, G. Schmid, *Angew. Chem. Int. Ed. Engl.* **1999**, 38, 2894–2896.

90 M. Rarey, M. Stahl, *J. Comput. Aided Mol. Des.* **2001**, 15, 497–520.

91 V. Gillet, P. Willett, J. Bradshaw, *J. Chem. Inf. Comput. Sci.* **2003**, 43, 338–345.

92 A. Schuffenhauer, P. Floersheim, P. Acklin, E. Jacoby, *J. Chem. Inf. Comput. Sci.* **2003**, 43, 391–405.

93 The ISIS public keys are also known as MACCS keys. ISIS/Base and MACCS are both products of MDL Information Systems, Inc., San Leandro, CA, http://www.mdl.com/ (**2003**).

94 L. H. Hall, L. B. Kier, *J. Chem. Inf. Comput. Sci.* **1995**, 35, 1039–1045.

95 N. Salim, J. Holliday, P. Willett, *J. Chem. Inf. Comput. Sci.* **2003**, 43, 435–442.

96 L. Xue, J. W. Godden, F. L. Stahura, J. Bajorath, *J. Chem. Inf. Comput. Sci.* **2003**, 43, 1218–1225.

97 A. K. Ghose, V. N. Viswanadhan, J. J. Wendoloski, *J. Comb. Chem.* **1999**, 1, 55–68.

98 M. J. McGregor, S. M. Muskal, *J. Chem. Inf. Comput. Sci.* **1999**, 39, 569–574.

99 M. J. McGregor, S. M. Muskal, *J. Chem. Inf. Comput. Sci.* **2000**, 40, 117–125.

100 D. Xie, A. Tropsha, T. Schlick, *J. Chem. Inf. Comput. Sci.* **2000**, 40, 167–177.

101 M. G. Grigorov, H. Schlichtherle-Cerny, M. Affolter, S. Kochhar, *J. Chem. Inf. Comput. Sci.* **2003**, 43, 1248–1258.

102 R. S. Pearlman, K. M. Smith, *J. Chem. Inf. Comput. Sci.* **1999**, 39, 28–35.

103 B. Pirard, S. D. Pickett, *J. Chem. Inf. Comput. Sci.* **2000**, 40, 1431–1440.

104 A. A. Patchett, R. P. Nargund, in *Annual Reports in Medicinal Chemistry*, Vol. 35, G. L. Trainor (Ed.), Academic Press, San Diego **2000**, pp. 289–298.

105 G. Mueller, *Drug Discov. Today* **2003**, 8, 681–691.

106 J. S. Mason, I. Morize, P. R. Menard, D. L. Cheney, C. Hulme, R. F. Labaudiniere, *J. Med. Chem.* **1999**, 42, 3251–3264.

107 R. P. Sheridan, *J. Chem. Inf. Comput. Sci.* **2003**, 43, 1037–1050.

108 M. F. M. Engels, *Ernst Schering Res. Found. Workshop.* **2003**, 42, 87–101.

109 J. W. Raymond, P. Willett, *J. Compu.-Aided Mol. Design.* **2002**, 16, 521–533.

110 C. A. Nicolaou, S. Y. Tamura, B. P. Kelly, S. I. Bassett, R. F. Nutt, *J. Chem. Inf. Comp. Sci.* **2002**, 42, 1069–1079.

111 G. Roberts, G. J. Myatt, W. P. Johnson, K. P. Cross, P. E. Blower Jr., *J. Chem. Inf. Comput. Sci.* **2000**, 40, 1302–1314.

112 G. W. Bemis, M. A. Murcko, *J. Med. Chem.* **1996**, 39, 2887–2893.

113 J. Fejzo, C. A. Lepre, J. W. Peng, G. W. Bemis, M. A. Murcko, J. M. Moore, *Chem. Biol.* **1999**, 6, 755–769.

114 M. Bohl, J. Dunbar, E. M. Gifford, T. Heritage, D. J. Wild, P. Willett, D. J. Wilton, *Quant. Struct.-Act. Relat.* **2002**, 21, 590–597.

115 X. Q. Lewell, A. C. Jones, C. L. Bruce, G. Harper, M. M. Jones, I. M. McLay, J. Bradshaw, *J. Med. Chem.* **2003**, 46, 3257–3274.

116 P. Ertl, *J. Chem. Inf. Comput. Sci.* **2003**, 43, 374–380.

117 http://www.cresset-bmd.com/ (**2003**).

118 R. Crossley, *Modern Drug Discov.* **2002**, 12, 18–22.

119 R. Crossley, M. Lipkin, M. J. Slater, P. De Zoysa, *Curr. Drug Discov.* **2003**, 4, 39–44.

120 X. Q. Lewell, R. Smith, *J. Mol. Graph. Model.* **1997**, 15, 43–48.

121 X. Q. Lewell, D. B. Judd, S. P. Watson, M. M. Hann, *J. Chem. Inf. Comput. Sci.* **1998**, 38, 511–522.

122 G. Schneider, M. L. Lee, M. Stahl, P. Schneider, *J. Comput. Aided Mol. Des.* **2000**, 14, 487–494.

123 J. Sadowski, H. Kubinyi, *J. Med. Chem.* **1998**, 41, 3325–3329.

124 J. Sadowski, *Curr. Opin. Chem. Biol.* **2000**, 4, 280–282.

125 Ajay, G. W. Bemis, M. A. Murcko, *J. Med. Chem.* **1999**, 42, 4942–4951.

126 K. V. Balakin, S. E. Tkachenko, S. A. Lang, I. Okun, A. A. Ivashchenko, N. P. Savchuk, *J. Chem. Inf. Comput. Sci.* **2002**, 42, 1332–1342.

127 S. A. Lang, A. V. Kozyukov, K. V. Balakin, A. V. Skorenko, A. A. Ivashchenko, N. P. Savchuk, *J. Comput. Aided Mol. Des.* **2002**, 16, 803–807.

128 D. T. Manallack, W. R. Pitt, E. Gancia, J. G. Montana, D. J. Livingstone, M. G. Ford, D. C. Whitley, *J. Chem. Inf. Comput. Sci.* **2002**, 42, 1256–1262.

129 R. Glen, *Chem. Comm.* **2002**, 23, 2745–2747.

130 R. L. Strausberg, S. Schreiber, *Science* **2003**, 300, 294–295.

131 D. Michalovich, J. Overington, R. Fagan, *Curr. Opin. Pharmacol.* **2002**, 2, 574–580.

132 J. Gunther, A. Bergner, M. Hendlich, G. Klebe, *J. Mol. Biol.* **2003**, 326, 621–636.

133 S. Schmitt, D. Kuhn, G. Klebe, *J. Mol. Biol.* **2002**, 323, 387–406.

134 http://www.eidogen.com/ (**2003**).

135 M. L. Lamb, K. W. Burdick, S. Toba, M. M. Young, A. G. Skillman, X. Zou, J. R. Arnold, I. D. Kuntz, *Proteins* **2001**, 42, 296–318.

136 X. Chen, C. Y. Ung, Y. Chen, *Nat. Prod. Rep.* **2003**, 20, 432–444.

137 C. Bissantz, P. Bernard, M. Hibert, D. Rognan, *Proteins* **2003**, 50, 5–25.

138 C. N. Cavasotto, A. J. Orry, R. A. Abagyan, *Proteins* **2003**, 51, 423–433.

139 E. D. Zanders, D. S. Bailey, P. M. Dean, *Drug Discov. Today* **2002**, 7, 711–718.

140 E. Jacoby, J. Davies, M. J. J. Blommers, *Curr. Topics Med. Chem.* **2002**, 2, 1279–1291.

141 H. J. Böhm, M. Boehringer, D. Bur, H. Gmünder, W. Huber, W. Klaus, D. Kostrewa, H. Kühne, T. Lübbers, N. Meunier-Keller, F. Müller, *J. Med. Chem.* **2000**, 43, 2664–2674.

142 C. W. Murray, M. L. Verdonk, *J. Comput. Aided Mol. Des.* **2002**, 16, 741–753.

143 E. Jacoby, A. Schuffenhauer, P. Floersheim, *Drug News Perspect.* **2003**, 16, 93–102.

6
Chemical Kinomics

Bert M. Klebl, Henrik Daub, and György Kéri

6.1
Introduction

This chapter on 'chemical kinomics' does not intend to add another 'omics' discipline to a constantly growing family. Rather, chemical kinomics is a discipline of chemical genomics, reflecting the appreciation of chemical genomics as a truly emerging direction in the world of drug discovery. Chemical genomics evolved because there is an urgent need for novel druggable targets and an even more urgent need for generating novel chemical entities against those targets, with the ultimate goal of elaborating new treatments for all kinds of diseases. Based on the wealth of information coming from genomics, chemical genomics tries to exploit the results from genomics and translate it into efficient drug development. The long-term goal is to generate a specific small-molecule ligand for every protein encoded by the genome, whether through combinatorial or organic chemistry or through natural-product screening. Such a ligand must not only bind to the protein of interest, but also modulate its biological activity, which usually means to inhibit its biological function. Only in a few instances do such ligands act through activating the target. Rationally, these genome-wide ligand–protein interactions represent an ideal starting point for drug development for any kind of disease. Given good knowledge of the physicochemical and pharmacological properties, these ligands are ideal tools for investigating the biological function of unknown proteins. Such ligands can validate the druggability of a protein of interest.

The human genome includes approximately 32 000 genes [1, 2], which are subsequently translated into the proteome. The human proteome is estimated to include from 100 000 to several million different protein molecules [3–5]. The proteome is much larger than the genome, because single genes can be translated into different proteins, and distinct proteins can be post-translationally modified through phosphorylation, glycosylation, ubiquitination, etc. Ideally, the chemical biology space should cover ≥ 100 000 specific ligands, one specific to each individual protein. But it is rather unrealistic to assume that ligands can become specific enough to discriminate between splice variants or even post-translationally modified

Chemogenomics in Drug Discovery: A Medicinal Chemistry Perspective.
Edited by Hugo Kubinyi and Gerhard Müller

ISBN: 3-527-30987-X

proteins. Therefore, 32 000 specific small-molecule ligands – each one designed for the products of one particular gene – would serve the ultimate goal of understanding the function of every gene by modifying the protein functions by using these ligands as tools for testing in biological models. Synthesizing 32 000 specific, selective ligands seems a gigantic task. Reaching the shear number is not so much a problem [6] as solving the specificity and selectivity issues of the compounds. Undoubtedly, the pharmaceutical and biotechnology industries have the most immediate interest in generating ligands that are as specific and selective as possible. Typically, such ligands are then used as starting points for chemical optimization in drug discovery programs. Drug development teaches us that the generation of a specific ligand and its translation into a lead compound is a time-consuming process. Despite this timing issue in generating selective and specific ligands, the central goal of chemical genomics is to accelerate the process of drug discovery [7]. Chemical genomics and chemical biology cannot be understood simply as the workhorse for generating a chemical toolbox. As was mentioned above, the compounds should also serve as ideal starting points for medicinal chemistry projects. Since the generation of such highly specific ligands through directed organic chemistry is a tremendously laborious task, it is wise to focus on achievable goals. For example, a target gene family approach is such an achievable goal, which allows keeping the focus not only on biology, but also on the underlying chemistry. Proteases, G-protein coupled receptors (GPCRs), nuclear receptors, and protein kinases, for example, represent such target families [8]. The common feature of the protein kinase family is their catalytic center, the kinase domain. All kinases accept adenosine triphosphate (ATP) as a cosubstrate so as to transfer the γ phosphate of ATP to a protein, peptide, or lipid substrate [9]. The ATP-binding site of kinases has been well described through numerous crystallization and cocrystallization efforts. In the direct vicinity of the ATP-binding site, kinases also bear special hydrophobic front and back pockets. These pockets are ideally suited to bind specifically designed chemical ligands [10]. The ligands exploit the hydrophobic pockets and the hinge region of the ATP-binding site to bind mostly in an ATP-competitive manner to the kinase domains [11]. Experience in drug discovery – especially since the past 10 years – teaches us that specific ATP-site specific protein kinase inhibitors can be generated, despite the high amino acid sequence homologies of kinase domains within the protein kinase family [12]. These ATP-site competitors, like Gefinitib (Iressa™) and Imatinib (Gleevec™), have made it into the clinic and onto the market [13, 14]. An urgent requirement for generating such a target-to-product success story is a strong integration between biology and chemistry, as well as their close interaction. This integrated approach to specialized kinase biology and kinase chemistry is referred to here as chemical kinomics. Researchers in the field of chemical kinomics intend to generate as many as possible specific kinase inhibitors and to implement them as tools in signal-transduction research and as leads for drug discovery. Subsequently, these leads need to be converted into marketable drugs, thereby providing help, especially in curing life threatening diseases.

6.2
Chemical Biology: The Hope

The chemical genomics and biology approaches have been implemented to accelerate drug discovery. In the pharmaceutical industry, we went through times of pharmacological profiling, which were then replaced by the paradigm of structure-based drug design. In the 1990s the advents of high-throughput screening and combinatorial chemistry peaked, leading to a high degree of automation in screening and chemistry. But still the number of new chemical entities did not increase significantly [15]. Just around 2000, the 'omics' popularity began in industrial research, providing companies with an endless list of targets, but no real solutions. There is a significant amount of criticism that all these 'omics' technologies have not delivered the expected results – new drugs [16]. Undoubtedly, it is still too early to be overcritical, especially with the more recent developments. The era of chemical biology and genomics is just now trying to apply these technologies in parallel, to shorten the duration of drug development and to gain from the resulting synergies. Clearly, this demands a seamless integration of chemistry, biology, and pharmacology. Traditionally, research and development in the pharmaceutical industry has been oriented along the value chain (Figure 6.1 a). A project starts with identifying and validating a potential target through biological means, which usually takes ~2 years. Within this period, an assay for compound screening is developed and adapted to high-throughput screening. Screening efforts typically result in a number of active compounds, which, after inspection and evaluation, turn into hits. Now medicinal chemistry comes into play and eventually turns these hits into leads. Further chemical optimization of leads from one or more classes is done with the intention of increasing the activity to the target. In modern drug development, lead compounds are not only improved in terms of their activity against the primary target, but also in terms of their pharmacological and physicochemical properties [17]. Ultimately, a compound with a balanced profile is nominated as a candidate and enters preclinical and clinical development. The process from hit to candidate usually takes approximately 4 to 5 years. In a typical chemical biology approach, all these activities are brought together in a parallel fashion to accelerate drug development (Figure 6.1 b) [7, 18].

6.3
Chemical Kinomics: A Target Gene Family Approach in Chemical Biology

This section focuses on the topic of this review – on the chemical genomics approach within the target family of protein kinases. First, we give a short introduction to the successful drug development activities in the field of protein kinases. Then, we describe some standard forward and reverse chemical genomics technologies, again specifically designed and adjusted for use with kinases.

Chemical genetics is a subdivision of the larger field of chemical genomics. Orthogonal chemical genetics turned into an important novel discipline in the

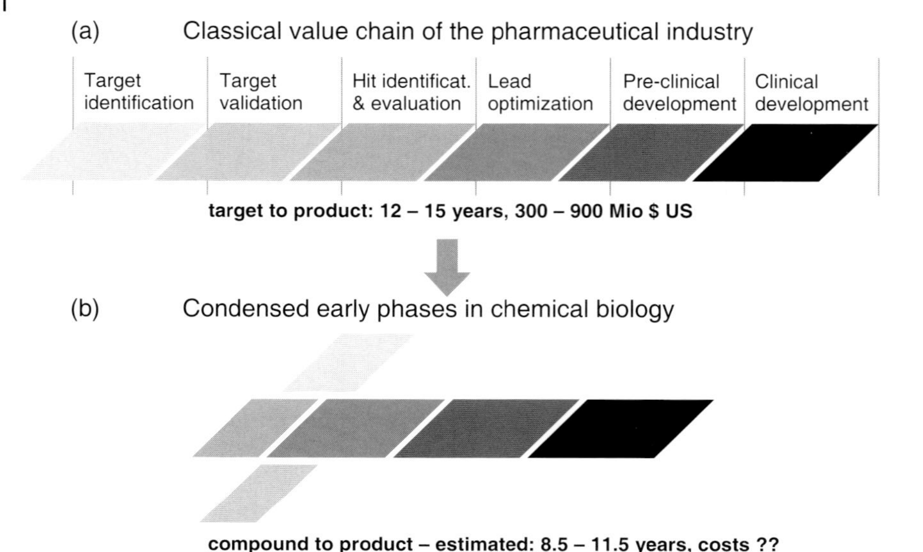

Figure 6.1 The classical pharmaceutical value chain extrapolated into the era of chemical biology.
(a) Until recently, drug discovery was rather linear in terms of both activities and decision making. In an exploratory phase, a project team identifies a target and subsequently tries to validate this target. This process usually takes about 2 years and also includes the development of an assay that can be used for high-throughput screening. The period of high-throughput screening, hit verification, hit optimization, lead nomination, and lead optimization takes on average 4.5 years until the nomination of a candidate. Preclinical and clinical development represent precisely defined steps in drug development and cannot be condensed much. A drug discovery program takes between 12 and 15 years from target to product.
(b) In the era of chemical kinomics (biology), a drug discovery program starts with screening a profiled and biased library for compounds active in a biological model for a disease of interest. The actives are then used for target identification and validation in a typical forward chemical genomics approach.
In parallel, hits are nominated from the group of actives, and chemical optimization is begun, to generate optimized leads. Ideally, the biased library consists of lead-like compounds and allows an efficient, short optimization program. We can anticipate that this parallel approach in chemical biology will lead to time reduction by ~ 3.5 years as compared to the classical linear approach.

area of kinase research and is a subdivision of chemical kinomics. Three related approaches, which now have great impact, are described. Another important subdivision of chemical kinomics is a kinase-directed, specialized chemical proteomics technology, the KinaTor^{TM} (see Section 6.3.4). In summary, the discipline of chemical kinomics is composed of chemical genomics, chemical genetics, screening, and chemical proteomics technologies. In the following four sections, we describe technologies as well as their applications.

6.3.1 Protein Kinase Inhibitor History

Protein kinases catalyze the phosphorylation of proteins, peptides, lipids, and sugars by using ATP, the energy store of the cell, as a phosphate donor:

$$\text{Protein kinase-catalyzed reactions:}$$
$$\text{ATP} + \text{substrate} \xrightarrow{\text{kinase}} \text{ADP} + \text{substrate-P}$$

Scheme 6.1

Until 1995, the family of protein kinases was regarded as undruggable. This was due to their high degree of sequence homology, identical catalytic mechanisms, highly identical protein folding topologies, and their common cosubstrate ATP. This attitude drastically changed when Novartis started to work on Bcr-Abl and its inhibitor STI571 (CGP57148B, Imatinib, now marketed as Gleevec™) [10, 13] and SmithKline initiated work on p38 inhibitors like SB203580 [19]. Bcr-Abl is a 'fused' mutant of the human tyrosine kinase c-Abl. Unlike c-Abl, the mutant Bcr-Abl is constitutively active and thus causes cells of a myeloid origin to proliferate, leading to chronic myeloid leukemia [13]. p38 is a member of the MAP kinase family and has been reported to be involved in a number of physiological processes such as cell survival, cell cycle, proliferation, apoptosis, cytoskeletal changes, and gene expression. Therefore, kinases play a central role in signal transduction in every kind of cell [20]. As a consequence, kinases are reported to be involved in a plethora of diseases. Basically, there is no therapeutic indication for which protein kinases can be excluded as potential targets, at least on the biological side. Genes for 518 kinases were originally discovered in the human genome [9] but, using our in-house bioinformatics tools, we have identified and cloned 534 protein kinases. In the past 8 to 9 years, the pharmaceutical industry has experienced an enormous paradigm shift: kinases are now regarded as druggable. This is mostly due to the launch of Gleevec™ in 2001 for chronic myeloid leukemia [13] and Iressa™, another small-molecule drug, acting on the EGF receptor tyrosine kinase [14], which was launched for non small cell lung cancer in 2002. In addition to small-molecule design, some of the tyrosine kinase receptors, which are accessible from the extracellular side, have also been used as targets for the development of biologicals such as trastuzumab. Trastuzumab is a humanized monoclonal antibody against the receptor tyrosine kinase ErbB2. Trastuzumab is effectively used in certain types of breast cancers and is marketed as Herceptin™ [21].

As of August 2003, at least 48 small-molecule kinase inhibitors are in various clinical trials (published in Pharmaprojects: http://www.pjbpubs.co.uk/pharmaprojects.htm). The R&D spending for small-molecule development is currently highest for protein kinase based research in the pharmaceutical and biotech industry [8]. Most of these small-molecule kinase inhibitors are designed as ATP-site competitors. Today, it is believed that specificity and selectivity can be achieved for small molecules even within such narrow spaces as highly homologous kinase domains. When protein kinases are addressed as molecular targets in a drug discovery program, small molecules are the preferred option, since most, if not all, kinase activities are located inside cells.

6.3.2 Chemical Kinomics: An Amenable Approach

Chemical kinomics implies a focused target gene family approach, dealing with protein kinases in the field of chemical biology. As mentioned above, the seamless integration of chemistry and biology is an important prerequisite for accelerating drug discovery. Focusing on a target class like protein kinases bears many advantages. When working within a target gene family, we face many commonalities with respect to biological methods such as purifying proteins, assaying activities, etc. The same is true in chemistry, especially with a focus on ATP-site competitors. Chemistry can use a limited number of core structures and differently decorate them to gain selectivity and specificity. In the following sections, we review several technologies that represent typical chemical biology/genomics tools. These tools have been developed especially for protein kinases, too.

6.3.2.1 Examples of Traditional Chemical Genomics Using Kinase Inhibitors

Among others, the biotechnology company Rosetta was a pioneer in applying target-specific ligands on a genomic scale. A breakthrough for these technologies was achieved when Hughes and colleagues [22] published their compendium of expression profiles. They published a strategy for exploiting gene expression profiles of known and unknown genes, to discern the function of the novel genes. They demonstrated that the role of novel genes can be functionally discovered by determining the gene expression profile of a knockout mutant of the gene of interest. The profile of the knockout strain or cell line is compared to the experimentally determined gene clusters of transcription profiles of knockouts or perturbations of known genes. Typically, the profile for a novel gene matches one of these clusters and therefore suggests and anticipates a role for the gene of interest. This gene profiling technique can be equally well applied to the analysis of perturbations caused by treatment with chemical ligands.

An example [22] illustrates how gene expression profiling can be used to assign a target pathway to a compound of interest. In an effort to identify the unknown target for the topical anesthetic dyclonine, a transcriptional response most similar to the knockout of *ERG2* in yeast was obtained after treating a wild-type yeast strain with dyclonine. Indeed, subsequent functional genomics experiments suggested that yeast Erg2p was a molecular target for dyclonine. Erg2p is the yeast homolog of the human sigma receptor, which is a neurosteroid-interacting protein that positively regulates potassium conductance. Thus, a potential mechanism for dyclonine is that it binds the sigma receptor and inhibits nerve conduction by reducing the potassium current [22]. Hence, gene expression profiling and transcriptome analysis represents a way of defining the mode of action of novel drugs and compounds.

A similar but less comprehensive approach was published by Gray et al. [23]. They, for the first time, implemented global gene expression profiling for kinase inhibitors. They determined the expression profiles for 2 different cyclin-dependent kinase (CDK) inhibitors, flavopiridol and compound 52, a purvalanol B analog. Since these inhibitors inhibited human and *Saccharomyces cerevisiae* CDKs equally well, yeast was used as a living 'computer' for understanding gene function. Importantly, they showed that the inhibitors indeed affect cell cycle genes as well as phosphate metabolism genes, which are known to be under the control of a yeast-specific CDK, Pho85p. As expected, the transcriptional profiles for the two structurally different CDK active-site inhibitors overlapped by only ~50%, suggesting that the two compounds may affect pathways involving CDKs to different degrees. The compound-induced profiles were also compared to a transcriptional analysis of a genetic disruption in one of the yeast CDKs. Again, significant overlap, but also differences, were observed between the genetic and chemical perturbations. Thus, this approach demonstrated a useful way of evaluating the selectivity of drug candidates in identifying proteins whose inhibition might specifically potentiate the effects of a primary drug. The lack of correspondence in the changes of mRNA transcript levels resulting from chemical and genetic inactivation underscores the intrinsic differences in these methods in modulating biological function [23]. On a genomic level these findings underscore the fact that chemical inactivation of the target affects only the CDK activity, whereas genetic inactivation affects the entire protein. There is an important and fundamental difference between applying a kinase inhibitor to a biological system of interest and using a knockout of a kinase gene in the same system. This finding has an important effect on classical target validation strategies. Typically, a gene product is defined as a valid drug target if the knockout of a gene of interest suppresses a relevant phenotype. In light of the above findings, it should be emphasized again that it is much more valuable to achieve effects through the use of a modulatory ligand. Knockouts are a nice add-on technology, but must not be used as a decision point in chemistry-driven drug discovery projects.

Another approach was used for the immunosuppressant rapamycin [24, 25]. Rapamycin binds to FKBP12, and this complex inactivates the kinase mTor as well as its yeast homologs Tor1p and Tor2p. mTor is a member of the phosphatidylinositol kinase-related kinases. Rapamycin treatment and transcriptional profiling in yeast were used to determine the role of Tor proteins in glucose activation and nitrogen discrimination pathways and in the pathways that respond to the diauxic shift [24]. Using epistasis experiments in the presence of chemical ligands ('chemical epistasis') and global expression analyses resulted in the transcriptional program induced by rapamycin being partitioned among five effectors of the Tor proteins. Clearly, a striking similarity between shifting yeast to low-quality carbon or nitrogen sources and treatment with rapamycin was demonstrated. Depending on which nutrient is limited in quality, the Tor proteins can modulate a given pathway differentially. Integrating the partition analysis of the transcriptional program of rapamycin with the biochemical data, a novel architecture of Tor protein signaling and of the nutrient-response network was proposed [25]. This work is a masterpiece in chemical biology/genomics, exploiting a compound to learn and understand more about biology and cellular physiology.

In the meantime, pioneering transcriptional profiling work has been extended from yeast to eukaryotes like *Caenorhabditis elegans* [26], *Drosophila melanogaster* [27], and higher eukaryotes like *Danio rerio* [28], and also to mammalian cells [29]. All the investigations show that it is possible to determine the effects of chemical ligands on the transcriptional profile. Practical applications of these approaches include the development of transcriptional profiles and gene-expression compendiums that can be used to predict drug toxicology and to detect drug-related 'off-target' effects. There is optimism that the development of cellular gene-expression fingerprints associated with compounds having known toxicological properties or mechanisms might lead to the ability to predict in-vivo toxicity of new compounds in advance of extensive animal toxicological testing [30–32]. This will be equally applicable to new kinase inhibitors.

6.3.2.2 **Forward Chemical Genomics Using a Kinase-biased Compound Library**

In this section, we refer to a broadly applicable method, which has been optimized for the use on kinases.

Our in-house library has been designed by exploiting existing public and proprietary knowledge in current small-molecule kinase chemistry. It consists of ~8500 ATP-site competitors, composed of more than 85 different chemical scaffolds, most of which have been described in the scientific literature and in patents, including patent applications. These kinase-biased inhibitors cover a broad spectrum of inhibition. The kinase inhibitors can be used in a typical forward chemical genomics approach [30]. An immediate requirement for a forward chemical genomics approach is a cell-based assay, ideally a phenotypical assay. These assays typically resemble a disease of interest on the cellular level. The assay is adapted to allow compound screening and screening is then performed without knowing the molecular target [33]. Compounds that elicit biological activity can be pharmacologically profiled or can be used immediately for target identification. Target identification is a straightforward approach especially within the family of kinases, since we have implemented a very efficient technology for identifying kinase targets, called KinaTor™. KinaTor™ is described in detail in Section 6.3.4.

6.3.2.3 **Chemical Validation**

This section also describes a kinase-specialized technology that could in theory have a much broader applicability. The technology was developed as a tool for the validation of druggable kinase targets in various biological models of relevant diseases. It resembles a reverse chemical genomics approach [30]. 'Chemical validation' is a tool for proving the druggability and validity of a presumed kinase target that has been proposed as a target for a given disease through other means.

A fraction of our in-house ~8500 ATP-site competitor compounds have been profiled in a number of cellular assays for toxicity, solubility, and permeability. Nontoxic, soluble, permeable compounds have been gathered into a subset of the main library, which we call the 'validation library'. The validation library currently consists of ~570 compounds. For chemical validation a potential kinase target is expressed and purified, a cell-free activity assay is developed, and in parallel a cellular disease model has to be established. The 570 compounds of the validation library were first tested in a biochemical assay. Ranking the IC_{50} values of the most potent inhibitors generated a chemical fingerprint for the kinase of interest. Inhibitors with IC_{50}s in the single-digit micromolar range or lower are suitable for chemical validation and subsequent testing in the cellular disease model. Typically, we need more than 5 different inhibitors, covering at least 3 orders of magnitude of inhibition with regard to their biochemically determined IC_{50}s. The inhibitors, which have defined the chemical fingerprint in the cell-free activity assay, are now used in the cellular assay representing the disease model. If the ranking of the IC_{50}s can be qualitatively reproduced in the cellular assay, the target is validated through this

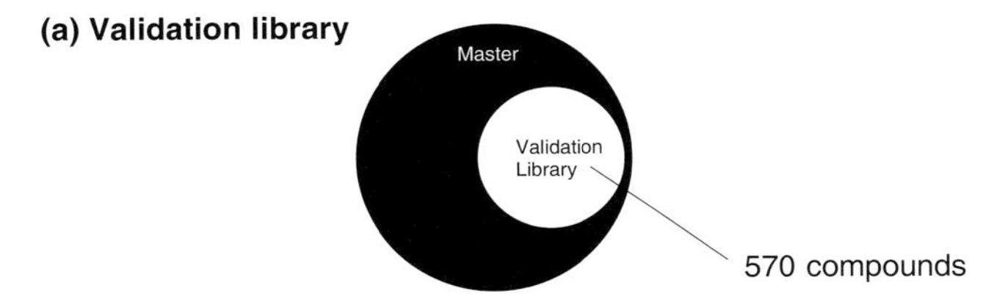

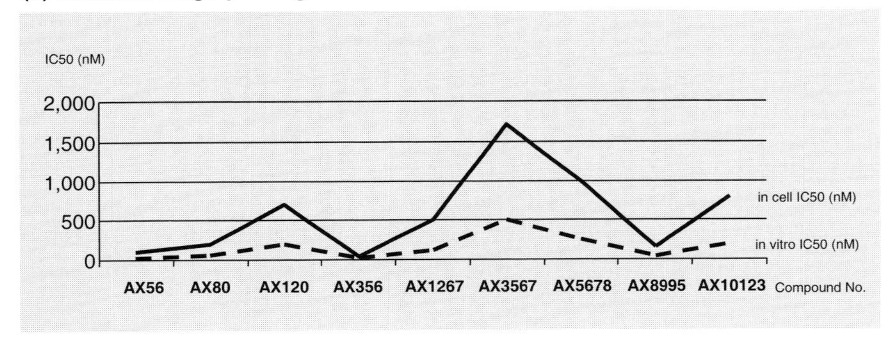

Figure 6.2 Chemical validation.
(a) Axxima's master library consists of 8500 ATP-site competitive small-molecule kinase inhibitors, and ~570 of these compounds have been sufficiently profiled in terms of toxicity (in cellular assays), permeability, and solubility. These are assembled in the so-called validation library. Compounds in the validation library have a purity of > 95% and are available in at least one-gram quantities.
(b) Compounds from the validation library are used for chemical fingerprinting. First, these compounds are screened against a kinase of interest at a concentration of 10 μM in an in-vitro kinase activity assay. Then the potent inhibitors are selected and promoted to the stage of IC_{50} determination. Inhibitors with IC_{50}s in the single-digit micromolar range or below can be used for chemical validation in a subsequent cellular assay. The graph schematically represents the IC_{50} values (y axis) for a number of potent compounds, indicated on the x axis. The solid line indicates the qualitatively matching values in a corresponding cellular assay, preferably a disease model. The in-vitro data clearly parallel the cellular data in this graph, demonstrating that the target kinase can be validated via chemical validation.

chemical fingerprint (Figure 6.2). We should mention that the absolute IC_{50} values do not have to match quantitatively. Rather, the chemical fingerprint needs to be reproduced qualitatively (Figure 6.2 b). We have used the chemical validation approach successfully for a number of kinase targets, which have further progressed significantly in our in-house pipeline. Chemical validation might be used also for other target genes.

6.3.3 Orthogonal Chemical Genetics

The benefits of combining orthogonal chemistry and genetics in drug discovery have been revealed through several studies in diverse areas of cell biology [34]. Each of the experiments has used the same fundamental approach of first modifying a small-molecule ligand (to make it 'orthogonal'), followed by changing the protein structure in a complementary way to accept the orthogonal ligand. This 'lock and key' approach takes place in a stepwise fashion: first, a small molecule (the 'key') that binds to the protein of interest is modified in a manner designed to eliminate its ability to bind to its native target. This modified compound is said to be orthogonal in normal cells, because it can no longer interact with its natural protein target or with any other target in the cell. Second, the individual protein (the 'lock') of interest is engineered to accept the orthogonal compound. Importantly, the mutation to the protein must affect only the binding of the orthogonal compound and not otherwise modify the protein's function [35]. Pioneering work was done by Hwang and Miller in 1987 for GTPases [36]. Kevan Shokat and coworkers implemented orthogonal chemical genetics successfully for protein kinases by introducing analog-sensitive kinase alleles (ASKA).

6.3.3.1 ASKAs: Analog-sensitive Kinase Alleles

The ASKA technology is based upon the discovery of analog-sensitive kinase alleles (the locks) and corresponding small-molecule analog compounds (the keys) that specifically modulate ASKA activity [37]. Unlike other orthogonal chemical genetic approaches, the ASKA system can be applied even to members of diverse protein kinase subfamilies [38]. It is highly modular and can be efficiently applied across the kinase superfamily. The key feature of the ASKA approach is the creation of a subtle but unique structural distinction between the catalytic domain of one kinase and all other kinases in the genome. This distinction is achieved by introducing a mutation in the ATP-binding pocket of the kinase. All kinases have a bulky amino acid residue at a conserved position in the ATP-binding pocket, the 'gatekeeper'. Mutation of the gatekeeper to an alanine or glycine allows access to the deep hydrophobic pocket. Importantly, this mutation is quiet in terms of kinase function and activity. This method is widely applicable and does not require 3D structural information. The gatekeeper is readily apparent from simple amino acid sequence alignments [38]. Next, a small modular set of functional ATP analogs and potent inhibitors are synthesized that fit only the engineered kinases Thus. the ASKA system works efficiently across the kinase superfamily to analyze and validate each kinase target. The ASKA technology has multiple applications to the drug discovery process, including [35].

- Identification of direct kinase substrates, facilitating the understanding of target function in the context of disease-relevant cellular pathways and facilitating assay development.
- Cell-based and in-vivo model systems for pharmacologically relevant target validation.
- Chemical genomic profiling – using microarrays to determine the effects of highly specific drug-mediated kinase inhibition of gene expression.
- Direct coupling of chemical genomic profiling to high-throughput and high-content drug screens.
- Preclinical in-vivo models with reference compounds to establish therapeutic indices and provide a source of biomarkers.

Shokat and colleagues demonstrated the versatility of the ASKA system by experimentally identifying novel kinase substrates through the use of functional ATP analogs. Implementing this technology, they recently determined that hnRNP-K is a direct substrate of Jnk-2 [39], Dok-1 for v-Src [40], and 181 different substrates for the *S. cerevisiae* cyclin-dependent kinase (Cdk1p) [41]. These were demonstrated to be direct substrates for Cdk1p – a kinase for which few substrates have been described despite its importance in cell-cycle control. Only 12 proteins from these 181 Cdk1p substrates had been previously described as Cdk1p substrates. In this study, 695 yeast ORFs were analyzed, allowing the total number of Cdk1p substrates in yeast to be estimated as greater than 500. Studies like the Cdk1p substrate identification approach may lead to the discovery of unforeseen regulatory connections in cell-cycle control. In a recent study, Eblen et al. [42] described direct substrates of ERK2, which also were identified through the ASKA method. After initial labeling with a radioactive cyclopentyl ATP analog and identification of the spots in 2D gels, the procedure was scaled up to detect the immunoprecipitated substrates by mass spectrometry. The ubiquitin ligase EDD and the nucleoporin Tpr were identified as two novel substrates in addition to the known substrate Rsk1. Elucidation of kinase pathways by direct substrate identification is important for pathway placement and functional annotation of orphan kinases [35] and also for assay development and screening.

ASKAs not only accept functional ATP analogs as cosubstrates, they are also inhibited by orthogonal kinase inhibitors, which are highly potent and cell permeable and show excellent bioavailability and low toxicity in mice. Thus, a specific kinase can be validated as a drug target by treating ASKAs in cells or whole animals with the orthogonal inhibitor and studying the genomic, proteomic, cellular, physiological, and/or phenotypic consequences of such inhibition [35]. For thorough validation of a kinase as a drug target for one or more diseases, homologous recombination in embryonic stem cells or traditional transgenic approaches (on kinase knockout backgrounds) have to be used to generate ASKA knockin cell lines or knockin mice. Thus, the ultimate validation of a kinase of interest can be conducted before or in parallel with lead identification and optimization, but well before entering costly preclinical or clinical development. The ASKA system, in combination with the orthogonal kinase inhibitor, can be also combined with gene expression profiling experiments (see Figure 3 in reference [35]). The complete set of genes that are up- or down-regulated in an inhibitor study represents a comprehensive genomic blueprint of specific kinase inhibition. This blueprint forms the basis of an array-based high-content assay for optimizing potency and specificity in a kinase inhibitor lead series (by comparing the ASKA/orthogonal ligand results with the lead series, see Figure 4 in reference [35]). We should mention that the blueprints generated by the ASKA approach are fundamentally different from those generated by knockdown or knockout technologies, as has been demonstrated in a rather straightforward analysis using *S. cerevisiae*. There, the role of Pho85p in a crucial metabolic pathway was revealed only upon acute chemical inhibition of the ASKA, and it was missed by pure knockout analysis [43].

Finally, ASKA mice should provide crucial information regarding the therapeutic index. ASKA-based in-vivo studies will be able to establish mechanism and target-based efficacy and toxicity for most protein kinases. Such information should prove useful in the preclinical testing of development candidates, because it will allow distinction between mechanism (target)-based and compound (off-target)-based toxicities [35]. At this point, we would like to challenge the scientific community and express our interest in the development of p38-ASKA mice, to either promote or discourage the numerous clinical trials of p38 for various indications and to clarify the associated liver toxicity upon treatment of patients with p38 inhibitors.

6.3.3.2 **Cohen's Inhibitor-insensitive p38 Mutants**

The MAP kinase p38 has been reported to play a crucial role in LPS- and TNFα-mediated inflammatory responses. Inhibition of p38 is supposed to inhibit the inflammatory response, such as secretion of TNFα and other cytokines. In addition, p38 has been reported to be involved in a number of other cellular responses, such as cell survival, cell cycling, proliferation, apoptosis, cytoskeletal changes, and gene expression. Almost as numerous as its potential cellular roles are the associations with a number of diseases, including inflammatory diseases like rheumatoid arthritis, Crohn's disease, inflammatory bowel diseases in general; cardiac hypertrophy; and Alzheimer's disease, where dysregulation of p38 has been reported to play a role [44]. SmithKline promoted the development of a p38-specific inhibitor for inflammation, for diseases such as rheumatoid arthritis and psoriasis. They generated the prototypical anti-inflammatory compound SB203580, a pyridinyl imidazole (Figure 6.6 a) and derivatives thereof to inhibit p38α and p38β [19]. SB203580 is a small-molecule ATP-site competitor of p38 [19]. SB203580 was the first p38 inhibitor under development [19], and SB242235 is its chemical analog, which has been promoted to the stage of clinical trials. As a consequence of the pioneering work with SB203580 and p38α, there are now at least 6 different clinical trials underway, having in common the inhibition of p38 and underscoring the importance of this kinase and its modulation with a chemical ligand. In the meantime, SB203580 became commercially available through a number of vendors.

Cocrystallization of p38α with pyridinyl imadazoles related to SB203580 [45, 46] pinpointed Thr-106 as being critical to inhibition by SB203580. Thr-106 or the conserved residue in other kinases was subsequently nominated as the so-called gatekeeper residue. As predicted, mutation of the gatekeeper to amino acids with larger sidechains made p38α insensitive to SB203580 [45–47]. By implication, the insensitivity of other MAP kinase family members to SB203580 was due to amino acid residues with larger sidechains at this position than the Thr-106 of p38α. Indeed, mutating these larger amino acids to smaller residues rendered the respective MAP kinases sensitive to the drug [47–49].

In the following paragraph, we describe the development of the technology, based on its first example, the kinase p38α.

Essentially, SB203580 and its analog SB202190 have been described more than 2000 times as specific and selective p38α/β inhibitors in the scientific literature. Some doubts arose with regard to the selectivity of the compound. Subsequently, the Cohen lab generated a drug-resistant mutant of p38α, $p38\alpha^{T106M,H107P,L108F}$. This triple mutant also includes a mutation in the above-described gatekeeper residue, Thr-106, and no longer binds SB203580. The mutant was reintroduced into the 293 cell line to generate a stable cell line that allows the inducible expression of $p38\alpha^{T106M,H107P,L108F}$. A corresponding stable cell line was generated with wild-type p38α as a control [50]. They anticipated that the downstream phosphorylation of bona fide p38α substrates would not be inhibited in the presence of the inhibitor-insensitive mutant and SB203580, whereas it should be inhibited in the presence of wild-type p38α and SB203580. This system was used to confirm that MAPKAP-K2, MSK1, HSP27, and CREB/ATF1 are indeed downstream, although not necessarily direct substrates of p38α. In contrast, the activation of c-Raf, induced by SB203580, is independent of the inhibition of p38α and presumably caused by direct binding of SB203580 to c-Raf itself [50].

This was the first time that a drug-insensitive protein kinase was used for the purpose of drug target validation. Eyers et al. also discussed the general applicability to the study of other protein kinases [50], as is described in the next section.

6.3.3.3 Active Inhibitor-insensitive Kinase Mutants (Orthogonal Protein Kinases)

Based on the work done by Shokat and colleagues on ASKAs, the work done in the Cohen lab, and clinical information coming from the analysis of Gleevec™-insensitive chronic myeloid leukemia patients [51], we learned that every protein kinase can be mutated into an active, inhibitor-insensitive mutant. The structural requirements for inhibitor binding are crucial determinants in designing the inhibitor-insensitive mutants. Often, changing the gatekeeper residue into a bulkier residue results in generation of an inhibitor-insensitive mutant (schematically illustrated in Figure 6.3). Underlying structural information from cocrystallization experiments facilitates generation of the mutants. In addition, we were also able to generate inhibitor-insensitive mutants even for kinases for which no structural information is available. As with the ASKAs, the mutation is quiet in terms of kinase function and activity. Since we have a large library of relatively specific kinase inhibitors, this method works efficiently across the entire kinase family. The only

Inhibitor-insensitive kinases

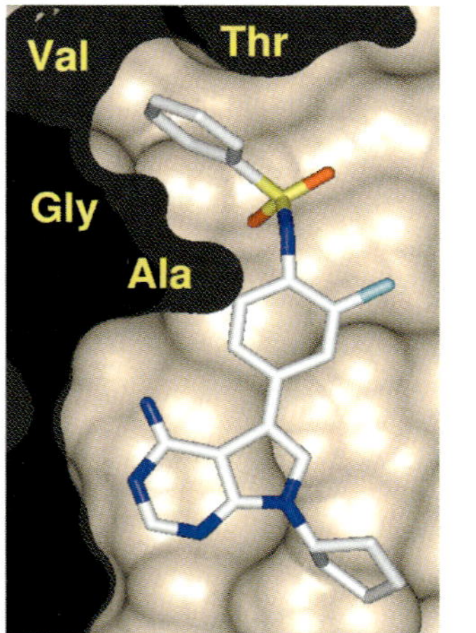

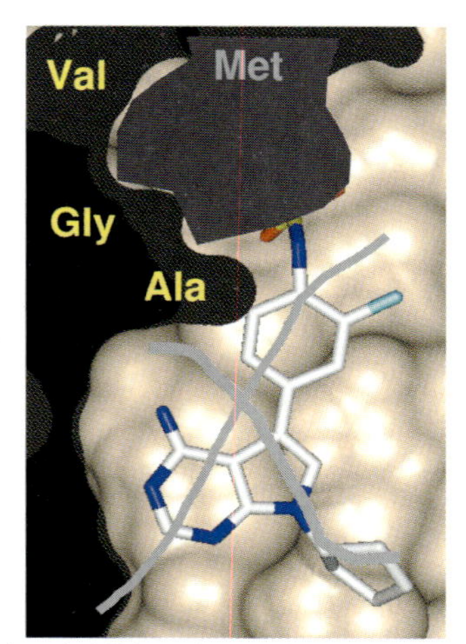

Figure 6.3 Schematic illustration of the generation of inhibitor-insensitive kinase mutants. The interaction of ATP-site competitors with kinase domains has been structurally characterized through the so-called Traxler model [10]. The part of the inhibitor that corresponds to the adenine ring binds to the hinge region of the kinase domain via H bonds. Next to the hinge region are the hydrophobic back pocket and the surface-exposed front pocket, which do not play a role in ATP binding. However, these pockets are extremely critical determinants in inhibitor binding, since the so-called gatekeeper is found in the back pocket. There is also a sugar-binding pocket and a phosphate-binding region, which do not have a major influence on inhibitor design. The gatekeeper and the residues around it in the back pocket are the important determinants for high-affinity binding. ATP-site competitors typically exploit the steric space in the hydrophobic back pocket. By mutating a small amino acid in the back pocket into one with a bulkier residue, inhibitor binding is abolished, typically without affecting the specific activity of the kinase reaction.

requirement is a specific kinase inhibitor for a novel kinase target, which is usually identified through library screening in a biochemical activity assay. Like the ASKA technology, this method is used for a highly stringent target validation. In contrast to the ASKA system, inhibitor-insensitive kinases do not need to be integrated into the genome as transgenes; a simple, efficient transfection method is sufficient. The active inhibitor-insensitive kinases and the specific inhibitor can be reintroduced into the cellular model system for a disease of interest. Only if the inhibitor looses its inhibitory potential in the presence of the kinase mutant, the kinase in question is also a proper druggable target for the disease of interest.

We have applied this method and initially validated the technology in a model system employing the EGF receptor (EGF-R) and its specific inhibitors PD153035 [52]. Figure 6.4 a shows the rationale for mutating threonine residue 766 of the EGF-R to a methionine (EGF-R^{T766M}). As expected, PD153035 very potently inhibits

(a) Rationale for choosing Thr766 of the EGF-R as a target for mutation

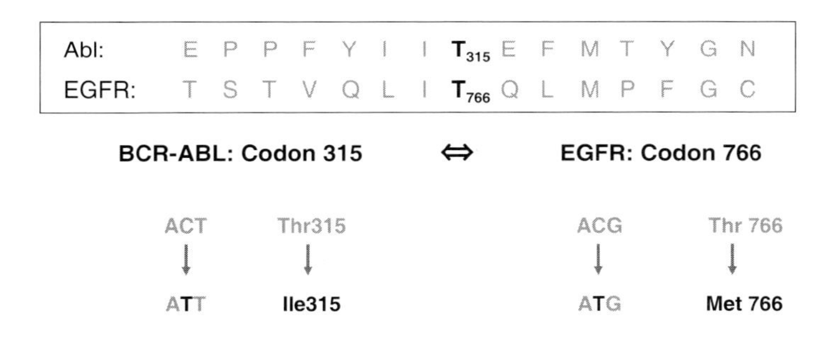

(b) Replacement of Thr-766 by methionine confers drug-resistance to EGFR

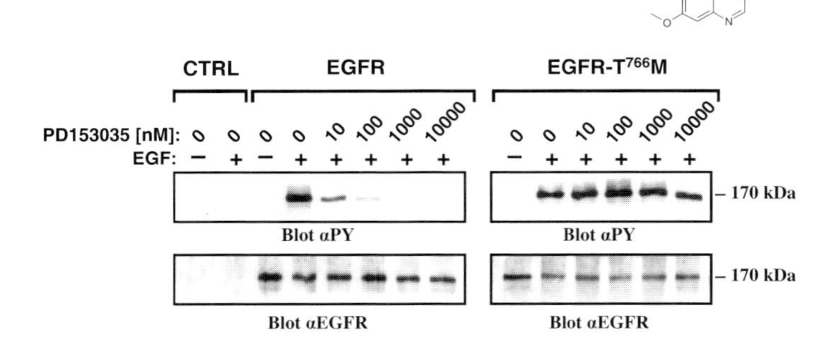

(c) EGF-R^{T766M} is inhibitor-insensitive, but active

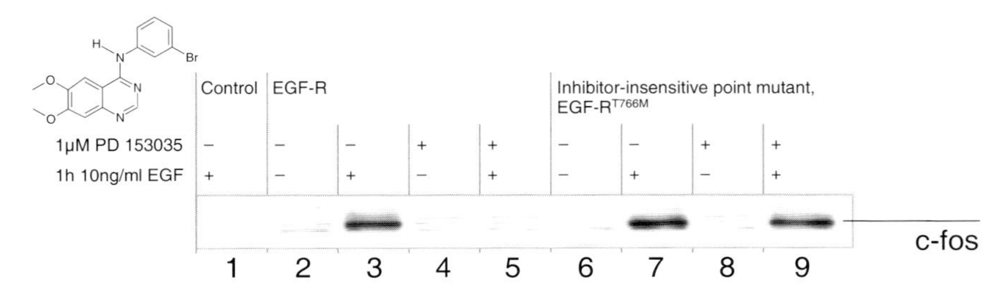

Figure 6.4 Legend see page 182.

EGF-R tyrosine autophosphorylation and the EGF-R^{T766M} mutant no longer responds to PD153035 treatment (Figure 6.4 b). It is known that EGF binds to and activates EGF-R. EGF-R induces phosphorylation of a MAP kinase cascade involving MEK1 and ERK1/2. Activation of this cascade leads to up-regulation of c-Fos [53]. For the purpose of validating our technology, EGF-R inhibitors like PD153035 can prove the known role (validity) of EGF-R in the regulation of c-Fos expression (Figure 6.4 c). Indeed, in the presence of the EGF-R^{T766M} mutation, the expression of c-Fos is no longer inhibited by specific EGF-R inhibitors. Thus, these active, inhibitor-insensitive kinase mutants are very potent tools for target validation. The parallel use of specific chemical ligands and genetic methods provides a maximum of specificity and accuracy in deciphering the roles of potential targets.

6.3.4
Chemical Proteomics for Kinases: KinaTor™

KinaTor™ was mentioned in Section 6.3.2.2. as a technological solution for forward chemical genomics approaches, for identifying the molecular targets of biologically active kinase inhibitors. The identification of a valid molecular target is necessary for a subsequent classical target-driven compound optimization program to generate a clinical candidate. The KinaTor™ technology is straightforward in theory. It relies on the coupling of specific, selective small-molecule kinase inhibitors via a linker to a matrix (Figure 6.5) to generate an affinity chromatography material that exploits the strong binding affinities of specific kinase inhibitors for their molecular targets.

Figure 6.4 Inhibitor-insensitive EGF-R mutants and their use for target validation. The EGF-R signaling pathway is well understood. It is used here to prove the validity of the Axxima technology.
(a) Thr-766 of EGF-R was selected as a target for mutation, based on the published information for the Gleevec- and PD180790-resistant Bcr-Abl mutants, for which Thr-315 has been described as the critical determinant for inhibitor binding [53]. Subsequent amino acid alignment led to the identification of Thr-766 in EGF-R as a potential residue for mutagenesis.
(b) EF1.1 –/– fibroblasts derived from EGF-R knockout mice were a generous gift from M. Sibilia and E. Wagner (Vienna, Austria). Retroviral infections of EF1.1 –/– cells, immunoprecipitations, and experimental details are described in [54], The upper-left panel shows inhibition of the wild-type EGF-R in the presence of increasing concentrations of PD153035, as measured by tyrosine autophosphorylation of EGF-R. Both lower panels represent control loading of the individual lanes with equal amounts of EGF-R, as determined with a polyclonal anti-EGFR antibody. The upper-right panel shows results for cells expressing EGF-R^{T766M} instead of wild-type EGF-R.
(c) EF1.1 –/–cells, stably expressing either EGF-R or EGF-R^{T766M} as described in B, were treated as outlined in B. Instead of monitoring EGF-R activity, total cell lysates were prepared and analyzed for c-Fos protein expression with a polyclonal anti-c-Fos antibody. c-Fos expression is induced through the EGF-R/MAPK cascade. Lane 1 represents a mock-transfected cell line treated with EGF.
Cell extracts in lanes 2 to 5 and lanes 6 to 9 represent cells transfected with wild-type EGF-R and EGF-R^{T766M}, respectively.
10 ng mL^{-1} EGF and 1 μM PD153035 were added where indicated [54].

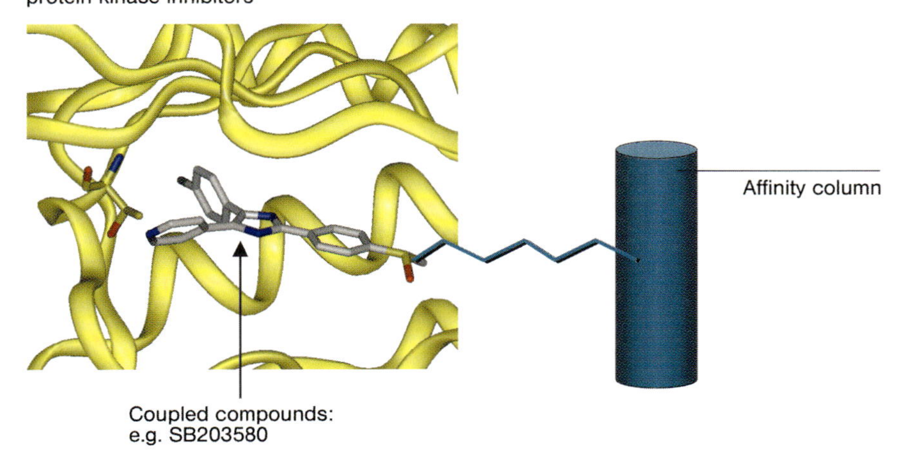

Figure 6.5 Scheme for developing a kinase inhibitor affinity matrix. Structural information from kinase–inhibitor cocrystals is valuable for determining the solvent-accessible sites of an inhibitor. The solvent-accessible site is subsequently functionalized by introducing an amino group, which is needed for coupling the inhibitor to the linker, which has been bound to Sepharose beads. For coupling details, refer to [58]. The length of the spacer is an important parameter for subsequent affinity chromatography with these matrices.

In the past, immobilization of compounds and subsequent affinity chromatography has been tried at many companies and at universities. So far, with a few exceptions coming from the field of the CDK inhibitors, purvalanol and paullones [54–56], the technology has not yet delivered, because the following factors were not considered:

- structural aspects of ligand binding to its receptor,
- physicochemical properties of the derivative for immobilization,
- feasibility of chemical synthesis of the derivatives.

Perhaps the most important points are the adsorption and elution conditions during the actual affinity chromatography. Cocrystallization of kinase domains with inhibitors, or at least a good binding model hypothesis, are important sources of information feeding into the generation of such affinity chromatography materials. To generate a functional affinity chromatography matrix, the solvent-accessible sites of the small-molecule inhibitor need to be assessed. Coupling of the linker material to the kinase inhibitor is ideally achieved through these solvent-accessible sites. Typically, kinase inhibitors with a known biological activity and one known in-vitro target are linked to a matrix. The free ligand, the functionalized derivatives, and the coupled material are routinely checked for activity in an in-vitro kinase assay.

Only active compounds are pursued. After coupling to the matrix, immobilized ligands are then processed in an affinity chromatography approach to identify all the relevant targets and off-targets that bind to the matrix. Typically, a crude extract is prepared from a cell line or a relevant tissue. The immobilized kinase inhibitor with biological activity is incubated under stringent conditions with this extract (conditions have to be optimized for each ligand) [57]. After this binding step, the beads are washed and the bound material is eluted either with free ligand or under denaturing conditions to identify all the bound material. The eluates are applied to gel electrophoresis. The 16-BAC/SDS/PAGE method is used for preparative gel electrophoresis [58]. Protein spots are excised, digested with trypsin, and identified by mass spectrometry. Thus, all qualitatively binding proteins can be identified (Figure 6.6 b, right). In a subsequent step, all the initially bound kinase and nonkinases (if any) are cloned, expressed, and purified. In-vitro enzyme assays are established for every bound protein. Using these in-vitro assays, the affinity of the free ligand for the various kinases and nonkinases is determined. Thus, quantitative binding can be determined and even the binding to nonkinases can be verified.

As a typical example and to illustrate the KinaTor™ technology, we immobilized the p38 inhibitor SB203580 (Figure 6.6 a), which we introduced in Section 6.3.3.2. In a large number of publications SB203580 and its closely related analog SB202190 have been used as chemical ligands to probe biological effects; that is, the application of SB203580 to cellular systems or in animals to cause biological responses. These responses have been assigned solely to the action of p38. But recently we learned that SB203580 is not exclusively selective for p38α and p38β. It also inhibits another serine/threonine kinase, GSK3β, to some extent [59]. We immobilized SB203580, relying on structural information obtained from pyridinyl-imidazole–p38 cocrystals

Figure 6.6 SB203580 KinaTor™.
(a) Chemical structures of SB203580 and its amino derivative used for immobilization.
(b) Frozen HeLa cells were lysed and applied to the immobilized SB203580 material. 16-BAC/SDS/PAGE was performed on the eluant as described [60]. The protein spots were excised and subjected to analysis by mass spectrometry. MALDI spectra were acquired with a Bruker Ultraflex time-of-flight (TOF)/TOF mass spectrometer with LIFT technology and anchor chip targets. Data analysis was performed with Bruker's Biotools and the MASCOT program. The kinase positions are indicated in the gel. Further experimental details are in [59]. Kinases known to bind to SB203580 have black labels, novel ones have grey labels.
(c) Kinases were identified by mass spectrometry as described in B. Active kinases were purchased or were cloned, expressed, and purified as described in [59]. The assay conditions for each single kinase were established. Subsequently, SB203580 was tested in in-vitro kinase assays initially at a concentration of 10 μM. When potent inhibition occurred, the IC_{50}s for SB203580 were determined (RICK, p38α, GAK, CK1δ: plot on the left). Kinase activities in the absence of inhibitors were set to 100%, and remaining activities at different concentrations of SB203580 are expressed relative to this value. Values for p38β and GSK3β were taken from [61]. Sequence alignments covering the hydrophobic back pockets [10] are shown on the right and suggest a similar mode of binding of SB203580 to the various kinases, via the so-called gatekeeper (shaded).
For further experimental details, refer to [59].

The p38 inhibitor, SB203580 — The amino-derivative of SB203580

(a)

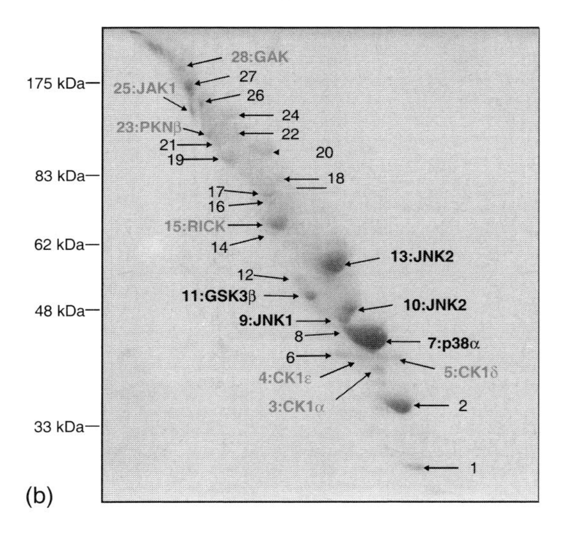

(b)

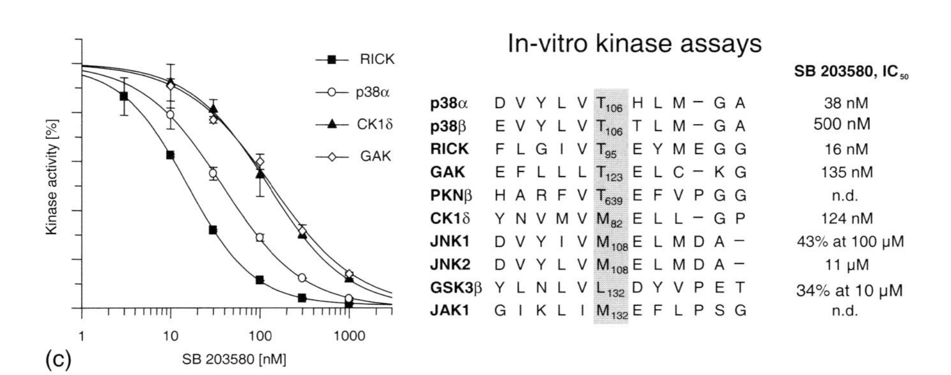

												SB 203580, IC_{50}
p38α	D	V	Y	L	V	T_{106}	H	L	M	–	G A	38 nM
p38β	E	V	Y	L	V	T_{106}	T	L	M	–	G A	500 nM
RICK	F	L	G	I	V	T_{95}	E	Y	M	E	G G	16 nM
GAK	E	F	L	L	L	T_{123}	E	L	C	–	K G	135 nM
PKNβ	H	A	R	F	V	T_{639}	E	F	V	P	G G	n.d.
CK1δ	Y	N	V	M	V	M_{82}	E	L	L	–	G P	124 nM
JNK1	D	V	Y	I	V	M_{108}	E	L	M	D	A –	43% at 100 μM
JNK2	D	V	Y	L	V	M_{108}	E	L	M	D	A –	11 μM
GSK3β	Y	L	N	L	V	L_{132}	D	Y	V	P	E T	34% at 10 μM
JAK1	G	I	K	L	I	M_{132}	E	F	L	P	S G	n.d.

(c)

Figure 6.6 Legend see page 184.

[45, 46]. First we generated an amino group containing the SB203580 derivative, which has an unchanged activity on p38α in an in-vitro kinase assay. Then a HeLa cell lysate was prepared and applied to the SB203580 affinity material. Following the above-described procedure for affinity purification, separation, and identification (see also the legend for Figure 6.6 b), we have been able to identify at least 7 novel kinases that bind to the SB203580 matrix (Figure 6.6 b) [58]. As a positive control, p38α was identified as well. In cell-free in-vitro kinase assays, we demonstrated that some of these novel interaction partners have a very high affinity for SB203580. For example, SB203580 is twice as potent as an inhibitor of RICK (RIP2, CARDIAK) than of p38α (IC_{50}s of 16 nM and 38 nM for RICK and p38α, respectively (Figure 6.6 c)). RICK is a serine/threonine kinase, which has been identified within the 7 novel binders. SB203580 inhibits the serine/threonine kinases GAK and CK1δ in the same range of potency as for p38α (Figure 6.6 c). These results raise the question as to the reality of the broadly described biological effects of p38α and p38β, which were defined as such by simply adding SB203580 to biological model systems and testing its effects. Typically, SB203580 has been applied at a concentration of 10 μM to the cellular models. Given a roughly equipotent activity of SB203580 against RICK, GAK, CK1α, CK1δ, and CK1ε, the SB203580-mediated effects in cells cannot be simply attributed to inhibition of p38. Rather, the combined inhibition of all these kinases or the inhibition of just one of them could be responsible for the compound-induced phenotype. In particular for inflammatory processes, we tend to question the assigned role of p38α, since RICK has also been reported to play an important role in the innate immune response [57, 60, 61]. Given this knowledge, the time has come to generate inhibitor-insensitive mutants (see Section 6.3.3.3) of p38α and RICK in order to define the individual roles of each kinase. The involvement of kinases like GAK, PKNβ, Jak1, CK1α, CK1δ, and CK1ε could be tested by using inhibitor-insensitive mutants as well, or by using other target-validation technologies, such as siRNA-mediated suppression of gene transcripts.

In summary, the KinaTor™ technology appears to be extremely potent for potentially revealing all binding partners of biologically active kinase inhibitors. Transmembrane domain proteins bind, as well as lipid kinases and other nucleotide-binding proteins, such as heat-shock proteins and oxidoreductases (unpublished data). So far, there are no limitations.

A very important and additional aspect of KinaTor™ is its use for lead optimization in kinase projects. Every pharmaceutical drug development program based on kinases suffers from the issue of selectivity. So far, the only solution has been to arbitrarily test lead compounds against as many kinases as possible in the available biochemical kinase assays. With KinaTor™, a lead can be immobilized on a matrix and the binding partners can be determined experimentally as outlined above for SB203580. Subsequently, the binding partners of KinaTor™ are assembled into a project-specific selectivity panel. Thus, this panel is not an arbitrarily assembled collection of available kinase assays, but rather an experimentally determined panel. Such a project-specific selectivity panel has huge implications for lead optimization, because the chemistry is now directed in a proper way to optimize for the target of interest and away from the off-targets, which have been identified through

KinaTor^{TM}. Typically, we suggest repeating the KinaTor^{TM} process at least one more time near the end of a lead optimization program.

Applying KinaTor^{TM} in lead optimization bears a further advantage: it is assumed that lead optimization is not a simple hunt for activity against the molecular target, but also includes optimization in terms of physicochemical and pharmacological properties. Thus, using KinaTor^{TM} for compounds in the late lead optimization stage relies on having a balanced lead compound. It is anticipated that this lead compound will not be monospecific to one target of interest, but rather will display a good therapeutic window of inhibition with regard to potential off-targets. Therefore, it is further anticipated that kinases will be identified that still bind to the inhibitor matrix. These kinases could very well be of further therapeutic interest for other indications. Consequently, such a combined application of KinaTor^{TM} to lead compounds can feed the so-called MasterKey approach [62], exploiting a well balanced compound class more than once within a target gene family for a number of different indications. We predict that the combination of KinaTor^{TM} with the MasterKey approach will significantly help to shorten drug development times in kinase programs.

6.4 Conclusions

The era of chemical genomics started out in by helping to significantly shorten drug development times. We are convinced that a radical and seamless integration of biology and chemistry is absolutely required to allow chemical genomics, or better, chemical biology to deliver on this promise. Importantly, chemical ligands per se cannot be regarded as lead compounds just because they can cause biological responses. A ligand or a biologically active compound usually needs to go through a number of optimization cycles to become a lead and finally a candidate. Therefore, chemical biology can aid drug development, but it cannot allow us to abandon classical medicinal chemistry. Keeping the needs for compound optimization in mind, we believe that chemical biology is most efficiently implemented within a target gene family approach, just as described in this article for the family of protein kinases. By applying technologies like inhibitor-insensitive kinases in combination with KinaTor^{TM} and the MasterKey approach, we were recently able to shorten drug development times, especially in the early phases of target identification/validation and lead optimization. We are looking forward to a period in which many novel drugs may result from kinase inhibitor projects.

Acknowledgments

We thank all our colleagues at Axxima and Vichem for their support and encouragement. Without their contributions this work would not have been possible. We gratefully acknowledge the intellectual support of Gerhard Müller and Matthias

Stein-Gerlach. Thanks also to Käthe Klebl for all the support. We are thankful for the financial support from the BMBF and our investors for establishing the chemical validation and KinaTor™ technologies.

References

1 E. S. Lander, L. M. Linton, B. Birren, C. Nusbaum, M. C. Zody, J. Baldwin, K. Devon, K. Dewar, M. Doyle, W. Fitz-Hugh et al., *Nature* **2001**, *409*, 860–921.

2 J. C. Venter, M. D. Adams, E. W. Myers, P. W. Li, R. J. Mural, G. G. Sutton, H. O. Smith, M. Yandell, C. A. Evans, R. A. Holt et al., *Science* **2001**, *291*, 1304–1341.

3 T. Laurell, G. Marko-Varga, *Proteomics* **2002**, *2*, 345–351.

4 G. L. Miklos, R. Maleszka, *Proteomics* **2001**, *1*, 169–178.

5 D. A. Jeffery, M. Bogyo, *Curr. Opin. Biotech.* **2003**, *14*, 87–95.

6 G. MacBeath, *Genome Biol.* **2001**, *2*, 1–6.

7 A. Sehgal, *Curr. Med. Chem.* **2003**, *10*, 749–755.

8 A. L. Hopkins, C. R. Groom, *Nat. Rev. Drug Discovery* **2002**, *1*, 727–730.

9 G. Manning, D. B. Whyte, R. Martinez, T. Hunter, S. Sudarsanam, *Science* **2002**, *298*, 1912–1934.

10 P. Traxler, G. Bold, E. Buchdunger, G. Caravatti, P. Furet, P. Manley, T. O'Reilly, J. Wood, J. Zimmermann, *Med. Res. Rev.* **2001**, *21*, 499–512.

11 P. Cohen, *Curr. Opin. Chem. Biol.* **1999**, *3*, 459–465.

12 D. Fabbro, S. Ruetz, E. Buchdunger, S. W. Cowan-Jacob, G. Fendrich, J. Liebetanz, J. Mestan, T. O'Reilly, P. Traxler, B. Chaudhuri, H. Fretz, J. Zimmermann, T. Meyer, G. Caravatti, P. Furet, P. W. Manley, *Pharm. Therapeutics* **2002**, *93*, 79–98.

13 R. Capdeville, E. Buchdunger, J. Zimmermann, A. Matter, *Nat. Rev. Drug Discov.* **2002**, *1*, 493–502.

14 M. Mushin, J. Graham, P. Kirkpatrick, *Nat. Rev. Drug Discov.* **2003**, *2*, 515–516.

15 G. M. Mine, *Annu. Rep. Med. Chem.* **2003**, *38*, 383–396 (see also http://www.phrma.org/).

16 D. E. Szymkowski, *Drug Discovery Today* **2003**, *8*, 157–159.

17 H. van de Waterbeemd, E. Gifford, *Nat. Rev. Drug Discov.* **2003**, *2*, 192–204.

18 A. Sehgal, *Curr. Opin. Drug Discovery Dev.* **2002**, *5*, 526–531.

19 J. C. Lee, J. T. Laydon, P. C. McDonnell, T. F. Gallagher, S. Kumar, D. Green, D. McNulty, M. J. Blumenthal, J. R. Keys, S. W. Landvatter, J. E. Strickler, M. M. McLaughlin, I. R. Siemens, S. M. Fisher, G. P. Livi, J. R. White, J. L. Adams, P. R. Young, *Nature* **1994**, *372*, 739–746.

20 P. Cohen, *Trends Biochem. Sciences* **2000**, *25*, 596–601.

21 K. Mokbel, D. Hassanally, *Curr. Med. Res. Opin.* **2001**, *17*, 51–59.

22 T. R. Hughes, M. J. Marton, A. R. Jones, C. J. Roberts, R. Stoughton, C. D. Armour, H. A. Bennett, E. Coffey, H. Dai, Y. D. He, M. J. Kidd, A. M. King, M. R. Meyer, D. Slade, P. Y. Lum, S. B. Stepaniants, D. D. Shoemaker, D. Gachotte, K. Chakraburtty, J. Simon, M. Bard, S. H. Friend, *Cell* **2000**, *102*, 109–126.

23 N. S. Gray, L. Wodicka, A. M. Thunnissen, T. C. Norman, S. Kwon, F. H. Espinoza, D. O. Morgan, G. Barnes, S. LeClerc, L. Meijer, S. H. Kim, D. J. Lockhart, P. G. Schultz, *Science* **1998**, *281*, 533–538.

24 J. S. Hardwick, F. G. Kuruvilla, J. K. Tong, A. F. Shamji, S. L. Schreiber, *Proc. Natl. Acad. Sci. USA* **1999**, *96*, 14866–14870.

25 A. F. Shamji, F. G. Kuruvilla, S. L. Schreiber, *Curr. Biol.* **2000**, *10*, 1574–1581.

26 N. Custodia, S. J. Won, A. Novillo, M. Wieland, C. Li, I. P. Callard, *Ann. N Y Acad. Sci.* **2001**, *948*, 32–42.

27 G. Le Goff, S. Boundy, P. J. Daborn, J. L. Yen, L. Sofer, R. Lind, C. Sabourault, L. Madi-Ravazzi, R. H. ffrench Constant, *Insect Biochem. Mol. Biol.* **2003**, *33*, 701–708.

28 P. R. Hoyt, M. J. Doktycz, K. L. Beattie, M. S. Greeley Jr., *Ecotoxicology* **2003**, *12*, 469–474.

29 F. W. Frueh, K. C. Hayashibara, P. O. Brown, J. P. Whitlock Jr., *Toxicol. Lett.* **2001**, *122*, 189–203.

30 F. R. Salemme, *Pharmacogenomics* **2003**, *4*, 257–267.

31 L. J. Browne, L. M. Furness, G. Natsoulis, C. Pearson, K. Jarnagin, *Targets* **2002**, *1*, 59–65.

32 J. E. Staunton, D. K. Slonim, H. A. Coller, P. Tamayo, M. J. Angelo, J. Park, U. Scherf, J. K. Lee, W. O. Reinhold, J. N. Weinstein, J. P. Mesirov, E. S. Lander, T. R. Golub, *Proc. Natl. Acad. Sci. USA* **2001**, *98*, 10787–10792.

33 G. E. Croston, *Trends Biotech.* **2002**, *20*, 110–115.

34 A. C. Bishop, O. Buzko, S. Heyeck-Dumas, I. Jung, B. Kraybill, Y. Liu, K. Shah, S. Ulrich, L. Witucki, F. Yang, C. Zhang, K. M. Shokat, *Annu. Rev. Biophys. Biomol. Struct.* **2000**, *29*, 577–606.

35 K. Shokat, M. Velleca, *Drug Discovery Today* **2002**, *7*, 872–879.

36 Y. W. Hwang, D. L. Miller, *J. Biol. Chem.* **1987**, *262*, 13081–13085.

37 A. C. Bishop, O. Buzko, K. M. Shokat, *Trends Cell Biol.* **2001**, *11*, 167–172.

38 A. C. Bishop, J. A. Ubersax, D. T. Petsch, D. P. Matheos, N. S. Gray, J. Blethrow, E. Shimizu, J. Z. Tsien, P. G. Schultz, M. D. Rose, J. L. Wood, D. O. Morgan, K. M. Shokat, *Nature* **2000**, *407*, 395–401.

39 H. Habelhah, K. Shah, L. Huang, A. L. Burlingame, K. M. Shokat, Z. Ronai, *J. Biol. Chem.* **2001**, *276*, 18090–18095.

40 K. Shah, K. M. Shokat, *Chem. Biol.* **2002**, *9*, 35–47.

41 J. A. Ubersax, E. L. Woodbury, P. N. Quang, M. Paraz, J. D. Blethrow, K. Shah, K. M. Shokat, D. O. Morgat, *Nature* **2003**, *425*, 859–864.

42 S. T. Eblen, N. V. Kumar, K. Shah, M. J. Henderson, C. K. W. Watts, K. M. Shokat, M. J. Weber, *J. Biol. Chem.* **2003**, *278*, 14926–14935.

43 A. S. Carroll, A. C. Bishop, J. L. DeRisi, K. M. S hokat, E. K. O'Shea, *Proc. Natl. Acad. Sci. USA* **2001**, *98*, 12578–12583.

44 S. Kumar, J. Boehm, J. C. Lee, *Nat. Rev. Drug Discovery* **2003**, *2*, 717–726.

45 K. P. Wilson, P. G. McCaffrey, K. Hsiao, S. Pazhanisamy, V. Galullo, G. W. Bemis, M. J. Fitzgibbon, P. R. Caron, M. A. Murcko, M. S. S. Su, *Chem. Biol.* **1997**, *4*, 423–431.

46 L. Tong, S. Pav, D. M. White, S. Rogers, K. M. Crane, C. L. Cywin, M. L. Brown, C. A. Pargellis, *Nat. Struct. Biol.* **1997**, *4*, 311–316.

47 P. A. Eyers, M. Craxton, N. Morrice, P. Cohen, M. Goedert, *Chem. Biol.* **1998**, *5*, 321–328.

48 R. J. Gum, M. M. McLaughlin, S. Kumar, Z. Wang, M. J. Bower, J. C. Lee, J. L. Adams, G. P. Livi, E. J. Goldsmith, P. R. Young, *J. Biol. Chem.* **1998**, *273*, 15605–15610.

49 T. Fox, J. T. Coll, X. Xie, P. J. Ford, U. A. Germann, M. D. Porter, S. Pahzanisamy, M. A. Fleming, V. Galullo, M. S. S. Su, K. P. Wilson, *Protein Sci.* **1998**, *7*, 2249–2255.

50 P. A. Eyers, P. van den Ijssel, R. A. Quinlan, M. Goedert, P. Cohen, *FEBS Letters* **1999**, *451*, 191–196.

51 P. La Rosee, A. S. Corbin, E. P. Stoffregen, M. W. Deininger, B. J. Druker, *Cancer Res.* **2002**, *62*, 7149–7153.

52 S. Blencke, A. Ullrich, H. Daub, *J. Biol. Chem.* **2003**, *278*, 15435–15440.

53 D. B. Chen, J. S. Davis, *Mol. Cell. Endocrinol.* **2003**, *200*, 141–154.

54 M. Knockaert, N. Gray, E. Damiens, Y. T. Chang, P. Grellier, K. Grant, D. Fergusson, J. Mottram, M. Soete, J. F. Dubremetz, K. Le Roch, C. Doerig, P. Schultz, L. Meijer, *Chem. Biol.* **2000**, *7*, 411–422.

55 M. Knockaert, K. Wieking, S. Schmitt, M. Leost, K. M. Grant, J. C. Mottram, C. Kunick, L. Meijer, *J. Biol. Chem.* **2002**, *277*, 25493–25501.

56 L. M. Schang, A. Bantly, M. Knockaert, F. Shaheen, L. Meijer, M. H. Malim, N. S. Gray, P. A. Schaffer, *J. Virol.* **2002**, *76*, 7874–7882.

57 K. Godl, J. Wissing, A. Kurtenbach, P. Habenberger, S. Blencke, H. Gutbrod, K. Salassidis, M. Stein-Gerlach, A. Missio, M. Cotten, H. Daub, *Proc. Natl. Acad. Sci. USA* **2003**, *100*, 15434–15439.
58 H. Daub, S. Blencke, P. Habenberger, A. Kurtenbach, J. Dennenmoser, J. Wissing, A. Ullrich, M. Cotten, *J. Virol.* **2002**, *76*, 8124–8137.
59 S. P. Davies, H. Reddy, M. Caivano, P. Cohen, *Biochem. J.* **2000**, *351*, 95–105.
60 A. I. Chin, P. W. Dempsey, K. Bruhn, J. F. Miller, Y. Xu, G. Cheng, *Nature* **2002**, *416*, 190–194.
61 K. Kobayashi, N. Inohara, L. D. Hernandez, J. E. Galán, G. Núnez, C. A. Janeway, R. Medzhitov, R. A. Flavell, *Nature* **2002**, *416*, 194–199.
62 G. Müller, *Drug Discovery Today* **2003**, *8*, 681–691.

7
Structural Aspects of Kinases and Their Inhibitors

Rogier Buijsman

7.1
Introduction

Kinases are pivotal regulators of the signal transduction pathways, thereby controlling cellular processes like metabolism, transcription, cell cycle progression, apoptosis, and differentiation. Kinases form by far the largest enzyme family [1] constituting ~1.7% of the human genome and currently having 518 different family members. It is now well established that many diseases, like cancer, diabetes, and rheumatoid arthritis, are modulated by kinases. In this respect it is no surprise that selective kinase inhibition has become an important goal for the pharmaceutical industry. Currently, over 20% of all research programs at major pharmaceutical companies are aimed at kinases. These research efforts have resulted in the recent approval of two new kinase inhibitors, Gleevec (Imatinib) and Iressa (Gefitinib) (Figure 7.1). Moreover, approximately 40 selective inhibitors have entered clinical trials, and new patent applications are published almost every day (Table 7.1) [2].

The advent of high-throughput technologies in structural biology provided a plethora of kinase X-ray structures. The structural information embedded in these structures gave profound insight into kinases at a molecular level and facilitated the design of potent and selective inhibitors [3]. In this chapter current insights into protein kinases as well as their small-molecule inhibitors are discussed according to their reported X-ray structures.

The first section describes the structure of kinases in general, the specific domains as well as the activation mechanism, and is followed by a section on the different kinase inhibition principles. In the next section, a detailed description of the structural features of all kinase inhibitors currently cocrystallized with their target kinase is given. Finally, some structural features that determine the selectivity of several kinase inhibitors are addressed.

The aim of this review is to give the reader a brief overview of the exciting research field of protein kinases. It is, however, by no means a comprehensive overview, and I would like to invite motivated readers to read one of the many excellent reviews on this research topic [4].

Chemogenomics in Drug Discovery: A Medicinal Chemistry Perspective.
Edited by Hugo Kubinyi and Gerhard Müller

ISBN: 3-527-30987-X

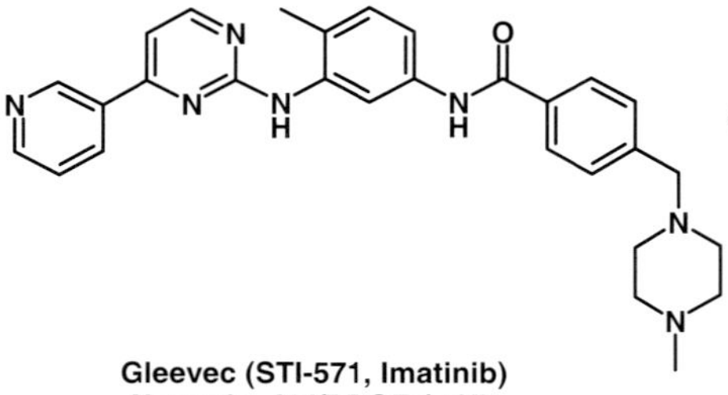

Gleevec (STI-571, Imatinib)
Novartis, Abl/PDGFr/c-Kit
Launched

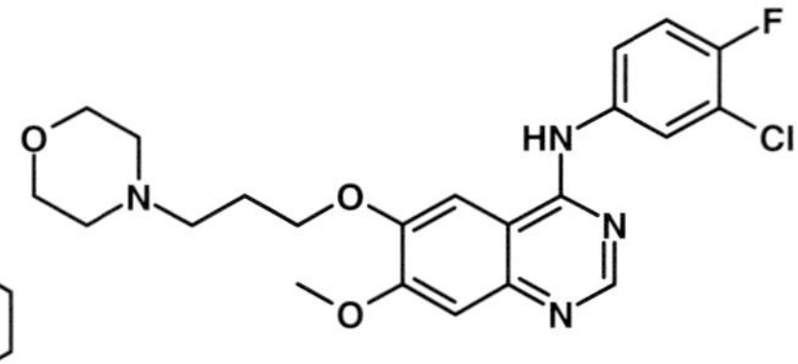

Iressa (ZD1839, Gefinitib)
Astra-Zeneca, EGFr
Launched

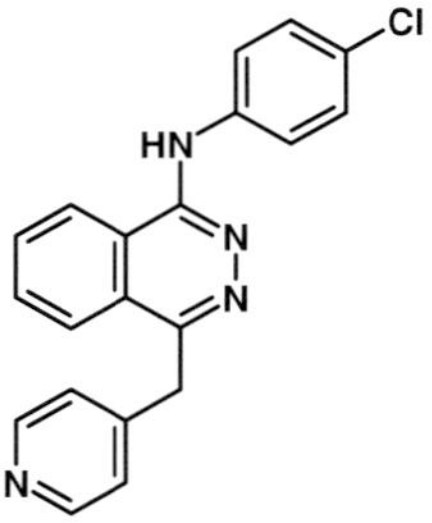

CPG-79787 (Vatalanib)
Novartis/Schering, VEGFr
Phase III

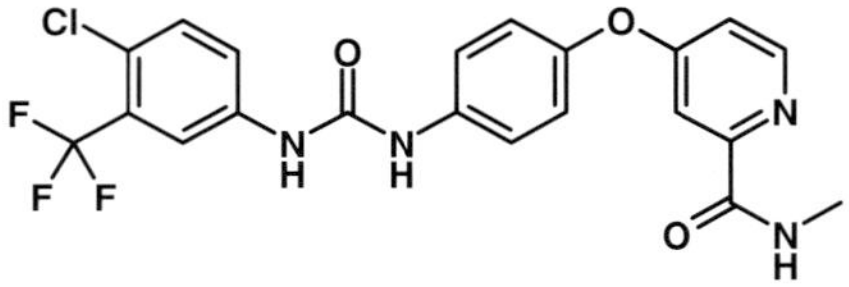

BAY-43-9006
Bayer/Onyx,cRaf
Phase III

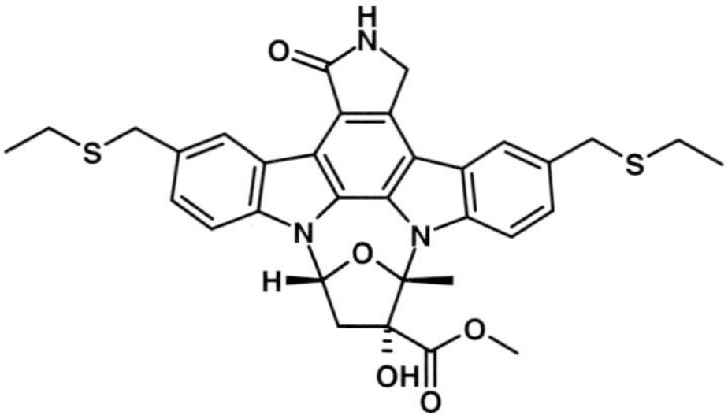

CEP-1347
Lundbeck/Cephalon/Kyowa, PKC/JNK/MLK
Phase III

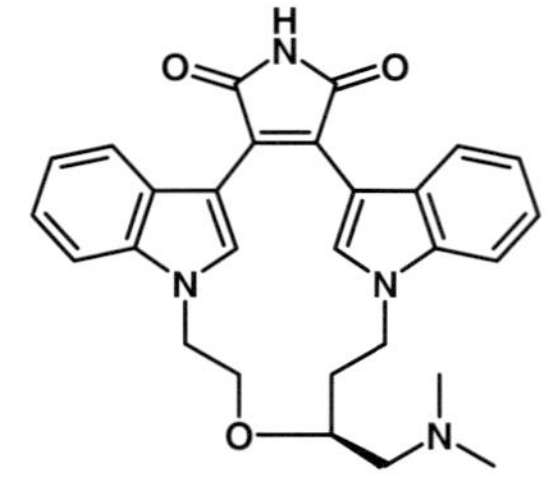

LY-333531(Ruboxistaurin)
Eli Lilly, PKCβ
Phase III

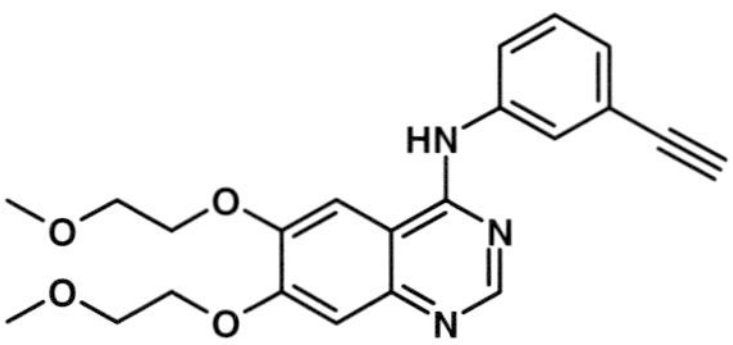

Tarceva (Erlotinib, OSI-774)
OSI/Roche/Genentech, EGFr
Phase III

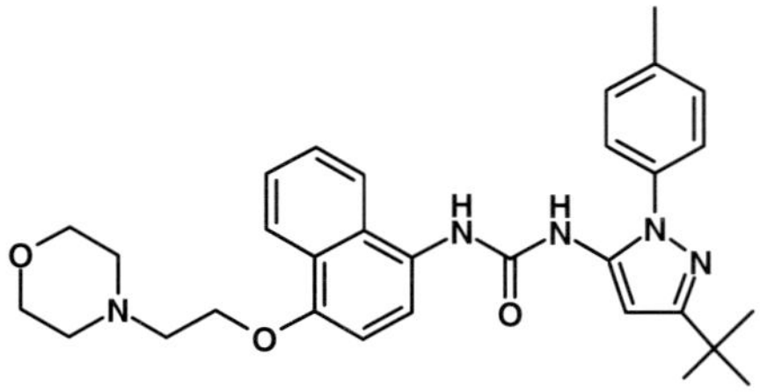

BIRB-796 (Doramapimod)
Boehringer Ingelheim, p38
Phase III

Figure 7.1 Structures of kinase inhibitors currently on the market or in phase 3 clinical trials.

Table 7.1 Presentation of all kinase projects currently in clinical trials (taken from http://www.iddb3.com).

Status	*Company*	*Drug*	*Disease*	*Kinase target*
Launched	Novartis	Gleevec (STI-571, Imantinib)	Cancer	Abl (cKit, PDGF)
Launched	AstraZeneca	Iressa (ZD1839, Gefinitib)	Cancer	EGFr
Phase 3	Novartis, Schering AG	CPG-79787 (Vatalanib)	Cancer	VEGF
Phase 3	OSI, Roche, Genentech	Erlotinib (OSI-774, Tarceva)	Cancer	EGFr
Phase 3	Lundbeck, Cephalon, Kyowa	CEP-1347	Alzheimer, Parkinson	PKC, JNK, MLK
Phase 3	Eli Lilly	LY-333531 (Ruboxistaurin)	Cancer	PKC-β
Phase 2/3	Boehringer Ingelheim	BIRB-796 (Doramapimod)	Inflammation, RA, Psoriasis, Crohn	p38
Phase 3	Bayer/Onyx	BAY-43-9006	Cancer	c-Raf
Phase 2	Pfizer	PD-184352 (CI-1040)	Cancer	MEK-1
Phase 2	Scios/J&J	SCIO-469	RA, Crohn	p38
Phase 2	Pfizer	CI-1033 (PD-183805, Canertinib)	Cancer	EGFr
Phase 2	GSK	GW-572016 (Lapatinib)	Cancer	EGFr, ErbB-2β
Phase 2	Wyeth	EKB-569	Cancer	EGFr, ErbB-2β
Phase 2	Aventis	ZD-6474	Cancer	VEGF, Kdr, EGFr
Phase 2	Kyowa	UCN-01	Cancer	CDK, Chk1, PKC
Phase 2	Cephalon	CEP-701	Cancer	Flt-3
Phase 2	Novartis	PKC-412 (Midostaurin)	Cancer	PKC
Phase 2	Pfizer	SU-6668	Cancer	FGF, VEGF, PDGF
Phase 2	Pfizer	SU-11248	Cancer	PDGF
Phase 2	Aventis	Alvocidib	Cancer	CDK
Phase 2	Cyclacel	CYC-202	Cancer	CDK
Phase 2[a)]	Schering, Asahi	ZK-258594 (Fasudil)	Angina	Rock
Phase 2	Vertex/Kissei	VX-702	Inflammation, Cardiovascular Disease	p38
Phase 2	Eli Lilly	LY-317615	Cancer	PKCβ
Phase 2	Rigel	R-112	Allergic rhinitis, Asthma	Syk
Phase 2	Pfizer	SC-80036	RA	p38
Phase 2	Pfizer, Agouron	AG-13736	Cancer	VEGF, PDGF

Table 7.1 (continued)

Status	*Company*	*Drug*	*Disease*	*Kinase target*
Phase 1	BMS	BMS-387032	Cancer	CDK2
Phase 1	Pfizer	CP-547632	Cancer	VEGF
Phase 1	Roche	Ro-320-1195	RA, inflammation	p38
Phase 1	Pfizer	CP-724714	Cancer	ErbB2, HER2
Phase 1	Millenium	MLN-518	Cancer	Flt-3
Phase 1	Novartis	AEE-788	Cancer	EGFr, ErbB-2β, VEGF
Phase 1	Pfizer	CP-690550	Psoriasis, Transplantation	JAK-3
Phase 1	Celgene	CC-401	Immune Disorder, Transplantation	JNK
Phase 1	Bayer	BAY-57-9352	Cancer	Kdr
Phase 1	Ariad	AP-23573	Cancer	mTOR
Phase 1	Roche	R-1487	RA	NR
Phase 1	Amgen	AMG-548	RA, inflammation	p38
Phase 1	GSK	681323	RA, COPD	p38
Phase 1	Scios/J&J	SCIO-323	RA	p38
Phase 1	Astra-Zeneca	AZD-2171	Cancer	VEGF
Phase 1	GSK	786034	Cancer	VEGF-2

[a] Fasudil is marketed in Japan.

7.2 Structural Aspects of Kinases

7.2.1 The General Structure of an Activated Kinase

In 1991 the structure of PKA was determined [5], which provided a complete view of the 3D arrangement of amino acids that constitute the protein kinase domain. The rapid growth of X-ray data on many other kinases that followed revealed that the folding topology of protein kinases is extremely well conserved. Protein kinases are folded into two subdomains or lobes: (1) the N-terminal lobe, which is composed of a five-stranded β sheet and one α helix (C helix) and (2) the C-terminal lobe, which is larger and predominantly helical (Figure 7.2) [6]. The two lobes are connected via a hinge domain, which allows rotation of the two lobes. A comparison between active and inactive CDK2 in complex with indirubin-5-sulfonate showed that the N lobe rotates by ~5° with respect to the C lobe upon activation [7]. The hinge domain is also involved in the binding of ATP, forming two hydrogen bonds. The first hydrogen bond is formed between the backbone carbonyl of Glu-81 (CDK2 numbering [8]) and the N^6 of ATP, and second between the backbone NH of Leu-83 and the N^1 of ATP (Figure 7.3).

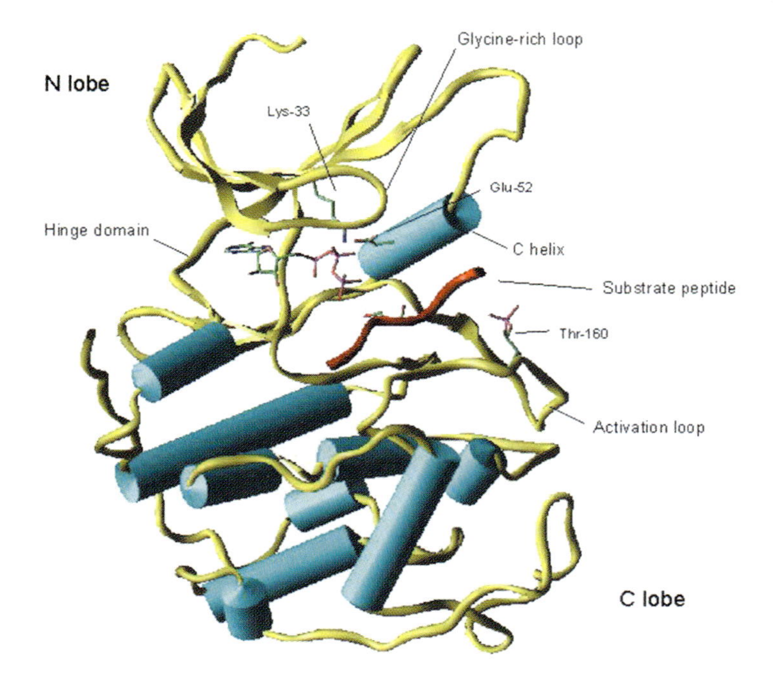

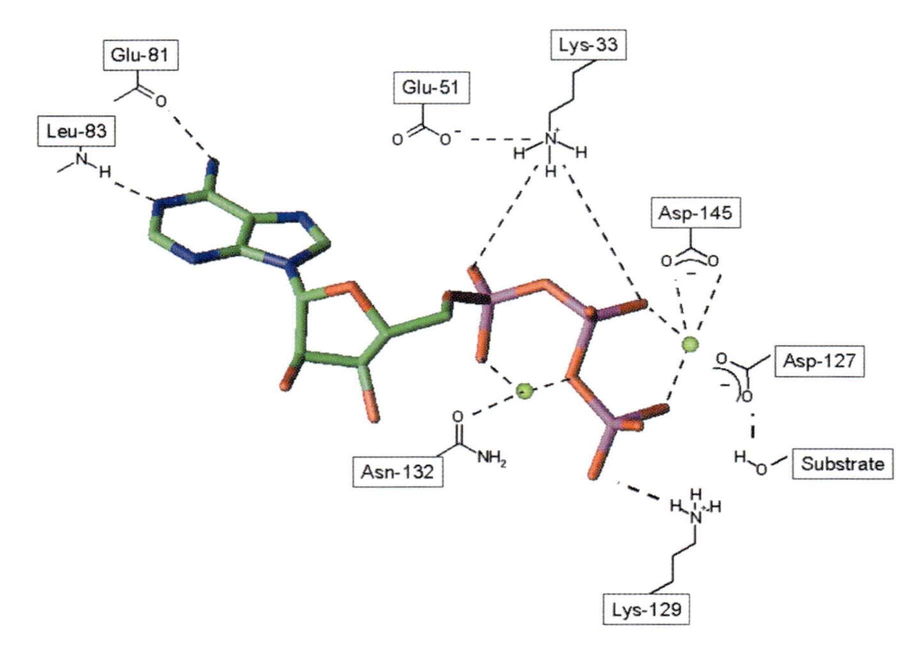

Figure 7.3 Key residue interactions with ATP in the active site of CDK2.

The C helix forms the back wall of the ATP-binding site [9]. It contains a conserved glutamic acid residue (Glu-52), which is of key importance in the phosphotransfer process, forming an ion pair with Lys-33 (Figure 7.3). Lys-33, which is buried deep in the ATP-binding cleft, makes a crucial contact with the α,β-phosphate oxygens, positioning them so as to facilitate the γ-phosphoryl transfer [4 (i)].

The phosphate-binding loop (or glycine-rich loop) forms the roof of the ATP-binding site and contains a conserved glycine-rich sequence motif (GXGXϕG), where ϕ is usually tyrosine or phenylalanine. The glycine residues make the loop very flexible, thus allowing the ATP-binding site to open and close during catalysis. Taylor et al. [9] suggested that, during the fast phosphoryl transfer step, the enzyme is in its closed conformation and opens again during the slower, rate-determining step to release ADP (Scheme 1).

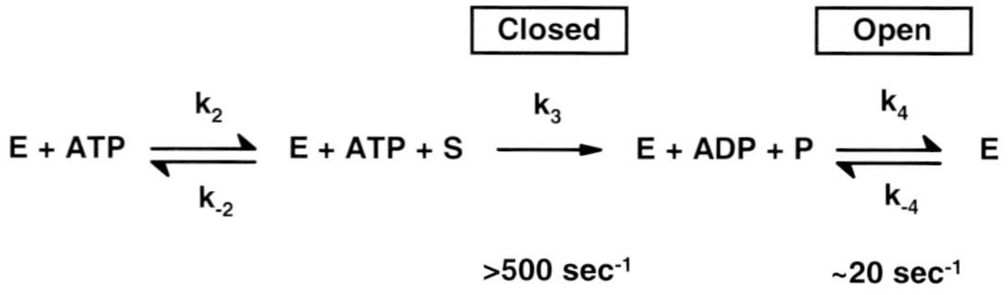

$$\text{E + ATP} \underset{k_{-2}}{\overset{k_2}{\rightleftharpoons}} \text{E + ATP + S} \xrightarrow{k_3} \text{E + ADP + P} \underset{k_{-4}}{\overset{k_4}{\rightleftharpoons}} \text{E}$$

Closed: $>500\ \text{sec}^{-1}$ (step k_3); Open: $\sim 20\ \text{sec}^{-1}$ (step k_4)

Scheme 7.1 Reaction pathway for catalysis (taken from [9]). S = substrate peptide, P = phosphopeptide product.

Another very important conserved domain in kinases is the activation loop. This 20–30 amino acid region is positioned between a highly conserved DFG and APE motif (IDA region, inter DFG–APE region). In its activated state the activation loop is in an open, extended conformation, which allows substrate binding to the kinase. The aspartate residue (Asp-145) of the DFG motif interacts with one of the two magnesium ions in the active site.

The presence of the two magnesium ions, which chelate the β- and γ-phosphate oxygens, positions the terminal phosphate group and reduces electrostatic repulsion of the incoming nucleophile [4 (i)].

Several other highly conserved key residues assist in the phosphotransfer catalysis. Asp-127 is located near the incoming nucleophile and may direct the hydroxyl function for the attack on the terminal phosphate. Asn-132 interacts with the second magnesium ion, and Lys-129 forms an ion pair with the terminal phosphate.

Although most of the conserved phosphate-binding residues in the ATP-binding site are crucial for the catalytic process, they do not contribute much to the free energy of binding of the ATP kinase, which is nicely reflected in the equipotent affinity of PKA for ATP, ADP, and adenosine [10].

7.2.2 Kinase Activation

The conformational changes that occur during the activation of a kinase are illustrated by four different crystal structures of CDK2 (1HCL [11], 1FIN [12], 1JST [13], and 1QMZ [14]), each representing a different stage in the activation process. The activation of CDK2 is triggered by its binding to cyclin-A. The crystal structure of inactive monomeric CDK2 (stage 1; Figure 7.4) shows the activation loop in its unphosphorylated state, located in the active-site area. In this conformation the activation loop prevents binding of ATP and the substrate.

In the absence of cyclin-A, the C helix is twisted and the conserved Glu-51 residue on its surface faces the solvent and is unable to coordinate with Lys-33, which instead coordinates with Asp-145. The torsion angles of Phe-146 and Asp-145 in the DFG motif are typical for an inactive kinase [15] (Table 7.2) and show that the orientation of Asp-145 is unfit for catalysis.

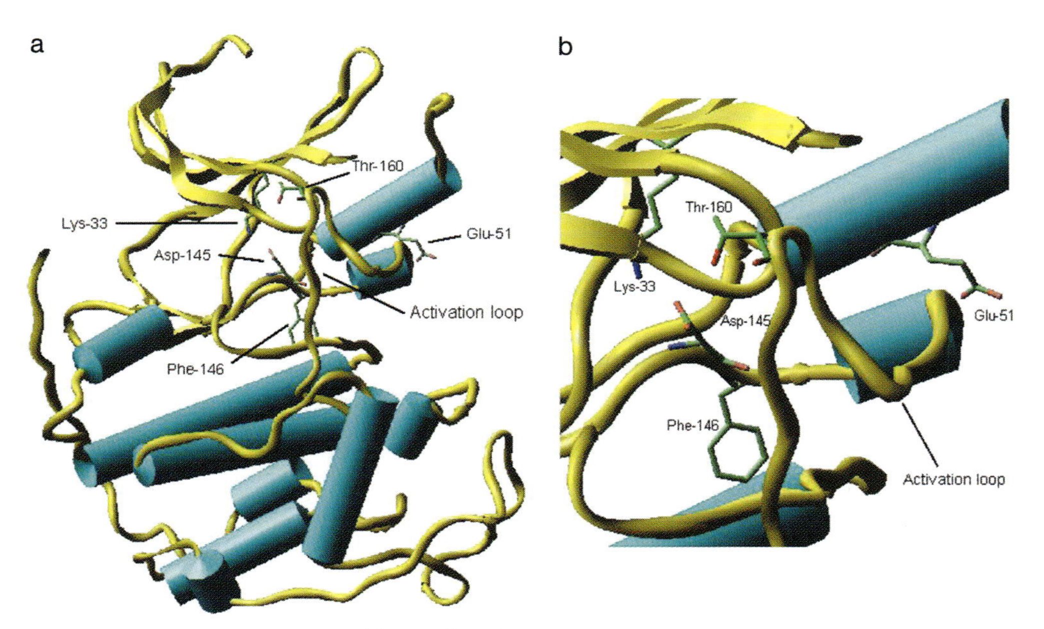

Figure 7.4 CDK2 in its inactivated form (pdb code: 1HCL).

Table 7.2 Torsion angles of Asp and Phe residues in the conserved DFG loop.

	Stage 1	*Stage 2*	*Stage 3*	*Stage 4*	*Hck inact. (1QCF)*	*Lck act. (3LCK)*
Asp ϕ	61.4	41.8	48.9	60.2	66.1	54.2
Asp ψ	30.4	80.5	78.8	70.4	14.5	81.3
Phe ϕ	−66.4	−45.5	−92.2	−88.6	−58.7	−95.0
Phe ψ	137.4	25.6	30.3	22.3	126.2	22.6

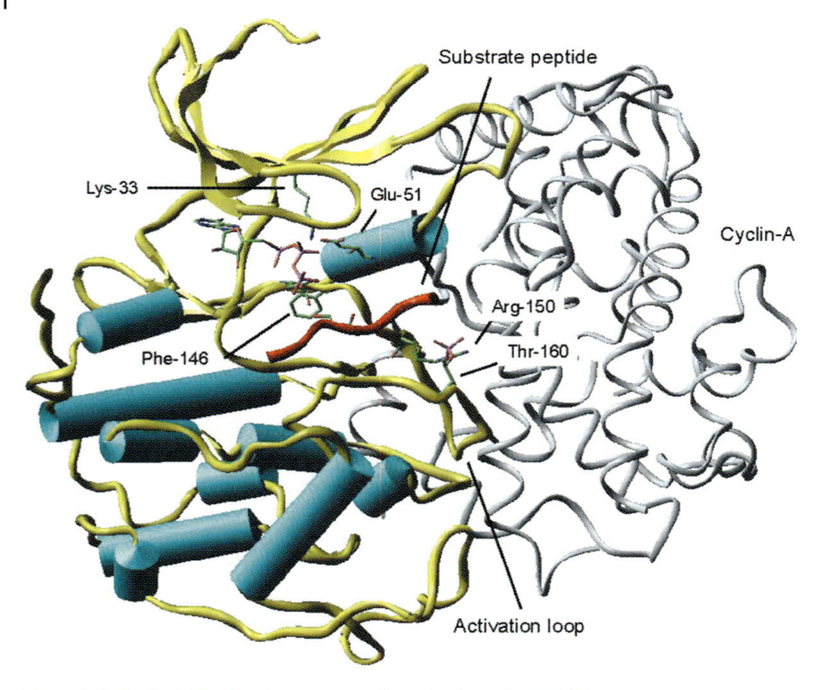

Figure 7.5 CyclinA-CDK2 in its activated form (pdb code: 1QMZ).

Upon binding cyclin-A (stage 2), the C helix rotates around its long axis, swinging Glu-51 inward and thus enabling it to contact Lys-33. The activation loop, which seems to be structurally coupled to the C helix, adopts a conformation that is similar to the activated form of CDK2. In stage three, Thr-160 has been phosphorylated by a CDK-activating kinase, and the conformation of the activation loop is stabilized by an ion pair between the phosphorylated Thr-160 and its conserved neighbor Arg-150. In stage four the substrate is bound to the kinase, which has all its catalytic residues lined up now for the γ-phosphotransfer (Figure 7.5). Similar activation mechanisms have been reported for the Src family of kinase and the insulin receptor kinase (IRK), although the trigger of activation is different [6].

7.3
Kinase Inhibition Principles

7.3.1
Substrate-competitive Inhibitors

The substrate-binding site seems to have obvious advantages over the ATP-binding site as a target for inhibiting kinase activity. First, substrate-binding inhibitors are not affected by the high ATP concentration found in cells. Second, the substrate-

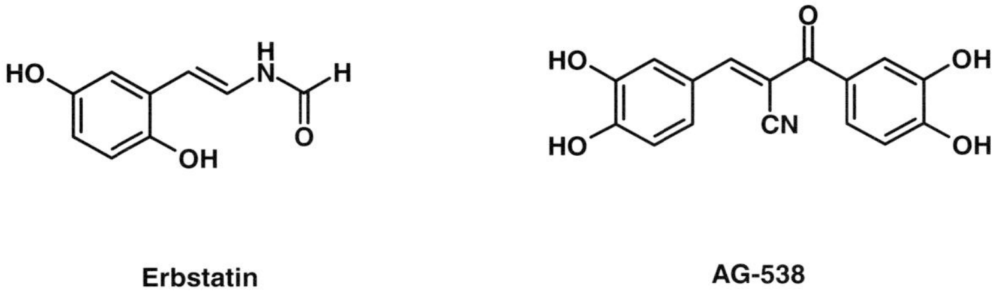

Erbstatin

AG-538

Figure 7.6 Structures of Erbstatin and AG-538.

binding site of a kinase controls selectivity, whereas the ATP-binding site is highly conserved throughout all kinase family members.

Substrate-competitive inhibition is a well known strategy for targeting enzymes, which has been applied successfully in enzyme classes such as the proteases. Nevertheless, its use for kinase inhibition has met with little success. One of the reasons is the rather stretched substrate pocket of kinases. Kinases are likely to use additional binding pockets, which are not located in the immediate environment of the active site [16, 17]. Therefore, kinases lack the specific hydrophobic pockets that could serve as targets for peptidomimetics, as occurs with HIV protease or thrombin.

One of the few examples of substrate-competitive inhibitors is the natural product Erbstatin (Figure 7.6). The phenolic moiety in Erbstatin is supposed to act as a tyrosine mimic, and this hypothesis led to the synthesis of Erbstatin [18] analogs called Tyrphostins. Levitski et al. [19] recently reported that one of their tyrphostins, AG-538, was selective for IGF-1R over Src and PKB. Promising as this may seem, the progress after a decade of research in the tyrphostin area has been rather disappointing.

Another approach toward effective substrate-competitive inhibitors was recently reported by Hubbard et al. [20]. Linking the known IRS-727 octadecapeptide substrate to a stable ATP mimic resulted in a combined ATP- and substrate-competitive inhibitor (compound **1**, Figure 7.7) having a K_i of 370 nM for IRK.

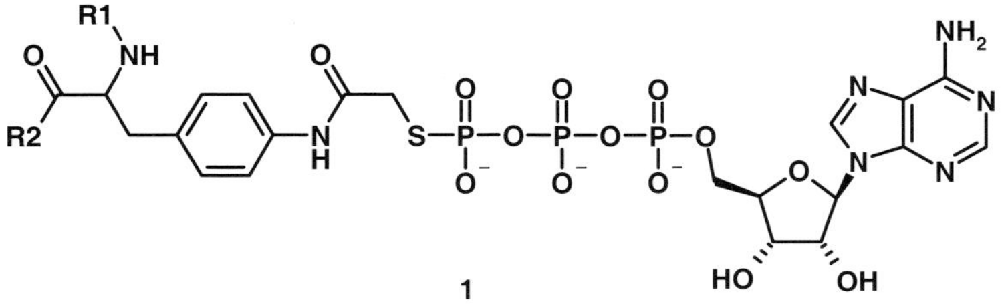

1

R1 = AcNH-Lys-Lys-Lys-Leu-Pro-Ala-Thr-Gly-Asp-

R2 = -Met-Asn-Met-Ser-Pro-Val-Gly-Asp-CO_2H

Figure 7.7 Structure of substrate-competitive inhibitor **1**.

An obvious drawback of this approach is that it involves large peptide-like molecules, which are notorious for problems with oral absorption, stability, and cell penetration.

7.3.2
ATP-competitive Inhibitors

The road toward potent, selective ATP-competitive inhibitors is burdened with an impressive amount of hurdles, and for a long time, finding these inhibitors was generally considered a mission impossible. The most prominent hurdle in developing ATP-competitive inhibitors is *specificity*. Apart from the fact that many proteins use purine-based cofactors [3], there are 518 kinase family members that share a highly conserved ATP-binding site. Another puzzling hurdle is *potency*. The intracellular concentration of ATP is overwhelming (up to 8 mM), which has a major impact on the potency of a competitive inhibitor. Concentrations required for an inhibitor to reach 50% inhibition are typically two to three orders of magnitude higher than its inhibition constant (K_i) [21].

Against all odds, the pharmaceutical industry began an intensive search for these 'magic bullets'. Fortunately, production and crystallization of kinases is relatively easy in comparison to other protein families, and an impressive number of crystal structures of many kinases and their small-molecule inhibitors has become available (see Figure 7.16, Section 7.4.1). This wealth of structural information has aided the design and synthesis of high-affinity ligands for kinases. An in-depth discussion of these structures is included in Section 7.4.

Today 23 ATP-competitive inhibitors for which structures have been disclosed are under clinical evaluation (5 in phase 1 clinical trials, 14 in phase 2, 4 in phase 3), and Iressa was recently approved by the U. S. FDA (see Figure 7.1 and Table 7.1).

Staurosporine analogs (5) and the 4-anilinoquinazolines and closely related structures (6) are the most represented kinase templates in these trials.

7.3.3
Activation Inhibitors/Allosteric Modulators

The true pearls of kinase inhibition are compounds that prevent activation of kinases rather than competing with the endogenous cofactor ATP. The first compound identified as having such a mechanism of action was Gleevec (see Figure 7.1). After determination of the 3D structure of Abelson kinase (c-ABL) in complex with Gleevec [22, 23], it became clear that the compound has an extended interaction with a binding site spatially distinct from the ATP-binding site. Access to this allosteric binding site is normally impeded by the phenyl moiety of the conserved DFG motif (DFG-in). Gleevec induces a structural transition that leads to movement of the Phe-382 residue of the DFG motif toward the ATP-binding site (DFG-out) (Figure 7.8). Consequently, the activation loop adopts a conformation that mimics substrate binding to the enzyme and prevents its activation by other kinases.

Conformational restriction of the activation loop clearly comes at the cost of binding free energy (negative entropy), which is compensated for by the numerous

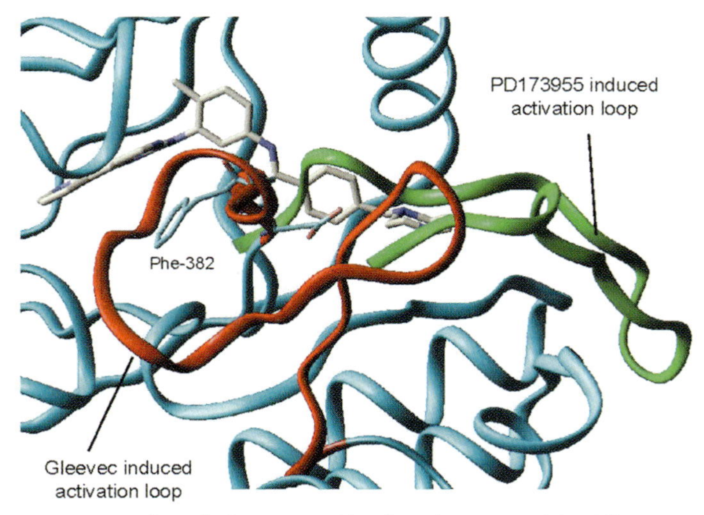

Figure 7.8 Binding of Gleevec to c-Abl (pdb code: 1IEP) and the difference in activation loop conformation between Gleevec and PD-173955 bound c-Abl.

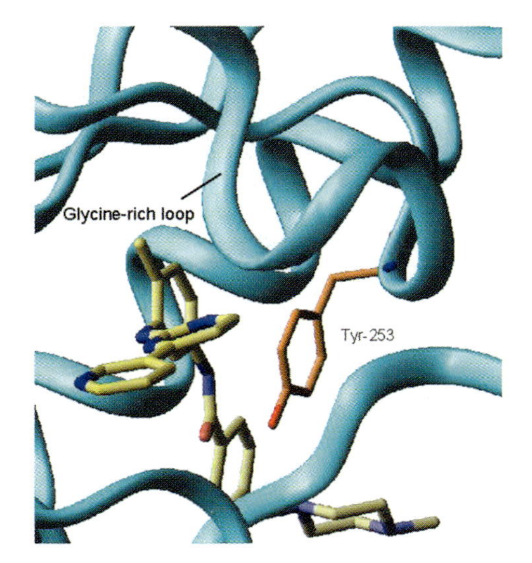

Figure 7.9 Glycine-rich loop in Gleevec-bound c-Abl.

van der Waals interactions of Gleevec with c-Abl. Apart from additional hydrophobic interactions in the allosteric binding region, Gleevec induces a typical [24, 25] conformational change in the glycine-rich loop of the Abelson kinase, moving Tyr-253 within van der Waals distance from its pyrimidine core (Figure 7.9). Nevertheless, the ATP-competitive inhibitor PD-173955 (Figure 7.10) shows greater potency against c-Abl than does Gleevec, despite the large difference in binding surface area (1251 $Å^2$ for Gleevec vs. 913 $Å^2$ for PD-173955), indicating that the entropy penalty due to the conformationally restricted activation loop is indeed considerable [26].

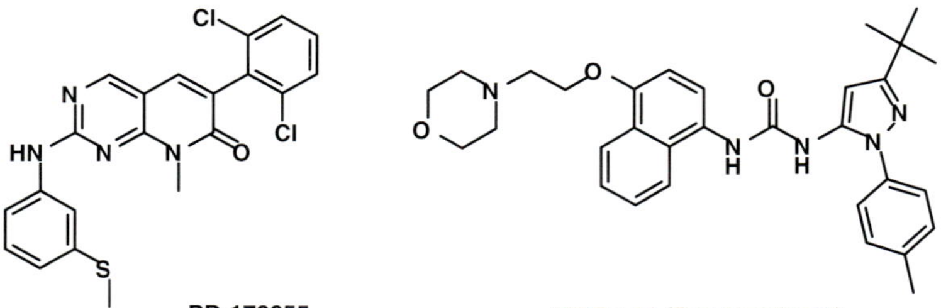

PD-173955 BIRB-796 (Doramapimod)

Figure 7.10 Structures of PD-173955 and BIRB-796.

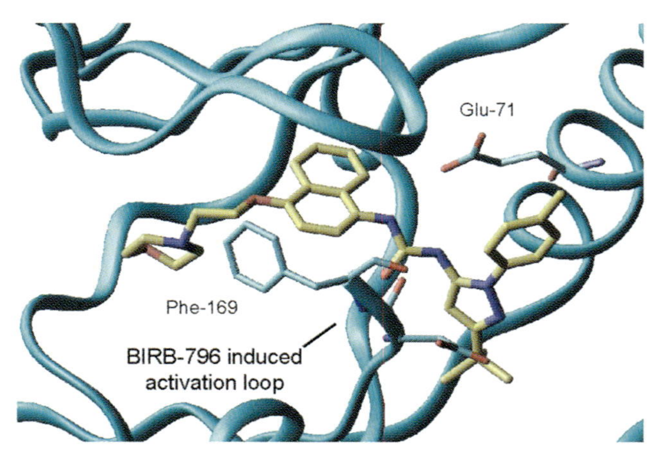

Figure 7.11 Binding of BIRB-796 to p38 MAP kinase (pdb code: 1KV2).

The unique binding mode of Gleevec with c-Abl is not a stand-alone case, as shown recently for the p38 MAPK inhibitor BIRB-796 (Doramapimod, see Figure 7.10). The slow binding kinetics of BIRB-796 indicates that p38 undergoes a rearrangement upon binding. X-ray studies [27] of BIRB-796 showed that the pyrazole moiety of this diaryl urea class inhibitor indeed binds in a similar allosteric binding pocket as Gleevec, which is exposed by a conformational change in the activation loop (DFG-out, Figure 7.11).

The *t*-butyl as well as the toluyl group in BIRB-796, which do not have an equivalent in Gleevec, contribute significantly to the free energy of binding. BIRB-796 has a dual mechanism of action, inhibiting p38 MAP kinase activity as well as p38MAP kinase activation by other kinases [28].

The last compound class known to bind allosterically to its target kinase is the 2-(4-iodo-phenylamino)benzhydroxamate esters. CI-1040 (Figure 7.12), which is a member of this class of inhibitors, has a potent binding to MEK1 and is currently in phase 2 clinical trials for cancer. An analog of CI-1040 (compound **2**, Figure 7.12) has recently been cocrystallized with MEK1 and ATP [29]. The ternary complex showed that the inhibitor occupies the same binding site as Gleevec and BIRB-796 and that the activation loop adopts an inactive conformation.

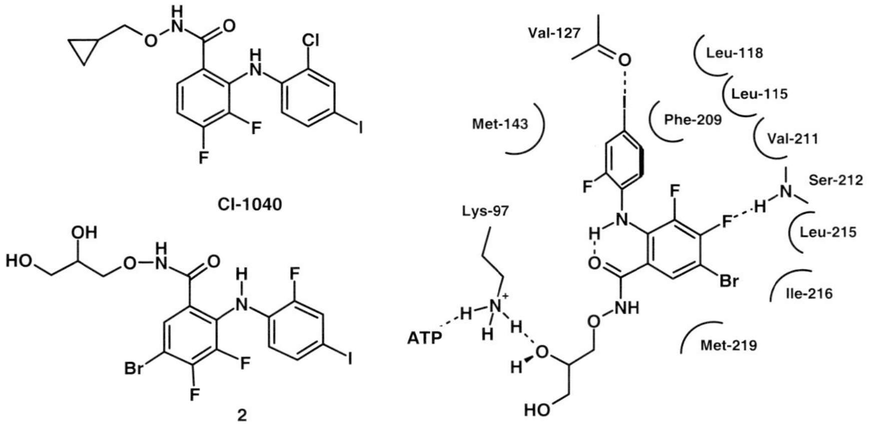

Figure 7.12 Structure of CI-1040 and the binding mode of a CI-1040 analog **2** (taken from [29]).

7.3.4
Irreversible Inhibitors

Irreversible inhibitors have a major advantage over their reversible counterparts in that their action is sustained after systemic clearance. However, the pharmaceutical industry usually does not make use of irreversible inhibition principles, because unspecific binding to proteins other than the target enzyme may lead to considerable toxicity. Since most of the current cancer therapies are unspecific and very toxic, the irreversible kinase inhibition principle may well be applied in this research area, provided that an improved risk/benefit ratio is observed [4e].

The first example of an irreversible inhibitor is the natural product Wortmannin, which was isolated from *Penicillium wortmannii* [30]. Wortmannin effectively inhibits PI3K at low nanomolar concentration and was shown to be specific across a large panel of kinases [31]. Covalent attachment to PI3K occurs after attack by Lys-883, which is essential for phosphate transfer (see Section 7.2), at the furan ring of Wortmannin [32]. Attack at this ring specifically occurs within the catalytic site of the PI3K kinase and is unaffected by nucleophiles in aqueous solution.

Apart from many hydrophobic interactions, which seem to account for most of the binding energy of Wortmannin, five possible hydrogen bonds can be observed in the crystal structure (Figure 7.13). Noteworthy is the formation of a hydrogen bond between Asp-964 and the enol function of the B ring, which is unchallenged by desolvation enthalpy prior to binding to the PI3K kinase and therefore contributes fully to the free energy of binding.

The 6-acrylamido-4-anilinoquinazolines constitute another class of irreversible inhibitor targeted at the EGFr tyrosine kinase. This type of inhibitor has been derived from the very potent (6 pM) and selective 4-anilinoquinazolines, e.g., PD-0153035 [4e]. A detailed description of the binding mode of the 4-anilinoquinazolines is given in Section 7.4.1. The Michael acceptor at the 6-position of the 4-anilino-

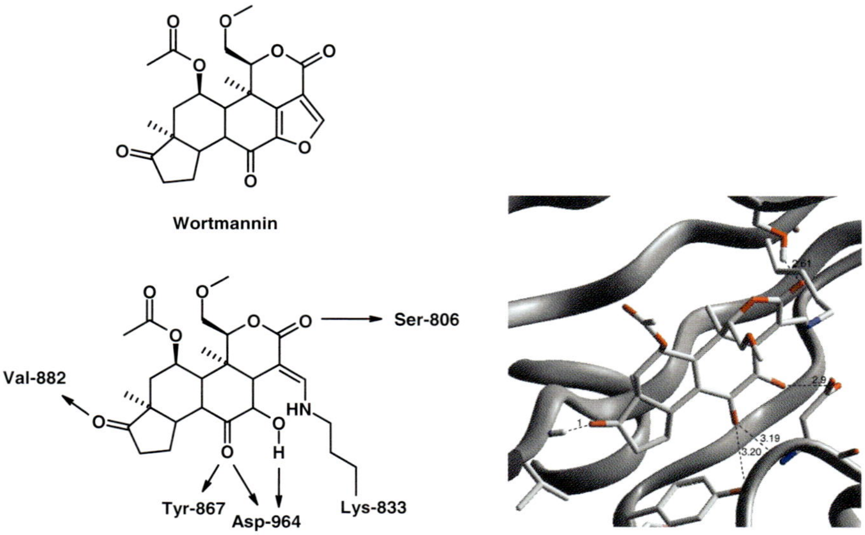

Wortmannin

Val-882 Ser-806 Tyr-867 Asp-964 Lys-833

Figure 7.13 Binding of Wortmannin to PI3K (pdb code: 1E7U).

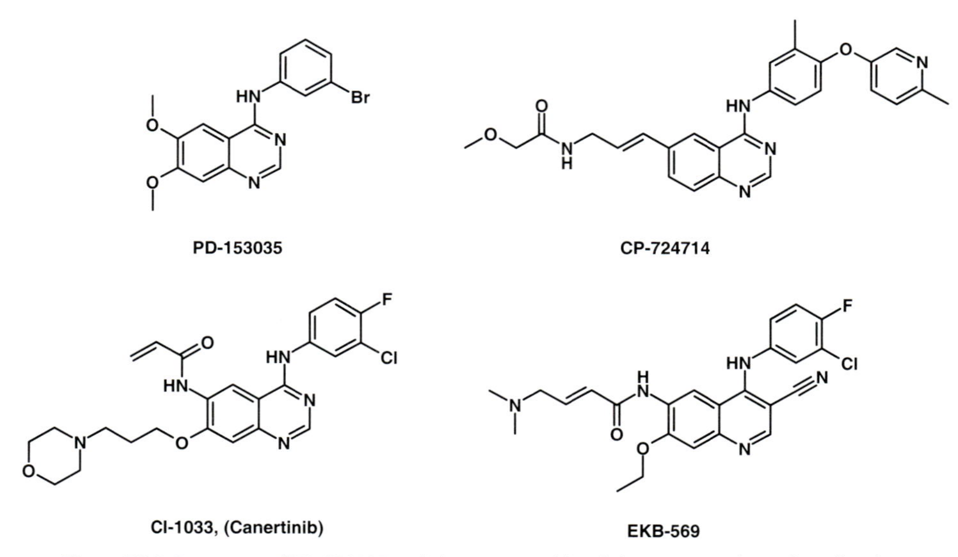

Figure 7.14 Structures of PD-153035 and three irreversible inhibitors currently in clinical trials.

quinazolines was rationally designed to target a cysteine (Cys-773) residue located in the ATP-binding site and found exclusively in EGFr kinase. The 6-acrylamido-4-anilinoquinazolines were unequivocally demonstrated to selectively bind to the catalytic domain of EGFr and alkylate Cys-773. Today, three compounds of this class are in clinical trials: CP-724714 (Pfizer, Phase 1), EKB-569 (Wyeth, Phase 2), and CI-1033 (Canertinib, Pfizer, Phase 2) (Figure 7.14 and Table 7.1).

7.4
Structural Aspects of Kinase Inhibitors

7.4.1
Kinase Inhibitor Scaffolds

According to Traxler's binding model [33] for ATP-competitive kinase inhibitors, there are five distinct subsites (Figure 7.15) within the ATP-binding site that have a distinct chemical environments and local sequence differences that enable the medicinal chemist to design potent, specific kinase inhibitors. The five subsites are:

- *Adenine-binding region (ABR)* – all ATP-competitive kinase inhibitors bind in this hydrophobic region and interact with the hinge domain via hydrogen bonds.
- *Ribose-binding pocket (RBR)* – this region is hydrophilic and is often exploited to accommodate solubilizing groups. This region is not highly conserved and contains unique residues, which could be used to direct selectivity (e.g., Cys-773 in EGFr, see Section 7.3.4).
- *Binding region I (BR-I)* – this pocket extends in the direction of the N^6 of ATP and is not involved in binding ATP. This region is not conserved and is used to improve affinity as well as selectivity. Access to this region is controlled by the so-called gatekeeper residue (see Section 7.4.2.2).
- *Binding region II (BR-II)* – this region is not accessed by ATP and could be used to obtain binding affinity and selectivity.
- *Phosphate-binding region (PBR)* – this hydrophilic region is highly solvent-exposed and seems unimportant for affecting affinity (see Section 7.2.1).

There are currently 82 different X-ray structures [34] deposited in the Protein Databank, which contain 20 different kinase inhibitor classes and 16 different kinases (Table 7.3). Figure 7.16 presents each of the 20 different kinase inhibitor classes and their respective binding modes in the kinase.

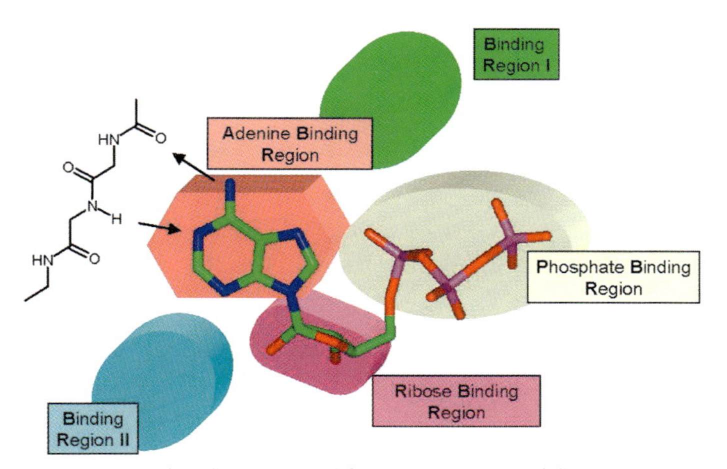

Figure 7.15 Traxler's binding model for ATP-competitive inhibitors.

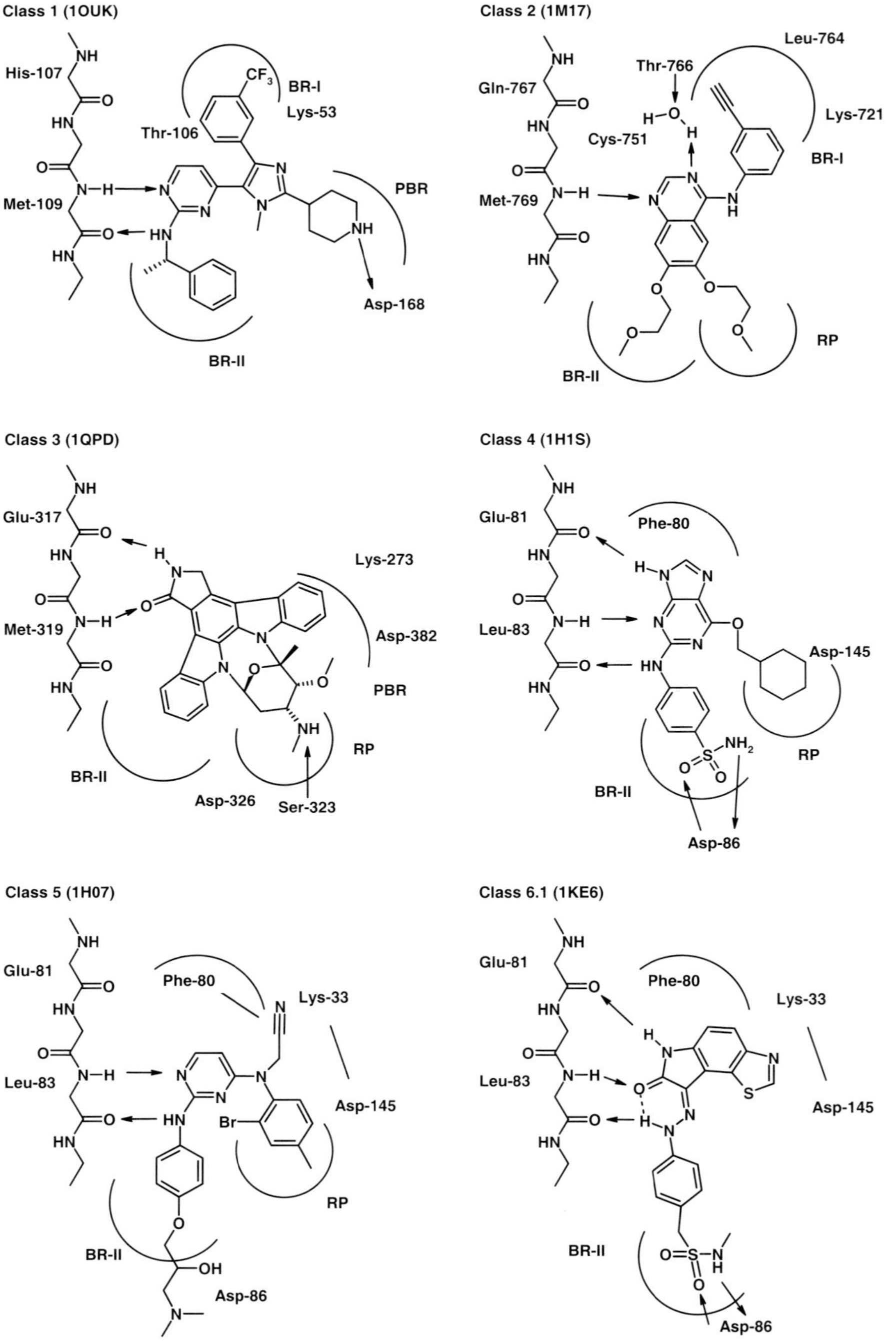

Class 1 (1OUK)

Class 2 (1M17)

Class 3 (1QPD)

Class 4 (1H1S)

Class 5 (1H07)

Class 6.1 (1KE6)

Figure 7.16 Binding modes of all known kinase inhibitor classes in their respective kinases.

Class 6.2 (1EH4)

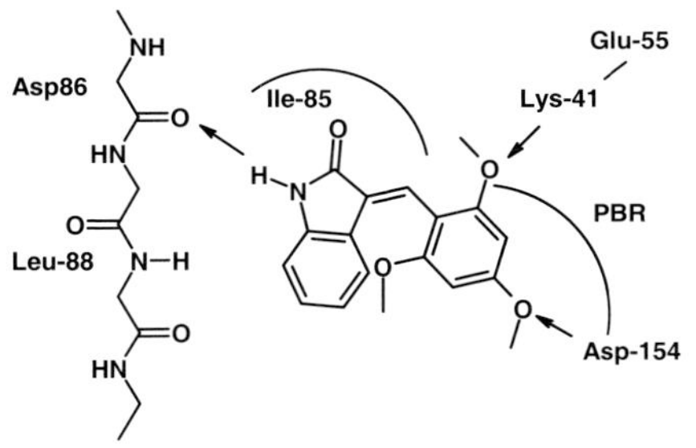

Class 7 (1GIH)

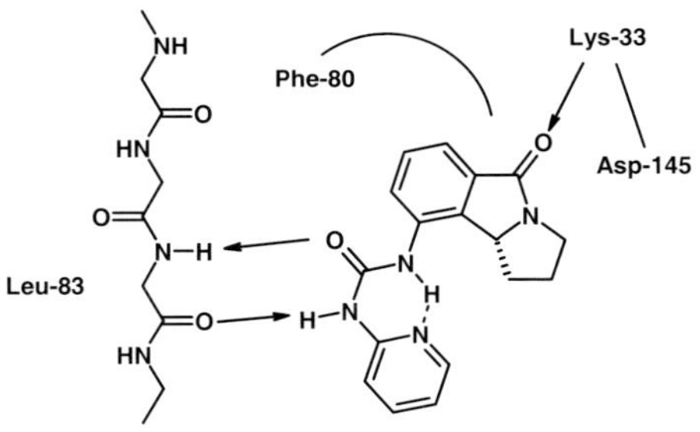

Class 8 (2FGI)

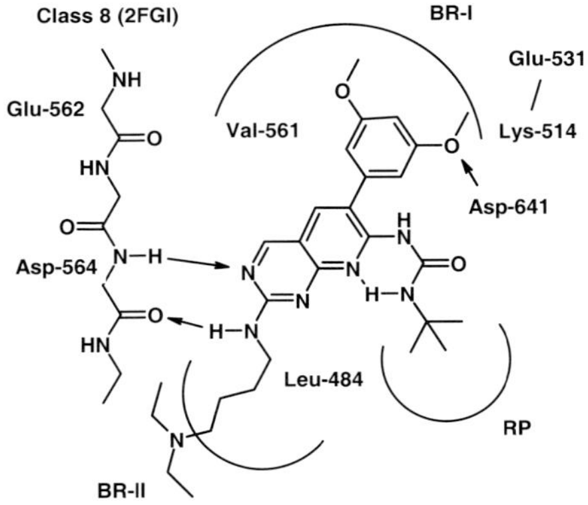

Class 9 (1QPE)

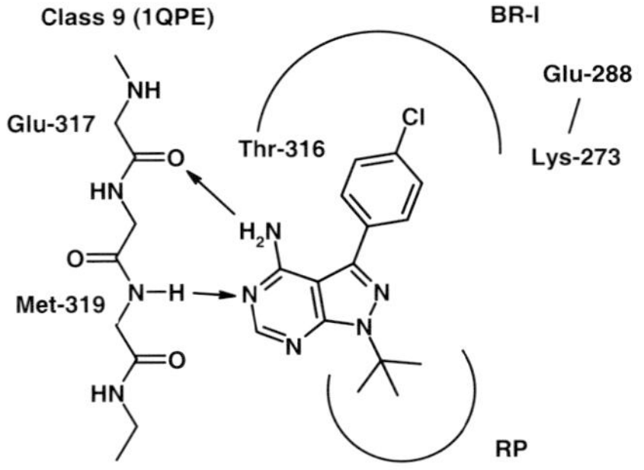

Class 10 (1PMV)

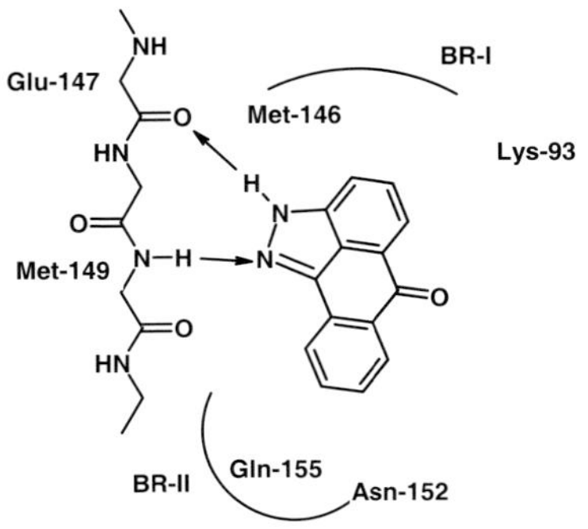

Class 11 (1DM2)

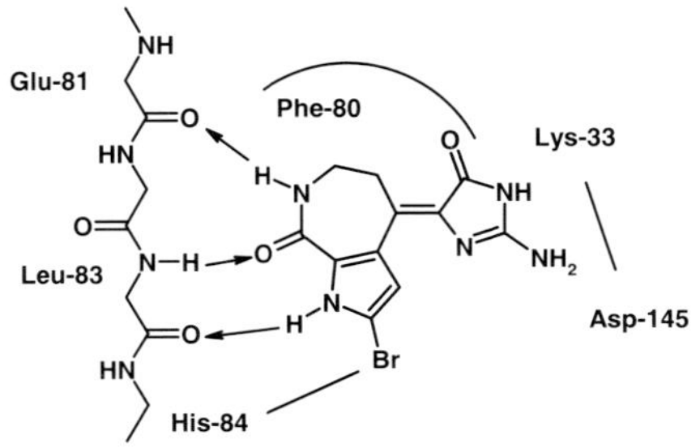

Figure 7.16 (continued)

Class 12.1 (1E90)

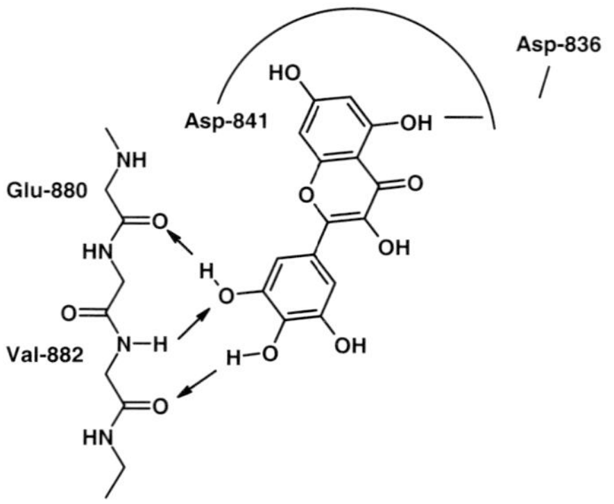

Class 12.3 (2HCK)

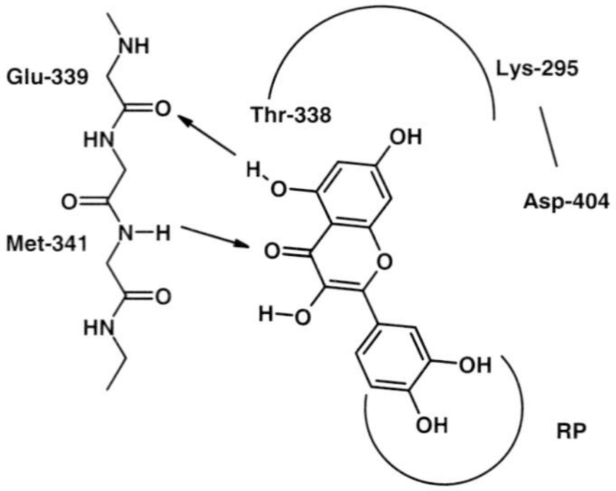

Class 12.3 (2HCK)

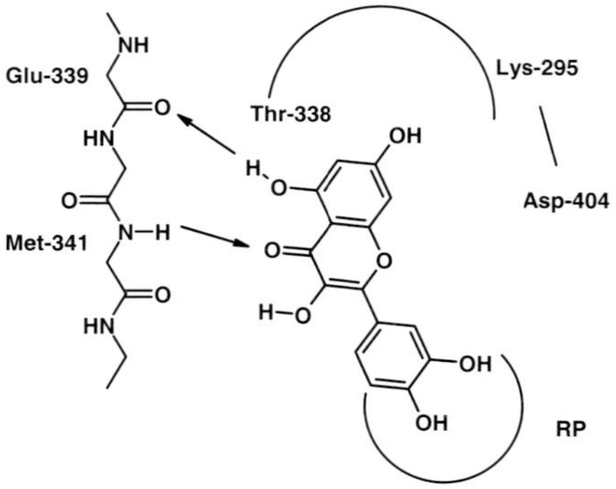

Class 13 (1PMU)

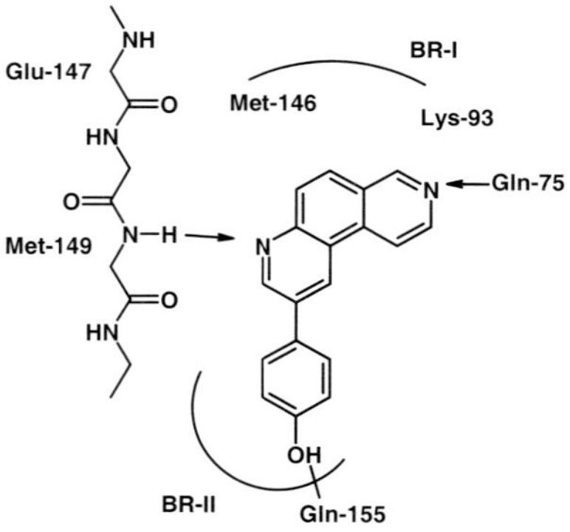

Class 14 (1G5S)

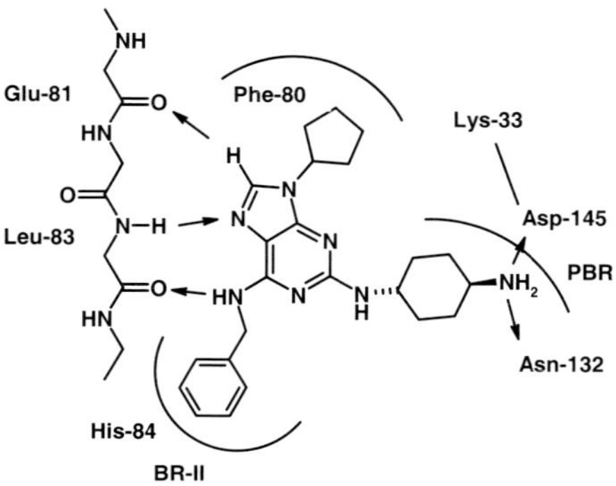

Class 15 (1BX6)

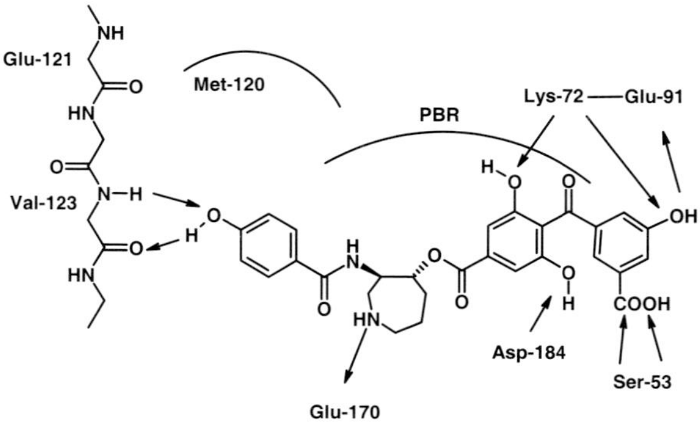

Figure 7.16 (continued)

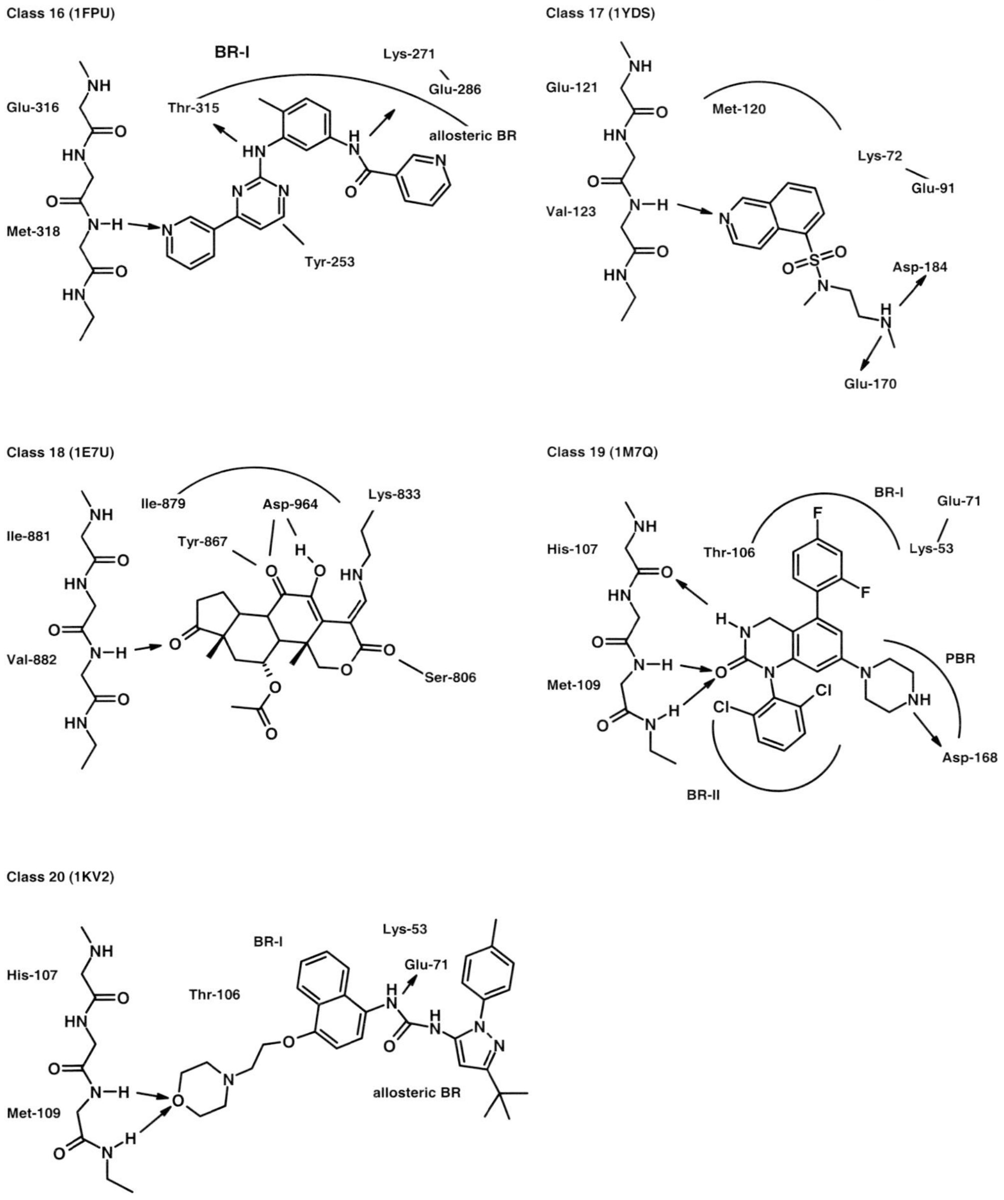

Class 16 (1FPU)

Class 17 (1YDS)

Class 18 (1E7U)

Class 19 (1M7Q)

Class 20 (1KV2)

Figure 7.16 (continued)

Table 7.3 Available kinase inhibitor classes in the Protein Databank.

No.	PDB code	Res.	Kinase	Species	Inhibitor class	IC50 (nM)	Ref.
1	1A9U	2.5	p38α	human	1	48	35
2	1BL6	2.5	p38α	human	1	160	35
3	1BL7	2.5	p38α	human	1	19	35
4	1BMK	2.4	p38α	human	1	25	35
5	1OUK	2.5	p38α	human	1	0.13	36
6	1OZ1	2.1	p38α	human	1	6	37
7	1PMN	2.2	JNK3	human	1	7.1	38
8	1PMQ	2.2	JNK3	human	1	1.6	38
9	1PME	2	ERK2	human (mutant)	1	0.76	39
10	3ERK	2.1	ERK2	rat	1	1,800	35
11	1DI8	2.2	CDK2	human	2	1,000	40
12	1DI9	2.6	p38α	human	2	5,000	40
13	1M17	2.6	EGFr	human	2	2	41, 42
14	1E8Z	2.4	PI3K	human	3	9,000	43
15	1QPD	2	LCK	human	3	40	44
16	1QPJ	2.2	LCK	human	3	40	44
17	1BYG	2.4	CSK	human	3	> 1,000	45
18	1NVQ	2	CHK1	human	3	5.6	46
19	1NVR	1.8	CHK1	human	3	7.8	46
20	1NVS	1.8	CHK1	human	3	15	46
21	1AQ1	2	CDK2	human	3	7	47
22	1PKD	2.3	CDK2	human	3	30	46, 48
23	1STC	2.3	cAPK	bovine	3	8*	49
24	1H0V	1.9	CDK3	human	4	35% @ 100 μM	50
25	1E1V	1.95	CDK2	human	4	12,000*	51
26	1GZ8	1.3	CDK2	human	4	31% @ 100 μM	50
27	1H0U	2	CDK2	human	4	40% @ 100 μM	50
28	1H1P	2.1	CDK2	human	4	12,000	52
29	1H1Q	2.5	CDK2	human	4	1,000	52
30	1H1R	2	CDK2	human	4	2,300	52
31	1H1S	2	CDK2	human	4	6	52
32	1H00	1.6	CDK2	human	5	38,000	53
33	1H01	1.79	CDK2	human	5	22,000	53
34	1H06	2.31	CDK2	human	5	35,000	53
35	1H07	1.85	CDK2	human	5	3,000	53
36	1H08	1.8	CDK2	human	5	> 100,000	53
37	1JSV	1.96	CDK2	human	5	2,000*	54
38	1E1X	1.85	CDK2	human	5	1,300	51

Table 7.3 (continued)

No.	PDB code	Res.	Kinase	Species	Inhibitor class	IC50 (nM)	Ref.
39	1E9H	2.5	CDK2	human	6.1	35	55
40	1FVT	2.2	CDK2	human	6.1	60	56
41	1FVV	2.8	CDK2	human	6.1	10	56
42	1KE5	2.2	CDK2	human	6.1	560	57
43	1KE6	2	CDK2	human	6.1	5.6	57
44	1KE7	2	CDK2	human	6.1	8.9	57
45	1KE8	2	CDK2	human	6.1	1,000	57
46	1KE9	2	CDK2	human	6.1	660	57
47	1P2A	2.5	CDK2	human	6.1	12	58
48	1AGW	2.4	FGFR1	human	6.1	10,000	24
49	1FGI	2.5	FGFR1	human	6.1	20,000	24
50	1EH4	2.8	CK1	fission yeast	6.2	1,000	59
51	1GIH	2.8	CDK2	human	7	96	60
52	1GII	2	CDK4 mimic CDK2	human	7	250	60
53	1GIJ	2.2	CDK4 mimic CDK2	human	7	1,600	60
54	1OPL	3.4	c-Abl	human	8	1	61
55	2FGI	2.5	FGFR1	human	8	29	62
56	1M52	2.6	c-Abl	mouse	8	1	23
57	1QCF	2	HCK	human	9	–	63
58	1QPE	2	LCK	human	9	20	65
59	1JVP	1.53	CDK2	human	10	1,600	64
60	1PMV	2.5	JNK3	human	10	150	65
61	1DM2	2.1	CDK2	human	11	70	66
62	1E90	2.7	PI3K	wild boar	12.1	1,800	43
63	1E8W	2.5	PI3K	wild boar	12.2	3,800	43
64	2HCK	3	HCK	human	12.3	–	67
65	1PMU	2.7	JNK3	human	13	590	65
66	1CKP	2.05	CDK2	human	14	–	68
67	1G5S	2.61	CDK2	human	14	48	69
68	1BX6	2.1	cAPK	mouse	15	4.7*	70, 71
69	1FPU	2.4	c-Abl	mouse	16	400	72, 75
70	1OPJ	1.8	c-Abl	mouse	16	38	73, 75
71	1IEP	2.1	c-Abl	mouse	16	38	74, 75
72	1YDR	2.2	cAPK	bovine	17	3,000	76
73	1YDS	2.2	cAPK	bovine	17	1,200	76
74	1YDT	2.3	cAPK	bovine	17	48	76
75	2CSN	2.5	CK1	fission yeast	17	8,500*	77
76	1E7U	2	PI3K	wild boar	18	4.2	43
77	1M7Q	2.4	p38α	human	19	2.6	78
78	1OVE	2.1	p38α	human	19	4.3	36
79	1OUY	2.5	p38α	human	19	0.74	36

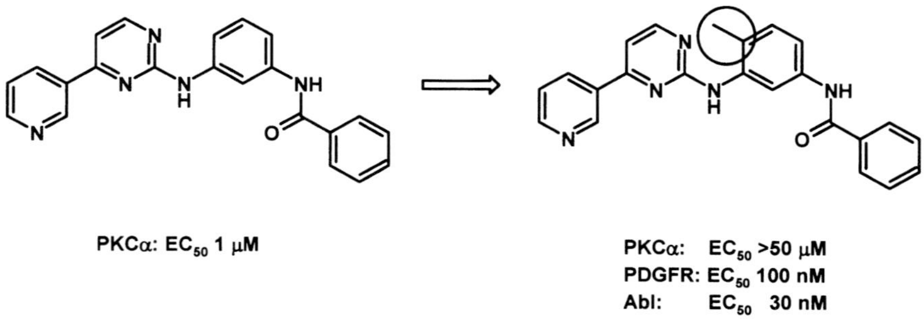

PKCα: EC_{50} 1 μM

PKCα: EC_{50} >50 μM
PDGFR: EC_{50} 100 nM
Abl: EC_{50} 30 nM

Figure 7.17 Selectivity of Gleevec is due to a single methyl group.

7.4.2 Selectivity Issues

7.4.2.1 The Selectivity Dogma

As long ago as the 1990s the possibility of selective kinase inhibition was demonstrated by medicinal chemists at Novartis. This group found that introduction of a single methyl group (Figure 7.17) changed an unselective PKCα inhibitor into a potent and selective PDGFR and c-Abl kinase inhibitor [33]. The underlying rationale for this induced selectivity is a forced change in the preferred conformation of the phenyl ring resulting in a steric clash with the ATP-binding site of PKCα kinase. The identification and, especially, the fast approval of Gleevec dispelled the dogmatic view of the pharmaceutical industry that selective kinase inhibition was impossible to achieve.

7.4.2.2 The Gatekeeper

As mentioned in Section 7.3.3, Gleevec binds to an extended form of the hydrophobic binding region I, access to which is controlled by a so-called 'gatekeeper' residue (Thr-315). The importance of this residue is demonstrated in Gleevec-resistant CML patients, who most commonly have a T315I mutation, thus locking up the hydrophobic pocket. [79]. By expressing the T315I mutant form of c-Abl kinase, it was subsequently shown that Gleevec was unable to bind to this mutant form [80].

Another well explored example of a gatekeeper residue determining selectivity is the binding of SB-203580 to p38α and p38β. MAP kinases p38α and β both contain a threonine (Thr-106) as a gatekeeper residue, which exposes the hydrophobic binding region I to the p-fluorophenyl group of SB-203580 (Figure 7.18 a). Other related kinases, such as p38γ, p38δ, and the JNK family are unaffected by SB-203580, because they lack the Thr gatekeeper and have a methionine residue instead blocking entrance to region I (Figure 7.18 b). The gatekeeper theory is supported by mutagenesis studies [81], which indicate that sensitivity to SB-203580 is reduced 10 fold with a T106M mutation. Moreover, binding to p38γ,δ and JNK-1 can be observed by a gatekeeper change to Thr. Selectivity of SB-203580 over other, less closely related kinases can be anticipated by the nature of their respective gatekeeper

Table 7.4 Correlation between activity of SB203580 on a kinase and identity of its gatekeeper residue.

Kinase	*Gatekeeper residue*	*SB203580 (IC_{50}, nM)*
p38α	Thr	48
p38β	Thr	50
p38γ	Met	> 10 000
p38δ	Met	> 10 000
UNK1	Met	~ 5 000
JNK2β1	Met	280
JNK2α2	Met	1 900
ERK2	Gln	> 100 000
MEK-1	Met	61 000
CDC2	Phe	> 50 000
PKA	Met	83 000
PKC-β2	Met	> 50 000
TGF-βI	Ser	20 000
TGF-βII	Thr	40 000
c-Raf	Thr	360
MAPKAP-K2	Met	> 10 000
ZAP-70	Met	> 20 000
LCK	Thr	20 000
EGFr	Thr	10 000

residues. Lee and coworkers [82] identified a nice correlation between the activity of SB203580 against a particular kinase and its corresponding gatekeeper residue (Table 7.4). In several instances, however, the gatekeeper residue is identical to p38α, and still a low activity is observed for these kinases. Furthermore, one should be very careful not to overemphasize the effect of the gatekeeper residue. A recent report [65] on a JNK3 crystal structure together with an analog of SB-203580 (i.e., compound **3**, Figure 7.19) showed that Met-146 no longer blocks the entrance for the lipophilic dichlorophenyl moiety and adopts a different conformation, moving 3 Å to the back (Figure 7.19). In this conformation the sulfur atom of Met-146 is engaged in a favorable [83] contact with the dichlorophenyl ring of the inhibitor.

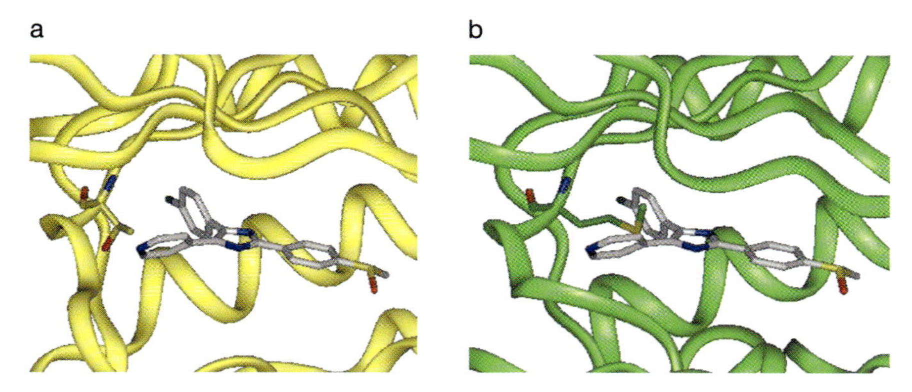

Figure 7.18
a) Binding of SB203580 to P38 Map kinase (pdb code 1A9U).
b) Binding of SB203580 to JNK-3.

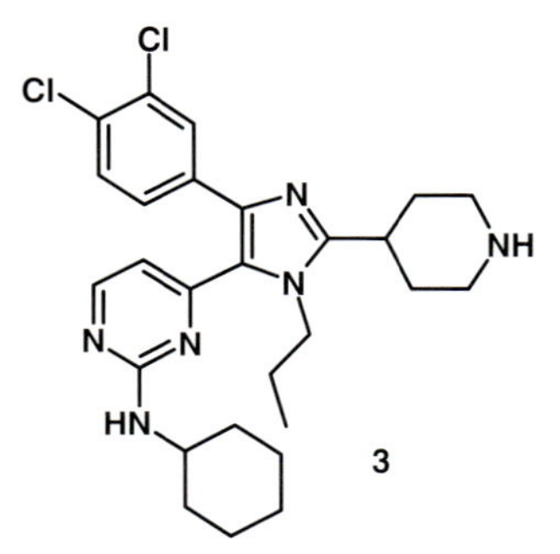

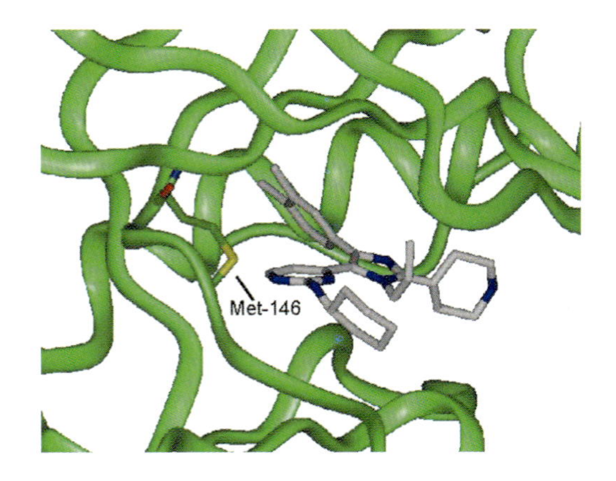

Figure 7.19 Structure of compound **3** and its binding to JNK-3 (pdb code: 1PMQ).

7.4.2.3 **Hinge-directed Selectivity**

As seen in Section 7.4.1, most kinase inhibitors share a hydrogen bond-acceptor and -donor pattern complementary to the hinge domain of their target kinase, which is necessary to avoid repulsion upon interaction. The dihydropyridopyrimidinone class of p38 MAP kinase inhibitor [78] is unique in this respect, having a carbonyl hydrogen bond-accepting group that is not complementary to the lower mainchain carbonyl group of the hinge domain. To avoid repulsion, a peptide flip between Met-109 and Gly-110 of the hinge domain in p38 occurs, thus allowing formation of two hydrogen bonds between the mainchain nitrogens of Met-109 and Gly-110 and the carbonyl group of the inhibitor (compound **4**, Figure 7.20). It is known that for this type of peptide flip to occur without a significant loss of binding energy, the hinge domain should have an X-Gly motif [84]. Glycine-110 is a residue that is conserved only in the p38 α,β, and γ isoforms, hence providing a perfect explanation for the observed selectivity of the dihydropyridopyrimidinone inhibitors for these isoforms. Constructed mutants of p38α (G110A and G110D) supported this hypothesis, showing a reduced inhibition by this type of inhibitor [36].

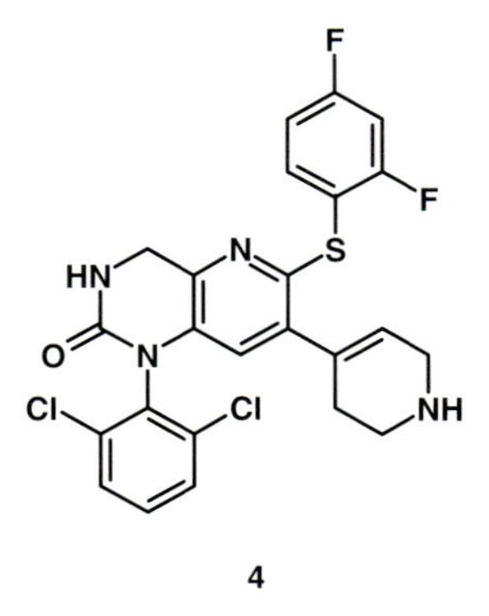

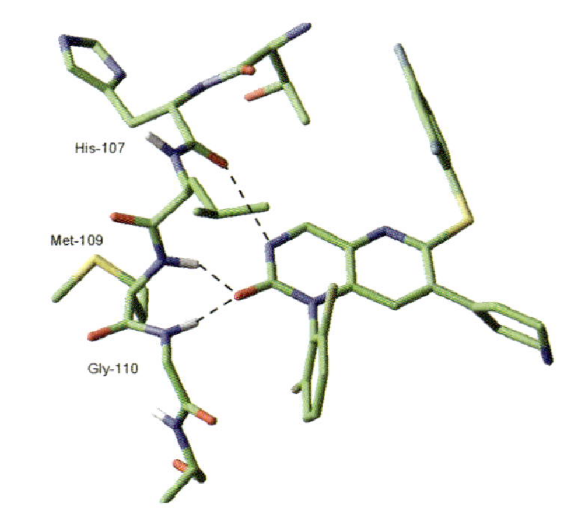

Figure 7.20 Structure of compound **4** and its binding to p38 (pdb code: 1OUY).

7.4.2.4 Binding Region II-directed Selectivity

Selective inhibition of the target kinase can also be obtained by addressing binding region II. The Banyu Tsukuba Research Institute has explored this particular strategy in their search for selective cyclin-dependent kinase 4 (CDK4) inhibitors [60, 85, 86], which are believed to suppress tumor growth by G_1 arrest.

A very elegant de-novo design protocol making use of CDK2 structural information enabled them to identify a very potent and selective CDK inhibitor (compound **5**, Figure 7.21). Subsequently, an attempt was made to improve the selectivity of the compound toward CDK4. By analyzing a sequence alignment of CDK4 and CDK2, they anticipated that CDK4 could accommodate larger substituents in the solvent-exposed binding region II, because of a difference in one residue (Thr-102 in CDK4 vs. Lys-89 in CDK2). Creating a steric repulsion in CDK2 by equipping the inhibitor with an aminochloroindanyl moiety at a suitable position enormously improved its selectivity toward CDK4 (compound **6**, Figure 7.21 and Table 7.5).

A reversed CDK selectivity profile could be obtained by targeting Lys-89, as illustrated by the discovery of Purvalanol B, which is a potent and selective CDK2 inhibitor having almost no activity against CDK4 [87]. The crystal structure of Purvalanol B in CDK2 revealed that Lys-89 makes a strong salt bridge with the carboxylate moiety of Purvalanol B. This interaction improved the potency towards CDK2 10 fold and the selectivity towards CDK4 more than 100 fold compared to Purvalanol A, which lacks the carboxylate function (Table 7.5).

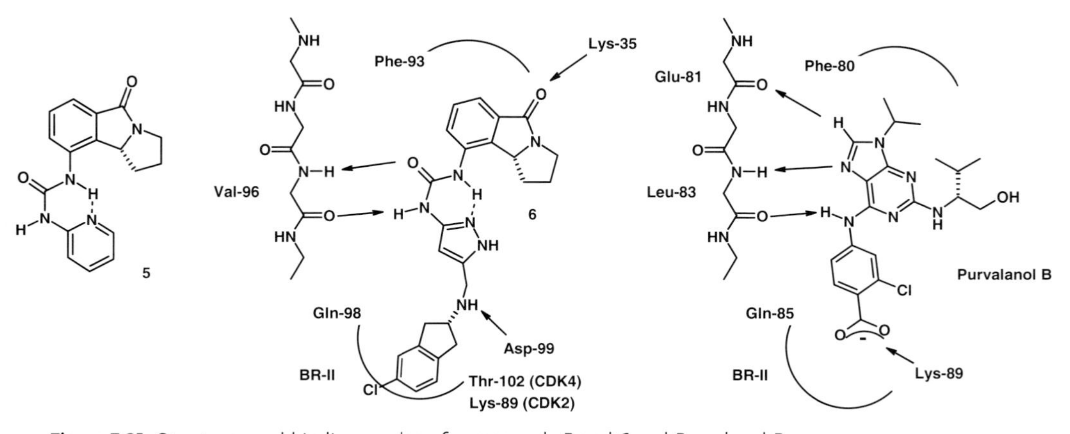

Figure 7.21 Structures and binding modes of compounds **5** and **6** and Purvalanol B.

Table 7.5 Activity and selectivity of CDK inhibitors **4** and **5** and Purvalanol A and B.

Compound	*CDK2 (nM)*	*CDK 4 (nM)*	*CDK2/CDK4*	*CDK4/CDK2*
5	78	42	1.9	
6	189	1.6	118.1	
Purvalanol A	70	850		12.1
Purvalanol B	6	> 10 000		> 1667

7.5 Outlook

This has been an exciting decade for researchers in the kinase field. High-throughput screening of various compound collections and structural insights into the human 'kinome' have enabled the development of an impressive repertoire of kinase inhibitors. As a result, a new generation of potent, selective kinase inhibitors is now under clinical evaluation, which will certainly emphasize the maturity of this research field and change the view that Gleevec and Iressa are 'just' exceptions to the rule.

Acknowledgments

I would like to thank G. Müller, C. A. A. van Boeckel, J. de Man, and R. Azevedo for their assistance and contributing discussions.

References

1 G. Manning, D. B. Whyte, R. Martinez, T. Hunter, S. Sudarsanam, *Science* **2002**, *298*, 1912–1934.
2 C. R. Groom, A. L. Hopkins, *Drug Discovery Today* **2002**, *7*, 801–802.
3 G. Scapin, *Drug Discovery Today* **2002**, *7*, 601–611.
4 (a) J. Dumas, *Exp. Opin. Ther. Patents* **2001**, *11*, 405–429; (b) S. E. Wilkinson, W. Harris, *Emerging Drugs* **2000**, *5*, 287–297; (c) P. Traxler, P. Furet, *Pharmacol. Ther.* **1999**, *82*, 195–206; (d) L. M. Toledo, N. B. Lydon, D. Elbaum, *Curr. Med. Chem.* **1999**, *6*, 775; (e) A. Bridges, *Chem. Rev.* **2001**, *101*, 2541–2571; (f) D. H. Williams, T. Michell, *Curr. Opin. Pharmacol.* **2002**, *2*, 1–7; (g) P. Cohen, *Nat. Rev. Drug Disc.* **2002**, *1*, 309–315; (h) J. Dancey, E. A. Sausville, *Nat. Rev. Drug Disc.* **2003**, *2*, 296–313; (i) J. A. Adams, *Chem. Rev.* **2001**, *101*, 2271–2290.
5 D. R. Knighton, J. Zheng, L. F. Ten Eyck, V. A. Ashford, N.-H. Xuong, S. S. Taylor, J. M. Sowadski, *Science* **1991**, *253*, 407.
6 M. Huse, J. Kuriyan, *Cell* **2002**, *109*, 275.
7 T. G. Davies, P. Tunnah, L. Meijer, D. Marko, G. Eisenbrand, J. A. Endicott, M. E. M. Noble, *Structure* **2001**, *9*, 389.
8 The residue numbering used in this section corresponds to the CDK2 numbering unless stated otherwise.
9 S. S. Taylor, E. Radzio-Andezelm, Madhusudan, X. Cheng, L. T. Eyck, N. Narayana, *Pharmacol. Ther.* **1999**, *82*, 133.
10 J. Hoppe, W. Freist, R. Marutzky, S. Shaltiel, *Eur. J. Biochem.* **1978**, *90*, 427.
11 U. Schulze-Gahmen, J. Brandsen, H. D. Jones, D. O. Morgan, L. Meijer, J. Vesely, S. H. Kim, *Proteins* **1995**, *22*, 378.
12 P. D. Jeffrey, A. A. Russo, K. Polyak, E. Gibbs, J. Hurwitz, J. Massague, N. P. Pavletich, *Nature* **1995**, *376*, 313.
13 A. A. Russo, P. D. Jeffrey, N. P. Pavletich, *Nat. Struct. Biol.* **1996**, *3*, 696.
14 N. R. Brown, M. E. Noble, J. A. Endicott, L. N. Johnson, *Nat. Cell Biol.* **1999**, *1*, 438.
15 F. D. Boehmer, L. Karagyozov, A. Uecker, H. Serve, A. Botzki, S. Mahboobi, S. Dove, *J. Biol. Chem.* **2003**, *7*, 5148.
16 R. D. Mitchell, D. B. Glass, C. Wong, K. Angelos, D. A. Walsh, *Biochemistry* **1995**, *34*, 528.
17 J. Hawkins, S. Zheng, B. Frantz, P. LoGrasso, *Arch. Biochem. Biophys.* **2000**, *382*, 310.

18 A. Gazit, P. Yaish, C. Gilon, A. Levitski, *J. Med. Chem.* **1989**, *32*, 2344.
19 G. Blum, A. Gazit, A. Levitski, *Biochemistry* **2000**, *39*, 15705.
20 K. Parang, J. H. Till, A. J. Ablooglu, R. A. Kohanski, S. R. Hubbard, P. A. Cole, *Nat. Struct. Biol.* **2001**, *8*, 37.
21 D. Lawrence, J. Niu, *Pharmacol. Ther.* **1998**, *77*, 81.
22 T. Schindler, W. Bornmann, P. Pellicena, W. T. Miller, B. Clarkson, J. Kuriyan, *Science* **2000**, *289*, 1938.
23 B. Nagar, W. Bornmann, P. Pellicena, T. Schindler, D. R. Veach, W. T. Miller, B. Clarkson, J. Kuriyan, *Cancer Res.* **2002**, *62*, 4236.
24 M. Mohammadi, G. McMahon, L. Sun, C. Tang, P. Hirth, B. K. Yeh, S. R. Hubbard, *Science* **1997**, *276*, 955.
25 G. Scapin, S. B. Patel, J. Lisnock, J. W. Becker, P. V. Lograsso, *Chem. Biol.* **2003**, *10*, 705.
26 F. D. Boehmer, L. Karagyozov, A. Uecker, H. Serve, A. Botzki, S. Mahboobi, S. Dove, *J. Biol. Chem.* **2003**, *7*, 5148.
27 C. Pargellis, L. Tong, L. Churchill, P. F. Cirillo, T. Gilmore, A. G. Graham, P. M. Grob, E. R. Hickey, N. Moss, S. Pav, J. Regan, *Nat. Struct. Biol.* **2002**, *9*, 268.
28 R. Schwartz, *Annu. Sci. Meeting Amer. Coll. Rheumatol.* Oct. 23–28 **2003**, Abstr. BR2.
29 H. Tecle, *Protein Kinases: New Therapies and New Technologies*, Amsterdam, Netherlands, 22–23 October **2003**, p. 15.
30 P. W. Brian, P. J. Curtis, H. G. Hemming, G. L. F. Norris, *Trans. Br. Mycol.* **1957**, *40*, 365.
31 S. P. Davies, H. Reddy, M. Caivano, P. Cohen, *Biochem. J.* **2000**, *351*, 95–105.
32 M. P. Wymann, G. Bulgarelli-Leva, M. J. Zvelebil, L. Pirola, B. Van Haesebroeck, M. D. Waterfield, G. Panayotou, *Mol. Cell. Biol.* **1996**, *16*, 1722.
33 P. Traxler, P. Furet, *Pharmacol. Ther.* **1999**, *82*, 195.
34 Kinase structures in their apo form or binary structures with ATP or stable analogs thereof have been omitted in this analysis.
35 Z. Wang, B. J. Canagarajah, J. C. Boehm, S. Kassisa, M. H. Cobb, P. R. Young, S. Abdel-Meguid, J. L. Adams, E. J. Goldsmith, *Structure (London)* **1998**, *6*, 1117.
36 C. E. Fitzgerald, S. B. Patel, J. W. Becker, P. M. Cameron, D. Zaller, V. B. Pikounis, S. J. O'Keefe, G. Scapin, *Nat. Struct. Biol.* **2003**, *10*, 764.
37 #. Trejo, H. Arzeno, M. Browner, S. Chanda, S. Cheng, D. D. Comer, S. A. Dalrymple, P. Dunten, J.-A. Lafargue, B. Lovejoy, J. Freire-Moar, J. Lim, J. McIntosh, J. Miller, E. Papp, D. Reuter, R. Roberts, F. Sanpablo, J. Saunders, K. Song, A. Villasenor, S. D. Warren, M. Welch, P. Weller, P. E. Whiteley, L. Zeng, D. M. Goldstein, *J. Med. Chem.* **2003**, *46*, 4702.
38 G. Scapin, S. B. Patel, J. Lisnock, J. W. Becker, P. V. Lograsso, *Chem. Biol.* **2003**, *10*, 705.
39 T. Fox, J. T. Coll, X. Xie, P. J. Ford, U. A. Germann, M. D. Porter, S. Pazhanisamy, M. A. Fleming, V. Galullo, M. S. Su, K. P. Wilson, *Protein Sci.* **1998**, *7*, 2249.
40 L. Shewchuk, A. Hassell, B. Wisely, W. Rocque, W. Holmes, J. Veal, L. F. Kuyper, *J. Med. Chem.* **2000**, *43*, 133.
41 J. Stamos, M. X. Sliwkowski, C. Eigenbrot, *J. Biol. Chem.* **2002**, *277*, 46265.
42 J. D. Moyer, E. G. Barbacci, K. K. Iwata, L. Arnold, B. Boman, A. Cunningham, C. DiOrio, J. Doty, M. J. Morin, M. P. Moyer, M. Neveu, V. A. Pollack, L. R. Pustilnik, M. M. Reynolds, D. Sloan, A. Theleman, P. Miller, *Cancer Res.* **1997**, *57*, 4838.
43 E. H. Walker, M. E. Pacold, O. Perisic, L. Stephens, P. T. Hawkins, M. P. Whymann, R. L. Williams, *Mol. Cell.* **2000**, *6*, 909.
44 X. Zhu, J. L. Kim, P. E. Rose, D. R. Stover, L. M. Toledo, H. Zhao, K. A. Morgenstern, *Structure (London)* **1999**, *7*, 651.
45 M. B. A. C. Lamers, A. A. Antson, R. E. Hubbard, R. K. Scott, D. H. Williams, *J. Mol. Biol.* **1999**, *285*, 713.
46 B. Zhao, M. J. Bower, P. J. McDevitt, H. Zhao, S. T. Davis, K. O. Johanson, S. M. Green, N. O. Concha, B. B. Zhou, *J. Biol. Chem.* **2002**, *277*, 46609.

47 A. M. Lawrie, M. E. Noble, P. Tunnah, N. R. Brown, L. N. Johnson, J. A. Endicott, *Nat. Struct. Biol* **1997**, *4*, 796.
48 L. N. Johnson, E. de Moliner, N. R. Brown, H. Song, to be published.
49 L. Prade, R. A. Engh, A. Girod, V. Kinzel, R. Huber, D. Bossemeyer, *Structure (London)* **1997**, *5*, 1627.
50 A. E. Gibson, C. E. Arris, J. Bentley, F. T. Boyle, N. J. Curtin, T. G. Davies, J. A. Endicott, B. T. Golding, S. Grant, R. J. Griffin, P. Jewsbury, L. N. Johnson, V. Mesguiche, D. R. Newell, M. E. Noble, J. A. Tucker, H. J. Whitfield, *J. Med. Chem.* **2002**, *45*, 3381.
51 C. E. Arris, F. T. Boyle, A. H. Calvert, N. J. Curtin, J. A. Endicott, E. F. Garman, A. E. Gibson, B. T. Golding, S. Grant, R. J. Griffin, P. Jewsbury, L. N. Johnson, A. M. Lawrie, D. R. Newell, M. E. M. Noble, E. A. Sausville, R. Schultz, W. Yu, *J. Med. Chem.* **2000**, *43*, 2797.
52 T. G. Davies, J. Bentley, C. E. Arris, F. T. Boyle, N. J. Curtin, J. A. Endicott, A. E. Gibson, B. T. Golding, R. J. Griffin, I. R. Hardcastle, P. Jewsbury, L. N. Johnson, V. Mesguiche, D. R. Newell, M. E. M. Noble, J. A. Tucker, L Wang, H. J. Whitfield, *Nat. Struct. Biol.* **2002**, *9*, 745.
53 J. F. Beattie, G. A. Breault, R. P. A. Ellston, S. Green, P. J. Jewsbury, C. J. Midgley, R. T. Naven, C. A. Minshull, R. A. Pauptit, J. A. Tucker, J. E. Pease, *Bioorg. Med. Chem. Lett.* **2003**, *13*, 2955.
54 P. M. Clare, R. A. Poorman, L. C. Kelley, K. D. Watenpaugh, C. A. Bannow, K. L. Leach, *J. Biol. Chem.* **2001**, *276*, 48292.
55 T. G. Davies, P. Tunnah, L. Meijer, D. Marko, G. Eisenbrand, J. A. Endicott, M. E. Noble, *Structure (Cambridge)* **2001**, *9*, 389.
56 S. T. Davis, B. G. Benson, H. N. Bramson, D. E. Chapman, S. H. Dickerson, K. M. Dold, D. J. Eberwein, M. Edelstein, S. V. Frye, R. T. Gampe Jr., R. J. Griffin, P. A. Harris, A. M. Hassell, W. D. Holmes, R. N. Hunter, V. B. Knick, K. Lackey, B. Lovejoy, M. J. Luzzio, D. Murray, P. Parker, W. J. Rocque, L. Shewch, *Science* **2001**, *291*, 134.
57 H. N. Bramson, J. Corona, S. T. Davis, S. H. Dickerson, M. Edelstein, S. V. Frye, R. T. Gampe, A. M. Hassell, L. M. Shewchuk, L. F. Kuyper, *J. Med. Chem.* **2001**, *44*, 4339.
58 J.-J. Liu, A. Dermatakis, C. M. Lukacs, F. Konzelmann, Y. Chen, U. Kammlott, W. Depinto, H. Yang, X. Yin, Y. Chen, A. Schutt, M. E. Simcox, K.-C. Luk, *Bioorg. Med. Chem.* **2003**, *13*, 2465.
59 N. Mashoon, A. J. Demaggio, V. Tereshko, S. C. Bergmeier, M. Egli, M. F. Hoekstra, J. Kuret, *J. Biol. Chem.* **2000**, *275*, 20052.
60 M. Ikuta, K. Kamata, K. Fukasawa, T. Honma, T. Machida, H. Hirai, I. Suzuki-Takahashi, T. Hayama, S. Nishimura, *J. Biol. Chem.* **2001**, *276*, 27548.
61 B. Nagar, O. Hantschel, M. A. Young, K. Scheffzek, D. Veach, W. Bornmann, B. Clarkson, G. Superti-Furga, J. Kuriyan, *Cell* **2003**, *112*, 859.
62 M. Mohammadi, S. Froum, J. M. Hamby, M. C. Schroeder, R. L. Panek, G. H. Lu, A. V. Eliseenkova, D. Green, J. Schlessinger, S. R. Hubbard, *EMBO* **1998**, *17*, 5896.
63 T. Schindler, F. Sicheri, A. Pico, A. Gazit, A. Levitzki, J. Kuriyan, *Mol. Cell.* **1999**, *3*, 639.
64 P. Furet, T. Meyer, A. Strauss, S. Raccuglia, J. M. Rondeau, *Bioorg. Med. Chem. Lett.* **2002**, *12*, 221.
65 G. Scapin, S. B. Patel, J. Lisnock, J. W. Becker, P. V. Lograsso, *Chem. Biol.* **2003**, *10*, 705.
66 L. Meijer, A. M. Thunnissen, A. W. White, M. Garnier, M. Nikolic, L. H. Tsai, J. Walter, K. E. Cleverley, P. C. Salinas, Y. Z. Wu, J. Biernat, E. M. Mandelkow, S.-H. Kim, G. R. Pettit, *Chem. Biol.* **2000**, *7*, 51.
67 F. Sicheri, I. Moarefi, J. Kuriyan, *Nature* **1997**, *385*, 602.
68 N. S. Gray, L. Wodicka, A. M. Thunnissen, T. C. Norman, S. Kwon, F. H. Espinoza, D. O. Morgan, G. Barnes, S. LeClerc, L. Meijer, S. H. Kim, D. J. Lockhart, P. G. Schultz, *Science* **1998**, *281*, 533.

69 M. K. Dreyer, D. R. Borcherding, J. A. Dumont, N. P. Peet, J. T. Tsay, P. S. Wright, A. J. Bitonti, J. Shen, S.-H. Kim, *J. Med. Chem.* **2001**, *44*, 524.

70 N. Narayana, T. C. Diller, K. Koide, M. E. Bunnage, K. C. Nicolaou, L. L. Brunton, N. H. Xuong, L. F. ten Eyck, S. S. Taylor, *Biochemistry* **1999**, *38*, 2367.

71 K. Koide, M. E. Bunnage, L. Gomez Paloma, J. R. Kanter, S. S. Taylor, L. L. Brunton, K. C. Nicolaou, *Chem. Biol.* **1995**, *2*, 601.

72 T. Schindler, W. Bornmann, P. Pellicena, W. T. Miller, B. Clarkson, J. Kuriyan, *Science* **2000**, *289*, 1938.

73 O. Hantschel, B. Nagar, S. Guettler, J. Kretzschmar, K. Dorey, J. Kuriyan, G. Superti-Furga, *Cell* **2003**, *112*, 845.

74 B. Nagar, W. Bornmann, P. Pellicena, T. Schindler, D. R. Veach, W. T. Miller, B. Clarkson, J. Kuriyan, *Cancer Res.* **2002**, *62*, 4236.

75 J. Zimmermann, E. Buchdunger, H. Mett, T. Meyer, N. B. Lydon, *Bioorg. Med. Chem. Lett.* **1997**, *7*, 187.

76 R. A. Engh, A. Girod, V. Kinzel, R. Huber, D. Bossemeyer, *J. Biol. Chem.* **1996**, *271*, 26157.

77 R.-M. Xu, G. Carmel, J. Kuret, X. Cheng, *Proc. Natl. Acad. Sci. U. S. A.* **1996**, *93*, 6308.

78 J. E. Stelmach, L. Liu, S. B. Patel, J. V. Pivnichny, G. Scapin, S. Singh, C. E. C. A. Hop, Z. Wang, P. M. Cameron, E. A. Nichols, S. J. O'Keefe, E. A. O'Neill, D. M. Schmatz, C. D. Schwartz, C. M. Thompson, D. M. Zaller, J. B. Doherty, *Bioorg. Med. Chem. Lett.* **2003**, *13*, 277.

79 M. E. Gorre, M. Mohammed, K. Ellwood, N. Hsu, R. Paquette, P. N. Rao, C. L. Sawyers, *Science* **2001**, *293*, 876.

80 P. L. Rosee, A. S. Corbin, E. P. Stoffregen, M. W. Deininger, B. J. Drucker, *Cancer Research* **2002**, *62*, 7149.

81 R. J. Gum, M. M. McLaughlin, S. Kumar, Z. L. Wang, M. J. Bower, J. C. Lee, J. L. Adams, G. P. Livi, E. J. Goldsmith, P. R. Young, *J. Biol. Chem.* **1998**, *273*, 15605.

82 J. C. Lee, S. Kassis, S. Kumar, A. Badger, J. L. Adams, *Pharmacol. Ther.* **1999**, *82*, 389.

83 E. A. Meyer, R. K. Castellano, F. Diederich, *Angew. Chem. Int. Ed.* **2003**, *42*, 1211.

84 S. Hayward, *Protein Sci.* **2001**, *10*, 2219.

85 T. Honma, K. Hayashi, T. Aoyama, N. Hashimoto, T. Machida, K. Fukasawa, T. Iwama, C. Ikeura, M. Ikuta, I. Suzuki-Takahashi, Y. Iwasawa, T. Hayama, S. Nishimura, H. Morishima, *J. Med. Chem.* **2001**, *44*, 4615.

86 T. Honma, T. Yoshizumi, N. Hashimoto, K. Hayashi, N. Kawanishi, K. Fukasawa, T. Takaki, C. Ikeura, M. Ikuta, I. Suzuki-Takahashi, T. Hayama, S. Nishimura, H. Morishima, *J. Med. Chem.* **2001**, *44*, 4628.

87 N. S. Gray, L. Wodicka, A. W. H. Thunnissen, T. C. Norman, S. Kwon, F. H. Espinoza, D. O. Morgan, G. Barnes, S. LeClerc, L. Meijer, S.-H. Kim, D. J. Lockhart, P. G. Schultz, *Science* **1998**, *281*, 533.

8
A Chemical Genomics Approach for Ion Channel Modulators

Karl-Heinz Baringhaus and Gerhard Hessler

8.1
Introduction

Ion channel modulators offer significant therapeutic opportunities in a number of areas, including arrhythmia, asthma, CNS disorders, coronary heart disease, hypertension, inflammation, and water retention. New ion channels are constantly being discovered and characterized in terms of their pharmacology, physiology, and structure [1]. In addition, more and more selective ion channel modulators are emerging, upon which drug discovery programs can be initiated.

The physiological effects of ion channels are based on the regulation of ion fluxes (e.g., K^+, Na^+, Ca^{2+}, Cl^-) across membranes, which affect, for example, osmotic pressure, nerve signal transmission, and muscle contraction. Ion permeation is extremely fast (up to 10^7 ions s^{-1}) and highly selective [1].

Drews et al. [2] classified ion channels as the fourth-most important target class for drug therapies after receptors, enzymes, and hormones, and a more recent analysis considered kinases, GPCRs, and cation channels to be the most interesting target classes for pharmaceutical research [3]. Currently, drugs targeting ion channels generate over 24 billion dollars in sales per annum.

Appropriate drug targets should meet several criteria, such as known biological functions, as well as robust assay systems for in vitro characterization and testing. Furthermore, they need to be accessible to low molecular weight compounds in vivo. Ion channels meet most of these 'druggability' criteria and can be viewed as suitable targets for small molecule drugs [3].

Potassium (K^+) ion channels, for example, are recognized as critical regulators of cellular activities and are linked to several disease indications, including ventricular arrhythmias, long QT syndrome, and atrial fibrillation, as well as to insulin secretion and T-cell activation [4]. The long QT syndrome, for instance, is associated with an inhibition of the hERG channel in the heart. hERG inhibition represents an important safety consideration in drug discovery. Due to their hERG blocking properties and subsequent QT interval prolongation, several diverse drugs such as Terfenadine, Cisapride, and Astemizole have been withdrawn from the

Chemogenomics in Drug Discovery: A Medicinal Chemistry Perspective.
Edited by Hugo Kubinyi and Gerhard Müller

ISBN: 3-527-30987-X

market [5]. In comparison, the voltage-gated Kv1.3 channel is of interest in therapeutic immune modulation in multiple sclerosis and other T-cell mediated autoimmune diseases, and Ca^{2+}-activated potassium channels are of interest for reducing hyperactive bladder by hyperpolarization of the smooth muscle in the bladder.

Calcium-channel blockers are used for treating cardiac arrhythmia and pulmonary hypertension and for prevention of reperfusion injury. Sodium channels have been linked to epilepsy and hyperkalemic periodic paralysis (Table 8.1).

Recently approved ion channel modulators include, for example, Nateglinide and Nimodipine. Nateglinide was approved in December 2000 as a blood glucose lowering agent. Nateglinide depolarizes pancreatic β cells by blocking the ATP-sensitive potassium (KATP) channel, whereby calcium channels are opened, resulting in calcium influx and insulin secretion. The extent of insulin release is glucose-dependent and decreases at low glucose levels. Nateglinide is highly tissue selective with low affinity for heart and muscle.

Nimodipine was approved in August 2000 for the improvement of neurological outcome by reducing the incidence and severity of ischemic deficits in patients with subarachnoid hemorrhage.

The sodium-channel inhibitor Amiloride is used for the treatment of chronic bronchitis, and the most frequently used anesthetic drug, Lidocain, inhibits voltage-gated sodium-channel α subunits, which mediate the pathophysiology of pain.

Despite their remarkable physiological value, ion channels are still an unexploited therapeutic target class, especially in comparison to G-protein coupled receptors. Hence, lead finding and lead optimization programs for ion channel modulators are becoming more and more interesting. As ion channels are strongly related to each other, a systematic exploration of this target family appears to be a promising way to accelerate drug discovery [6]. Chemical genomics refers to such systematic and in-depth exploration of a target family and fosters a knowledge-driven drug design approach [7]. This method is especially feasible for ion channel modulators, since considerable knowledge of pharmaceutically active structural classes and structure–activity relationships exists. In this chapter we summarize our current

Table 8.1 Pathophysiological conditions related to ion channels.

Channel	*Disease*
Calcium channels	Arrhythmia, diabetes, epilepsy, hypertension, migraine, stroke
Chloride channels	Cystic fibrosis, myotonia, muscoviscidos
Potassium channels	Arrhythmia, asthma, blood pressure, cardiac ischemia, cell proliferation, diabetes, epilepsy, cancer, immune suppression
Sodium channels	Epilepsy, migraine, myotonia, pain, stroke
Ligand-gated ion channels	Allergy, asthma, epilepsy, gastroesophageal reflux, inflammation, ischemia, learning and memory, migraine, neurodegenerative diseases, stroke

chemical genomics knowledge-based strategies for drug discovery of ion channel modulators. This includes structural information about ion channels, as well as lead-finding strategies in this field. The impact of this strategy is outlined by several successful examples.

We consider the highest impact of this strategy to be in lead finding, although such a target family-related approach offers further obvious advantages in the field of assay development, HTS technology, and compound optimization. In particular, the selectivity of ion channel modulators can be addressed by appropriate profiling of these compounds and by building channel-specific models applicable to lead optimization.

8.2 Structural Information on Ion Channels: Ion Channel Families

Ion channels form a large, diverse family of membrane proteins that can be grouped according to various criteria, such as the gating behavior or the ion selectivity, as shown in Table 8.2.

Classification according to such a scheme is not always simple, since ligand-gated channels like the NMDA-activated ion channel may show voltage dependence, and, on the other hand voltage-gated channels have ligand-binding sites [1]. Voltage-gated sodium channels can be activated by drugs like Veratridine, whereas the MaxiK channel is gated by calcium ions [1].

The classification of ion channels by their topology is exemplified for potassium channels in Figure 8.1 [1]. Potassium channels can be classified into 2TM/P channels, which contain two transmembrane helices (TM) with one P loop (P) between them, 6TM/P channels, 7TM/P channels, 8TM/2P channels, and 4TM/2P channels. The 4TM/2P family is called leakage channels and is targeted by numerous anesthetics [8].

The 6TM/P channel family contains six transmembrane helices, labeled S1 to S6. The S4 helix contains four to seven positively charged amino acids, which are responsible for sensing the membrane potential. Therefore, the S4 helix is called the voltage sensor. The S5 and S6 helices form the ion-conducting pore by tetramerization.

Table 8.2 Classification schemes for ion channels.

Gating	*Ion selectivity*
Voltage	Na^+
Ligand	Ca^{2+}
Mechanical	K^+
Thermal	Cl^-
	any ion, any cation, any monovalent cation

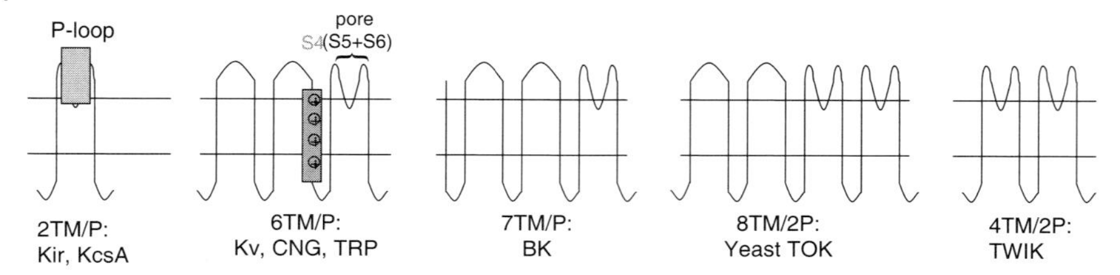

Figure 8.1 Architectures of potassium channels: different transmembrane topologies shown together with potassium channels as examples. In the 6TM family, the voltage-sensor helix S4 is highlighted, together with the pore-forming helices S5 and S6.

A number of different X-ray structures of bacterial potassium channels reveal the detailed atomic picture of the pore-forming part, helices S5 and S6 [9]. KcsA, which is crystallized in the closed conformation, has an overall structure similar to an inverted teepee [9a]. Four identical subunits surround the ion-conducting pathway (Figure 8.2). Each subunit contains two full transmembrane helices, S5 and S6, as well as the P loop. The S6 helices line the central cavity, whereas the S5 helices are involved in interactions with the lipid environment. In the closed channel conformation the transmembrane helices meet at the cytosolic side to block the ion conduction path. In the open conformation of the channel, the S6 helix kinks at a conserved glycine residue to open the ion conduction path, as shown in the structure of the bacterial channel MthK [10]. The ion conduction path is formed by the selectivity filter and the large water-filled central cavity.

The solution of potassium channel X-ray structures has significantly contributed to the understanding of mechanistic questions like the amazing selectivity of potassium channels. Although the atomic radii of potassium (1.33 Å) and of sodium (0.95 Å) differ only slightly, potassium channels select potassium over sodium ion by a factor of 1000. This tremendous selectivity is achieved by the coordination geometry of eight amide carbonyl groups in the selectivity filter, optimized for the coordination sphere of potassium ions [11].

Structural data on potassium channels has also improved the understanding of the gating mechanism. Gating comprises a signaling step and the opening of the ion conduction path. The elucidation of the structure of the bacterial voltage-gated potassium channel KvAP [9a], crystallized by using monoclonal antibody Fab fragments, yielded some unexpected insights into the design of the voltage-sensor helix S4. Mutational data demonstrated the role of the S4 helix in voltage sensing, and fluorescence labeling has shown movement of the S4 helix during gating [12, 13]. The prevailing model so far suggests a movement of helix S4 from one side of the membrane to the other upon changes in the membrane potential, although the findings of MacKinnon imply a different model. First, helix 3 is actually split into two different helices: the second part of helix 3 – called helix 3b – forms a helix–turn–helix motif with helix S4. This unit, called the 'voltage-sensor paddle', is actually oriented perpendicular to the pore unit and moves to the outer membrane side when the channel is opened.

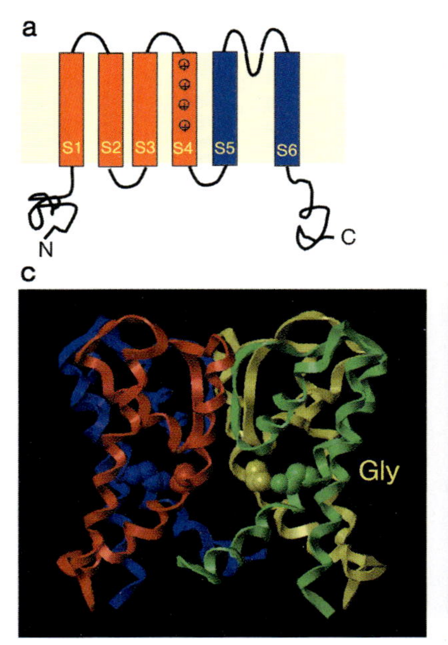

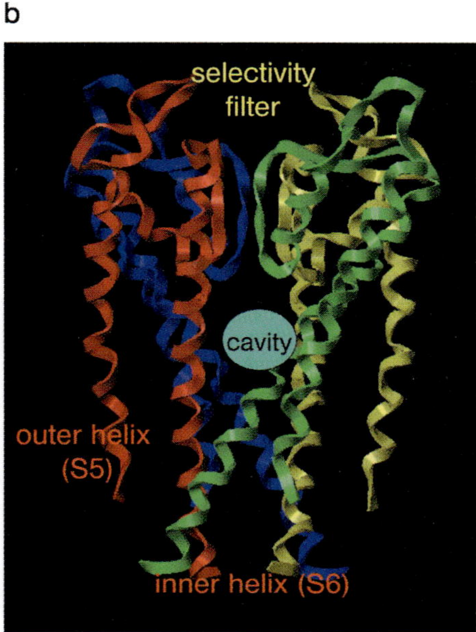

Figure 8.2
(a) Topology of 6TM/P potassium channels.
(b) X-ray structure of KcsA (PDB code: 1j95). The four monomers tetramerizing to form the functional channel are shown in different colors. Important structural features such as the S5 and S6 helices, central cavity, and selectivity filter are indicated.
(c) Structure of MthK pore (PDB code: 1LNQ). The structure is in the open channel conformation. The glycine residues that serve as a hinge for the bending of helix S6 are indicated.

Nevertheless, it is still unknown how the movement of the voltage-sensor paddle is linked to the opening of the ion conduction pathway, which is achieved by an outward bending of the S6 helix at the position of a conserved glycine. This helix movement opens the inner cavity to the cytosol, as shown by a comparison of the KcsA and MthK structures (Figure 8.2). For the inward-rectifying potassium channel family, a glycine residue in a different position could be the hinge position for formation of the opening pathway [14].

Chloride channels have a completely different structure from potassium channels [15]. The dimeric structure has two ion pathways, one formed by each monomer. The ion pathway does not run straight through the membrane, but is U-shaped. Amino acids stabilize the ion in the pathway by forming direct interactions with the chloride atom via hydrogen bond donors, just as the carbonyl groups in the selectivity filter of potassium channels stabilize the potassium cation.

Although, as described above, a couple of structures of ion channels have been solved, it is still a challenging task to express, purify, and crystallize these membrane

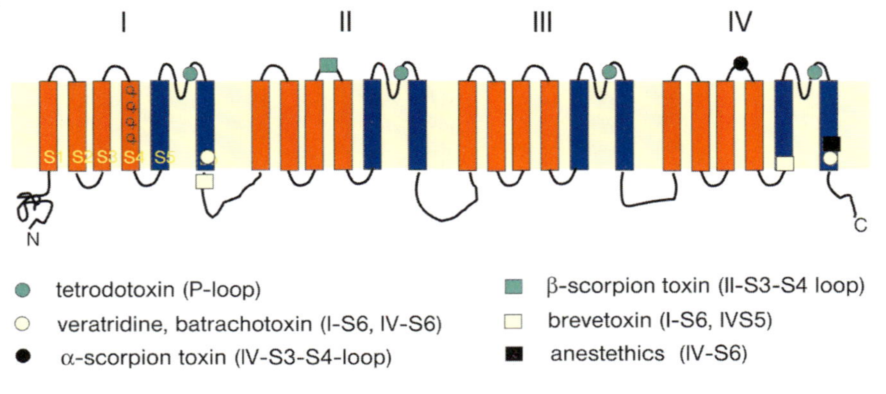

Figure 8.3 Topology of voltage-gated sodium channels. Known binding sites of peptides and drugs are marked. Voltage-gated sodium channels possess four 6TM/P domains.

proteins. Chang et al. [16], for example, state that approximately 24 000 crystallization conditions were tested to solve the structure of the MscL homolog from *Mycobacterium tuberculosis*, a mechanosensitive ion channel. Therefore, the number of 3D structures of ion channels is still very small compared to the number of enzyme structures. Most importantly, no crystal structure of a ligand–ion channel complex has been obtained so far.

Thus, structure-based drug design in the field of ion channels still has to rely on homology models of ion channels, which can be combined with conventional methods to map the ligand binding site, such as site-directed mutagenesis or photoaffinity labeling [17]. A number of different binding sites have thus been recognized on ion channels. For voltage-gated sodium channels, at least six different binding sites for toxins or drugs are known and are schematically depicted in Figure 8.3 [18].

Voltage-gated potassium channels also have a number of different binding sites. Similar to sodium channels, there is a binding site for peptide toxins at the outer vestibule of the pore. This binding site has been identified by site-directed mutagenesis for different peptide toxins, e.g., for the toxin ShK from a sea anemone, which blocks Kv1.3, or for Charybdotoxin, which blocks various potassium channels [19, 20].

Potassium channels also have binding sites within the ion conduction pore, as has been demonstrated for example for Kv1.3, Kv1.5, and hERG [21, 22]. Within the central cavity, there might be distinct but possibly overlapping binding sites. Ammonium ions bind in the upper part of the cavity, close to the selectivity filter; whereas for the hERG channel or for Kv1.3, mutational data indicate drug binding sites closer to the cytosolic part of the cavity. A recent study by Milnes et al. [23] raises the question of whether there is a 'nonaromatic' binding site within the hERG channel, since the binding affinity of Fluvoxamine is only partially attenuated by mutations of Tyr652 and Phe656 [23, 24].

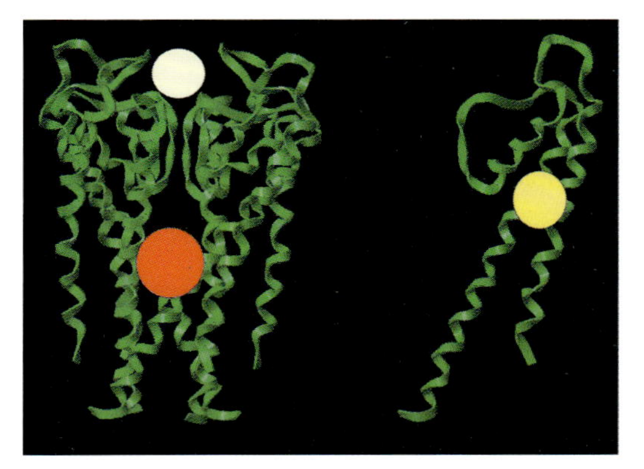

toxin binding site
cavity binding site
activator binding site

Figure 8.4 Structure of potassium channels with different binding sites in the pore domain marked. To show the binding location of R-L3 between helices S5 and S6, only the monomer is shown.

Recently, another binding site has been identified, the first binding site for a potassium channel activator [25]. The benzodiazepine derivative R-L3, a partial agonist of KCNQ1, binds between the S5 and S6 helices as indicated in Figure 8.4. Interestingly, the structurally related compound L-7 blocks KCNQ1 binding in the central cavity [26].

This example illustrates the difficulty of drug design in the absence of detailed structural knowledge, since even slight modifications can have a tremendous effect on the binding site and mode of action. In the field of ion channels, rational design is even more hampered by the fact that voltage-gated ion channels cycle through at least three different states – a resting state, an open state, and an inactivated state. Electrophysiological studies give evidence that blockers can interact with open channels as well as with closed channels. Vesnarinone or MK-499 require channel opening to bind [27, 28]. Other drugs, like Ketoconazole, bind to a closed state of the hERG channel, but Bertosamil binds to it in both its open and inactivated states [29, 30].

8.3
Lead-finding Strategies for Ion Channel Modulators

Appropriate lead-finding strategies for ion channel modulators make use of as much information as possible [31]. This includes information on modulators of closely related ion channels and presumably some 3D information about the particular target, either a homology model or available X-ray or NMR structures. The ligand information can be used for a ligand-based lead finding approach, whereas 3D structures are applicable to structure-based design. This section illustrates both lead-finding techniques through their application in several case studies.

8.3.1
Ligand-based Lead Finding

Ligand-based lead finding is based solely on information about putative ligands for a particular target or for a closely homologous target. This ligand information is then applied to select compounds that are closely linked to the reference molecules. This is achieved by 2D or 3D techniques. The 2D approach consists mainly of similarity and substructure searching, whereas the 3D method makes use of 3D pharmacophores built from a set of diverse compounds [32].

For similarity searching, all molecules are described by an appropriate binary descriptor (consisting of only zeros and ones). Such a binary fingerprint contains all structural information for a particular molecule and was applied at Aventis to identify new Kv1.5 inhibitors in the compound collection.

The Kv1.5 channel is a member of the voltage-gated K^+ channel family (which belongs to the 6TM/P family), whose functional form consists of four α subunits each containing 6 transmembrane segments [33]. The Kv1.5 pore domain is formed by four S5 and S6 segments from four different α subunits. In the human atrium, Kv1.5 is the molecular component of the repolarizing K^+ current Ikur, which contributes to the falling part of the cardiac action potential. Since Ikur has been found only in the human atrium, blockade of Kv1.5 has emerged as a promising approach for developing new atrial-selective antiarrythmics devoid of undesired effects observed with the currently available antiarrythmics [34]. When the Kv1.5 project was started at Aventis, no high-throughput screening assay was available, and our lead-finding strategy relied on database searching.

We used a compound from an Icagen patent as a query and identified two structurally different molecules that showed almost identical Kv1.5 activity as our reference molecule (Figure 8.5 a). Additionally, a Kv1.5 pharmacophore was derived from a lead series of Kv1.5 inhibitors. This pharmacophore, consisting of three hydrophobic features in a specific spatial orientation, was used to identify new putative Kv1.5 inhibitors in our corporate compound collection. The 12 most-promising compounds were selected based on their fit to the Kv1.5 pharmacophore. Subsequent biological profiling revealed one new lead structure (Figure 8.5 b).

Known side effects of a lead compound or drug can become an interesting opportunity to turn the side effect into the main pharmacological action of the compound. For the calcium antagonist Nifedipine, weak blocking of the calcium dependent potassium channels IK_{Ca} has been reported. IK_{Ca} is assumed to be involved in several diseases such as sickle cell anemia, immune disorders, and ischemic events [35–37]. Blocking of this channel was also proposed to be beneficial in traumatic brain injury [38]. Therefore, the calcium channel blocker Nifedipine was used as a starting point for developing a selective IK_{Ca} blocker with beneficial properties in a traumatic brain injury model [39]. Since the NH group of the dihydropyridine ring is a prerequisite for calcium antagonistic activity, it was replaced by the isoelectronic oxygen (Figure 8.6), leading to phenylpyrans that showed significant IK_{Ca} blocking activity [40]. Added electron-withdrawing substituents at the para position of the phenyl ring were able to further increase the

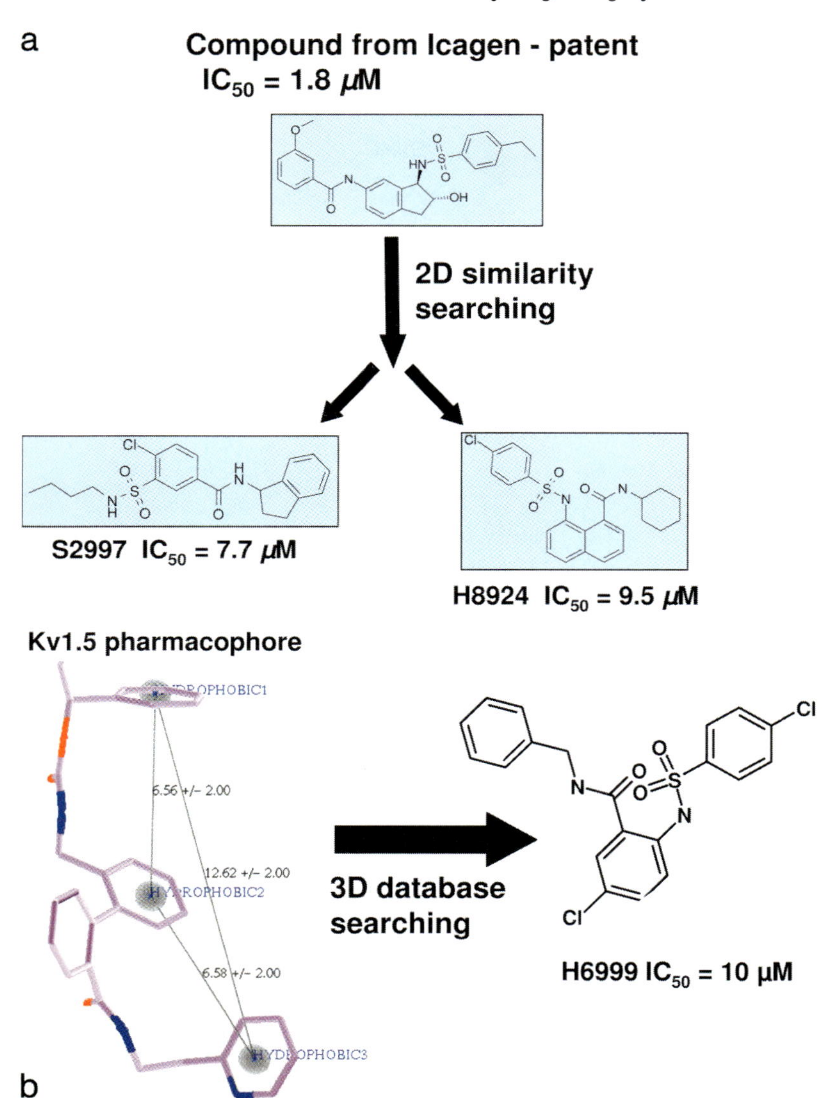

Figure 8.5
(a) Similarity-based 2D database searching for Kv1.5 inhibitors.
(b) Ligand-based Kv1.5 pharmacophore and its application in 3D database searching.

potassium channel activity. Overall, the SAR for the phenylpyrans on the IK_{Ca} channel was found to be orthogonal to the SAR of the dihydropyridines on the L-type calcium channel, allowing for the identification of IK_{Ca}-selective compounds.

Thus, this recent example nicely demonstrates that ion channel ligands can be valuable starting points for the identification of drugs acting against other members of the ion channel protein family.

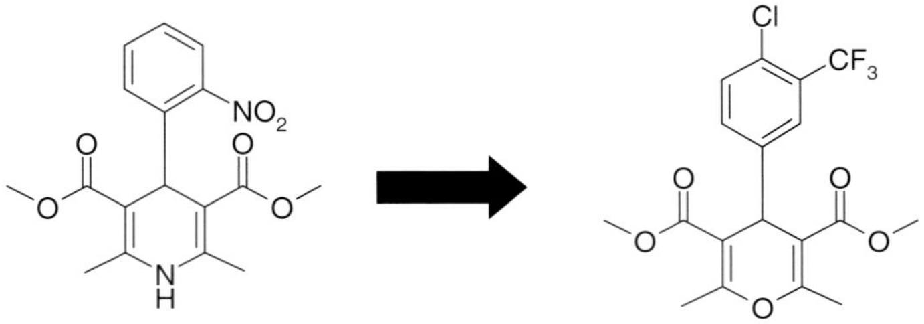

Figure 8.6 Nifedipine (left) provided a good starting point to obtain a selective blocker of IK_{Ca} (right), by slight changes in the central scaffold and the substitution pattern.

8.3.2
Structure-based Lead Finding

Structure-based lead finding requires a target 3D structure to start with. However, experimental elucidation of ion channel structures either NMR or X-ray crystallography is extremely difficult to achieve. Nevertheless, homology modeling of closely homologous channels to KcsA or MthK, for which 3D structures are available (see Section 8.2), makes this approach feasible.

In an early structure-based design effort, a combinatorial library was designed by using LUDI for Kv1.3 [41, 42]. Kv1.3 is involved in regulation of the membrane potential of human T cells, controlling calcium influx into the cell by voltage-dependent calcium channels [43]. Calcium influx ultimately results in cytokine release and cell proliferation. Therefore, Kv1.3 blockers might be interesting immunosuppressive compounds [44]. Various peptide toxins are known to block Kv1.3. Chandy and coworkers have used these structures in combination with mutant cycle analysis to derive a model of the outer vestibule of the Kv1.3 channel [45].

Within this outer vestibule model, LUDI calculations were focused on three amino acids from each subunit, which are known to be important for toxin binding: His404, Gly380, and Asp386 [46]. LUDI was used to suggest fragments interacting with these key amino acids. Fragment linking and modifications of the whole molecules, followed by molecular mechanics calculations, resulted in a phenylstilbene scaffold, which in the next step was varied in a combinatorial library comprising 400 compounds. The most active compound showed an IC_{50} of 2.9 μM (Figure 8.7).

This study was based on a model of the outer vestibule, which was developed using indirect evidence like the structure of known ligands and data from mutational analysis. At that time, the KcsA crystal structure or other potassium channel X-ray structures were not available. Meanwhile, more detailed knowledge of the atomic details of potassium channels allows the development of homology models [47] that can be successfully used in drug design, as demonstrated by the following example.

A recent structure-based lead-finding strategy was used for Kv1.5 inhibitors. The pore-forming domain of Kv1.5 exhibits 54% sequence homology with the bacterial K^+ channel KcsA from *Streptomyces lividans*, for which a crystal structure of the

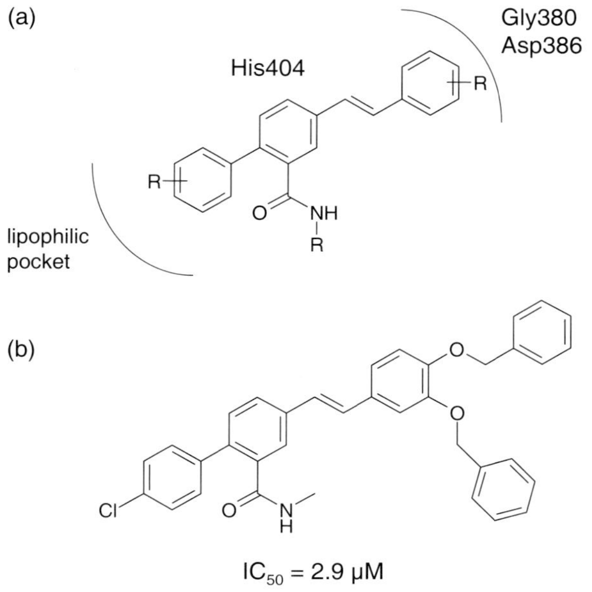

(a)

Gly380
Asp386
His404
R
R
O
NH
R
lipophilic pocket

(b)

O
O
Cl
O
N
H

IC_{50} = 2.9 µM

Figure 8.7
(a) Schematic representation of the proposed interactions of the phenylstilbene scaffold with residues from the outer vestibule.
(b) The shown structure has the highest activity for Kv1.3 found in this library.

closed channel is available. This structure (PDB code: 1bl8) was subsequently used as a template to build a homology model of the Kv1.5 pore-forming domain, using the Composer module of Sybyl 6.6 [48]. Starting with the α subunit of KcsA, a model of the S4 and S6 segments of Kv1.5 was built. Four of these segments were assembled according to the arrangement of the four α subunits of KcsA, representing the pore domain of Kv1.5. This model was refined by a two-step minimization protocol, involving minimization of the protein sidechains while keeping the backbone rigid, followed by minimization of the whole protein. The Sybyl 6.6 implementation of the AMBER forcefield was used to evaluate the energy of the system. The minimized structure was submitted to several tests for its quality and internal consistency, which included both geometric and profile analyses.

A computational elucidation of putative binding sites using the PASS algorithm revealed an internal site as most interesting for small organic molecules to interact with [49]. Subsequent more detailed analysis resulted in the derivation of a protein-based pharmacophore (Figure 8.8).

Use of this particular pharmacophore in subsequent 3D database searching of about 1 million compounds resulted in 244 interesting compounds, from which 19 compounds showed IC_{50} values below 10 µM.

The alignment of three of these hits is shown in Figure 8.9 a. Of course, these compounds exhibit high spatial similarity and fit remarkably well into the Kv1.5 binding site.

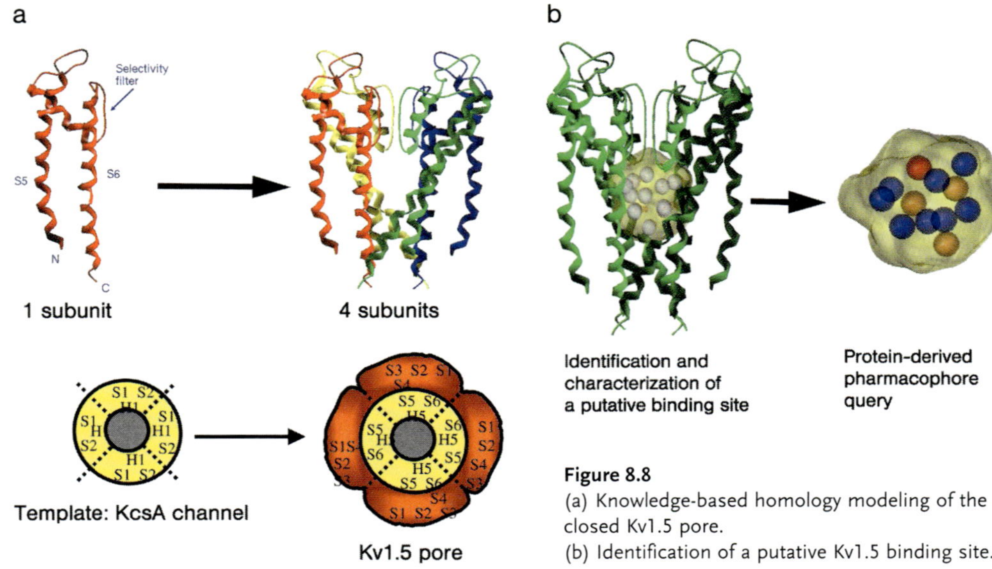

Figure 8.8
(a) Knowledge-based homology modeling of the closed Kv1.5 pore.
(b) Identification of a putative Kv1.5 binding site.

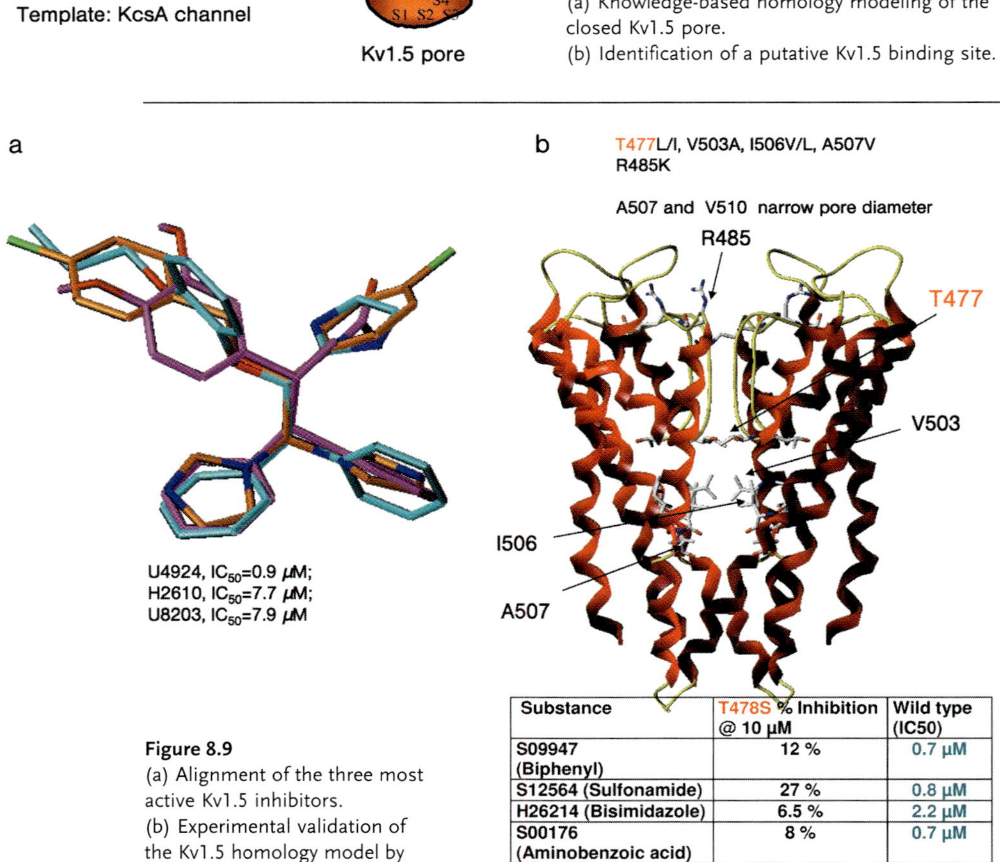

Substance	T478S % Inhibition @ 10 μM	Wild type (IC50)
S09947 (Biphenyl)	12 %	0.7 μM
S12564 (Sulfonamide)	27 %	0.8 μM
H26214 (Bisimidazole)	6.5 %	2.2 μM
S00176 (Aminobenzoic acid)	8 %	0.7 μM
U54924	33 %	1 μM

Figure 8.9
(a) Alignment of the three most active Kv1.5 inhibitors.
(b) Experimental validation of the Kv1.5 homology model by mutational data.

We proposed several mutations to the Kv1.5 channel to validate our model. Docking of our inhibitors in the Kv1.5 binding site revealed Thr477 as very important for binding. Indeed, mutation of Thr477 to serine left the Kv1.5 channel fully functional, but the activity of all four types of Kv1.5 inhibitors significantly decreased [50].

8.4 Design of Ion Channel Focused Libraries: Chemical Genomics

The consideration of all available ligand information concerning ion channel modulators, as well as the use of 3D structural information is required for designing appropriate ion channel focused libraries. This can be achieved by matching chemical and biological information in the target family of ion channels. The intersection of biological structures and functionalities with chemical structures and properties is derived to perform a knowledge-driven biased library design. This allows extraction of common structural features for ion channel modulators out of a practically infinite chemical space [51]. Applying such design criteria leads to chemical libraries that are enriched in preferred features of ion channel modulators. This section covers design principles and their application.

8.4.1 Design Principles

Matching chemical and biological information in the field of ion channels requires combined 2D and 3D analysis (Figure 8.10). The 2D approach is based on a collection

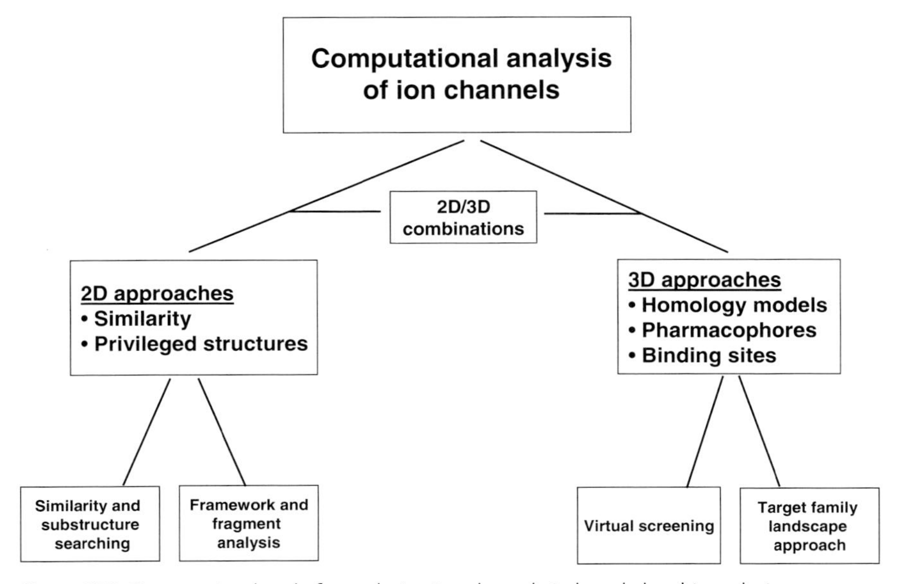

Figure 8.10 Computational tools for analyzing ion channels in knowledge-driven design.

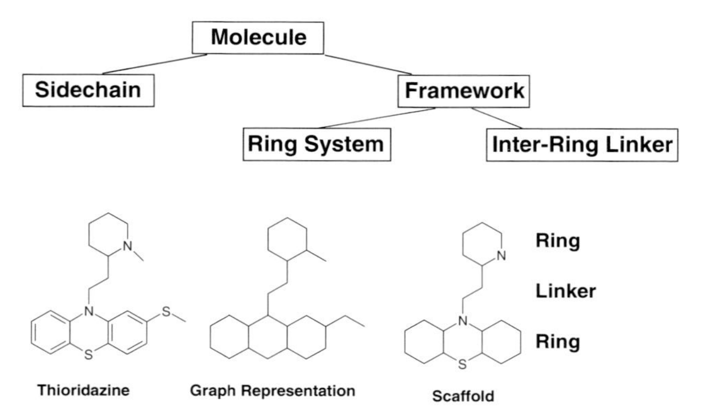

Figure 8.11 Topological framework analysis.

of biologically active compounds and consists mainly of similarity and substructure searching and of analysis of common frameworks and fragments to identify privileged chemotypes. Applicable 3D techniques are either ligand- or structure-based. The ligand-based method requires biologically active ion channel modulators to derive 3D pharmacophores, and the structure-based technique uses a 3D structure of ion channels for subsequent virtual screening.

Substructure searching is often used in drug design and needs no further clarification. Similarity searching is also a very well known technique described in more detail elsewhere [52]. We usually use MACCS keys, Unity fingerprints, CATS descriptors, and feature trees for similarity searching [53]. Each technique has its own strengths and weaknesses, so we favor parallel application of two or three of them.

Framework analysis was published by Bemis and Murcko in 1996 [54]. They analyzed shapes of existing drugs in a commercial database to extract drug-related molecular frameworks by following a graph theoretical approach to decompose molecules into rings and noncyclic sidechains. Linkers and rings together form the framework of a molecule, whereas sidechains are omitted (Figure 8.11). For example, framework analysis of Thioridazine starts with removal of the acyclic sidechains and leaves a framework composed of two rings and one inter-ring linker.

Application of this topological framework analysis to ion channel modulators yields access to privileged ion channel chemotypes. Conversion of these frameworks into appropriate scaffolds for synthesis allows subsequent building of ion channel focused libraries.

Fragment analysis is based on the RECAP algorithm, published in 1998 [55]. This retrosynthetic combinatorial analysis starts with a collection of active molecules and then fragments these molecules using any set of retrosynthetic reactions (Figure 8.12). For example, Cisapride is cleaved into four fragments based on three different bond cleavage types.

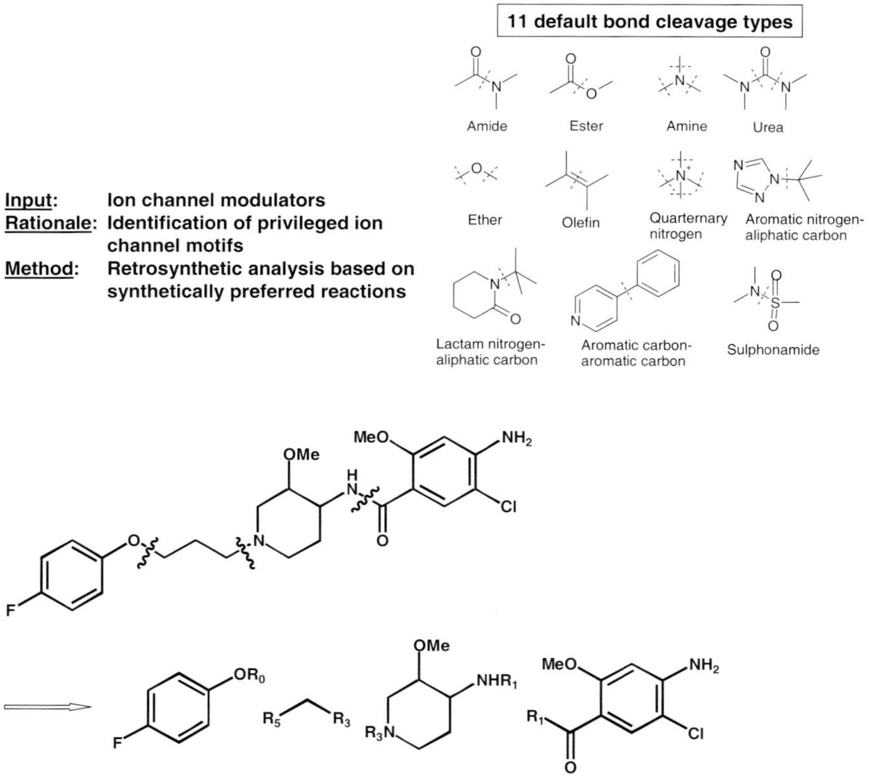

Input: Ion channel modulators
Rationale: Identification of privileged ion channel motifs
Method: Retrosynthetic analysis based on synthetically preferred reactions

11 default bond cleavage types

Amide, Ester, Amine, Urea, Ether, Olefin, Quarternary nitrogen, Aromatic nitrogen-aliphatic carbon, Lactam nitrogen-aliphatic carbon, Aromatic carbon-aromatic carbon, Sulphonamide

Figure 8.12 Retrosynthetic combinatorial analysis procedure (RECAP).

Resulting fragments are clustered and reclassified into sets of monomers for subsequent library design. The RECAP procedure derives not only suitable chemotypes but also appropriate building blocks for scaffold decoration. Since the monomers are extracted from biologically active compounds, there is a high likelihood that new molecules derived from them will contain biologically important motifs.

The 3D approach makes use of ion channel-specific pharmacophores, ion channel X-ray structures, and homology models (see Section 8.3). Ion channel X-ray and homology models are not as precise as structures of smaller proteins. The uncertainty regarding the binding mode of ion channel modulators also adds additional complexity to structure-based virtual screening. However, valid 3D pharmacophores can be derived from these structures and subsequently used to identify privileged ion channel chemotypes by virtual screening in proprietary and public databases [56]. Needless to say, both ligand- and structure-based pharmacophores require in-depth validation prior to their use in virtual screening.

8.4.2
Example: Building the Aventis Ion Channel Library

Ion channels are of potential interest in several therapeutic areas. However, appropriate high-throughput assays to test several hundred thousand compounds against a particular ion channel still lack sufficient signal-to-noise ratios [57]. Therefore, a biased ion channel library is of high interest for lead finding.

We performed a combined 2D and 3D analysis of chemical and biological space to identify ion channel privileged chemotypes. The 2D approach was based on a collection of biologically active compounds and consisted mainly of similarity and substructure searching and of analysis of common frameworks and fragments. Our 3D approach relied on multiple ion channel pharmacophores and homology models, which were used for virtual screening.

Our iterative ion channel library design process is outlined in Figure 8.13. Retrieval and critical review of literature and in-house data on ion channel modulators resulted in a collection of valuable lead compounds, suitable for 2D and 3D database mining in internal and external compound collections. Among others, we took into account calcium channel blockers like Clonidine, chloride channel blockers, potassium channel openers, K(ATP) channel blocker and openers (e.g., Glibenclamide), and NHE-1 inhibitors.

Scaffold proposals were collected and reviewed according to privileged ion channel motifs, chemical feasibility, and fit to our multiple pharmacophores. Building block selection, virtual library design, and filtering yielded small virtual libraries suitable for automated solution-phase synthesis. All synthesized compounds were finally purified and characterized prior to addition to our focused library.

Picked and purchased compounds, as well as all new designed chemotype focused libraries, were assessed in terms of quality, diversity, and drug-likeness [58]. The

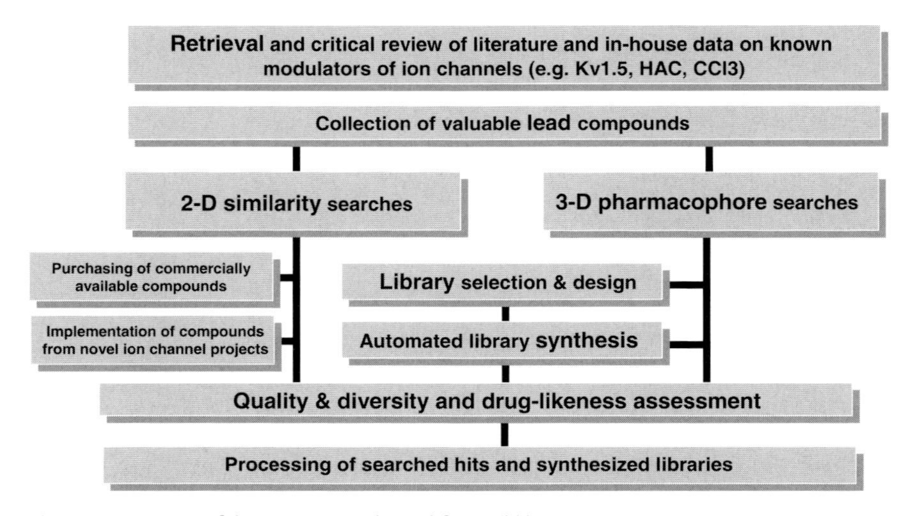

Figure 8.13 Design of the Aventis ion channel focused library.

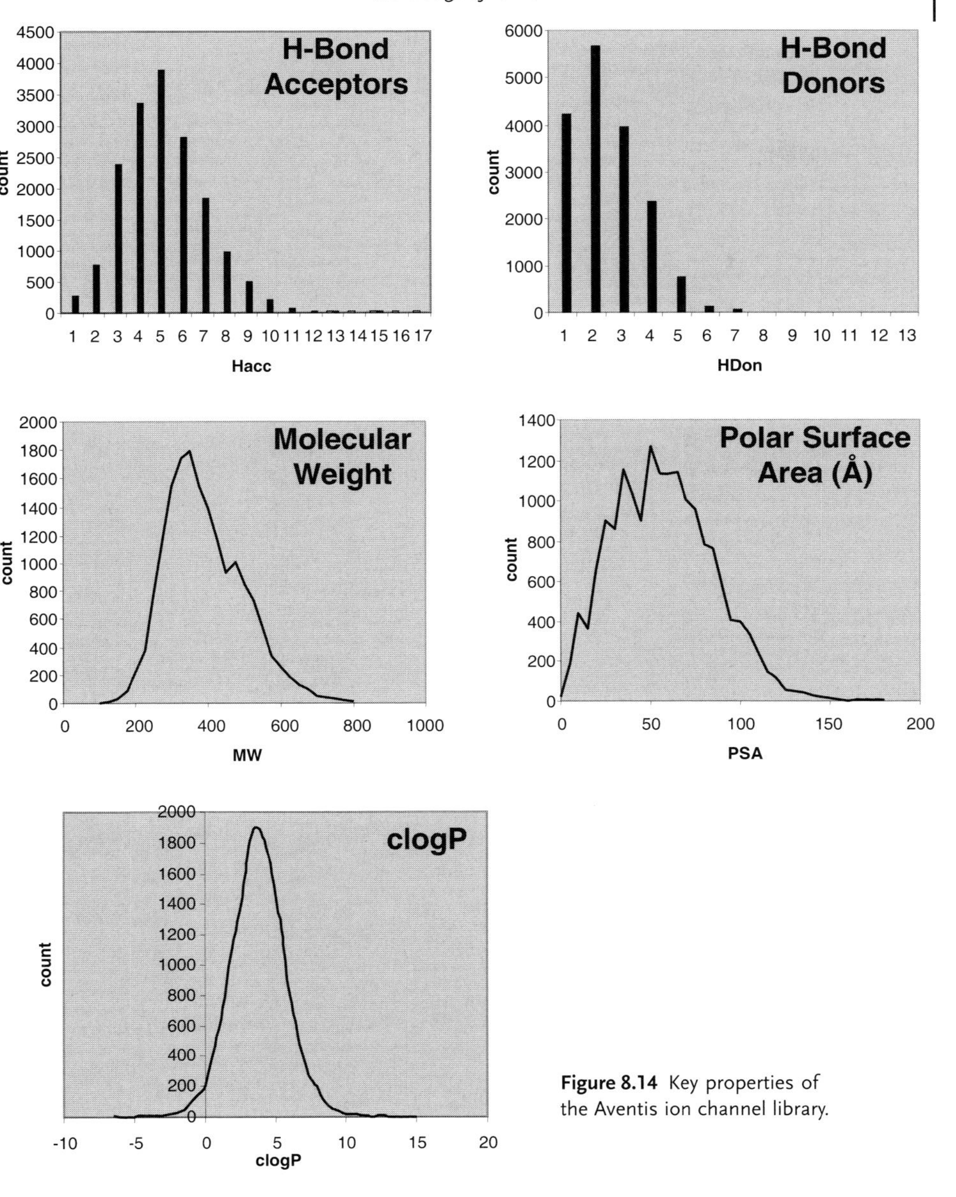

Figure 8.14 Key properties of the Aventis ion channel library.

remaining compounds were plated in our ion channel library, which is frequently used for screening of ion channels. New compounds and chemotypes from novel ion channel projects are continuously added to this focused library, and thus we constantly increase the value of our ion channel ligand collection.

The key properties of our ion channel library, namely molecular weight, polar surface area, clogP, and number of hydrogen bond acceptors and donors, are within the lead-like range of compounds (Figure 8.14) [59]. A purity analysis of a small

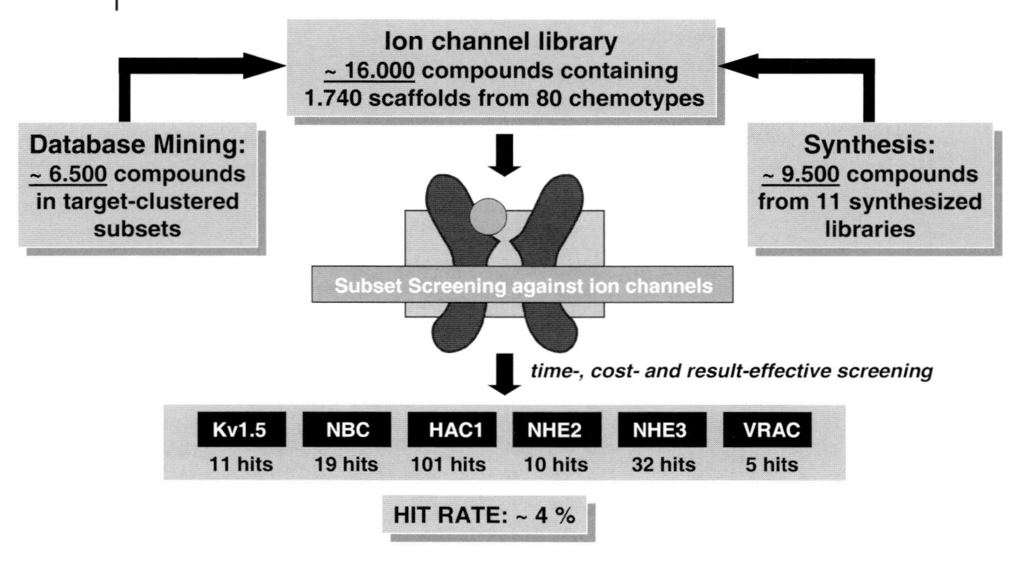

Figure 8.15 Aventis ion channel library: current status and hit rates.

subset of our liquid ion channel collection revealed that more than 75% of the compounds had acceptable purity.

Currently, the Aventis ion channel library contains approximately 16 000 compounds related to 1740 scaffolds and 80 chemotypes. Almost 6500 compounds have emerged from database mining, and 9500 compounds were synthesized (Figure 8.15).

Screening of this library or subsets of it against new ion channels revealed hit rates of approximately 4% (Figure 8.15), which is substantially higher than typical hit rates from high-throughput screening (~0.01%–0.1%). In addition, this focused screening identified highly valuable hits, since the library contains primarily drug-like or lead-like compounds. Profiling this library against several ion channels not only reveals channel-subtype specific chemotypes, but also offers the opportunity to build early structure–activity relationships on scaffolds, which is very helpful especially for optimizing activity and selectivity. Hence, our chemical genomics approach has yielded improved screening hit rates and better starting points for subsequent compound optimization, thus reducing the cycle times for screening and optimization.

Although almost all our screening efforts using this focused library against new ion channel targets have resulted in good hit rates, we were recently disappointed by finding a hit rate of only 0.8% against a specific potassium channel, and we thought about opportunities for improvement. Current antiarrhythmic agents, for example Sotalol, quite often show adrenoceptor inhibition and potassium- or multi-channel blockade [60]. Hence, a future prospect for our ion channel library is to identify common GPCR and ion channel chemotypes by profiling our GPCR library against ion channels and vice versa.

8.5 Conclusions

Identification of suitable lead compounds for subsequent optimization is one of the key needs in drug discovery today. Several techniques have been successfully applied, while a chemical genomics-driven approach, by building target family related compound libraries, seems to be a promising future strategy for lead finding. The high complexity of these efforts motivates a knowledge-driven design strategy, taking into account as much information as possible from targets and ligands. An in-depth scientific understanding of the intersection of biological and chemical information is crucial for enabling higher productivity in early compound identification. Hence, the relationship between certain chemotypes and biological targets within target families should drive this lead identification strategy.

The effort to bridge the chemical and biological space is called chemical genomics or chemogenomics. So far, no unique definition of chemical genomics has emerged from the literature, but the systematic exploration of target families is a common goal, based on the assumption that similar compounds bind to similar targets by similar mechanisms. This assumption is one of the foundations for the selection of compounds for our ion channel biased screening collection. Another important principle of chemical genomics is the integrated use of state-of-the-art computational tools to derive new ion channel binding motifs.

This chapter has discussed such a chemical genomics approach for ion channel modulators. Some case studies for ion channel lead finding illustrate the opportunities of target- and ligand-related strategies. Target family-related knowledge is of course mandatory for this process. However, limited accessibility to ion channel 3D structures and uncertainties in homology models mean that structure-based approaches are feasible only after thorough validation of the underlying models.

However, identification of ion channel modulators by screening compound libraries enriched with ion channel privileged chemotypes offers rapid, efficient access to lead compounds. Flexible data-driven building and optimization of such an ion channel focused library will enable better and faster lead identification of ion channel modulators.

The increasing information in biological and chemical space and its effective transformation into a knowledge-driven ion channel focused library may foster a paradigm shift in lead identification.

Acknowledgements

We thank Anna Gorokhov, Holger Heitsch, Bernard Pirard (Aventis Pharma), Gabriele Cruciani (Univ. Perugia), and Gisbert Schneider (Univ. Frankfurt) for stimulating discussions on several aspects of this chapter and Guiscard Seebohm for providing preprints of his recent work.

References

1 B. Hille, *Ionic Channels of Excitable Membranes*, 3rd ed., Sinauer Associates, Sunderland, MA **2001**.
2 (a) J. Drews, S. Ryser, *Nat. Biotechnol.* **1997**, *15*, 1318–1319; (b) J. Drews, *Science* **2000**, *287*, 1960–1964.
3 A. L. Hopkins, C. R. Groom, *Nat. Rev. Drug Discovery* **2002**, *1*, 727–730.
4 M. E. Curran, *Curr. Opin. Biotechnol.* **1998**, *9*, 565–572.
5 R. A. Pearlstein, R. Vaz, D. Rampe, *J. Med. Chem.* **2003**, *46*, 2017–2022 and references therein.
6 (a) G. Wess, M. Urmann, B. Sickenberger, *Angew. Chem.* **2001**, *113*, 3443–3453; (b) G. Wess, *Drug Discov. Today* **2002**, *7*, 533–535.
7 (a) E. Jacoby, A. Schuffenhauer, P. Floersheim, *Drug News Perspect.* **2003**, *16*, 93–102; (b) K. Shokat, M. Velleca, *Drug Discov. Today* **2002**, *7*, 872–879.
8 S. A. Goldstein, D. Bockenhauer, I. O'Kelly, N. Zilberberg, *Nature Rev. Neuroscience* **2001**, *2*, 175–184.
9 (a) D. A. Doyle, J. M. Cabral, R. A. Pfuetzner, A. Kuo, J. M. Gulbis, S. L. Cohen, B. T. Chait, R. MacKinnon, *Science* **1998**, *280*, 69–77; (b) Y. Jiang, A. Lee, J. Chen, V. Ruta, M. Cadene, B. T. Chait, R. MacKinnon, *Nature* **2003**, *423*, 33–41.
10 Y. Jiang, A. Lee, J. Chen, M. Cadene, B. T. Chait, R. MacKinnon, *Nature* **2002**, *417*, 515–522.
11 (a) Y. Zhou, J. H. Morals-Cabral, A. Kaufman, R. MacKinnon, *Nature* **2001**, *414*, 43–48; (b) B. Roux, R. MacKinnon, *Science* **1999**, *285*, 100–102.
12 (a) W. Stühmer, F. Conti, H. Suzuki, X. Wang, M. Noda, N. Yahagi, H. Kubo, S. Numa, *Nature* **1989**, *339*, 597–603; (b) D. M. Papazian, L. C. Timpe, Y. N. Jan, L. Y. Jan, *Nature* **1991**, *349*, 305–310; (c) E. R. Liman, P. Hess, *Nature* **1991**, *353*, 752–756.
13 (a) L. M. Mannuzzu, M. M. Moronne, E. Y. Isacoff, *Science* **1996**, *271*, 213–215; (b) A. Cha, F. Bezanilla, *Neuron* **1997**, *19*, 1127–1140.
14 A. Kuo, J. M. Gulbis, J. F. Antcliff, T. Rahman, E. D. Lowe, J. Zimmer, J. Cuthbertson, F. M. Ashcroft, T. Ezaki, D. A. Doyle, *Science* **2003**, *300*, 1922–1926.
15 (a) R. Dutzler, E. B. Campbell, M. Cadene, B. T. Chait, R. MacKinnon, *Nature* **2002**, *415*, 287–294; (b) R. Dutzler, E. B. Campbell, R. MacKinnon, *Science* **2003**, *300*, 108–112.
16 G. Change, R. H. Spencer, A. T. Lee, M. T. Barclay, D. C. Rees, *Science* **1998**, *282*, 2220–2226.
17 J. Striessnig, M. Grabner, J. Mitterdorfer, S. Hering, M. J. Sinneger, H. Glossman, *Trends Pharmacol. Sci.* **1998**, *19*, 108–115.
18 N. Ogata, Y. Ohishi, *Jpn. J. Pharmacol.* **2002**, *88*, 365–377.
19 K. Kalman, M. W. Pennington, M. D. Lanigan, A. Nguyen, H. Rauer, V. Mahnir, K. Paschetto, W. R. Kem, S. Grissmer, G. A. Gutmann, E. P. Christian, M. D. Cahalan, R. S. Norton, K. G. Chandy, *J. Biol. Chem.* **1998**, *273*, 32697–32707.
20 P. Stampe, L. Kolmakova-Partensky, C. Miller, *Biochemistry* **1994**, *33*, 443–450.
21 M. Hanner, B. Green, G. Ying-Duo, W. A. Schmalhofer, M. Matyskiela, D. J. Durand, J. P. Felix, L. Ana-Rosa, C. Bordallo, G. J. Kaczorowski, M. Kohler, M. L. Garcia, *Biochemistry* **2001**, *40*, 11687–11697.
22 J. S. Mitcheson, J. Chen, M. Lin, C. Culberson, M. C. Sanguinetti, *Proc. Natl. Acad. Sci. USA* **2000**, *97*, 12329–12333.
23 J. T. Milnes, O. Crociani, A. Arcangeli, J. C. Hancox, H. J. Witchel, *Br. J. Pharmacol.* **2003**, *139*, 887–898.
24 J. S. Mitcheson, *Br. J. Pharmacol.* **2003**, *139*, 883–884.
25 G. Seebohm, M. Pusch, J. Chen, M. C. Sanguinetti, *Cir Res.* **2003**, *93*, 941–947.
26 G. Seebohm, J. Chen, N. Strutz, C. Culberson, C. Lerche, M. C. Sanguinetti, *Mol. Pharmacol.* **2003**, *64*, 70–77.
27 K. Kamiya, J. S. Mitcheson, K. Yasui, I. Kodama, M. C. Sanguinetti, *Mol. Pharmacol.* **2001**, *60*, 244–253.

28 P. S. Spector, M. E. Curran, M. T. Keating, M. C. Sanguinetti, *Cir. Res.* **1996**, *78*, 499–503.

29 R. Dumaine, M.-L. Roy, A. M. Brown, *J. Pharmacol. Exp. Ther.* **1998**, *286*, 727–735.

30 E. Zitron, C. A. Karle, G. Wendt-Nordahl, S. Kathöfer, W. Zhang, D. Thomas, S. Weretka, J. Kiehn, *Br. J. Pharmacol.* **2002**, *137*, 221–228.

31 K. H. Bleicher, H.-J. Böhm, K. Müller, A. I. Alanine, *Nat. Rev. Drug Discovery* **2003**, *2*, 369–378, and references therein.

32 O. F. Güner, *Pharmacophore Perception, Development, and Use in Drug Design*, International University Line, La Jolla, CA **2000**.

33 (a) S. M. Cobbe, *Pacing Clin Electrophysiol.* **1994**, *17*, 1005–1010; (b) D. Fedida, B. Wible, Z. Wang, B. Fermini, F. Faust, S. Nattel, A. M. Brown, *Circ Res.* **1993**, *73*, 210–216; (c) D. J. Snyders, M. M. Tamkun, P. B. Bennett, *J. Gen. Physiol.* **1993**, *101*, 513–543; (d) M. J. Coghlan, W. A. Carroll, M. Gopalakrishnan, *J. Med. Chem.* **2001**, *44*, 1627–1653.

34 S. Peukert, J. Brendel, B. Pirard, A. Brüggemann, P. Below, H.-W. Kleemann, H. Hemmerle, W. Schmidt, *J. Med. Chem.* **2003**, *46*, 486–498.

35 N. J. Logson, J. Kang, J. A. Togo, E. P. Christian, J. Aiyar, *J. Biol Chem.* **1997**, *272*, 32723–32727.

36 W. J. Joiner, L. Y. Wang, M. D. Tang, L. Kaczmarek, *Proc. Natl. Acad. Sci. USA* **1997**, *94*, 11013–11018.

37 G. Gardos, *Biochem. Biophys. Acta* **1958**, *30*, 653–654.

38 B. S. Jensen, L. Teuber, D. Strobaek, P. Christophersen, S. P. Olesen (Neurosearch A/S), WO 2000069794, 2000.

39 K. Urbahns, E. Horvath, J.-P. Stasch, F. Mauler, *Bioorg. Med. Chem. Lett.* **2003**, *13*, 2637–2639.

40 (a) S. Goldmann, J. Stoltefuss, *Angew. Chem.* **1991**, *103*, 1587–1605; (b) D. J. Triggle, D. R. Langs, R. A. Janis, *Med. Res. Rev.* **1989**, 9, 123–180.

41 (a) H. J. Boehm, *J. Comp. Aided. Mol. Design* **1992**, *6*, 69–78; (b) H. J. Boehm, *J. Comp. Aided. Mol. Design* **1992**, *6*, 593–606.

42 A. Lew, A. R. Chamberlin, *Bioorg. Med. Chem. Lett.* **1999**, *9*, 3267–3272.

43 C. S. Lin, R. C. Boltz, J. T. Blake, M. Nguyen, A. Talento, P. A. Fischer, M. S. Springer, N. H. Sigal, R. S. Garcia, M. L. Kaczorowski, G. C. Koo, *J. Exp. Med.* **1993**, *177*, 637–645.

44 M. D. Chalan, K. G. Chandy, *Curr. Opin. Biotech.* **1997**, *8*, 749–756.

45 (a) J. Aiyar, J. P. Rizzi, G. A. Gutman, K. G. Chandy, *J. Biol. Chem.* **1996**, *271*, 31013–31017.

46 (a) P. Hidalgo, R. MacKinnon, *Science* **1995**, *268*, 307–310; (b) R. Ranganathan, J. H. Lewis, R. MacKinnon, *Neuron* **1996**, *16*, 131–139.

47 A. Giorgetti, P. Carloni, *Curr. Opin. Chem. Biol.* **2003**, *7*, 150–156.

48 Sybyl version 6.6, Tripos Inc., St. Louis, MO, USA.

49 G. P. Brady, P. F. W. Stouten, *J. Comput. Aided Mol. Design* **2000**, *14*, 383–401.

50 N. Decher, B. Pirard, F. Bundis, S. Peukert, K.-H. Baringhaus, A. E. Busch, K. Steinmeyer, M. C. Sanguinetti, *J. Biol. Chem.* **2003**, 278, 43564–43570.

51 K.-H. Baringhaus, T. Klabunde, H. Matter, T. Naumann, B. Pirard, in *Molecular Informatics: Confronting Complexity, Proceedings of the International Beilstein Workshop* May 2002 (Eds. M. G. Hicks, C. Kettner), Logos Verlag, Berlin **2003**, pp. 167–178.

52 P. M. Dean, *Molecular Similarity in Drug Design*, Chapman & Hall, Glasgow **1995**.

53 (a) MACCS-II, MDL Information Systems Inc.: San Leandro, CA **1992**; (b) *UNITY Reference Manual*, Tripos Inc., St. Louis, MO **1995**; (c) D. R. Flower, *J. Chem. Inf. Comput. Sci.* **1998**, *38*, 379–386; (d) G. Schneider, W. Neidhart, T. Giller, G. Schmid, *Angew. Chem.* **1999**, *111*, 3068–3070; (e) M. Rarey, J. S. Dixon, *J. Comput. Aided Mol. Design* **1996**, 10, 41–54.

54 M. A. Murcko, G. A. Bemis, *J. Med. Chem.* **1996**, *39*, 2887–2893.

55 X. Q. Lewell, D. B. Judd, S. P. Watson, M. M. Hann, *J. Chem. Inf. Comput. Sci.* **1998**, *38*, 511–522.

56 W. P. Walters, M. T Stahl, M. A. Murcko, *Drug Discov. Today* **1998**, *3*, 160–178, and references therein.

57 J. Xu, X. Wang, B. Ensign, M. Li, L. Wu, A. Guia, J. Xu, *Drug Discov. Today* **2001**, *6*, 1278–1287.

58 (a) K.-H. Baringhaus, *Bioforum International* **2001**, *5*, 126–129; (b) W. P. Walters, M. A. Murcko, *Adv. Drug Delivery Rev.* **2002**, *54*, 255–271, and references therein.

59 T. I. Oprea, A. M. Davis, S. J. Teague, P. D. Leeson, *J. Chem. Inf. Comput. Sci.* **2001**, *41*, 1308–1315.

60 M. G. Cimini, J. K. Gibson, *Annu. Rep. Med. Chem.* **1992**, *27*, 89–98.

9
Phosphodiesterase Inhibitors: A Chemogenomic View

Martin Hendrix and Christopher Kallus

9.1
Introduction

The superfamily of mammalian cyclic nucleotide phosphodiesterases is responsible for the degradation of the nucleotide 3′,5′-cyclic phosphates cGMP and cAMP to their respective hydrolysis products 5′-GMP and 5′-AMP (Figure 9.1) [1]. This seemingly simple change has far-reaching consequences, because it switches off the biological signal transmitted by the second messengers cAMP and cGMP. Not surprisingly, the enzymes that control this pathway are found in all animal tissues and cell types, and there are generally several PDE isoenzymes in any given cell.

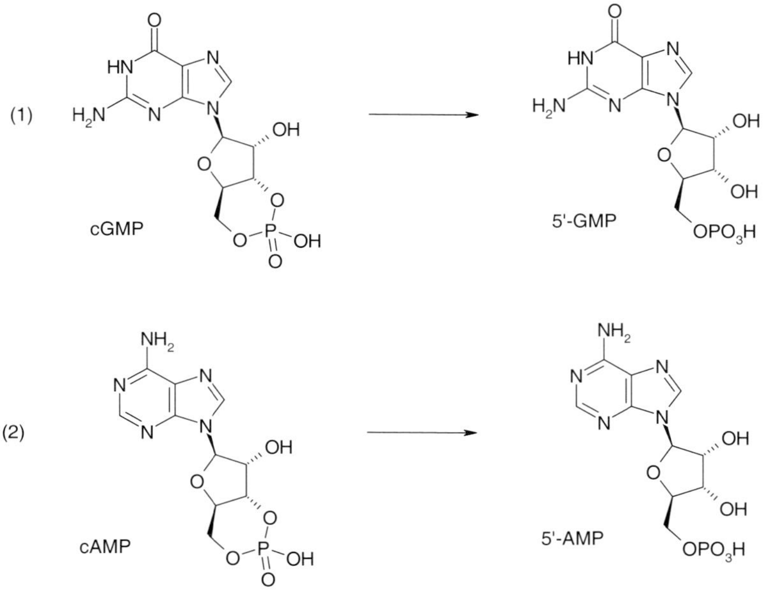

Figure 9.1 Reactions catalyzed by cyclic nucleotide phosphodiesterases.

Chemogenomics in Drug Discovery: A Medicinal Chemistry Perspective.
Edited by Hugo Kubinyi and Gerhard Müller

ISBN: 3-527-30987-X

For quite some time now it has likewise been recognized that these key enzymes also present an important opportunity for pharmacological intervention, and efforts in this area have begun to pay off. There is now a considerable body of information available on a multitude of PDE inhibitors. The purpose of this chapter is to take a unifying view of this complex field and try to extract the now apparent general structural features useful in PDE inhibitor design [2].

9.2
PDE Isoenzymes and Subtypes

Today 11 members of the human PDE superfamily are known, all of which are class I phosphodiesterases and all of which are intracellular or membrane-bound enzymes. Several of the isoenzymes are encoded by more than one gene which, in combination with the presence of different splice variants, brings the number of different PDE proteins to well over 50. The different isoenzymes are characterized according to their substrate specificity, sequence homology, kinetic properties, and sensitivity to certain known PDE inhibitors. Table 9.1 shows these properties together with the predominant tissue expression of the various PDEs.

The different PDE isoenzymes share a common domain structure, which comprises N-terminal regulatory domains and a C-terminal 250–300 amino-acid catalytic domain (Figure 9.2). All the different functional domains appear to be of ancient origin; gene duplication and domain shuffling seem to account for the multitude of contemporary isoenzymes. Among the different parts of the enzyme, the catalytic domains show the highest sequence identity among members of a PDE family (up to 80%). They contain two conserved metal-binding sites in a histidine-rich region with a HD(X_2)H(X_4)N motif, similar to metalloproteases such

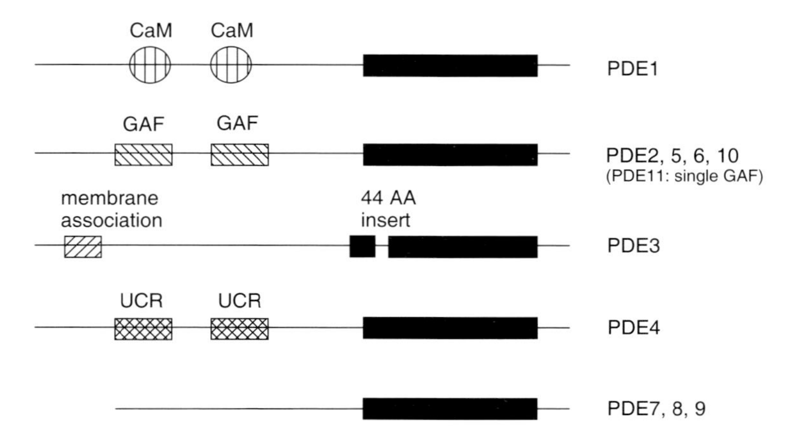

Figure 9.2 Schematic representation of the domain structure of PDEs. The conserved catalytic domain is in the C-terminal portion (solid bars), whereas various regulatory domains make up the N terminus.

Table 9.1 Genomic organization of human phosphodiesterases.

Name	*Characteristics*	*cAMP*	*cGMP*	*Number of genes*	*Primary tissue/ cellular distribution*
		Km (μM)			
PDE1	Ca^{2+}-CaM-stimulated	110 (PDE1A) 25 (PDE1B) 1 (PDE1C)	5 (PDE1A) 3 (PDE1B) 1 (PDE1C)	3	VSMC, brain, lung, heart
PDE2	cGMP-stimulated	30–100	10–30	1	adrenal cortex, brain, heart, liver, olfactory bulbus
PDE3	cGMP-inhibited	0.1–0.5	0.1–0.5	2	heart, lung, liver, adipocytes, immunocytes, pancreas
PDE4	cAMP-specific	0.5–4	> 50	4	immunocytes, lung, brain
PDE5	cGMP-specific	> 40	1.5	1	VSMC, SMC, lung, corp. cav., platelets
PDE6	photoreceptor	2000	60	3	retina (rods and cones)
PDE7	cAMP high-affinity	0.2	> 1000	2	skeletal muscle, T-cells
PDE8	cAMP high-affinity	0.7	> 100	2	broadly expressed, testis, ovary, intestine, colon
PDE9	cGMP high-affinity	> 100	0.07	1	broadly expressed, liver, kidney, spleen, brain
PDE10	dual substrate	0.5	3	1	broadly expressed, in mice most abundant in brain, testis
PDE11	dual substrate	1	0.5	1	testis, brain, corp. cav., skeletal muscle, prostate

VSCM: vascular smooth muscle cells
SMC: smooth muscle cells
CaM: calmodulin
Corp. cav.: corpus cavernosum

as thermolysin [3]. Five histidines, designated H-1 to H-5, are invariant in all Class I PDEs. H1, H2, H3 and H4 are part of the two metal-binding domains. Overall, 21 amino acids are invariant in the catalytic domain of 10 of the 11 human PDEs.

Amino acid residues of both metal-coordinating sites seem to be important for catalytic activity [4]. Zn, Mg, Co, and Mn all constitute possible catalytically active metals, although the catalytic activity varies. Based on recent publications, there seems to be a consensus now that a combination of Zn and Mg is the physiologically relevant catalytic species.

The N-terminal protein sequences show considerably more variation than the C terminus, reflecting the broad range of regulatory stimuli that they transmit to the enzyme (e.g., phosphorylation, calcium concentration, substrate activation, interaction with regulatory proteins). Thus, PDE1 contains a pair of calmodulin binding sites, making the enzyme sensitive to changes in calcium concentration.

The cAMP-specific PDE4 contains a pair of upstream conserved regions (UCR) that regulate catalytic activity. Perhaps the most prominent N-terminal sequence motif is the presence of so-called GAF domains in several cGMP-hydrolyzing enzymes (PDE2, 5, 6, 10, 11). The name GAF refers to the three protein types in which this homology sequence has been found so far: cGMP-PDEs, adenylyl cyclases, and a protein called FhIA. These are, at least in principle, allosteric cGMP-binding sites, although in some instances (e.g., the second GAF domain of PDE2) they remain unoccupied [5]. In addition to these regulatory functions, the N-terminal protein domains can also mediate dimerization. Most PDE enzymes appear *in vivo* as dimers or oligomers, although there does not seem to be cooperativity between the catalytic domains of the monomers.

Structural information on the catalytic domain of PDEs is now available for two isoenzymes (PDE4 and PDE5) by means of X-ray structural analyses including bound ligands. This has made it possible to understand the molecular basis of substrate recognition and features of the catalytic mechanism. In the initial landmark study of Xu et al. [6], the substrate cAMP was modeled into the active site of the experimentally determined PDE4B structure (Figure 9.3). The adenine is located in a hydrophobic core binding pocket made up of Leu393, Pro396, Ile410, Phe414, and Phe446. The nucleotide is recognized by a bidentate H-bond motif involving Gln443, a residue conserved across all PDE families. The rotation of this critical sidechain amide is fixed by a network of additional H bonds. If guanine (cGMP) instead of adenine needs to be recognized, as in cGMP-specific PDE5, the rotation of the same sidechain amide is reversed to meet the inverted H-bond requirements of the lactam of guanine. In this model, the charged phosphodiester is placed close to the binuclear metal center that is responsible for catalysis.

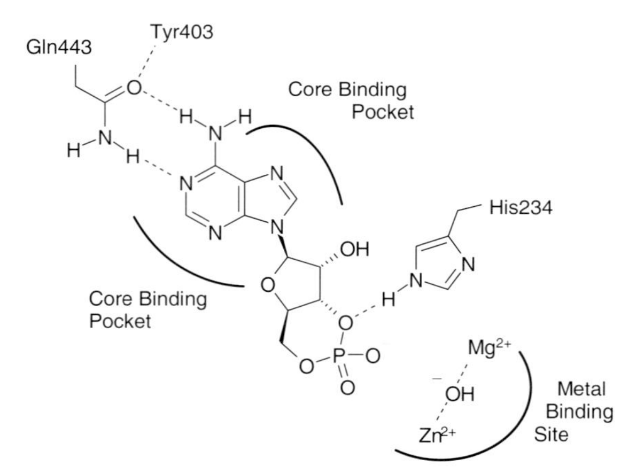

Figure 9.3 Recognition of cAMP by PDE4. A bidentate interaction between Glu443 and two of the adenine nitrogens is key to nucleotide discrimination for this cAMP-specific PDE. The core binding pocket contributes hydrophobic interactions with the heterocycle. The cyclic phosphate is thought to interact with the binuclear metal center, which may deliver an activated water molecule.

Hydrolysis of the phosphodiester bond proceeds with inversion of configuration at phosphorus, hence suggesting a standard nucleophilic substitution reaction by a water molecule [7]. A mechanism consistent with this finding has recently been suggested based on the structure of cAMP-bound PDE4: the water molecule or, more likely, the deprotonated hydroxide ion is activated by the Zn metal, whereas the other metal (Mg) could function as a charge-neutralizing Lewis acid [8]. Consistent with an in-line nucleophilic displacement reaction, the O3′ leaving group can be protonated by His234 acting as a proton donor.

9.3 Potential Therapeutic Applications of PDE Inhibitors

Medical applications of PDE inhibitors are numerous, and the subject is beyond the scope of this chapter. For detailed considerations, the reader is referred to in-depth reviews detailing the individual PDEs. Theophylline and other nonselective inhibitors have had a long history of medical use. However, the first widely adopted clinical use was that of selective PDE5 inhibitors for the treatment of erectile dysfunction (ED). As far as PDE4 is concerned, FDA approval of the first selective PDE4 inhibitor is still eagerly awaited. Such agents would have broad potential applications in inflammatory diseases such as asthma or COPD (chronic obstructive pulmonary disease) and possibly beyond (e.g., cognition disorders). On the other hand, PDE3 inhibitors have been available as approved drugs for some time, but have only seen limited clinical use, for example, in patients with heart failure. This may reflect particular issues with the limited PDE3 inhibitors available to date (e.g., lack of subtype selectivity for PDE3A vs. 3B) or may result from a more principal pharmacological feature of PDE3 as a target. Other PDEs have received far less attention. This is due to two main reasons: (1) suitable inhibitors have not been available, (2) basic understanding of the underlying biological mechanisms is still in development. The latter argument is particularly true for the 'young' PDEs 7–11.

9.4 Nonspecific PDE Inhibitors

The prototypical structural class of nonselective PDE inhibitors is represented by the methylxanthines (Figure 9.4), a family of plant-derived alkaloids that includes theophylline (**1**), caffeine (**2**), and theobromine (**3**) [9]. Although limited in potency, these simple naturally occurring xanthines were the 'parents' in the later discoveries of more potent synthetic derivatives such as pentoxyfylline (**4**) and isobutylmethylxanthine (IBMX, **5**). In particular, the latter compound has been widely used and has been regarded for decades as the 'gold standard' nonselective inhibitor of all PDEs. Only recently has it become clear that some of the newer PDEs (8 and 9) are not inhibited by IBMX. Derivatives of IBMX carrying substituents at the 8 position confer increased potency [10]. An example is compound **6**, which retains most of

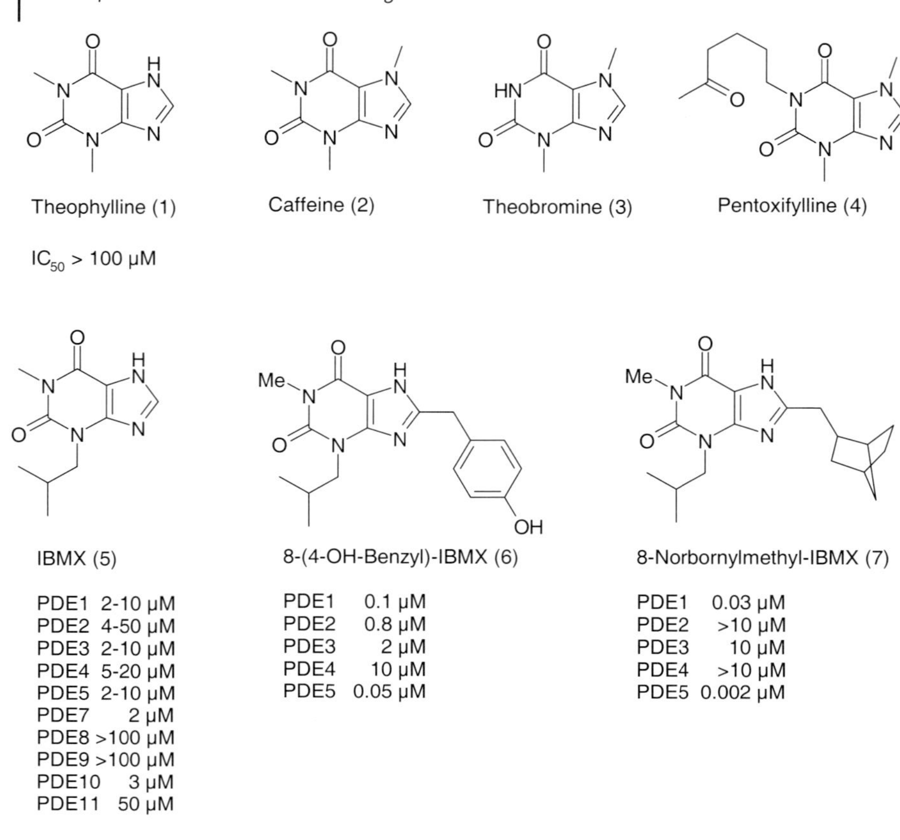

Figure 9.4 Nonselective xanthine derivatives. IBMX is considered the prototypical nonselective PDE inhibitor (even though two of the 'new' PDEs are insensitive to it).

the nonselectivity of IBMX. However, depending on the nature of the substituent, selectivity may start to emerge, as occurs for the bulky 8-norbonylmethyl-IBMX (**7**) which is indeed a selective PDE1/5 inhibitor. This approach of variation of a common core motif to introduce different PDE isoenzyme selectivity represents a recurring theme in the rest of this chapter.

The nonselective xanthines have served as a historical starting point and the first pharmacological tools in the area of PDE inhibitors. Isoenzyme-selective PDE inhibitors have been obtained by modification of the xanthine substituents and, more importantly, by turning to different core structures. These are discussed in the following sections, starting with inhibitors of cGMP-specific PDE5 and PDE6, followed by inhibitors of cAMP-specific PDE4 and the related dual-substrate PDE3. These two areas represent the bulk of available information regarding isoenzyme-selective inhibitors. We then discuss the knowledge on inhibitors of the remaining PDEs, with particular focus on structural themes that were introduced during the discussion of PDE5 and PDE4.

9.5 Inhibitors of the cGMP-specific PDE5 and PDE6

PDE5 is known as the cGMP-binding, cGMP-specific PDE. Only one gene product has been found, with four different splice variants. PDE5 contains two allosteric cGMP-binding domains in the amino-terminal region (GAF A and GAF B) [11]. PDE5 activation occurs upon binding of cGMP to the GAF A domain. Low catalytic activity has been observed in the nonactivated state. Both the catalytic domain (residues 537–860) and the allosteric binding site of PDE5 are highly specific for cGMP, and cGMP binding to the catalytic site seems to influence its binding to the allosteric site [12, 13]. As for many other PDE family members, PDE5 is a homodimer of 99-kDa subunits.

PDE5 was the first PDE for which the presence of metal-binding domains was demonstrated. Zn^{2+} is the most suitable metal ion for catalytic activity and leads to higher catalytic activity than Mn^{2+} and Mg^{2+}. In addition, a phosphorylation site is found on Ser92, thus giving rise to different regulatory pathways. Phosphorylation increases the activity of the enzyme at substrate concentrations below the K_M. Occupation of both allosteric cGMP binding sites is required for phosphorylation at Ser92. However, the catalytic domain alone seems to suffice for hydrolytic activity, since a truncated version containing only this region demonstrates catalytic activity [14].

Inhibition of PDE5 is a very broad field, and several reviews covering PDE5 inhibitors have appeared [15]. Instead of trying to cover all known inhibitors, a classification according to structural type is attempted. At least three main classes can be recognized:

- Substrate analogs characterized by an amide moiety, preferably tied into a pyrimidinone ring.
- Inhibitors carrying a methoxychlorobenzyl moiety.
- Indole-type PDE5 inhibitors.

9.5.1 Substrate-analogous PDE5 Inhibitors

The substrate-analogous PDE5 inhibitors are the broadest class of PDE5 inhibitors and include two marketed products, sildenafil and vardenafil (Figure 9.5). The cGMP analog zaprinast (**8**) served as starting point in this class. Importantly, zaprinast, which was originally in clinical trials as a potential treatment for asthma, introduced the important *ortho*-alkoxyphenyl moiety into the PDE5 inhibitors. At the time, zaprinast, with a K_i of 130 nM [16], constituted an important tool for evaluation of the role and function of PDE5.

A next step towards more potent inhibitors is exemplified by DMPPO (**9**), featuring a more hydrophobic right-hand portion of the core and the addition of a polar sulfonamide to the aromatic ring. These pyrazolopyrimidinones have been used for the construction of PDE inhibitors since the mid 1980s [17]. DMPPO in particular

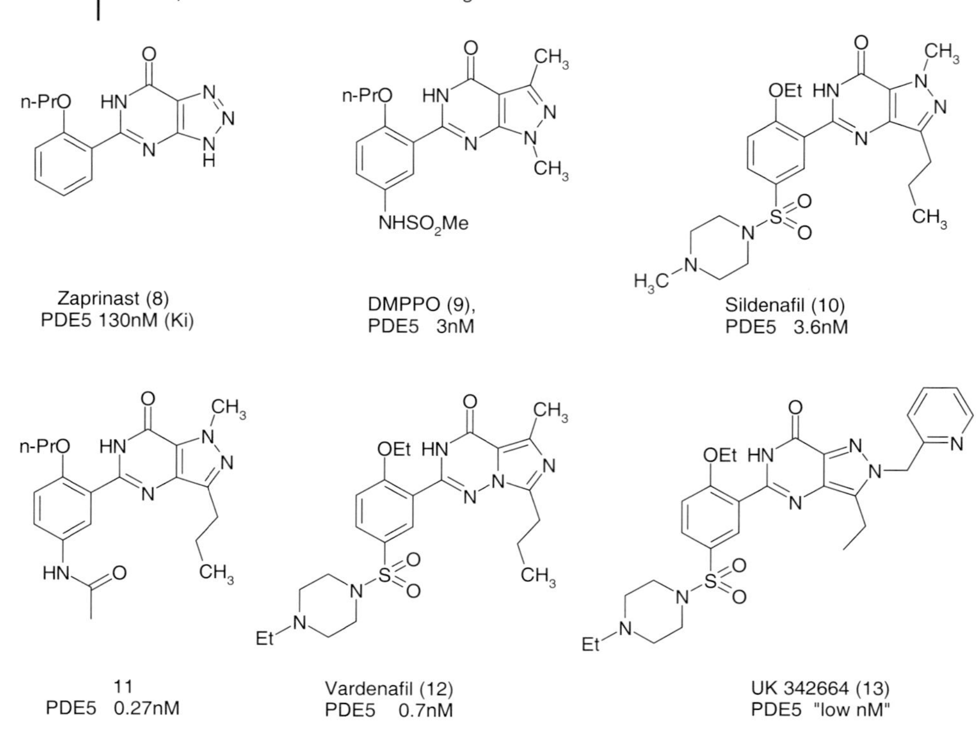

Figure 9.5 Substrate-analogous PDE5 inhibitors.

demonstrates the flexibility of the binding pocket on the right side of the molecule and shows an IC_{50} of 3 nM [18]. The next step in development was marked by the introduction of sildenafil (Viagra, UK 92,480, **10**). Considerable variability is allowed in the polar group in these structures, and a sulfonamide can be replaced by, e.g., ordinary amides (**11**) or ureas. These variations had been made so as to increase selectivity for PDE5 vs. PDE6. But, although the potency of such inhibitors compared to sildenafil was found to be increased, selectivity vs. PDE6 was not profoundly improved [19].

Vardenafil (Levitra, **12**), an imidazotriazinone PDE5 inhibitor, also falls into the substrate-analogous PDE5 inhibitor class. It demonstrates the most potent PDE5 inhibition of the marketed products in this class (IC_{50} = 0.7 nM). Compound **13** exemplifies isomeric pyrazolopyrimidinones with comparable activity to sildenafil [20]. Considering the observed binding of sildenafil with a water-mediated H bond between N2 of the pyrazole portion and Tyr612, the potent inhibitory activity suggests either a different binding mode or considerable distortion in the binding pocket.

As has been demonstrated through X-ray crystallographic analysis [21] with sildenafil (**10**) and vardenafil (**12**), the amide moiety forms a bidentate H-bond interaction with Gln 817, analogous to the mode in which the lactam portion of cGMP itself is expected to bind (Figure 9.6). The orientation of the sidechain amide

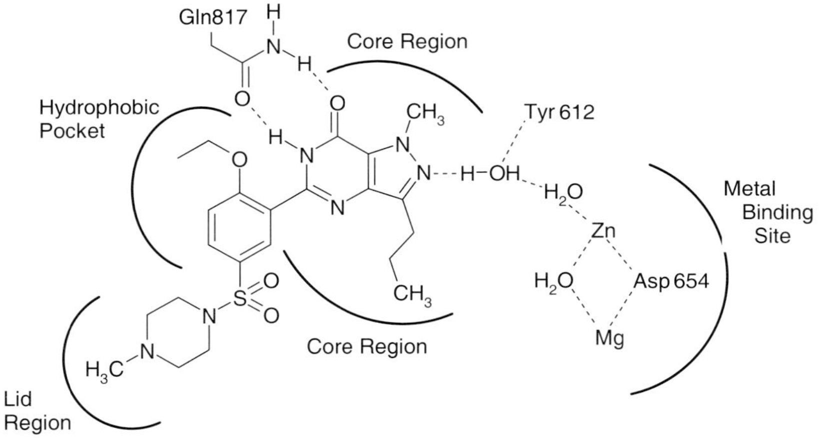

Figure 9.6 Binding of sildenafil to the PDE5 active site. The key glutamine residue is engaged in a bidentate interaction, which is also postulated for the recognition cGMP itself.

is fixed by further H-bond interactions within the protein. Interestingly, the analogous residue in PDE4B (Gln443) can reverse its orientation and thereby fit the inverted donor–acceptor requirements of the adenine nucleotide. Despite the H-bond acceptor role of the pyrimidinone oxygen, the sulfur analog of sildenafil, which should be a weaker H-bond acceptor, still demonstrates potent PDE5 inhibition [22]. The alkoxyphenyl sidechain fits into a hydrophobic pocket that differs from the PDE4 binding site by the substitution of a leucine by a methionine, which may be in part responsible for the only weak interaction of sildenafil with PDE4. Substantial binding energy is likely derived from this contact in particular, and it was even noted that the ethoxy group does not fill the entire pocket. Surprisingly, the very polar sulfonamide fragment does not seem to be involved in H-bond interactions, and the methyl piperazine residue points toward the protein surface and is engaged in some hydrophobic contacts.

Apart from the recognition of the lactam, the core region is largely engaged in hydrophobic contacts to the heterocycle. One exception is the pyrazole nitrogen, which makes a water-mediated contact to the metal binding site, which contains two metal ions. One of them, a zinc ion, makes contacts to several aspartate and histidine residues, whereas the other – presumably a magnesium ion – shares one of these contacts (Asp654) and is otherwise surrounded by five water molecules. In the binding model of cGMP in the active site, both metals can contact the cyclic phosphate moiety.

Figure 9.7 demonstrates the substantial flexibility in the right side of the substrate-analogous inhibitors. Compound **14** shows subnanomolar PDE5 inhibition combined with improved selectivity (90-fold) over PDE6 compared to sildenafil [23, 24]. The derivatives **15** and **16** demonstrate the effect of a bulky substituent in the region of N1 of sildenafil: potency and PDE6 selectivity are improved simulta-

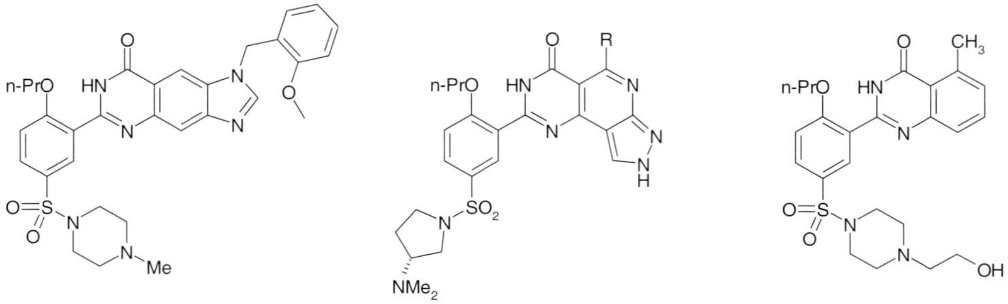

14
PDE5 0.62 nM

R= H 15 PDE5 2.1 nM
R= NHCH$_2$(4-F-Ph) 16 PDE5 0.3 nM

17
PDE5 6.5 nM

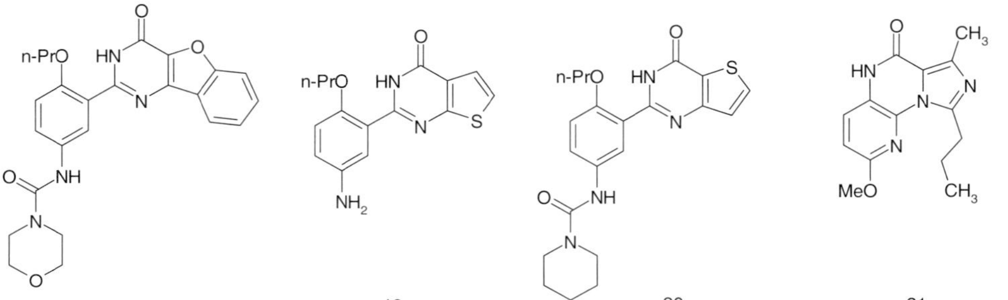

18
PDE5 18 nM

19
PDE5 1.7 nM

20
PDE5 3.5 nM

21
PDE5 10 nM
PDE3 20 nM

Figure 9.7 Structural variability within the substrate-analogous inhibitor family.

neously, although the molecular weight of the benzyl-carrying molecule approaches a value that may be problematic for good absorption from the GI tract [25]. The fused polycyclic derivatives **17–20** demonstrate that highly lipophilic heterocycles fused to the pyrimidinone are also accepted by the enzyme and are sufficient for high potency [26]. A comprehensive study has compared different heterocycle-fused pyrimidinone PDE5 inhibitors and demonstrated considerable influence of the heterocycle part [27]. The left side of the molecule likewise shows some flexibility, and even quite simple, low molecular weight derivatives such as **21** can serve as potent PDE inhibitors, although **21** lacks selectivity against PDE3 [28]. In keeping with the notion that the NH in such substrate analogs is directly involved in an H-bond, the N-alkylated derivative of **21** shows a much reduced PDE5 inhibitory potency.

Compound **22** (Figure 9.8) is a representative of Schering-Plough's tetracyclic PDE inhibitors that also include dual PDE1/5 inhibitors and that are characterized by the additional fused ring attached to the pyrimidinone [29]. This type of compound must be considered as extended xanthine analogs. Indeed, xanthines such as **23** can be potent, selective PDE5 inhibitors. Another variation of the xanthine theme is represented by **24** [30]. With respect to classification, the xanthines pose an interesting problem. As has been discussed, typical xanthine-based PDE

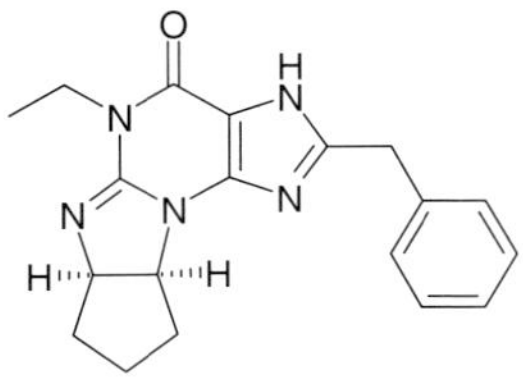

22
PDE5 19 nM

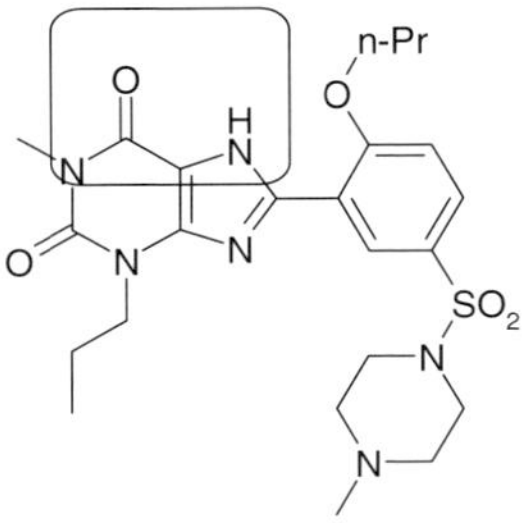

23
PDE5 14 nM

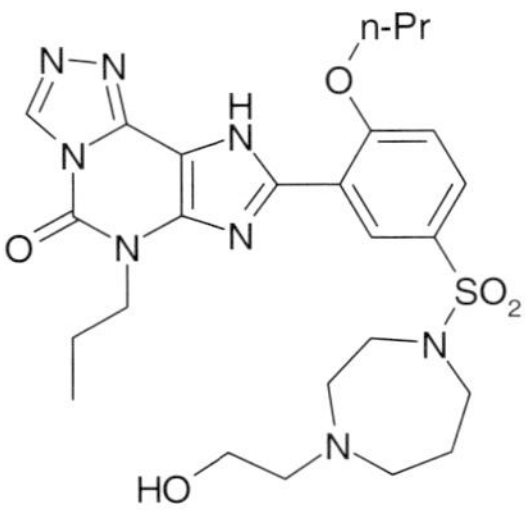

24
PDE5 0.34 nM

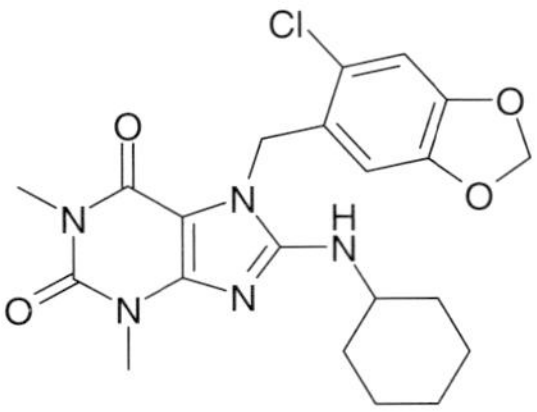

25
PDE5 <5 nM

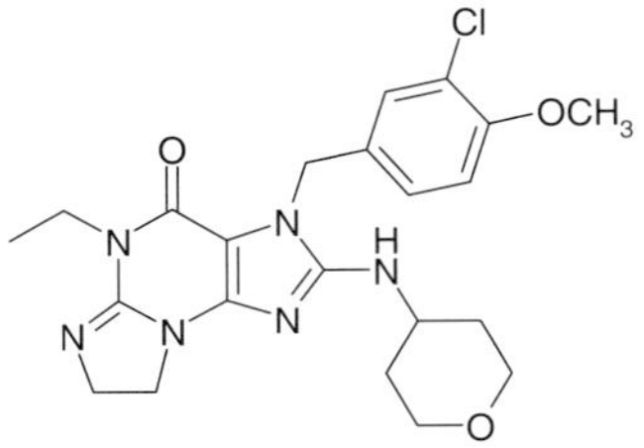

SCH 444877 (26)
PDE5 1.5 nM

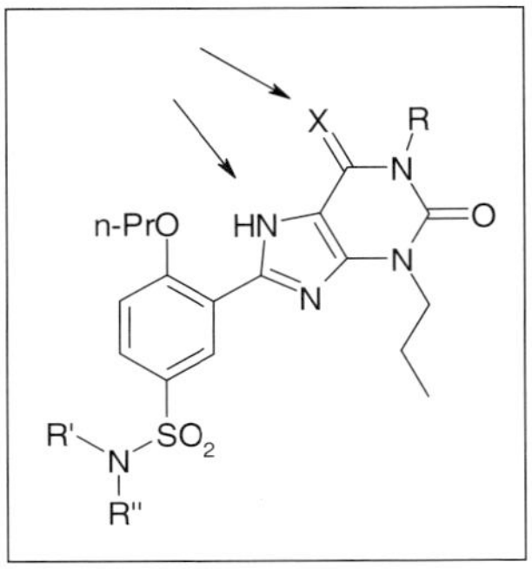

Figure 9.8 Xanthine-type PDE5 inhibitors.

inhibitors such as IBMX carry an alkyl substituent on N1 of the pyrimidinedione ring and thus technically lack the signature lactam moiety. However, the structural commonalities of the depicted examples to the inhibitors already discussed, including the presence of an NH close to the carbonyl oxygen (or equivalent H-bond acceptor), justifies their classification with the other substrate-analogous pyrimidinone inhibitors (Figure 9.8, inset).

Nonetheless, greater structural variation is allowed within the xanthine group which does not easily fit the above rationale. This is demonstrated by **25**, which can be considered a hybrid of **13** or **14** and a xanthine core [31]. By transitioning into the related tetracyclic guanine series, very potent inhibitors of PDE5 with very high selectivity for PDE6 were created, e.g., **26** [32]. Indeed, **25** and **26** represent the transition between substrate analogs and the series of chloromethoxybenzyl-type compounds discussed below.

9.5.2 Inhibitors Carrying a Chloromethoxybenzyl Substituent

Classification according to a substituent is of course unusual, since medicinal chemists conventionally group structures by their common core, thought to define a pharmacophoric element, and a set of substituents which, by definition, do not

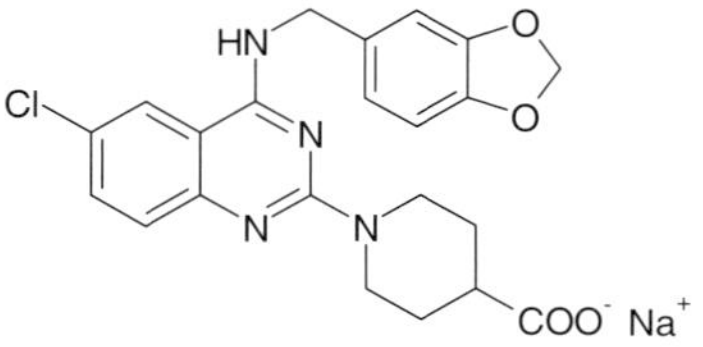

E4021 (27)
PDE5 3.7 nM

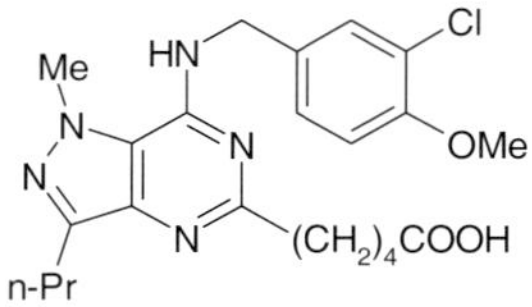

28

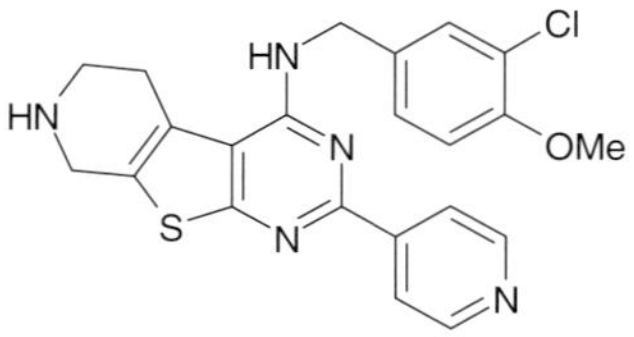

29
PDE5 2.6 nM

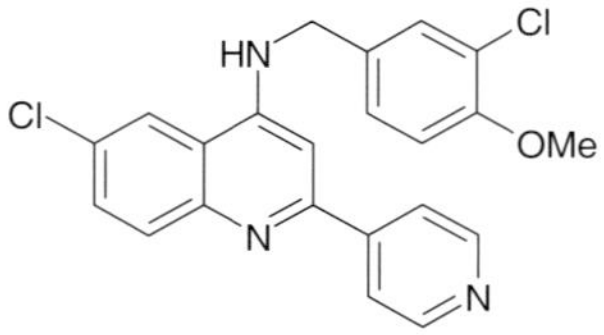

30
PDE5 0.24 nM

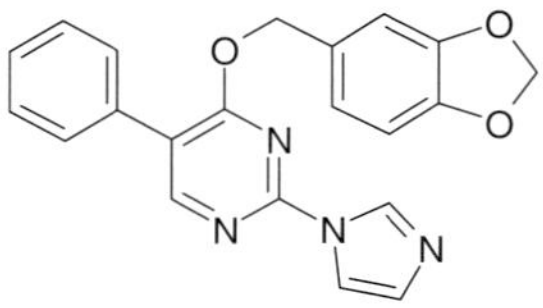

31
PDE5 14 nM

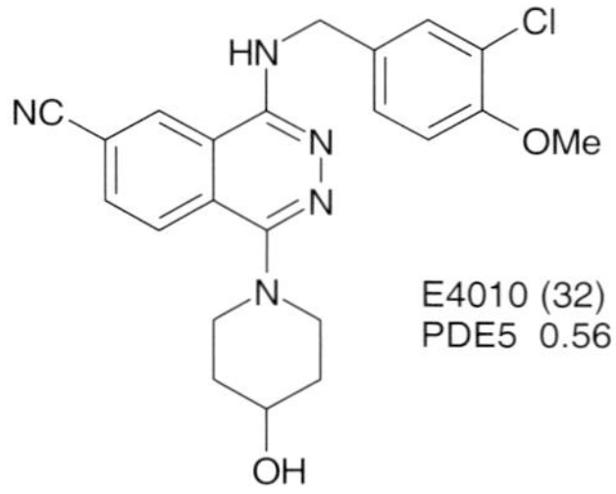

E4010 (32)
PDE5 0.56 nM

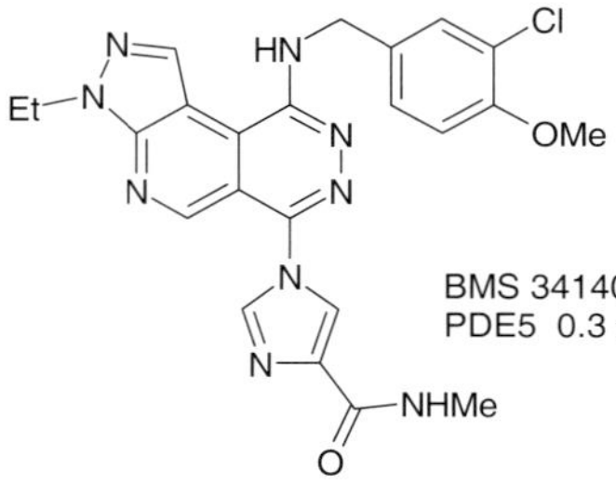

BMS 341400 (33)
PDE5 0.3 nM

Figure 9.9 Chloromethoxybenzyl-containing PDE5 inhibitors.

belong to the pharmacophore and allow some variability. Nevertheless, a chloromethoxy substituent (or isosteres thereof, e.g., methylenedioxy or dimethoxyphenyl) has been incorporated in enough interrelated PDE5 inhibitors that this classification seems justified. This includes a very diverse set of scaffolds and – together with the tadalafil series – seems to be an area of intense current patenting activity.

A number of 6,6 or 5,6 fused-ring systems fall into this category. A typical example of a quinazolinone-based inhibitor is E4021 (**27**) (Figure 9.9) [33]. This series of diaminoquinazolines has also spawned potent kinase inhibitors. Compound **28**, bearing a pyrazole fragment that resembles the right side of sildenafil, and the tricyclic derivative **29** demonstrate the broad variability of cores in this class [34]. In addition, analogs of **29** have been disclosed, in which the ring fused to the thiophene can also be an aromatic or a simple cyclohexyl ring [35]. Example **30** shows the transition from quinazolines to quinolines with retained PDE5 inhibitory activity [36]. In **31** the lipophilic part of the quinazoline has been replaced with a phenyl

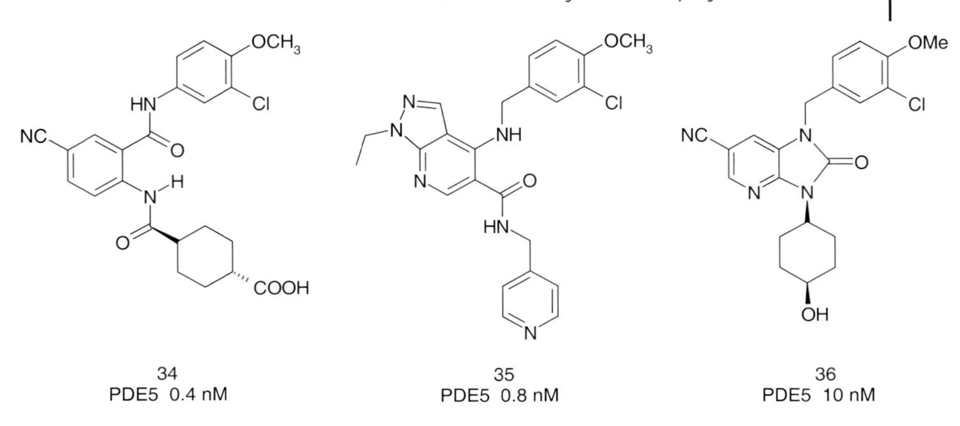

Figure 9.10 Different scaffolds for presentation of the chloromethoxyphenyl group.

substituent at the 5 position of the pyrimidine [37]. In addition, nitrogen is exchanged for oxygen as a linker to the benzyl group. The changes result in a somewhat reduced activity; however, PDE5 inhibition is still in the low double-digit nanomolar range.

Next to the quinazolines, phthalazines represent the most important heterocyclic series in this class of PDE5 inhibitors. At least two compounds from this category have reached clinical trials: E4010 (**32**) [38] and BMS341400 (**33**) [39]. Both compounds show subnanomolar PDE5 inhibition. Again, a chloromethoxybenzyl substituent seems to be among the preferred substituents in the top part of the molecule; polar substituents in the bottom part also characterize this series.

Phthalazines were discovered as scaffolds for both PDE5 inhibitors and kinase inhibitors. Similarly, anthranilic acid amides were disclosed for both pharmacological activities (Figure 9.10). Compared to the phthalazines, putatively an internal H-bond keeps the two substituents in a comparable spatial arrangement. Extremely potent PDE5 inhibitors can be found in this class, e.g., **34** [40]. A similar internal H-bond can be speculated for **35**, for which potent PDE5 inhibitory activity was likewise reported [41]. Compound **36** represents the transition between phthalazines and anthranilic acid amides [42].

9.5.3
Indole-type PDE5 Inhibitors

This series of PDE5 inhibitors is rapidly expanding, in part in response to the market entry of tadalafil (**37**, Figure 9.11). Tadalafil itself is a potent PDE5 inhibitor (IC_{50} = 2 nM), with improved selectivity for PDE6 compared to sildenafil and vardenafil but demonstrating a much lower selectivity for PDE11 (IC_{50} = 37 nM) [43]. The binding mode of tadalafil shows considerable differences from the substrate-analogous PDE5 inhibitors. The indole NH is involved in a single, not bidentate, H bond with Gln817 (Figure 9.11, inset). The methylenedioxy aromatic

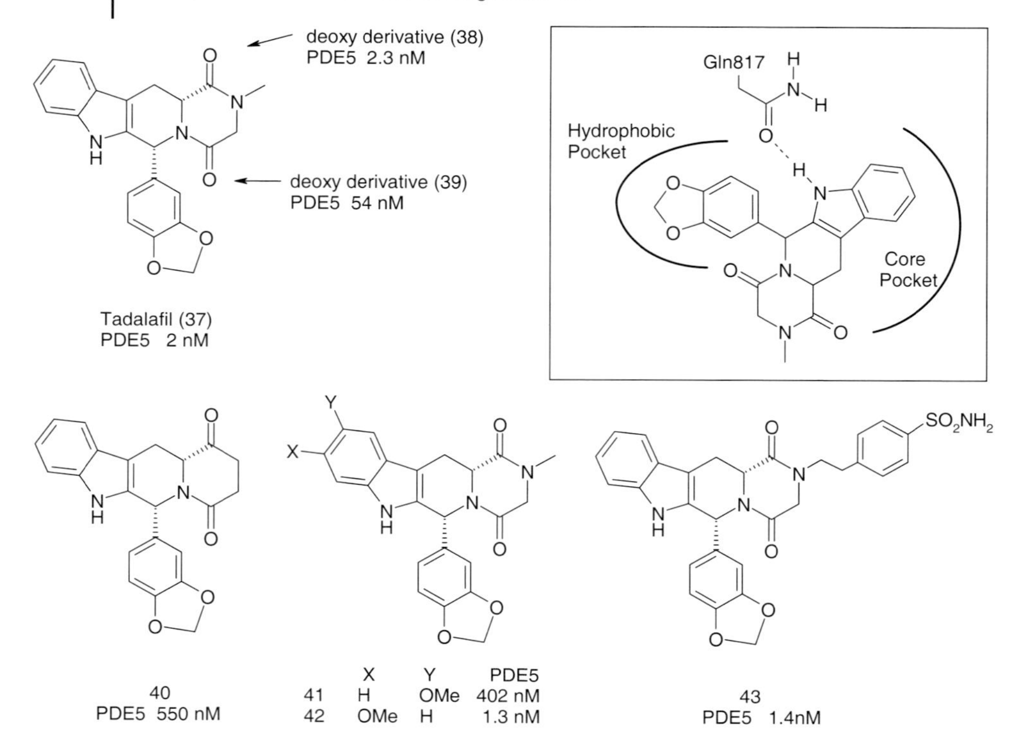

Figure 9.11 Indole-type PDE5 inhibitors. In contrast to the substrate-analogous inhibitors, the key glutamine residue is engaged in only a single – not a bidentate – H bond to the indole NH of tadalafil (see inset).

portion occupies space that is filled by the alkoxyaromatic group in vardenafil and sildenafil, i.e., the hydrophobic pocket. Only one of the two carbonyl oxygens in the diketopiperazine ring is essential for potent PDE5 inhibitory activity (compare **38** and **39**) [44]. The N-methyl group also seems to be important for high activity (**40**) [45]. Compounds **41** and **42** demonstrate the sensitivity of the system to substitution in the aromatic ring [46]. This highly regiospecific effect is reminiscent of the phthalazine class of PDE5 inhibitors. The bulky sulfonamide derivative **43** illustrates the room for large substituents on N1 of the diketopiperazine ring system [47].

The diketopiperazine system itself is not mandatory for potent PDE5 inhibition, as exemplified by compounds **44** and **45** (Figure 9.12) [48]. Compound **45** also demonstrates that the methylenedioxy fragment in tadalafil can be replaced by a methoxy group. Other compounds (**48**) show a dihydrofuran in this region. Although **47** completely lacks the diketopiperazine ring, the urethane carbonyl group substitutes for the more important carbonyl functionality and assures good activity [49]. As shown in **43**, the right side of the molecule shows some structural flexibility and can accommodate structures such as **47** and **48** [50].

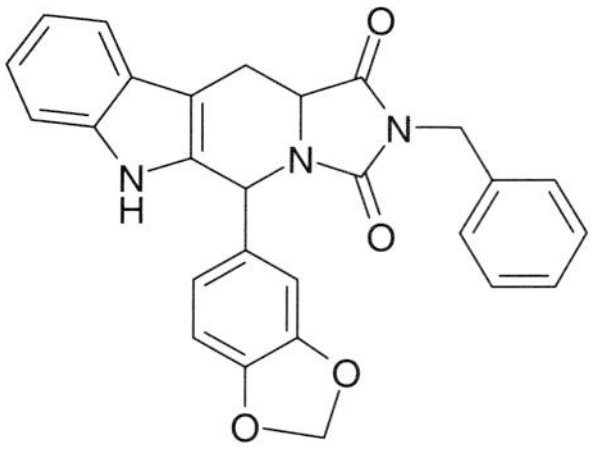

44
PDE5 <10 nM

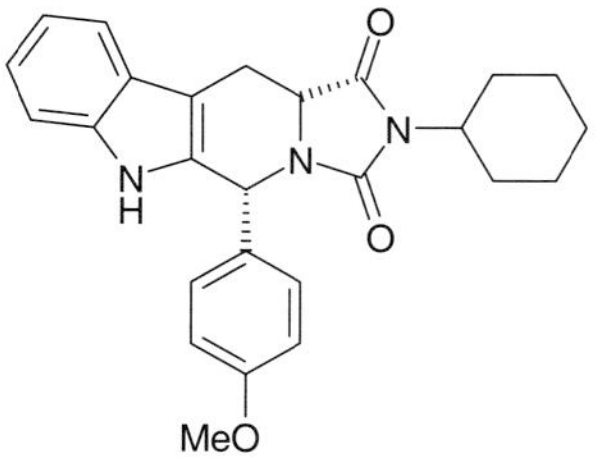

45
PDE5 7 nM

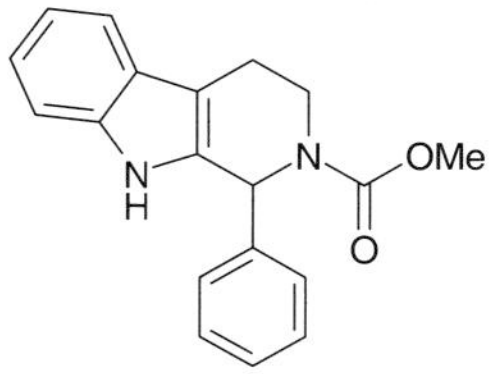

46
PDE5 1 nM

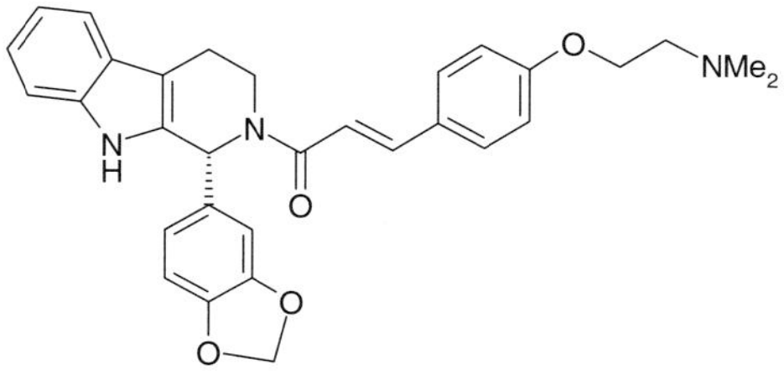

47

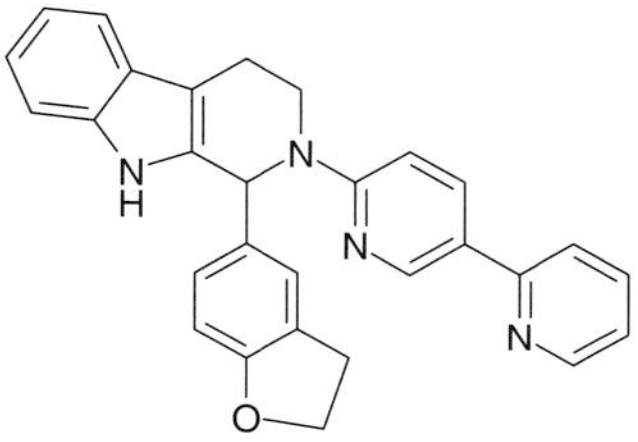

48
PDE5 0.69 nM (Ki)

Figure 9.12 Indole-type PDE5 inhibitors lacking the diketopiperazine ring.

An indole is not strictly required in this type of compound (Figure 9.13). An example is given by the quinolinones **49** and **50**, which show the transition from the tadalafil 6,5,6 fused system to an analogous 6,6,5 system. This quinolinone structure preserves the ability of the heterocycle NH to form the crucial H-bond to the glutamine oxygen of Gln817. The inhibitory potency of these compounds is excellent. Again, the diketopiperazine system is not essential, and compounds with only the requisite lower carbonyl functionality also show very potent PDE5 inhibition (**50**). Other compounds, such as **51** and **52**, seem to also bear some resemblance to the structural topology found in tadalafil, although the analogy is much less clear [51]. Although the trimethoxyphenyl part shows some similarity to the methylenedioxyphenyl group in tadalafil and both are connected to a hydrophobic aromatic core residue, the H-bond donor function is lacking. On the other hand, for the analogs of **51**, the presence of the carbonyl groups was also found to be important, another parallel to the SAR in the tadalafil series.

Since most standard medicinal chemistry issues (potency, selectivity, pK, etc.) have been addressed with various inhibitor classes and since three compounds are already on the market, new PDE5 inhibitors in the future may be especially relevant in indications distinct from male erectile dysfunction, as dual (or multiple)-specificity PDE inhibitors, or as templates for the synthesis of other PDE or kinase inhibitors.

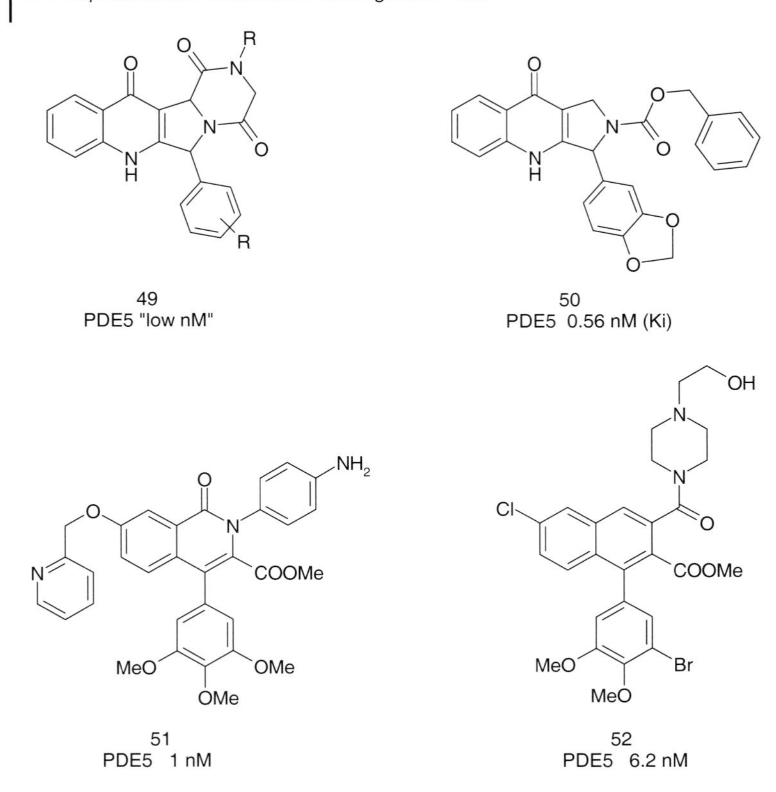

Figure 9.13 Tadalafil analogs without an indole ring, and related compounds.

9.6 PDE6 Inhibitors

PDE6 shows high selectivity for cGMP and also contains two allosteric cGMP binding sites in the N-terminal half of the protein. The occupation of these allosteric sites seems to play an important role in the control of catalytic activity, although the catalytic function does not seem to be directly influenced by these sites. Rather, occupation seems to affect the affinity for other regulatory proteins binding to PDE6 and thus indirectly influence catalytic activity [52].

Three different gene products are known for PDE6, which consists of hetero-tetramers formed from two homodimers. Historically, the monomeric subunits of PDE6 have been known as α (99 kDa), α‘ (94 kDa) and β (98 kDa) [53]. In addition, a much smaller γ subunit (9.7 kDa) acts as an inhibitory regulator of PDE6 activity and binds with very high affinity. The less frequently occurring soluble form of PDE6 also contains a δ subunit [54].

PDE6 is by far the dominant PDE occurring in rod and cone cells of the vertebrate retina. It plays a key role in visual signal transduction, which is unique among the

PDE family members. Absorption of light first activates rhodopsin which in turn activates the G protein transducin. Light activation of PDE6 then occurs through interaction of the activated (GTP-bound) α subunit of transducin with the inhibitory γ subunit of PDE6, thus leading to removal of PDE6 inhibition and rapid decline in cGMP concentration. This results in closure of cGMP-gated cation channels, inducing a hyperpolarized state in the receptor cells.

Since several vision impairments have been traced to defects in proper PDE6 function, and since side effects such as blue vision that occur in patients taking sildenafil have been speculated to be linked to a lack of sufficient PDE6 selectivity, PDE6 inhibitors have not been the target of intensive medicinal chemistry endeavors and so far have not been reported in their own right. The catalytic domain of PDE6 is very similar to the catalytic site of PDE5, which may be the reason for overlapping activity. Vardenafil and sildenafil inhibit PDE6 in the double-digit nanomolar range, and tadalafil at a considerably higher IC_{50}. As already mentioned, recent publications disclose PDE5 inhibitors with reduced PDE6 inhibitory activity.

9.7 Inhibitors of cAMP-metabolizing PDE4 and PDE3

PDE4 is often referred to as the cAMP-selective PDE, due to its high substrate specificity. Four different gene products are known, and differential splicing leads to more than 15 isoenzymes that fall into three categories: long (85–110 kDa), short (68–75 kDa), and super-short ($<$ 68 kDa) [55]. In addition, both membrane-bound and cytosolic variants are known. The most unusual feature of the PDE4 long forms is the presence of two upstream conserved regions (UCR1 and UCR2) which are 55 and 75 amino acids long and separated by a linker. Both UCRs are highly conserved between different species, implying an important role in PDE4 function. Putatively, the UCRs are involved in regulatory events, with the UCR2 domain acting to reduce catalytic activity. Variants with deleted portions of UCR1 and/or UCR2 show different sensitivity to the inhibitor rolipram (**53**, Figure 9.14). PDE4 can be activated by PKA phosphorylation at two sites, one of which is located in the UCR1. The phosphorylated enzyme has a lower IC_{50} for rolipram [56].

Pharmacologically, PDE4 is one of the best-investigated members of the PDE family, and many inhibitors of this enzyme are already known [57]. In recent years, insights into the atomic structure of PDE4 and into the interactions of substrate and inhibitors with the enzyme's active site have been acquired. From a chemogenomic point of view, understanding the design principles of PDE4 inhibitors with the help of structural information has become a valuable basis for extrapolation of these principles to other PDE members with similar conserved catalytic domains.

The most prominent group of structurally related PDE4 inhibitors was built on the structure of the archetypical inhibitor rolipram (**53**). Rolipram also serves to illustrate a particular difficulty in the field of PDE4 inhibitors, namely, the confusion that has surrounded the different behaviors of PDE4 preparations toward this (and other) inhibitors. This has led to the concept of low-affinity rolipram binding (LARB,

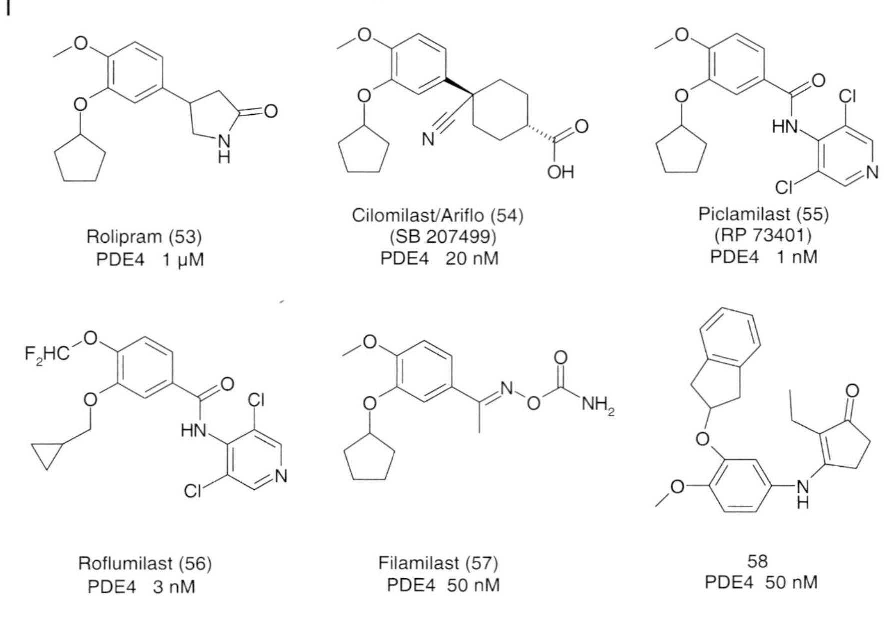

Figure 9.14 Catechol-type PDE4 inhibitors.

IC_{50} = 100–1500 nM for rolipram) and high-affinity rolipram binding (HARB, IC_{50} = 1–50 nM for rolipram) states. Today, most investigators believe that these represent two different conformational states of the same enzyme and that interactions with other proteins and/or phosphorylation of the enzyme can control the particular state. Much of the work discussed in this section was concerned with trying to identify molecules that preferably inhibit the LARB form, since the HARB form is thought to mediate dose-limiting inhibitor side effects such as nausea and vomiting. However, in a significant portion of the available literature and patent reports, it is unclear which state the experimentally employed PDE4 enzyme was in.

The defining structural element found in rolipram and its relatives is the dialkoxyphenyl (catechol) moiety. Although the development of rolipram (then for depression) was terminated early, many companies have since disclosed PDE4 inhibitors intended primarily for indications such as asthma and COPD (chronic obstructive pulmonary disease) [58]. For a long time, cilomilast (Ariflo, **54**) has appeared to be ahead in this race [59]. In November 2003, GSK received an 'approvable' letter from the FDA, despite the FDA advisory panel's earlier negative opinion of the compound. The extent to which new studies may have been requested by the FDA and will delay approval remains to be seen.

A second prominent structural feature found in many catechol type inhibitors, the 3,5-dichloropyridine residue, was introduced first in piclamilast (**55**) [60]. In a modified form, this combination of pyridine and catechol was taken up by many companies. Most notably, it is found in roflumilast (**56**) which, apart from cilomilast, must be considered the most advanced clinical candidate to date [61].

The pharmacophoric pattern of the catechols allows for substantial variability. Thus, acyclic derivatives such as the oxime derivative found in filamilast (**57**) can replace the lactam in rolipram [62]. The size of the hydrophobic group can also be increased, as seen in **58** [63].

How do the catechol inhibitors interact with the enzyme active site? The first insight into this question came from the crystal structure of PDE4D complexed with the mixed PDE4/3 inhibitor zardaverine (**59**, Figure 9.15) [64]. Zardaverine displays an overall flat molecular architecture combining the typical dialkoxyaryl motif of many PDE4 inhibitors with the pyridazinone ring carrying an amidic N–H, a fragment found in a number of PDE3 inhibitors. In contrast to other PDE4-specific inhibitors, the bulky substitution at the alkoxy group is not present [65]. Interestingly, the main H-bonding interaction found in the structure is the bidentate complexation of Gln466 by the catechol ether oxygens. This is the very same, highly conserved glutamine residue that is thought to interact in the key bidentate recognition of the substrate cAMP. In the structure of PDE5 the corresponding residue (Gln817) is also the one that provides the sole direct H bond to the substrate-analogous inhibitors sildenafil and vardenafil. The aryl moiety spans the core pocket of the active site. In particular, a stacking interaction is observed with Phe469. Furthermore, zardaverine partially extends into the metal binding region containing the two metal centers (probably Zn^{2+} and Mg^{2+}) that coordinate residues of the highly conserved PDE sequence motifs His-Asn-X-X-His and His-Asp-X-X-His. In the crystal structure, zardaverine also appears to make H bonding contact to a cacodylate group (dimethyl arsenate, a phosphodiester mimic), which cocrystallized with the metal centers. Whether this implies similar interaction under physiological conditions (e.g., with a phosphate) is unclear.

Zardaverine is small enough that it occupies only part of the active site. Inhibitors such as rolipram, containing the larger cyclopentyl group instead of methyl, would be expected to fill the S1 pocket of PDE4D (corresponding to the hydrophobic pocket of PDE5). Modeling studies on rolipram itself have come to differing conclusions. In a recent publication, Houslay and Adams [1e] suggested the binding mode depicted in Figure 9.15, whereas an earlier modeling study made a different proposal [66]. Based on the zardaverine structure, however, it seemed likely that the catechol fragment would be positioned in an analogous manner. Indeed, this was shown to be so in the recently published X-ray structure of rolipram bound to PDE4D [67].

A distinct variation within the catechol series is the introduction of an additional phenyl group branching from the variable connection between the catechol itself and the pyridine or equivalent polar fragment. As shown in Figure 9.16, this gave rise to Celltech's CDP-840 (**60**), a compound that was codeveloped with Merck/Frosst but was stopped in phase 2 [68]. The additional hydrophobic feature may be a contribution to the hydrophobic S3 pocket identified in the PDE4D structure (see above). Later developments by the Merck/Frosst group included compounds **61** and **62**, eventually leading to **63**. A variation with a different spacing for the phenyl group is demonstrated in compounds **64** and **65** [69]. The latter compound once again underscores the exchangeability of an amide (as in rolipram) and a pyridine ring (roflumiast, CDP-840) in this structural environment. Interestingly, the deoxygenated (i.e., indane instead of indanedione) derivative of **64** is a much less potent inhibitor. Finally, a classical scaffold approach is shown in **66**, in which morpholine serves as a template to present the three important PDE inhibition motifs in a correct topology [70].

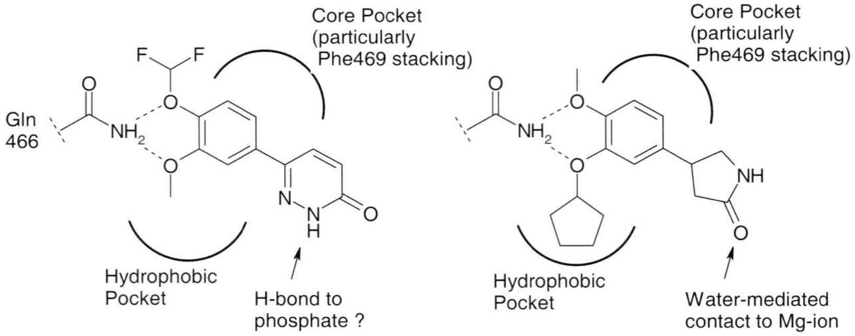
Core Pocket
(particularly
Phe469 stacking)
Gln
466
Hydrophobic
Pocket
H-bond to
phosphate ?
Core Pocket
(particularly
Phe469 stacking)
Hydrophobic
Pocket
Water-mediated
contact to Mg-ion

Zardaverine (59)
PDE4 160 nM
PDE3 580 nM

Rolipram
(PDE4 selective)

Figure 9.15 Binding of catechol Inhibitors to the active site of PDE4. The key glutamine residue (corresponding to Gln817 in PDE5) is involved in an H bonding contact with the catechol moiety.

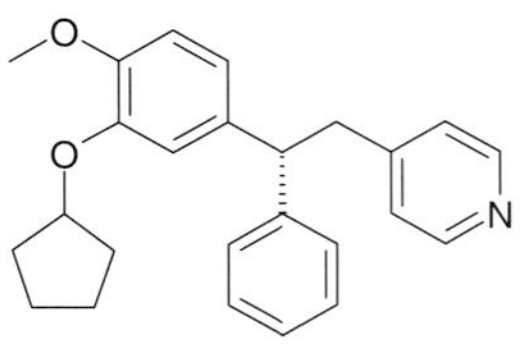

CDP-840 (60)
PDE4 4 nM

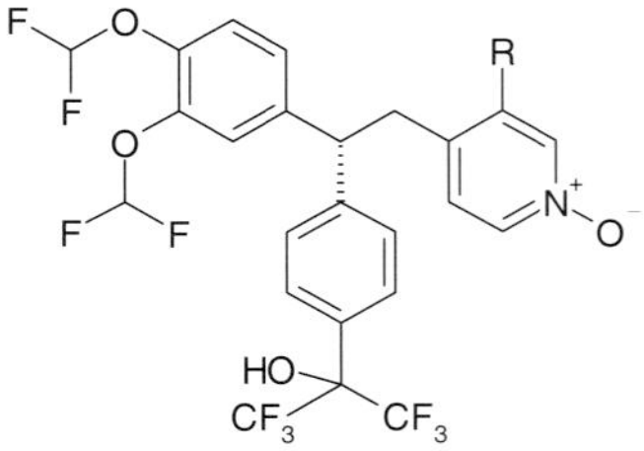

L-791,943 (61, R=H) PDE4 2 nM
L-826,141 (62, R=Me)

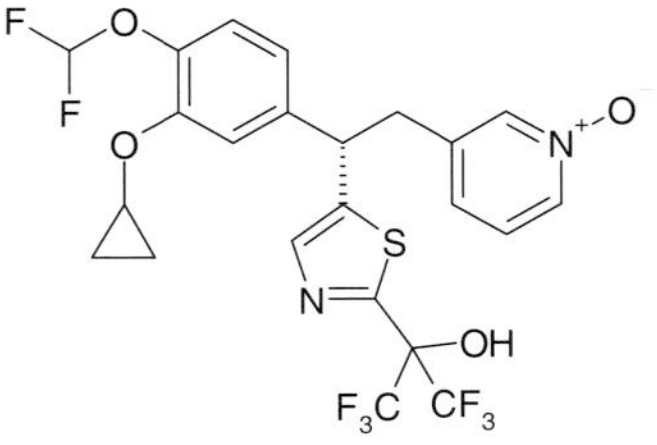

L-869,298 (63)
PDE4 0.5 nM

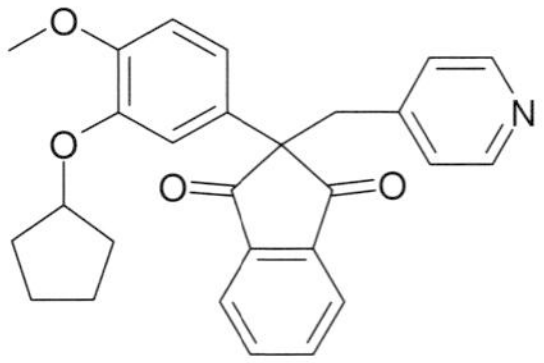

64
PDE4 30 nM

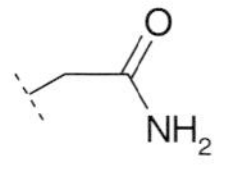

CDC-801 (65)

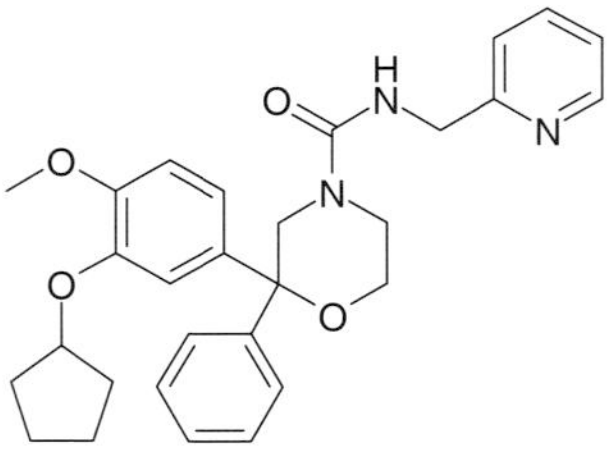

66
PDE4 28 nM

Figure 9.16 CDP-840 and related catechol derivatives featuring an additional aromatic residue.

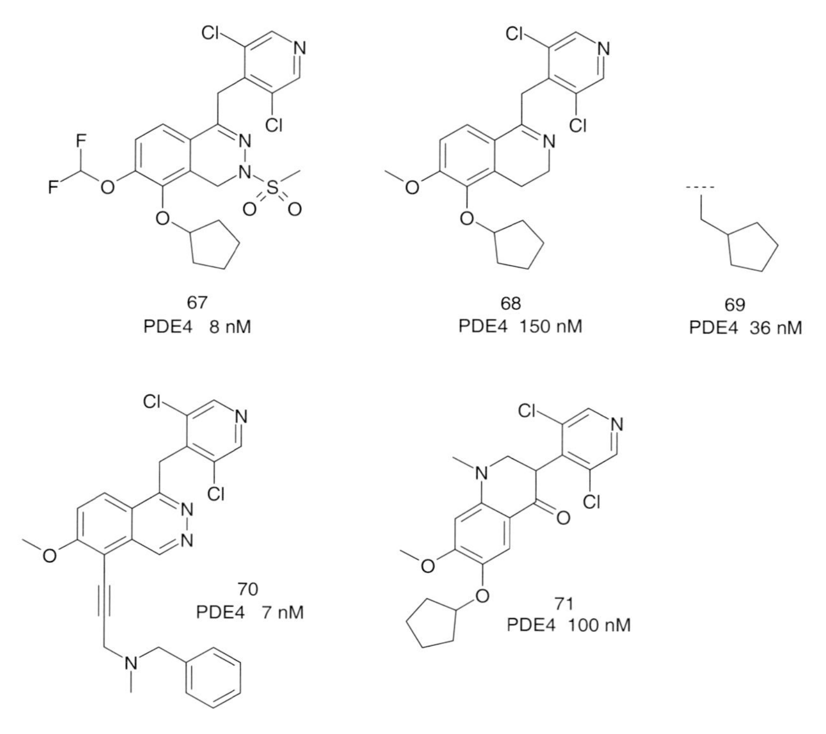

Figure 9.17 Fused catechol-type PDE-4 inhibitors where the catechol moiety is part of a bicyclic framework.

Figure 9.17 displays a related subfamily of PDE4 inhibitors, in which the typical catechol fragment is itself part of condensed bicyclic system as a chemical scaffold. For example, a variety of dialkoxy-substituted dihydrophthalazine (**67**) and dihydro-isoquinoline (**68**) derivatives have appeared in the patent literature [71]. Again, the combination of these scaffolds with a pyridyl moiety is preferred. Only one catechol oxygen is required, and replacing the second one with a methylene group (**69**) even results in enhanced activity. Interestingly, even an elaborate alkine moiety can replace the *cyclo*-pentyloxy fragment (**70**).

Another chemical subtype for inhibition of PDE4 comprises benzo-condensed heterocycles (Figure 9.18). In these structures one alkoxy group is typically placed in an α position to the ring fusion. As part of the heterocycle, an endocyclic heteroatom acting as an H-bond acceptor is placed ortho to the alkoxy group. Of course, from a pharmacological point of view these motifs are essentially masked catechol derivatives. These rigid chemical scaffolds define further series of privileged structures for PDE4 inhibition. Examples include alkoxy-substituted benzoxazoles (**72**), benzofuranes (**73**, **75**, **76**), and quinolines (**74**) and also extend to structures lacking one acceptor heteroatom, such as benzimidazoles (**77**) and indoles [72–75]. Known privileged fragments, which were already introduced in combination with catechols, are recurring in this constellation: we can notice the dichloropyridine/

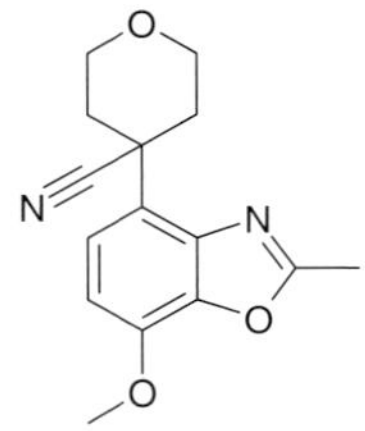

72
PDE4 135 nM

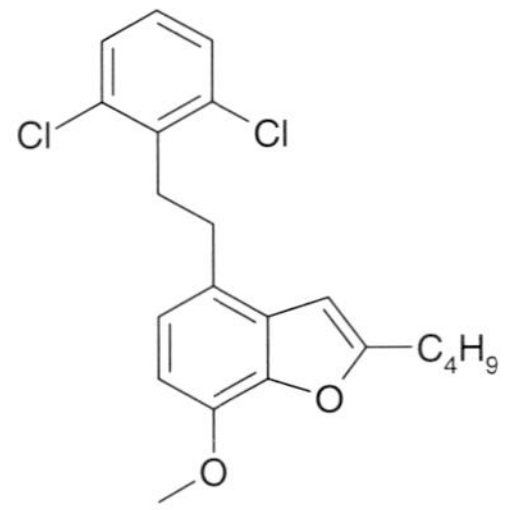

73
PDE4 1 nM

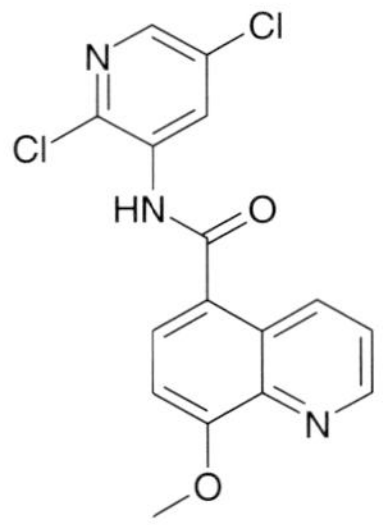

D-4418 (74)
PDE4 200 nM

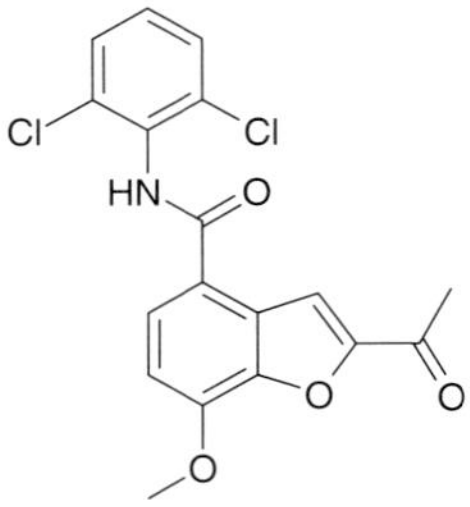

75
PDE4 1.6 nM

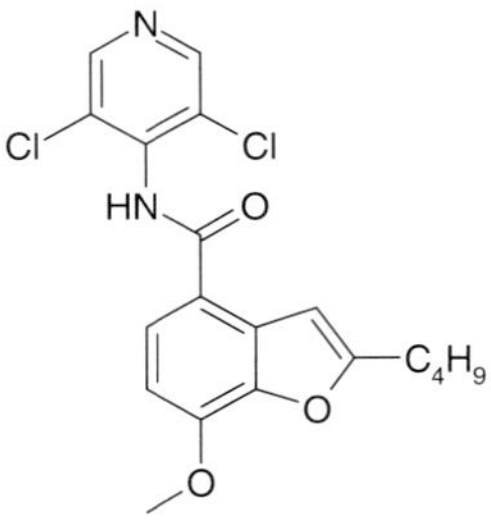

76

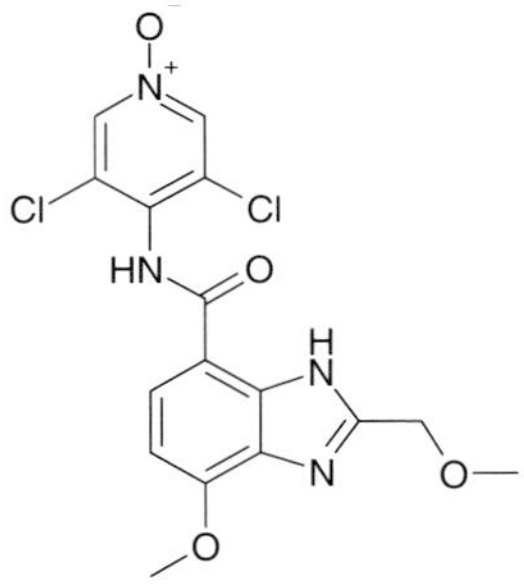

77

Figure 9.18 Masked catechols.

phenyl motif in **73–76**, also allowed as the N-oxide variant (**77**), or its replacement by a donor-substituted cyclohexyl nitrile as in **72** (compare cilomilast **54**). Furthermore, an additional lipophilic residue on the five-membered ring system is clearly preferred.

A particularly interesting heterocyclic replacement for the catechol group is the indazole ring system and related heterocycles (Figure 9.19). The relationship can be illustrated in a series of PDE4 inhibitors from ICOS beginning with IC-197 (**78**). As seen in **79** and **80**, the catechol moiety can be replaced with an indazole without much loss in potency [76]. This is in agreement with the previous finding that only one of the alkoxy donor moieties in the catechols is strictly required (compare **69**, **70**, **77**). The notion of this particular bioisosterism has also been utilized by scientists at Pfizer, who reported pyrazolopyridinones **81** and **82** [77]. Not surprisingly, the concept has also been extended to compounds **83** and **84**, which are the respective indazole analogs of roflumilast and cilomilast, although no enzyme inhibition data have been reported [78].

An extension of the scaffold concept introduced with compound **66** is given in the following examples (Figure 9.20). Typically, up to three substituents are orientated by using a heterobicylic template like benzoxazole **85**, benzimidazole **86**, or adenines **87**, **88** [79–81]. Since one of these substituents is always a catechol

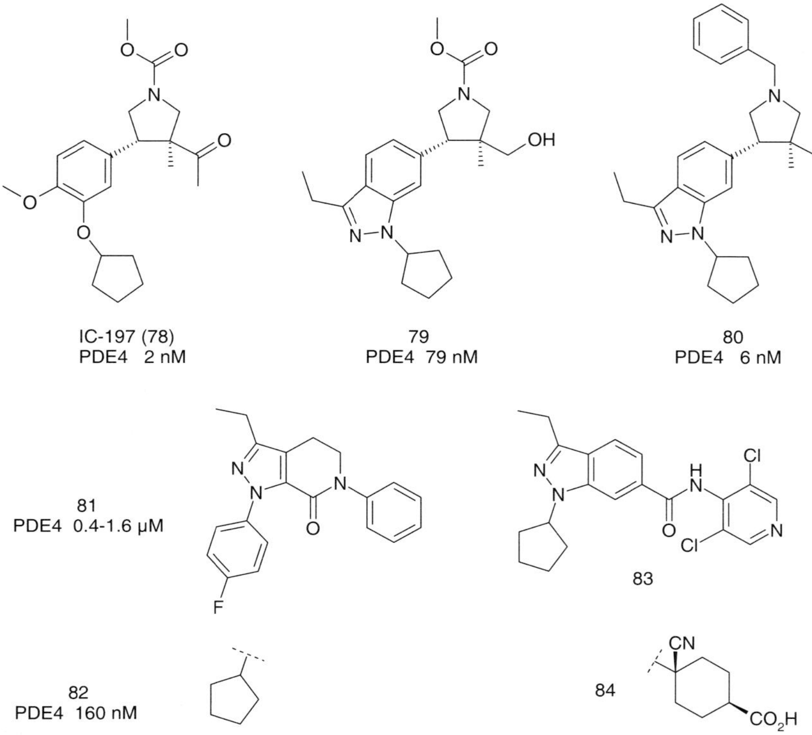

IC-197 (78)
PDE4 2 nM

79
PDE4 79 nM

80
PDE4 6 nM

81
PDE4 0.4-1.6 μM

83

82
PDE4 160 nM

84

Figure 9.19 Indazoles as catechol mimics.

derivative, one can speculate as to whether the heterocyclic scaffold serves not only as a template but simultaneously as a pyridine replacement.

A very interesting series of derivatives related to the early PDE4 inhibitor nitraquazone (**89**) [82] is shown in Figure 9.21 [83]. These lack the catechol moiety and may be more related to the xanthines like arofylline (**99**), but they are structurally distinct, and perhaps their most distinguishing feature is the presence of a lipophilic phenyl ring appended to a heterobicyclic core, resulting in a L-shaped molecular structure. Another common feature is the presence of a carbonyl group or analog thereof opposite the phenyl ring attachment. Examples that illustrate the scope of heterocycles are the azabenzopyrimidinones **90** and **91** from Yamanouchi and the condensed triazole **92** from Almirall [84]. Interestingly, in analogy to the catechols, the familiar pyridine motif can be combined with these core structures as well, giving rise to the highly potent PDE4 inhibitors **93** and **94** [85]. It is therefore tempting to speculate that these inhibitors likewise act in a substrate-like manner addressing the core adenosine binding pocket of PDE4. Possibly related to this structural motif is a series of rigidified benzodiazepinones from Pfizer/Parke Davis exemplified by **95** [86]. This eventually gave rise to two clinical candidates, of which CI-1044 (**96**) is the most recent [87].

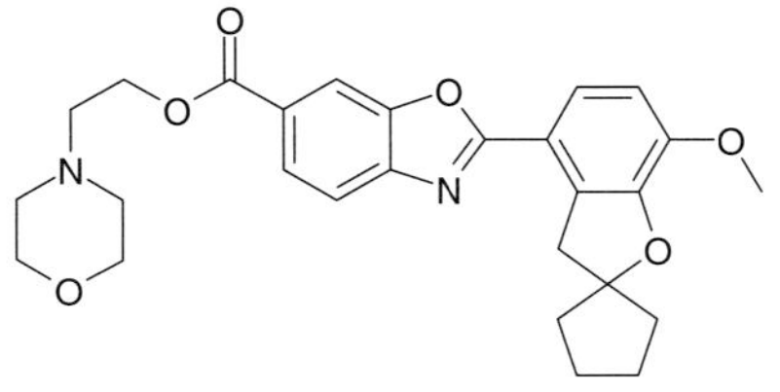

85
PDE4 125 nM

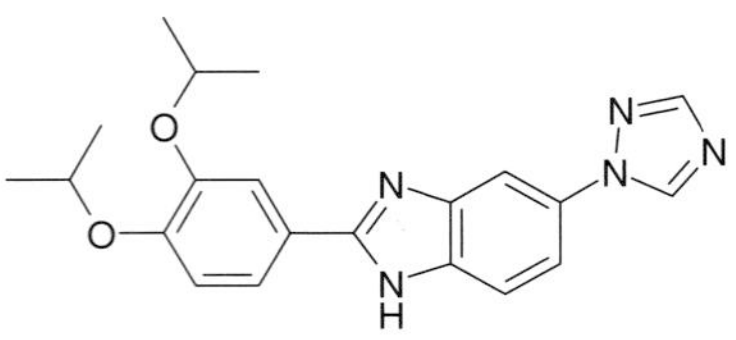

86
PDE4 20 nM

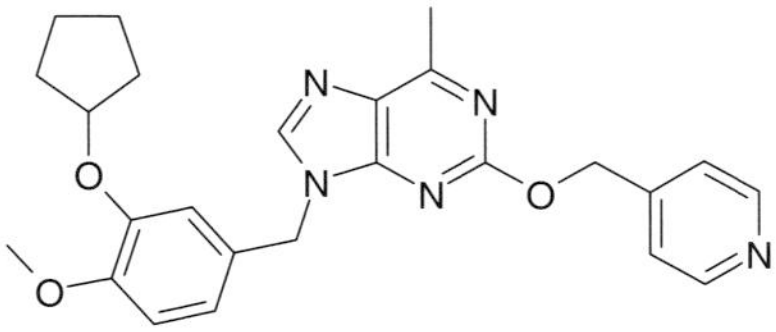

87
PDE4 0.5 nM

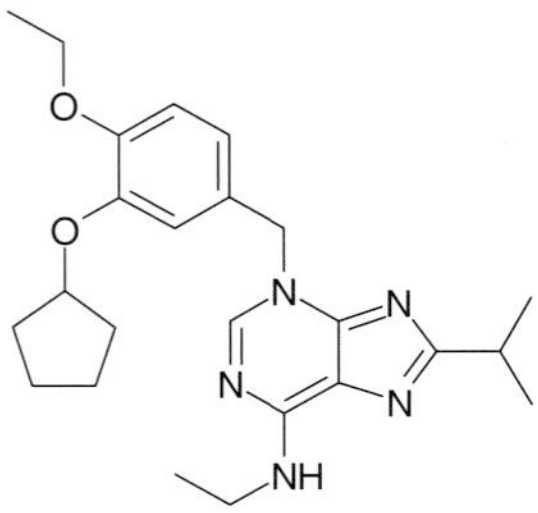

V-11294A (88)
PDE4 200 nM

Figure 9.20 Heterocyclic scaffolds for PDE4 privileged structures.

In comparison to the numerous compounds that fall into the catechol or related class, far fewer compounds belong to the other 'classical' scaffold types of PDE inhibitors. The xanthine core, which was the basis of several selective, highly potent PDE5 inhibitors, has been far less successful in the PDE4 arena by comparison (Figure 9.22). Denbufylline (**97**) was an early example with a structure close to IBMX [88]. This compound is representative of the trend toward more lipophilic and space-filling substituents at both pyrimidine nitrogens. The same holds true for cipamfylline **98** [89] and arofylline **99** [90]. The latter compound demonstrates that an aromatic moiety rather than the usual aliphatic residue can act as one of the lipophilic groups.

Beyond these classical examples regarding structural motifs, which fall into more or less sizeable clusters, there are some inhibitors that are not readily categorized into any of the above classes. We will discuss only two of these that have become significant because they both have proceeded into advanced stages of clinical development (Figure 9.23). BAY 19-8004 (**100**) has been developed by Bayer for the treatment of asthma and COPD [91]. This compound falls into a class of benzofuran derivatives which seemingly do not fit into the catechol-related group. Another compounds, AWD 12-281 (**101**), was discovered by AWD (formerly Asta Medica) and is currently under development by GSK [92]. Here, the familiar chloropyridine is presented without the catechol attachment, at least if one discounts the potential phenolic acceptor.

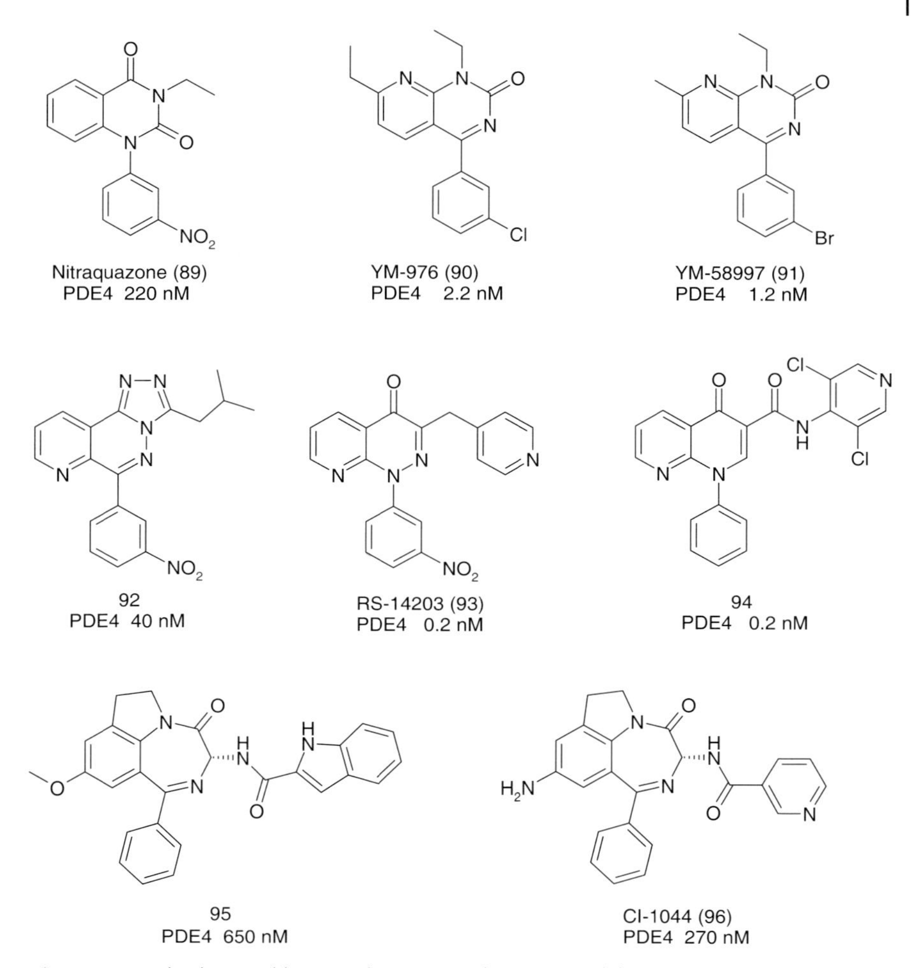

Figure 9.21 Aryl substituted heterocyclic compounds as PDE4 inhibitors.

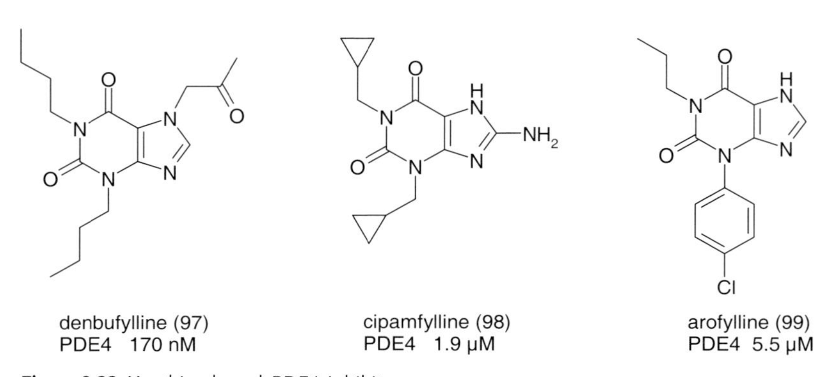

Figure 9.22 Xanthine-based PDE4 inhibitors.

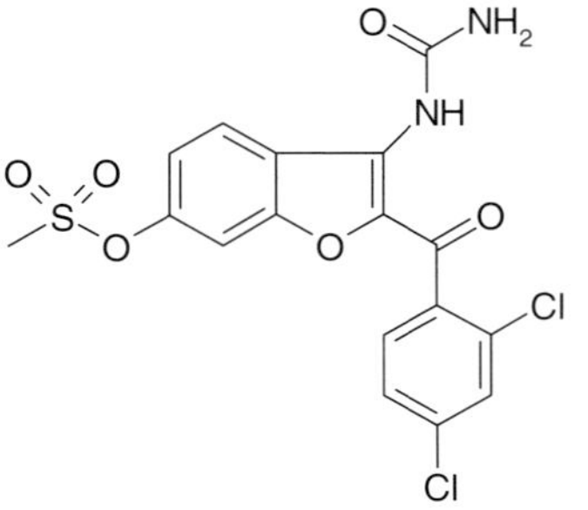

BAY 19-8004 (100)
PDE4 67 nM

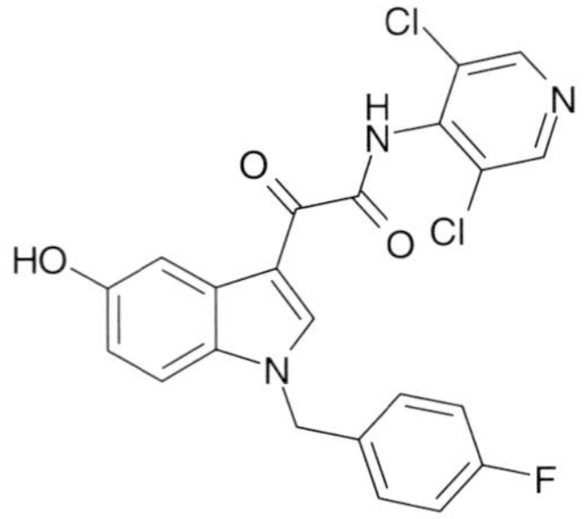

AWD 12-281 (101)
PDE4 26 nM

Figure 9.23 Other structural motifs for PDE4 inhibitors in clinical development.

9.7.1
Dual PDE4/3 Inhibitors

For several reasons, dual inhibitors of PDE4 and PDE3 have been of interest to pharmaceutical companies. Such inhibitors have been found, in particular, with the dihydroisoquinoline core (Figure 9.24) [93]. These inhibitors feature the classical catechol fragment embedded in a rigid tricyclic framework with an additional aromatic substituent. Examples include tolafentrine (**102**) and pumafentrine (**103**). Interestingly, it appears that, upon transitioning from the aza series to the carba analog, PDE3 inhibition is lost and selective PDE4 inhibitors are obtained, such as compound **104**.

tolafentrine (102)
PDE4 60 nM
PDE3 90 nM

pumafentrine (103)
PDE4 7 nM
PDE3 28 nM

104
PDE4 0.56 nM

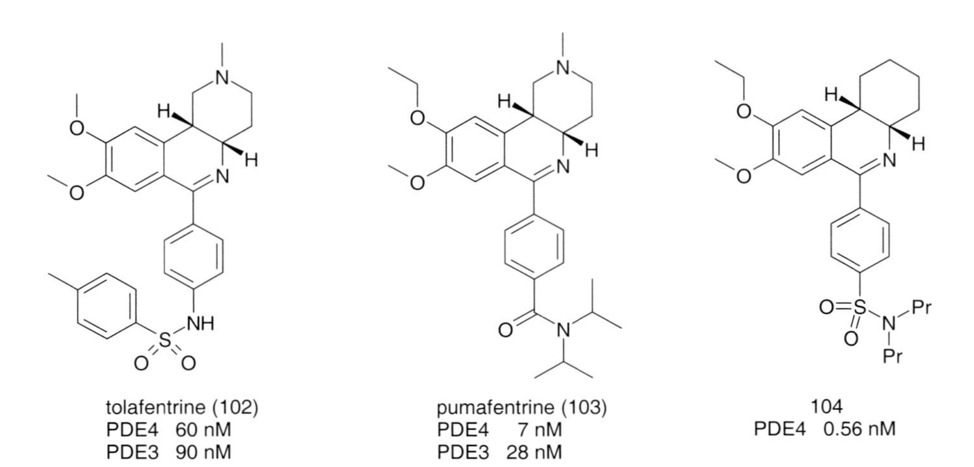

Figure 9.24 Dihydroisoquinoline derivatives as PDE4/3 inhibitors.

9.7.2
PDE3 Inhibitors

PDE3 hydrolyses cAMP with 10-fold higher activity than cGMP and is inhibited by high cGMP concentrations. Both nucleotides compete for the same catalytic site, thus creating crosstalk between cAMP- and cGMP-regulated pathways. Two gene products, PDE3A and PDE3B, have been found encoding proteins with a molecular mass of approximately 122 kDa. A membrane-anchoring domain is an additional feature of the particulate PDE3s that are found, e.g., in adipocytes and hepatocytes. A unique structural feature of the PDE3 sequence is a 44 amino acid insertion sequence at the amino-terminal end of the catalytic domain that is essential for activity [94]. Regulation of this dimeric enzymes occurs mainly via interaction with hormones such as leptin, glucagon, and insulin, and it plays a major role in the regulation of glycogen levels [95]. In addition, phosphorylation by PKA also leads to activation. Both ways of activation act independently and seem to work super-additively [96].

The known PDE3 inhibitors present a good example of a chemogenomic principle: the recognition of a conserved chemical pattern by a genetically conserved biologically active domain. A careful investigation shows that many PDE3 inhibitors are constructed according to a common blueprint, consisting of up to three modular building blocks (Figure 9.25). A constituent element of a PDE3 inhibitor is a lactamic structure which may appear in different hetero-substituted forms, e.g., as a

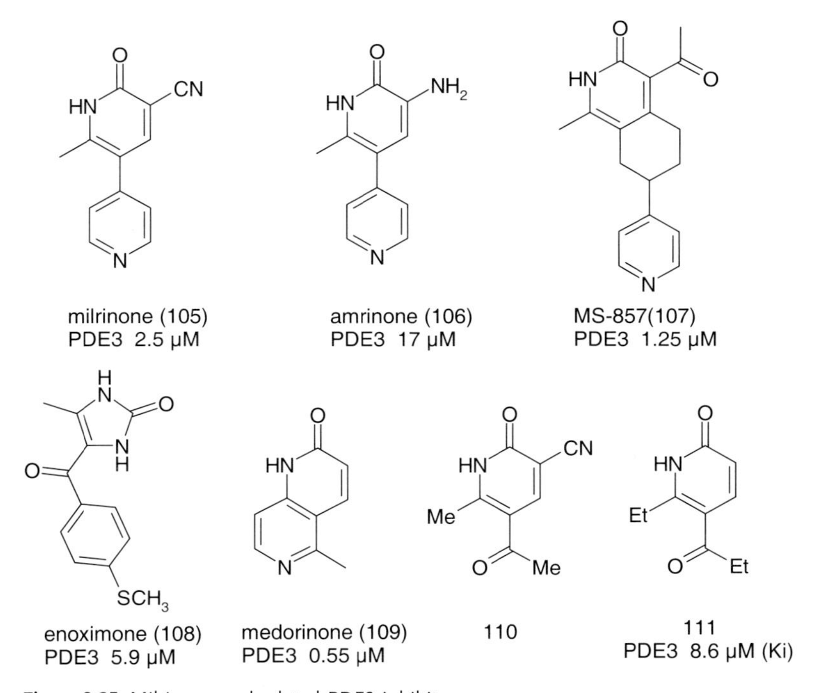

Figure 9.25 Milrinone and related PDE3 inhibitors.

semicarbazone or as a thiourea. In every structure, an amidic NH motif is essential for good binding. The second module is an aromatic ring, which may be a direct substituent of the lactamic ring (**105–108**) or part of a condensed lactamic polycycle (**109**) [97–100]. Both framework alternatives frequently contain a pyridyl substructure (as in the classical PDE3 inhibitors amrinone **105** and milrinone **106**) or a benzimidazole substructure (e.g., meribendan **112**, pimobendan **113**, Figure 9.26)

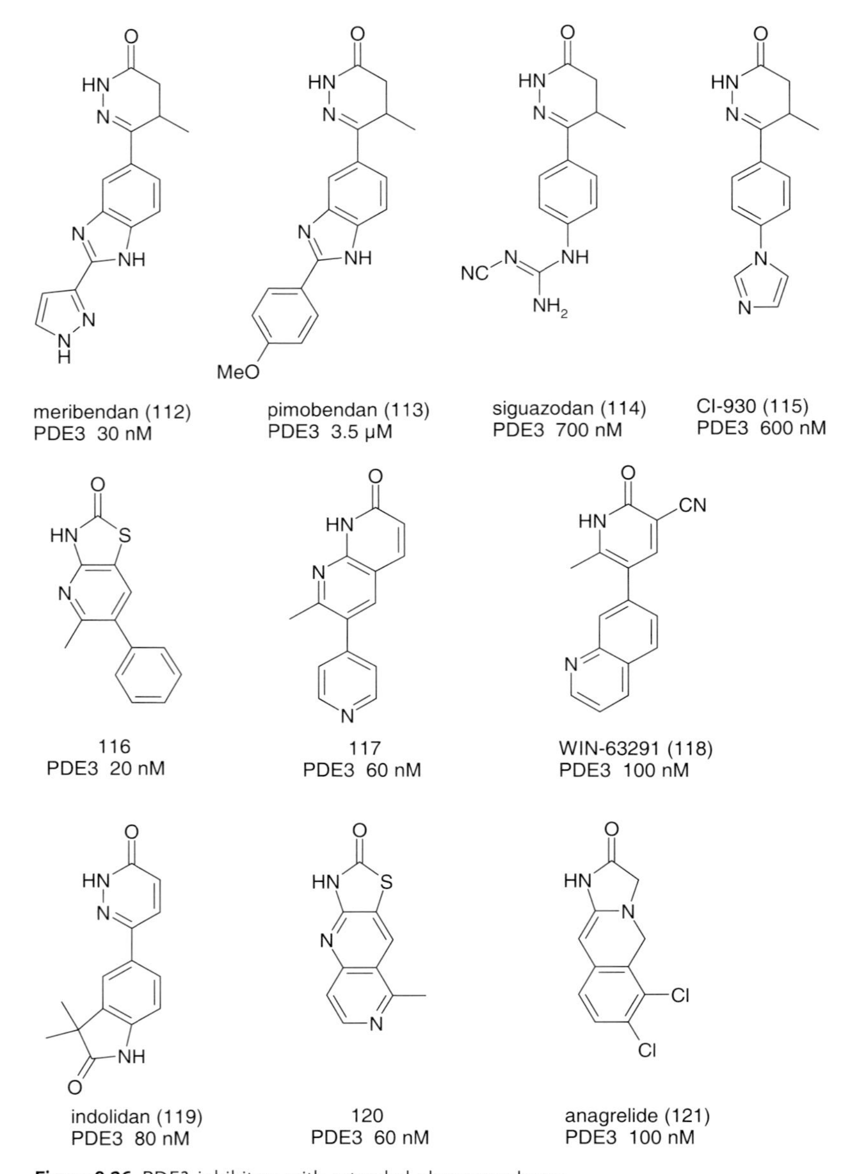

Figure 9.26 PDE3 inhibitors with extended pharmacophores.

that matches the common hydrophobic pharmacophore. These two-module inhibitors constitute a first subfamily of PDE3 inhibitors. Interestingly, in the simplest example the second motif on the lactam ring can be a ketone, as shown in **110** and **111** in which an methyl or ethyl carbonyl moiety replaces the pyridyl residue of milrinone [101].

The addition of a lipophilic ring pharmacophore to this bicyclic aryl or biaryl structure, e.g., phenyl, methoxyphenyl, ethylthiophenyl, imidazolyl, pyrazolyl, or pyridyl, generates the second inhibitor subfamily: full-length inhibitors consisting of three modules, such as meribendan or CI-930 (**112–115**, Figure 9.26) [102, 103]. Again, the middle pharmacophore can be incorporated into the lactam module (**116**, **117**) or into the additional lipophilic part of the inhibitor (**118**, **119**). Extreme condensation of the pharmacophores, so as to incorporate all three modules into one tricyclic system, resulted in inhibitors **120** and **121**.

Alternatively, the second of these two modules can be left out, as in **122** and **123** (Figure 9.27). Here, the lactam module and the lipophilic ring module on the other end are connected by a flexible alkyl linker, leading to structures that are reminiscent to those constructed by fragment-based approaches, thus forming a third inhibitor subfamily (Figure 9.27).

It is of great interest for the design of new PDE3 inhibitors to gain deeper insights on the 3D structures of inhibitors and their interaction with the enzyme. Earlier investigations have shown that a planar constellation of all three inhibitor parts is crucial for selective PDE inhibition [104]. More recently, studies based on a PDE3 homology model constructed from the X-ray structure of PDE4B2B shed some light on the structure–function relationship of this inhibitor type [105]. The lactam part of the inhibitors is suggested to occupy a hydrophobic pocket above the phosphate-binding region of cAMP formed by Thr908, Leu910, His913, Lys947, and Ile951; the major interaction of the lactam motif takes place between the backbone carbonyl of Thr908 and the ε-amino group of Lys947. Following the model further, the middle pharmacophore occupies the ribose position and blocks the catalytic activity of Tyr751 and His752. The third recognition pocket is proposed to be formed of Ile951, Ile968, Phe972, Leu1000, and in particular Phe1004, where

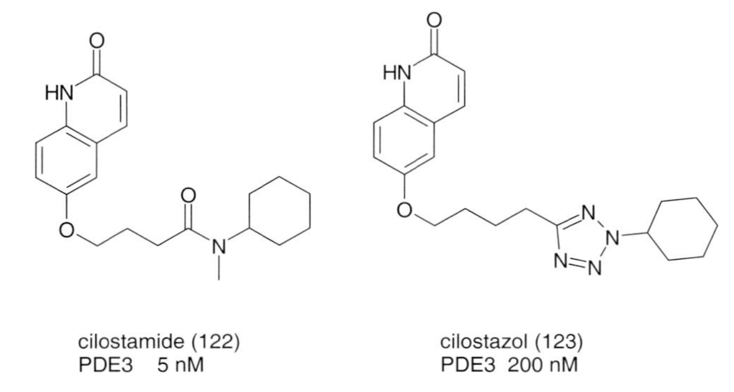

Figure 9.27 Cilostamide and cilostazol.

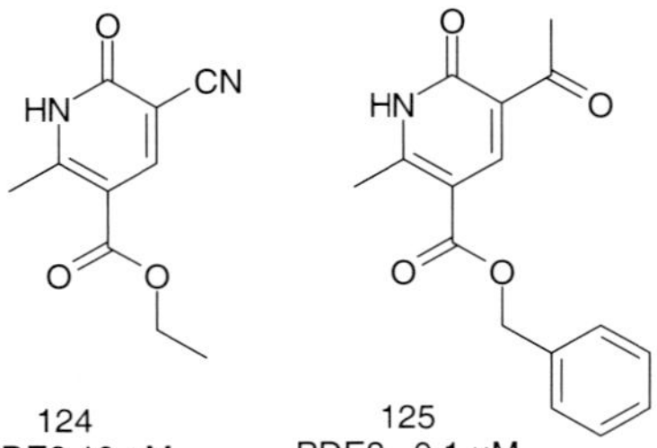

124
PDE3 10 μM

125
PDE3 0.1 μM

Figure 9.28 Influence of potential hydrophobic interaction with the core pocket.

the adenine ring of cAMP comes to reside. It might be addressed by the lipophilic fragments of the inhibitors equipped with the third module. Since these inhibitors are the most effective ones, the importance of this lipophilic adenine-binding pocket must be emphasized, especially in the context of new inhibitor and library design. The outstanding role of Phe1004 for high affinity has been shown by chemical site-directed mutagenesis experiments and is demonstrated by the de-novo designed inhibitor **125** [106]. Although the potency of **124** is almost exclusively based on the lactam pocket affinity, introduction of a phenyl group (presumably capable of π interactions with Phe1004) resulted in a dramatic increase in PDE3 inhibition (Figure 9.28).

9.8 Inhibitors of Other Phosphodiesterases

9.8.1 PDE1

The 75-kDa PDE1 isoenzyme is characterized by its dual specificity and its activation by Ca^{2+}/calmodulin. Three different gene products, PDE1A, PDE1B and PDE1C, are known. In addition, several C- and N-terminal splice variants have been characterized [107]. Among the different PDE1 isoenzymes, the substrate selectivity and the activation efficiency by Ca^{2+}/calmodulin varies. Although PDE1C hydrolyses both cyclic nucleotides with equal effectiveness, PDE1A and 1B prefer cGMP as the substrate.

The regulatory domain near the N terminus contains two calmodulin-binding domains (domain A and domain B), although only one calmodulin seems to bind to PDE1. Close to the calmodulin-binding domain A, an autoinhibitory domain has been found. Truncation of this region leads to activation of the enzyme with retained calmodulin-binding capacity. In addition to activation by Ca^{2+}/calmodulin, PDE1 can also be regulated by phosphorylation; the phosphorylation appears close to the calmodulin-binding domain B and leads to decreased sensitivity to activation by calmodulin [108].

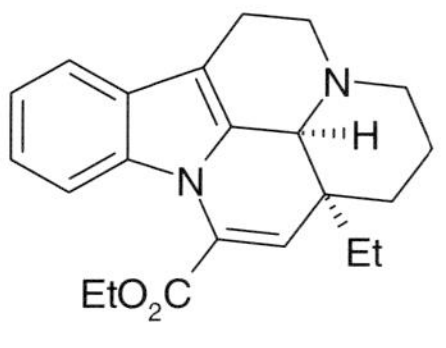

Vinpocetine (126)
PDE1 20 μM

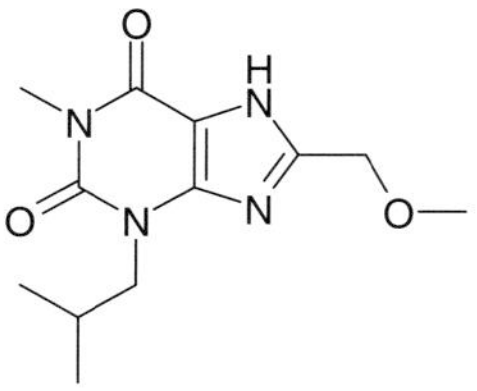

8-MeOMe-IBMX (127)
PDE1 8 μM
PDE5 10 μM

Figure 9.29 Classical inhibitors of PDE1.

PDE1 modulation has not been firmly tied pharmacologically to the treatment of a particular disease. PDE1's broad tissue distribution and dual substrate specificity make it an important player in the regulation of cellular cyclic nucleotide second messengers. In contrast to PDEs 5 and 4/3, which have received considerable attention from medicinal chemists, there is also much less information available on inhibitors of PDE1. For many years the alkaloid vinpocetine (**126**) could be cited as the most significant selective, albeit weak, PDE1 inhibitor (Figure 9.29) [109].

Several xanthine derivatives are potent inhibitors of PDE1. In this regard, the dual PDE1/5 inhibitor 8-methoxymethyl-IBMX (**127**) can be regarded as a first lead. In an effort to identify more potent PDE1/5 inhibitors, a team at Schering Plough has focused on the structurally related tetracyclic guanines (Figure 9.30) such as the benzyl-substituted **128** or the even more potent alkyl analog **129** [110]. In contrast, the isomeric 3-substituted benzyl derivative **130** is less potent. Nonetheless, if both substitutions are combined, they produce highly potent and selective PDE1 inhibition, as in **131** and **132**. The latter represents the most potent and specific PDE1 inhibitor published to date. Another hint as to the specific requirements of PDE1 can be gained from the tricyclic compound **133**, which has reduced hydrophobicity and steric bulk compared to **134** and is both more potent and more selective toward PDE1.

Much earlier, a structurally different inhibitor type bearing the heterocyclic core of sildenafil appeared in the patent literature (Figure 9.31). Compound **135** is a micromolar PDE1 inhibitor, which also shows affinity for the adenosine receptor [111]. By tuning the substituents, the more potent and selective inhibitor **136** was obtained some 10 years later by the Pfizer UK group [112].

A different approach was taken by medicinal chemists at Sterling Winthrop (Figure 9.32) [113]. In an attempt to combine the structural features of the PDE5 inhibitor zaprinast and the PDE3 inhibitor milrinone, they arrived at WIN 61626 (**137**) and WIN 61691 (**138**), which are dual PDE1/3 inhibitors with slight selectivity for PDE1.

Completely unrelated structures have appeared in the patent literature disclosed by Japanese inventors (Figure 9.33). The two imidazole derivatives **139** and **140** are micromolar PDE1 inhibitors [114]. More potent, and bearing some structural resemblance to the alkaloid papaverine, are the quinazolines **141** and **142** from

128
(R=Ph)
PDE1 100 nM
PDE5 80 nM

129
(R=$(CH_2)_2CHMe_2$)
PDE1 6 nM
PDE5 10 nM

130
PDE1 205 nM
PDE5 225 nM

131
(R=H)
PDE1 2 nM
PDE5 180 nM

132
(R=Ph)
PDE1 0.07 nM
PDE5 305 nM

133
(R1, R2 = Me)
PDE1 18 nM
PDE5 140 nM

134
(R1, R2 = $(CH_2)_4$)
PDE1 60 nM
PDE5 75 nM

Figure 9.30 Tetracyclic guanine inhibitors of PDE1 and PDE5.

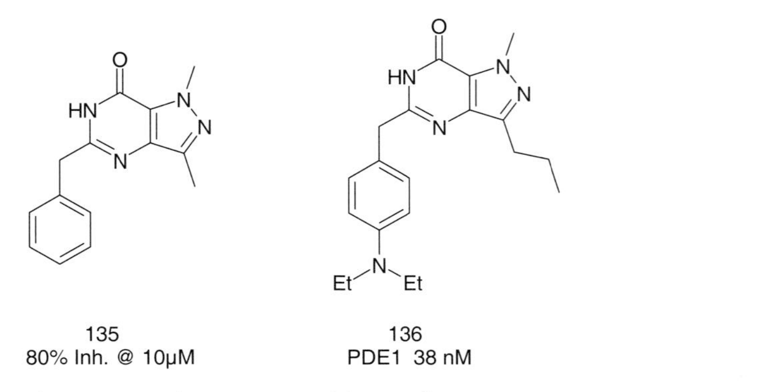

Figure 9.31 Imidazotriazinone Inhibitors of PDE1.

Eisai [115]. We should note, however, that no selectivity data have been published for these compounds.

In contrast to the other members of the PDE family, PDE1 is unique in its ability to interact with calmodulin. As would be expected therefore, this interaction can also be the target of potential inhibitors. This appears to be so for some natural-product inhibitors of PDE1 isolated from ginseng root. The ginsenosides Rb, Rc, and Re are moderately active (5–15 μM) steroidal inhibitors of CaM PDE isolated

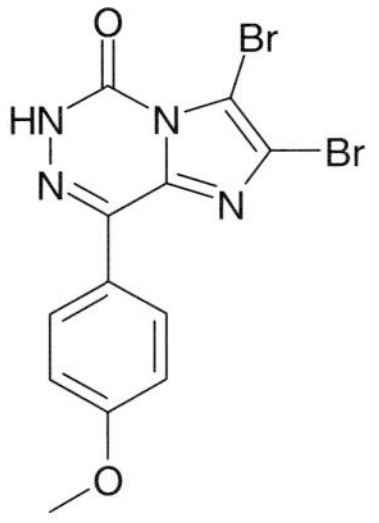

137 (WIN 61626) PDE1 450 nM PDE3 1.1 μM

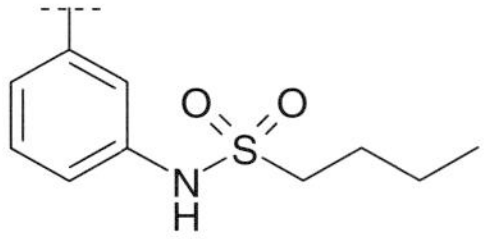

138 (WIN 61691) PDE1 85 nM PDE3 290 nM

Figure 9.32 Dual PDE1/3 Inhibitors.

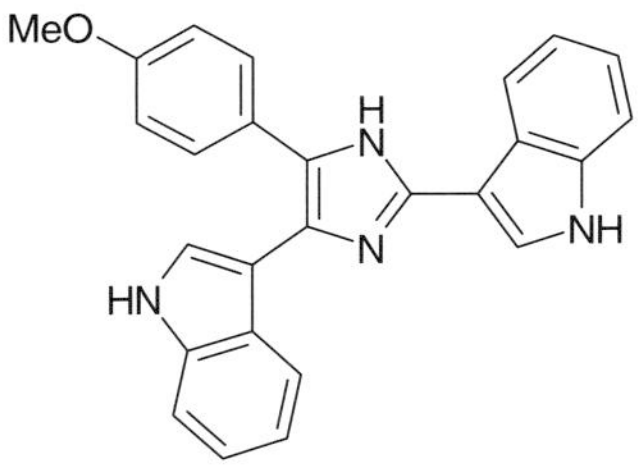

139 78% Inhib. @ 10μM

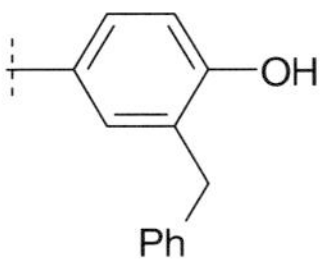

140 81% Inhib. @ 10μM

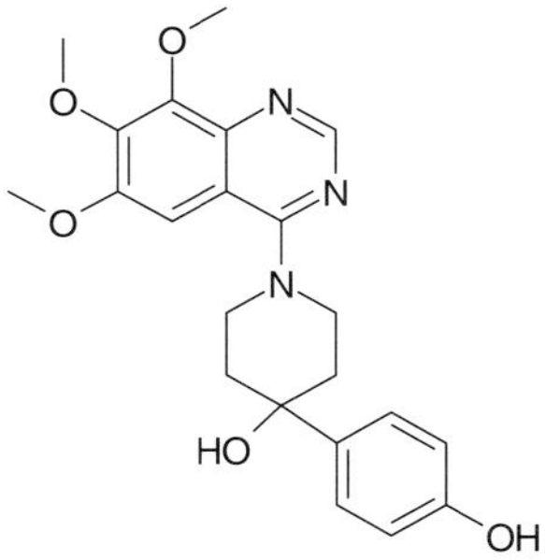

141 PDE1 22 nM

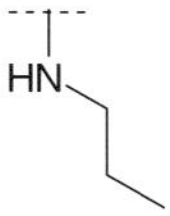

142 PDE1 170 nM

Figure 9.33 Other PDE1 inhibitors from the patent literature.

from bovine heart [116]. Interestingly, they cannot inhibit the CaM PDE from bovine brain, suggesting that the calmodulin interaction may be different for different PDE1 isoforms. Another steroid that may share the same inhibitory mechanism and possesses somewhat higher potency (60 nM) is the Streptomyces metabolite KS-505a.

9.8.2 PDE2

The dual-substrate PDE2 is a unique 105-kDa enzyme that is characterized by the ability to have its catalytic activity stimulated by cGMP binding to the allosteric regulatory site consisting of the tandem GAF domains GAF A and GAF B. The activation can be as high as 50 fold and is due to increased affinity of the enzyme for the substrate. The X-ray structure of cGMP bound to GAF B has recently been solved and revealed: in the cGMP-bound state both GAF domains form a heterodimer [117]. Only a single PDE2 gene product has been characterized, with several splice variants [118].

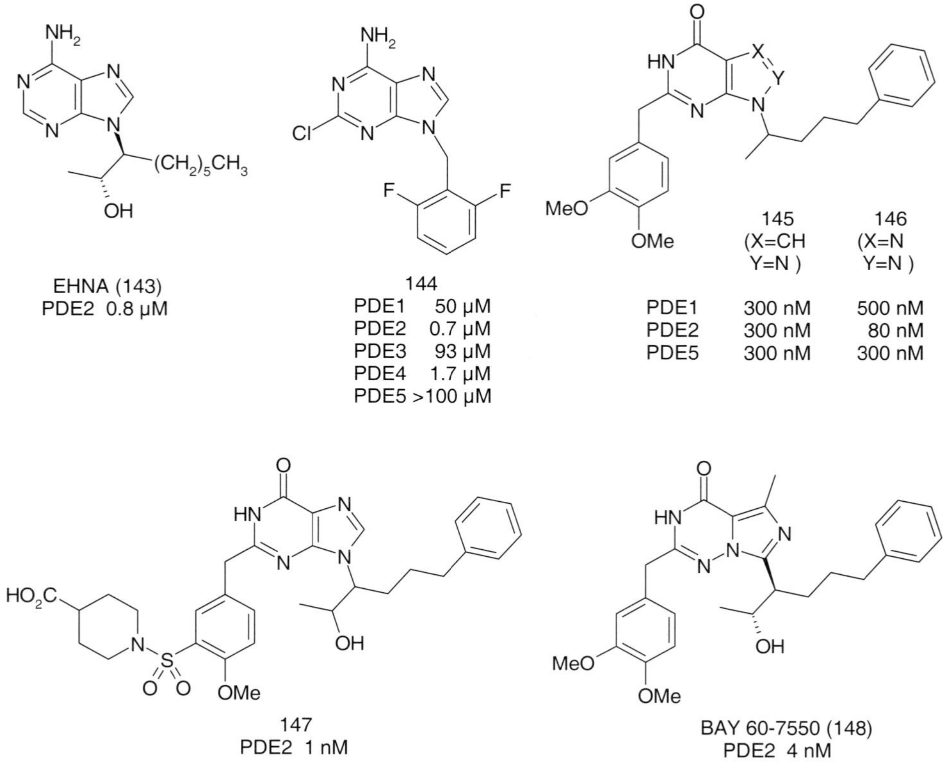

EHNA (143)
PDE2 0.8 μM

144
PDE1 50 μM
PDE2 0.7 μM
PDE3 93 μM
PDE4 1.7 μM
PDE5 >100 μM

	145 (X=CH Y=N)	146 (X=N Y=N)
PDE1	300 nM	500 nM
PDE2	300 nM	80 nM
PDE5	300 nM	300 nM

147
PDE2 1 nM

BAY 60-7550 (148)
PDE2 4 nM

Figure 9.34 Development of selective PDE2 inhibitors.

PDE2 shows broad tissue distribution in cytosolic as well as membrane-bound forms with a particularly high level of expression in the central nervous system. It has been speculated that the high capacity of PDE2 in the activated state makes it a possible emergency system in situations of uncontrolled increasing cyclic nucleotide levels. However, the biochemical and pharmacological characterization of PDE2 has clearly been hampered by a lack of potent, selective inhibitors. For many years, the only PDE2 inhibitor available was erythro-9-(2-hydroxy-3-nonyl)adenine (EHNA, **143**; Figure 9.34), a potent adenosine deaminase inhibitor (IC50 = 1 nM); its much less pronounced PDE inhibitory effects (IC50 = 0.8 μM) were not discovered until 1992 [119]. EHNA is selective for PDE2 over other PDEs, but its dual pharmacological activity presents a problem for its use as a biochemical tool. Another purine derivative (**144**) was reported to be a mixed PDE2/4 inhibitor [120].

More recently, Bayer disclosed potent, selective PDE2 inhibitors in a series of patents (Figure 9.34) [121]. A key feature of these inhibitors, which belong to the substrate-analog series, is a benzylic rather than a phenylic substitution at position 2 (using the common numbering system of purine nomenclature) of the heterocycle. Thus, inhibitors **145** and **146** demonstrated moderate potency toward PDE2 and show a mixed PDE1/2/5 inhibitor profile. In contrast, the purinone **147**, bearing a more elaborate substitution pattern, with a 2-hydroxy-6-phenylhexyl sidechain on one side and a sulfonamide residue on the other, is a very potent, selective inhibitor of PDE2. Bayer has also disclosed that selective PDE2 inhibitors can improve cognitive performance and thus could be used in the treatment of Alzheimer's disease and related disorders [122]. To this end, BAY 60-7550 (**148**) was shown to be active in animal models of learning and memory.

9.8.3
PDE7

Two splice-variants of PDE7A and a second gene product PDE7B with variations in the N-terminal portion of the protein are known; the two different gene products show 70% identity [123]. The molecular weight varies from 50 kDa (PDE7A2) to 52 kDa (PDE7B) and 57 kDa (PDE7A2). PDE7 is highly specific for cAMP and is inhibited neither by rolipram, the prototypical PDE4 inhibitor, nor by zaprinast, the classical PDE5 inhibitor. It is however inhibited by the broad-spectrum PDE inhibitor IBMX.

Inhibition of PDE7 is still an upcoming topic, and published reports including activity and SAR data are the exception. Nevertheless, patent filing has increased considerably during the past few years, giving a rough impression of the privileged pharmacophoric patterns for PDE7 inhibition. Reassuringly, classical substrate analogs such as purinone **149** (Figure 9.35) from Celltech are among the range of known PDE7 inhibitors, thus demonstrating that the principles discussed in the above sections can be transferred to this family of PDEs as well [124]. However, there is much heterogeneity within the group of patented/published PDE7 inhibitors. The group includes benzopyrrolidine **150** (Ono [125]), perhaps reminiscent of the prominent catechol-type scaffolds among PDE4 inhibitors, to spirocyclic urea **151** and thiadiazoles like **152** (both from Warner Lambert [126]), to thienopyrimidine **153** and related condensed heterocycles from Bayer [127]. Another company with interest in this field of inhibitors, Altana, successfully converted their previous PDE4/3-selective dihydroisoquinoline derivatives (see pumafentrine, **103**) into a lead series with PDE7 activity, as demonstrated by compound **154** [128].

The increasing number of patents in this area clearly suggests that substantial pharmaceutical interest is directed toward additional PDE7 inhibitors and that the disclosure of additional selective and potent compounds is only a matter of time. Inhibitors of PDE7 might offer a novel strategy in the treatment of T-cell mediated diseases and in modulating inflammatory and immunological responses. This thesis is supported by the clearly shown involvement of PDE7 in T-cell activation [129]. Since PDE7 is also present in airway epithelial cells, it may be an attractive target in the area of airway and inflammatory diseases as well [130]. More recently, it was discovered that PDE7 mRNA expression levels are altered in the brains of Alzheimer's patients, thus suggesting that inhibition of PDE7 could modulate the up-regulated cAMP pathways in Alzheimer's disease [131].

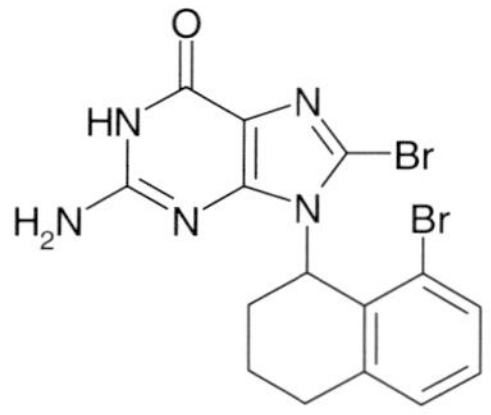

149
PDE7 1.3 μM

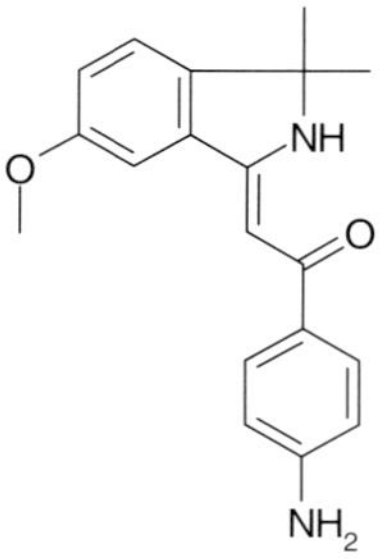

150
PDE7 23 nM

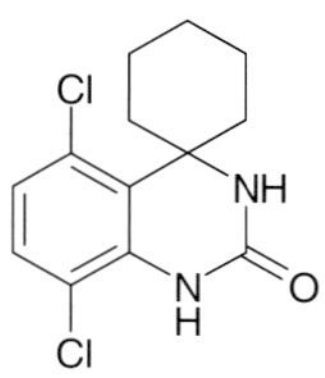

151
PDE7 14 nM

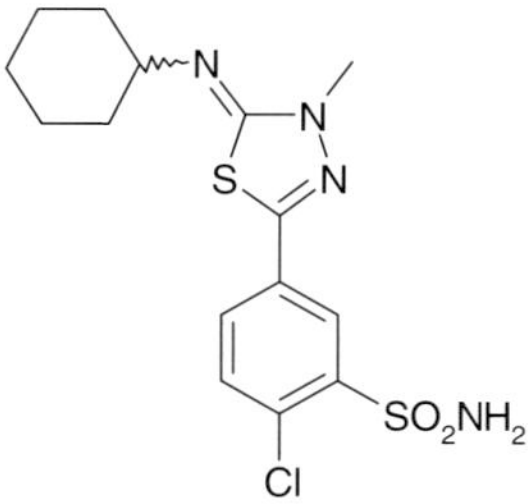

152
PDE7 70 nM

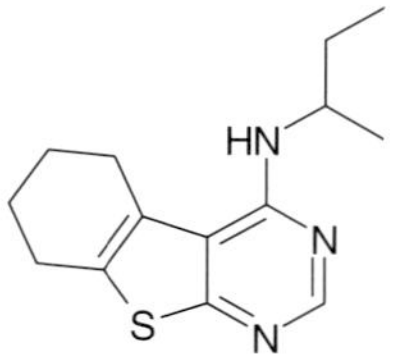

153

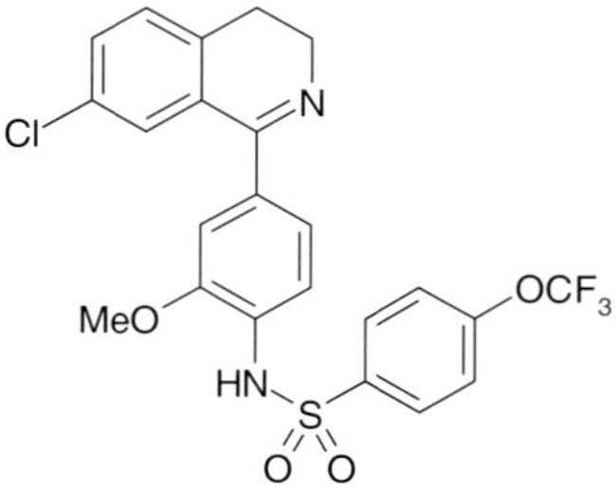

154
PDE7 32 nM

Figure 9.35 Diverse structures of PDE7 inhibitors.

9.8.4
Recently Discovered PDEs 8–11

In contrast to the numerous PDE inhibitors reported for the 'early' PDEs 1–5, PDEs 8–11 have as yet somewhat of an orphan status. The scattered information that is available was mostly disclosed in the patent literature, although a significant surge in activity can be expected. The available data suggest that these PDEs can be addressed with scaffolds that are related to those known for other PDE inhibitor classes. PDE8 (currently two subtypes 8A and 8B are known) shows high affinity for cAMP, even surpassing that of PDE4 (called the 'cAMP-specific PDE'), although its turnover is much lower than that of PDE4. PDE8 is not inhibited by the prototypical PDE4 inhibitor rolipram but shows some sensitivity to dipyridamole [132]. Only one isoform (encompassing several splice variants) is known for the cGMP-specific PDE9, the smallest PDE known to date (between 465 and 593 amino acids). The catalytic domain shows all 21 conserved amino acids common to PDE1–PDE8. The dual-specificity PDE10 occurs as a single gene product with at least two splice variants of approximately 88 kDa. The catalytic domain shows relatively low sequence identity with other PDEs. In common with several other PDEs, a pair of

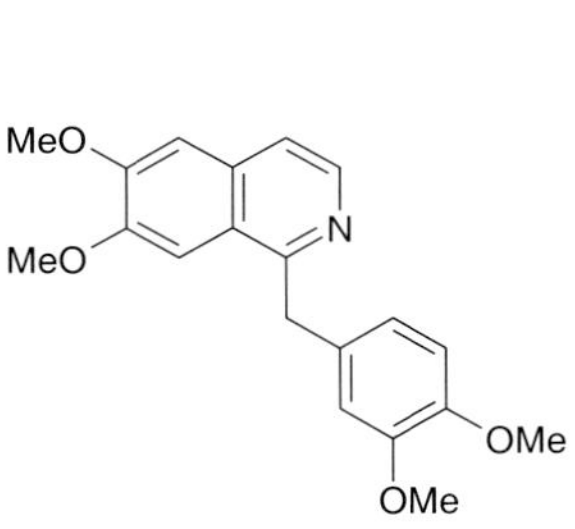

Papaverine (155)
PDE10 18 nM

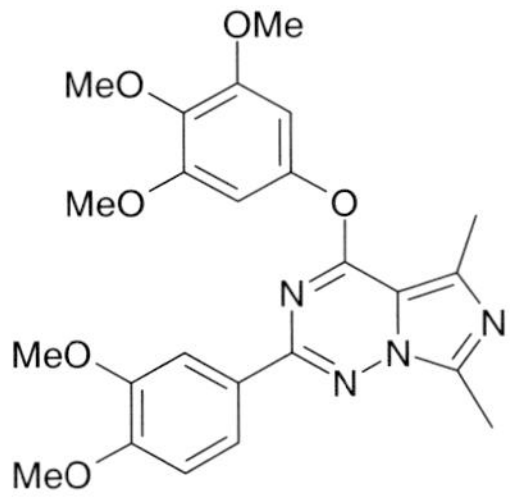

156
PDE10 35 nM

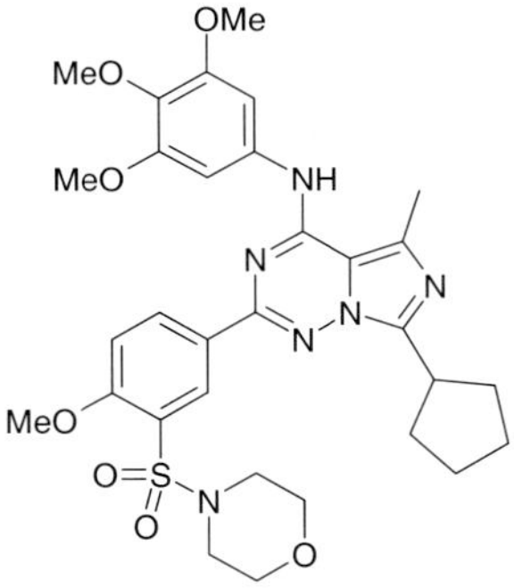

157
PDE10 13 nM

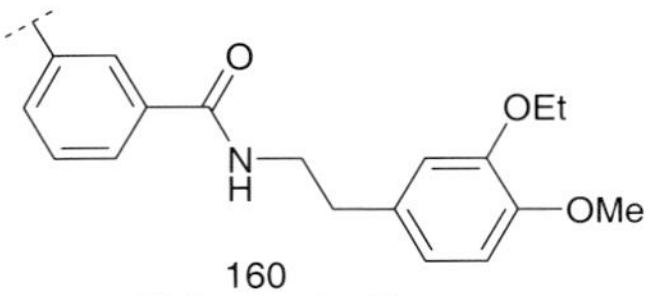

160
PDE10 <30 nM

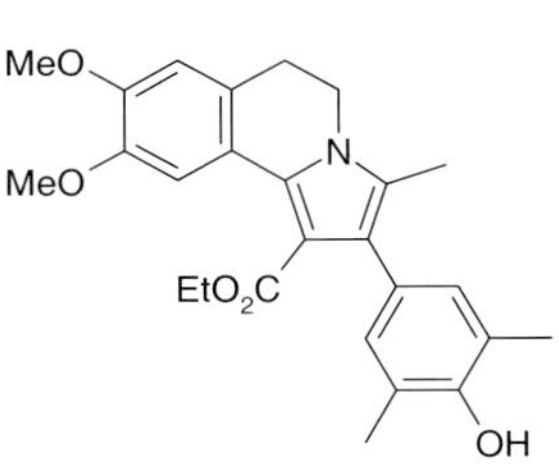

158
PDE10 30 nM

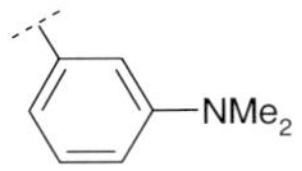

159
PDE10 42 nM

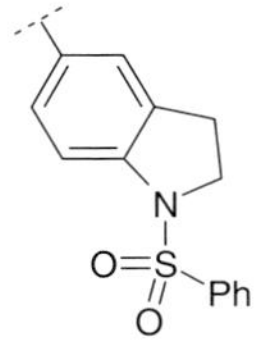

161
PDE10 <30 nM

Figure 9.36 Papaverine and related inhibitors of PDE10.

cGMP-binding GAF domains has been identified in the N terminus of the protein [133]. PDE11 is the most recently identified member of the PDE superfamily and likewise demonstrates dual specificity for cAMP and cGMP. Similar to PDE10, this rather small (490 amino acid residues, ~56 kDa) isoenzyme contains a GAF domain of unknown function. The most potent inhibitor for PDE11 found to date is tadalafil (IC_{50} = 36 nM), an unintentional circumstance that was discovered during the clinical development of this PDE5 inhibitor. Zaprinast and IBMX also inhibit PDE11, albeit on a micromolar level.

Interest in PDE 10 has been stimulated by the distinct high level of expression of the enzyme in the striatum region of the brain [134]. Scientists at Pfizer who were working with PDE10 found that papaverine (**155**, Figure 9.36), a classical PDE inhibitor that was previously thought to be unspecific, is a reasonably selective and potent inhibitor of the enzyme (the next-closest IC_{50} values are those for PDE4D and 4C, with values of 320 nM and 800 nM, respectively, and for PDE6 with 860 nM) [135]. Apart from this isolated example, Bayer has disclosed several PDE10 inhibitors

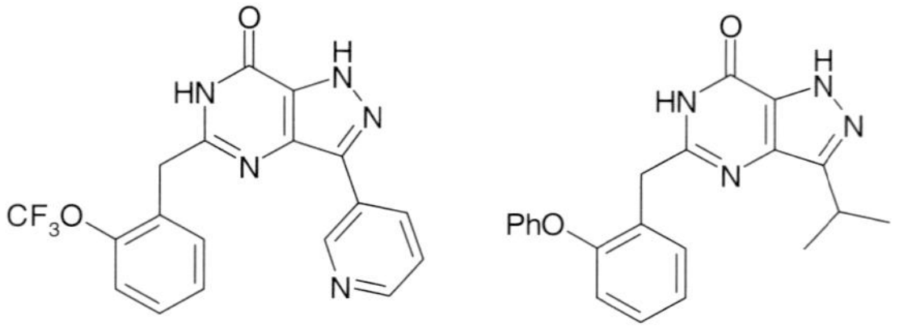

162
PDE9 141 nM

163
PDE9 126 nM

Figure 9.37 First examples of PDE9 inhibitors.

that fall into two different chemical series. Compounds **156** and **157** are imidazotriazine-based inhibitors that, on the one hand share the core of the PDE5 inhibitor vardenafil; on the other hand one could argue that they bear at least some structural resemblance to papaverine as well [136]. The other series of dihydropyrroloisoquinolines **158** to **161** are characterized by a dialkoxy substitution in the benzene ring and by an aromatic residue replacing the pyrrole [137]. These compounds likewise show some intriguingly similar features to papaverine, and thus all the inhibitors may loosely reflect a distinct structural motif in the PDE inhibitor family, leading to emphasis on the PDE10 isoform but yet general enough to inhibit other family members, as has been demonstrated with papaverine.

Still more elusive have been inhibitors of the remaining PDEs – types 8, 9, and 11. In particular, PDE8 and PDE9 are not sensitive to the prototypical inhibitor IBMX and thus appear to form a distinct class of their own. However, in a recent patent from Pfizer, moderately potent PDE9 inhibitors were disclosed that featured the 'classical' purinone-type heterocyclic core having a benzylic substituent in position 2 (again using purine numbering) [138]. In position 9 either aromatic, e.g., **162**, or aliphatic, e.g., **163**, residues appear to be tolerated (Figure 9.37). This again demonstrates that the same design principles can be carried over from one isoenzyme within a family to the next. Thus we may reasonably anticipate that potent specific inhibitors of PDE8 and PDE11, which have so far remained elusive, may soon be discovered within one of the known structural classes of inhibitor.

9.9
Summary: A Chemogenomic View of PDE Inhibitors

This chapter has illustrated the numerous advances made toward identifying selective PDE inhibitors. Despite the heterogeneity that may appear at first glance, several common structural features have become clear. The majority of compounds can be grouped into families of common substructures that address conserved features of the PDE active site. Most notably, the substrate-like cGMP-PDE5 inhibitors, the structurally unrelated indoline derivatives, and the cAMP-PDE4 inhibitors from the catechol class all form crucial H bonds to the same conserved glutamine residue that guides nucleotide recognition. Furthermore, in all instances, the hydrophobic interaction within the core-binding domain plays a key role in the energetics of inhibitor binding. This set of shared binding interactions mediated by completely different structures can be considered a chemical equivalent of convergent evolution, by which a common result is reached from different starting points. The design principles that have been at the root of this progress can be transferred from one PDE class to another, making us confident that an accelerated pace of discovery of new inhibitors may be near. This would be highly welcome from a pharmacological point of view, because new selective inhibitors are needed to explore and understand biological functions and, ultimately, to identify new drugs.

References

1 For general reviews covering the biology of PDEs see: (a) S. H. Francis, I. V. Turko, J. D. Corbin, *Prog. Nucleic Acid Res.* **2001**, *65*, 1–52; (b) J. A. Beavo, L. L. Burton, *Nature Rev. Mol. Cell Biol.* **2002**, *3*, 710–718; (c) S. H. Soderling, J. A. Beavo, *Curr. Opin. Cell Biol.* **2000**, *12*, 174–179; (d) M. Conti, S. L. C. Jin, *Prog. Nucleic Acid Res. Mol. Biol.* **2000**, *63*, 1–38; (e) M. A. Houslay, D. R. Adams, *Biochem. J.* **2003**, *370*, 1–18; (f) M. A. Houslay, W. Kolch, *Mol. Pharmacol.* **2000**, *58*, 659–668; (g) M. Conti, *Mol. Endocrinol.* **2000**, *14*, 1317–1327; (h) M. D. Houslay, G. Milligan, *Trends Biochem. Sci.* **1997**, *22*, 217–224; (i) S. D. Rybalkin, J. A. Beavo, *Biochem. Soc. Trans.* **1996**, *24*, 10005–10009; (j) J. A. Beavo, *Physiol. Rev.* **1995**, *75*, 725–748; (k) D. H. Maurice, D. Palmer, D. G. Tilley, H. A. Dunkerley, S. J. Netherton, D. R. Raymond, H. S. Elbartany, S. L. Jimmo, *Mol. Pharmacol.* **2003**, *64*, 533–546; (l) S. D. Rybalkin, C. Yan, K. E. Bornfeldt, J. A. Beavo, *Circ. Res.* **2003**, *93*, 280–291.

2 For reviews covering the early literature on PDE inhibitors see: (a) R. E. Weishaar, M. H. Cain, J. A. Bristol, *J. Med. Chem.* **1985**, *28*, 537–545; (b) C. D. Demoliou-Mason, *Exp. Opin. Ther. Patents* **1995**, *5*, 417–430; (c) *Phosphodiesterase Inhibitors* (Eds. C. Schudt, G. Dent, K. F. Rabe), Academic Press, London **1996**; (d) M. J. Perry, G. A. Higgs, *Curr. Opin. Chem. Biol.* **1998**, *2*, 472–481; (e) E. Sybertz, M. Czarniecki, *Exp. Opin. Ther. Patents* **1997**, *7*, 631–639.

3 S. H. Francis, J. L. Colbran, L. M. McAllister-Lucas, J. D. Corbin, *J. Biol. Chem.* **1994**, *269*, 22477–22480.

4 I. V. Turko, S. H. Francis, J. D. Corbin, *J. Biol. Chem.* **1998**, *273*, 6460–6466.

5 S. E. Martinez, A. Y. Wu, N. A. Glavas, X.-B. Tang, S. Turley, W. G. J. Hol, J. A. Beavo, *Proc. Natl. Acad. Sci. USA* **2002**, *99*, 13260–13265.

6 R. X. Xu, A. M. Hassell, D. Vanderwall, M. H. Lambert, W. D. Holmes, M. A. Luther, W. J. Rocque, M. V. Milburn, Y. Zhao, H. Ke, R. T. Nolte, *Science* **2000**, *288*, 1822–1825.

7 (a) N. D. Goldberg et al., *J. Biol. Chem.* **1980**, 255, 10344–10347; (b) P. M. Burgers et al., *J. Biol. Chem.* **1979**, *254*, 9959–9961.

8 Q. Huai, J. Colicelli, H. Ke, *Biochemistry* **2003**, *42*, 13220–13226.

9 G. Dent, K. F. Rabe, in *Phosphodiesterase Inhibitors* (Eds. C. Schudt, G. Dent, K. F. Rabe), Academic Press, London **1996**, pp. 41–64.

10 K. R. Sekhar, P. Grondin, S. H. Francis, J. D. Corbin, in *Phosphodiesterase Inhibitors* (Eds. C. Schudt, G. Dent, K. F. Rabe), Academic Press, London **1996**, pp. 135–146.

11 S. D. Rybalkin, I. G. Rybalkina, M. Shimizu-Albergine, X. Tang, J. A. Beavo, *EMBO J.* **2003**, *23*, 469–478.

12 M. K. Thomas et al., *Biochem. J.* **1990**, *265*, 14964–14970.

13 I. V. Turko et al., *Biochem. J.* **1998**, *329*, 505–510.

14 T. H. Fink et al., *J. Biol. Chem.* **1999**, *274*, 34613–34620.

15 For reviews see: (a) Z. Sui, *Exp. Opin. Ther. Pat.* **2003**, *13*, 1373–1388; (b) H. Haning, U. Niewoehner, E. Bischoff, *Prog. Med. Chem.* **2003**, *41* 249–306; (c) A. Stamford, *Ann. Rep. Med. Chem.* **2002**, *37*, 53–64; (d) G. N. Maw, *Ann. Rep. Med. Chem.* **1999**, *34*, 71 ff.; (e) D. P. Rotella, *Nat. Rev. Drug Discov.* **2002**, *1*, 674–682; (f) D. P. Rotella, *Drugs Future* **2001**, *26*, 153–162.

16 I. V. Turko et al., *Mol. Pharmacol.* **1999**, *56*, 124–130.

17 H. W. Hamilton, EP0201188 (**1985**).

18 (a) B. A. Dumaitre, N. Dodic, EP0636626 (**1995**); (b) B. Dumaitre, N. Dodic, *J. Med. Chem.* **1996**, *39*, 1635–1644.

19 (a) D.-K. Kim et al., *Bioorg. Med. Chem.* **2001**, *9*, 1895–1899; (b) D.-K. Kim et al., *Bioorg. Med. Chem.* **2001**, *9*, 3013–3021.

20 M. E. Bunnage, J. P. Mathias, S. D. A. Street, A. Wood, WO98/49166 (**1998**, Pfizer).

21 (a) B.-J. Sung, K. Y. Hwang, J. H. Jeon, J. I. Lee, Y.-S. Heo, J. H. Kim, J. Moon, M. Yoon, Y.-L. Hyun, E. Kim, S. J. Eum, S.-Y. Park, J.-O. Lee, T. G. Lee, S. Ro, J. M. Cho, *Nature* **2003**, *425*, 98–102; (b) D. G. Brown, C. R. Groom, A. L. Hopkins, T. M. Jenkins, S. H. Kamp, M. M. O'Gara, H. J. Ringrose, C. M. Robinson, W. E. Taylor, WO03/038080 (**2003**, Pfizer).

22 J.-H. Kim, Y. Kim, K. I. Choi, D. H. Kim, G. Nam, J. H. Seo, WO02/102802 (**2002**, Korea Institute of Science and Technology).

23 (a) D. P. Rotella, Z. Sun, Y. Zhu, J. Krupinski, R. Pongrac, L. Seliger, D. Normandin, J. E. Macor, *J. Med Chem.* **2000**, *43*, 5037–5043; (b) D. P. Rotella, Z. Sun, Y. Zhu, J. Krupinski, R. Pongrac, L. Seliger, D. Normandin, J. E. Macor, *J. Med Chem.* **2000**, *43*, 1257–1263.

24 J. E. Macor, D. P. Rotella, H. N. Weller, D. W. Cushman, J. P. Yevich, WO99/64004 (**1999**).

25 J. Krupinski, R. Pongrac, L. Seliger, D. Normandin, J. E. Macor, Y. Bi, P. Stoy, L. Adam, B. He, A. Watson, Z. Sun, *Bioorg. Med. Chem. Lett.* **2001**, *11*, 2461–2464.

26 (a) N. K. Terrett, WO 93/12095 (**1993**); (b) T. Oota, Y. Kawashima, K. Hatayama, JP07330777 (**1995**); (c) T. Oota, Y. Kawashima, K. Hatayama, JP07267961 (**1995**).

27 B. Dumaitre, N. Dodic, *J. Med. Chem.* **1996**, *39*, 1635–1644.

28 N. Hoefgen, S. Szelenyi, M. Degenhard, U. Egerland, WO00/43392 (**2000**).

29 (a) H.-S. Ahn, A. Bercovici, G. Boykow, A. Bronnenkant, S. Chackalamannil, J. Chow, T. Cleven, J. Cook, M. Czarniecki, C. Domalski, A. Fawzi, M. Green, A. Gündes, G. Ho, M. Laudicina, N. Lindo, K. Ma, M. Manna, B. McKittrick, B. Mirzai, T. Nechuta, B. Neustadt, C. Puchalski, K. Pula, L. Silverman, E. Smith, A. Stamford, R. P. Tedesco, H. Tsai, D. Tulshian, H. Vaccaro, R. W. Watkins, X. Weng, J. T. Witkowski, Y. Xia, H. Zhang, *J. Med. Chem.* **1997**, *40*, 2196–2210; (b) S. Vemulapaelli, R. W. Watkins, M. Chintala, H. Davis, H.-S. Ahn, A. Fawzi, D. Tulshian, P. Chiu, M. Chatterjee, C.-C. Lin, E. J. Sybertz, *Cardiovasc. Pharmacol.* **1996**, *28*, 862–869; (c) M. Czarniecki, H.-S. Ahn, E. J. Sybertz, *Ann. Rep. Med. Chem.* **1996**, *31*, 61–70; (d) G. D. Ho, L. Silverman, A. Bercovici, P. Puchalski, Y. Xia, M. Czarniecki, M. Green, R. Cleven, H. Zhang, A. Fawzi, *Bioorg. Med. Chem. Lett.* **1999**, *9*, 7–12.

30 (a) J. Gracia-Ferrer, J. Feixas-Gras, J. M. Prieto-Soto, A. Vega-Noverola, B. Vidal-Juan, WO01/07441 (**2001**); (b) B. Vidal-Juan, C. Esteve-Trias, J. Gracia-Ferrer, J. M. Prieto Soto, WO02/12246 (**2002**).

31 S. Chackalamannil, Y. Wang, C. D. Boyle, A. W. Stamford, WO02/24698 (**2002**, Schering Corp).

32 T. Asberom, Y. Hu, D. A. Pissarnitzki, R. Xu, Y. Wang, S. Chackalamannil,

J. W. Clader, A. W. Stamford, WO03/020724 (**2003**, Schering Corp).

33 (a) M. Miyahara, M. Ito, H. Itoh, T. Shiraishi, N. Isaka, T. Konishi, T. Nakano, *Eur. J. Pharmacol.* **1995**, *284*, 25–33; (b) T. Saeki, H. Adachi, Y. Takase, S. Yoshitake, S. Souda, I. Saito, *J. Pharmacol. Exp. Ther.* **1995**, *272*, 825–831.

34 (a) Y. Onoda, H. Takami, T. Seishi, D. Machii, Y. Nomoto, H. Takai, H. Okumura, T. Ohno, K. Yamada, M. Ichimura, WO99/43674 (**1999**); (b) H. Yamada, N. Umeda, S. Uchida, Y. Shiinoki, H. Horikoshi, N. Mochizuki, WO00/59912 (**2000**); (c) R. Jonas, P. Schelling, M. Christadler, F.-W. Kluxen, WO98/17668 (**1998**); (d) R. Jonas, H.-M. Eggenweiler, P. Schelling, M. Christadler, N. Beier, WO02/18389 (**2002**); (e) R. Jonas, H.-M. Eggenweiler, P. Schelling, M. Christadler, N. Beier, DE19942474 (**2001**); (f) R. Jonas, P. Schelling, M. Christadler, N. Beier, WO02/00660 (**2002**).

35 (a) R. Jonas, P. Schelling, M. Christadler, N. Beier, WO02/00664 (**2002**, Merck KGaA); (b) M. Brändle, T. Ehring, C. Wilm, WO01/64192 (**2001**, Merck KGaA).

36 N. Umeda, K. Ito, S. Uchida, Y. Shiinoki, WO01/12608 (**2001**).

37 S. J. Lee, J. Konishi, O. T. Macina, K. Kondo, D. T. Yu, T. Miskowski, EP0640599 (**1995**).

38 (a) N. Watanabe, Y. Kabasawa, S. Abe, M. Shibazaki, H. Ishihara, K. Kodama, H. Adachi, WO98/07430 (**1998**); N. Watanabe, N. Karibe, K. Miyazaki, F. Ozaki, A. Kamada, S. Miyazawa, Y. Naoe, T. Kaneko, I. Tsukada, T. Nagakura, H. Ishihara, K. Kodama, H. Adachi, WO99/42452 (**1999**).

39 G. Yu, J. Macor, H.-J. Chung, M. Humora, K. Katipally, Y. Wang, S. Kim, WO00/56719 (**2000**).

40 F. Ozaki, K. Ishibashi, H. Ikuta, H. Ishihara, S. Souda, WO95/18097 (**1995**).

41 G. Yu, H. J. Mason, X. Wu, J. Wang, S. Chong, G. Dorough, A. Henwood, R. Pongrac, L. Seliger, B. He, D. Normandin, L. Adam, J. Krupinski, J. E. Macor, *J. Med. Chem.* **2001**, *44*, 1025–1027.

42 K. Sawada, T. Inoue, Y. Sawada, T. Mizutani, WO01/05770; (**2001**).

43 E. Gebekor, S. Bethell, L. Fawcet, M. Mount, S. Phillips, *Eur. J. Urol.* **2002**, *63*, Suppl. 1, Abstr. 244.

44 M. W. Orme, J. S. Sawyer, L. M. Schultze, WO02/36593 (**2002**, Lilly Icos LLC).

45 M. W. Orme, J. S. Sawyer, A. C.-M. Daugan, A. Bombrun, F. Gellibert, L. M. Schultze, R. F. Brown, R. L. M. Gosmini, WO03/00691 (**2003**, Lilly Icos LLC).

46 M. W. Orme, J. S. Sawyer, L. M. Schultze, WO01/94345 (**2001**, Lilly Icos LLC).

47 M. W. Orme, J. S. Sawyer, L. M. Schultze, A. C.-M. Daugan, F. Gellibert, WO02/00656 (**2002**, Lilly Icos LLC).

48 (a) A. Daugan, R. F. Labaudiniere, WO96/32003 (**1996**); (b) A. Daugan, WO95/19978 (**1995**); (c) A. Daugan, F. Gellibert, WO97/03985 (**1997**); (d) J. S. Whitaker, I. Saenz de Tejada, K. M. Ferguson, WO01/80860 (**2001**); (e) P. L. Oren, N. R. Anderson, M. A. Kral, WO01/08686 (**2001**); (f) L. L. Allemeier, D. L. Brashear, K. M. Ferguson, W. E. Pullman, WO00/66114 (**2000**); (g) M. W. Orme, J. S. Sawyer, L. M. Schultze, WO02/10166 (**2002**); (h) M. W. Orme, J. S. Sawyer, A. C.-M. Daugan, R. Brown, L. M. Schultze, WO02/28858 (**2002**).

49 J. S. Sawyer, M. W. Orme, A. Bombrun, A. Bouillot, M. Dodic, M. Sierra, R. L. M. Gosmini, WO02/064590 (**2002**, Lilly Icos LLC).

50 (a) A. Bombrun, WO97/43287 (**1997**); (b) A. Bombrun, US6306870 (**2001**); (c) A. Bombrun, US6043252 (**2000**).

51 T. Ukita, Y. Nakamura, A. Kubo, Y. Yamamoto, M. Takahashi, J. Kotera, T. Ikeo, *J. Med. Chem.* **1999**, *42*, 1293–1305.

52 H. Mou et al., *J. Biol. Chem.* **1999**, *274*, 18813–18820.

53 N. O. Artemyev et al., *J. Biol. Chem.* **1996**, *271*, 25382–25388.

54 N. Li et al., *Genomics* **1998**, *49*, 76–82.
55 For a short introduction to the biology of PDE4 see M. Conti, W. Richter, C. Mehats, G. Livera, J.-Y. Park, C. Jin, *J. Biol. Chem.* **2003**, *273*, 5493–5496; for further information see also reference 1(e).
56 R. Hoffmann, I. R. Wilkinson, J. F. McCallum, P. Engels, M. D. Houslay, *Biochem. J.* **1998**, *333*, 139–149.
57 For reviews see: (a) P. Norman, *Expert Opin. Ther. Patents* **2002**, *12*, 93–111; (b) P. Norman, *Expert Opin. Ther. Patents* **2000**, *10*, 1415–1427; (c) P. Norman, *Expert Opin. Ther. Patents* **1999**, *9*, 1101–1118; (d) J. A. Karlsson, D. Aldous, *Expert Opin. Ther. Patents* **1997**, *7*, 989–1003; (e) T. J. Martin, *IDrugs* **2001**, *4*, 312–338; (f) H. J. Dyke, J. G. Montana, *Expert Opin. Investig. Drugs* **2002**, *11*, 1–13; (g) H. J. Dyke, J. G. Montana, *Expert Opin. Investig. Drugs* **1999**, *8*, 1301–1325; (h) Z. Huang, Y. Ducharme, D. MacDonald, A. Robichaud, *Curr. Opin. Chem. Biol.* **2001**, *5*, 432–438; (i) A. M. Doherty, *Curr. Opin. Chem. Biol.* **1999**, *3*, 466–473; (j) T. J. Torphy, *Am. J. Respir. Crit. Care Med.* **1998**, *157*, 351–370; (k) C. Burnouf, M.-P. Pruniaux, *Curr. Pharm. Des.* **2002**, *8*, 1255–1296.
58 (a) *Drugs Future* **1995**, *20*, 793–804; (b) N. Copper, M. M. Teixeira, J. Warneck, J. M. Miotla, R. E. Wills, D. M. T. Macari, R. W. Gristwood, P. G. Hellewell, *Br. J. Pharmacol.* **1999**, *126*, 1863–1871; D. W. P. Hay, *Curr. Opin. Chem. Biol.* **2000**, *4*, 412–419.
59 (a) S. B. Christensen, A. Guider, C. J. Forster, J. G. Gleason, P. E. Bender, J. M. Karpinski, W. E. DeWolf Jr., M. S. Barnette, D. C. Underwood, D. E. Griswold, L. B. Cieslinski, M. Burman, S. Bochnowicz, R. R. Osborn, C. D. Manning, M. Grous, L. M. Hillegas, J. O'Leary Bartus, M. D. Ryan, D. S. Eggleston, R. C. Haltiwanger, T. J. Torphy, *J. Med. Chem.* **1998**, *41*, 821–835.
60 (a) G. Fenton, P. N. Majid, M. N. Palfreyman, WO95/04045 (**1995**); (b) P. Agathangelou, *IDrugs* **1999**, *2*, 523–525.
61 (a) A. Hatzelmann, C. Schudt, *J. Pharmacol. Exp. Ther.* **2001**, *297*, 267–279; (b) D. S. Bundschuh, M. Eltze, J. Barsig, *J. Pharmacol. Exp. Ther.* **2001**, *297*, 280–290; W. Timmer, V. Leclerc, G. Birraux, M. Neuhäuser, A. Hatzelmann, T. Bethke, W. Wurst, *Am. J. Respir. Crit. Care Med.* **2000**, *161*, A505.
62 R. J. Heaslip, D. Y. Evans, B. D. Sickels, *FASEB J* **1994**, *8* (4 Part 1), Abs 2146.
63 M. Ine, K. Yamana, S. Noda, T. Akiyama, A. Akahama, JP-11189577 (**1999**).
64 M. E. Lee, J. Markowitz, J.-O. Lee, H. Lee, *FEBS Lett.* **2002**, *530*, 53–58.
65 C. Schudt, S. Winder, B. Muller, D. Ukena, *Biochem. Pharmacol.* **1991**, *42*, 153–162.
66 O. Dym, I. Xenarios, H. Ke, J. Colicelli, *Mol. Pharmacol.* **2002**, *61*, 20–25.
67 Q. Huai, H. Wang, Y. Sun, H.-Y. Kim, Y. Liu, H. Ke, *Structure* **2003**, *11*, 865–873.
68 B. Hughes, D. Howat, H. M. Lisle, T. James, N. Gozzard, K. Blease, P. Hughes, R. Kingaby, G. Warrellow, R. Alexander, J. Head, E. Boyd, M. Eaton, M. Perry, M. Wales, B. Smith, R. Owens, C. Catterall, S. Lumb, A. Russell, R. Allen, M. Merriman, D. Bloxham, G. Higgs, *Br. J. Pharmacol.* **1996**, *118*, 1183–1191.
69 W. He, F.-C. Huang, B. Hanney, J. Souness, B. Miller, G. Liang, J. Mason, S. Djuric, *J. Med. Chem.* **1998**, *41*, 4216–4223.
70 T. Akiyama, M. Ina, K. Yamana, A. Takahama, JP11322730 (**1999**).
71 (a) M. Napoletano, G. Norcini, G. Grancini, F. Pellacini, G. M. Leali, G. Morazzoni, WO00/005218 (**2000**); (b) M. Napoletano, G. Norcini, G. Grancini, F. Pellacini, G. M. Leali, G. Morazzoni, WO00/005219 (**2000**); (c) M. Napoletano, G. Norcini, G. Grancini, F. Pellacini, G. M. Leali, G. Morazzoni, WO99/32449 (**1999**); (d) M. Napoletano, G. Norcini, G. Grancini, F. Pellacini, G. M. Leali, G. Morazzoni, WO00/026218 (**2000**).
72 (a) T. Martin, W. R. Ulrich, H. Amschler, T. Baer, D. Flockerzi, B. Gutterer, U. Thibaut, A. Hatzelmann, H. Boss, R. Beume, H. P. Kley, D. Häfner, WO00/01695

(**2000**); (b) D. G. McGarry, J. R. Regan, F. A. Volz, C. Hulme, K. J. Moriarty, S. W. Djuric, J. E. Souness, B. E. Miller, D. M. Sweeney, *Bioorg. Med. Chem.* **1999**, *7*, 1131–1139; (c) G. Buckley, N. Cooper, H. J. Dyke, F. Galleway, L. Gowers, J. C. Gregory, D. R. Hannah, A. F. Haughan, P. G. Hellewell, H. J. Kendall, C. Lowe, R. Maxey, J. G. Montana, R. Naylor, C. L. Picken, K. A. Runcie, V. Sabin, B. R. Tuladhar, J. B. H. Warneck, *Bioorg. Med. Chem. Lett.* **2000**, *10*, 2137–2140.

73 J. Montana, N. Cooper, H. Dyke, J. Oxford, R. Gristwood, C. Lowe, H. Kendall, G. Buckley, R. Maxey, J. Warneck, J. Gregory, L. Gowers, F. Galleway, R. Wills, R. Naylor, B. Tudhalar, K. Broadley, H. Danahay, *Am. J. Respir. Crit. Care Med.* **1999**, *159*, A624.

74 C. Hulme, R. Methew, K. Moriarty, B. Miller, M. Ramanjulu, P. Cox, J. Souness, K. M. Page, J. Uhl, J. Travis, R. Labaudiniere, F. Huang, S. W. Djuric, *Bioorg. Med. Chem. Lett.* **1998**, *8*, 3053–3058.

75 J. Regan, J. Bruno, D. McGarry, G. Poli, B. Hanney, S. Bower, J. Travis, D. Sweeney, B. Miller, J. Souness, S. Djuric, *Bioorg. Med. Chem. Lett.* **1998**, *8*, 2737–2742.

76 J. J. Gaudino, WO01/47915 (**2001**, ICOS).

77 A. J. Duplantier, C. J. Andresen, J. B. Cheng, V. L. Cohan, C. Decker, F. DiCapua, K. G. Kraus, K. L. Johnson, C. R. Turner, J. P. UmLand, J. W. Watson, R. T. Wester, A. S. Williams, J. A. Williams, *J. Med. Chem.* **1998**, *41*, 2268–2277.

78 (a) A. Marfat, WO98/09961 (**1998**); (b) A. Marfat, WO99/23076 (**1999**).

79 T. Bär, T. Martin, W. R. Ulrich, H. Amschler, B. Gutterer, D. Flockerzi, A. Hatzelmann, H. Boss, R. Beume, R. Kilian, H. P. Kley, WO99/57115 (**1999**).

80 S. Yamada, T. Hosoya, K. Kitagawa, S. Inoue, M. Kiniwa, T. Asao, WO99/24425 (**1999**).

81 (a) T. Tanaka, E. Iwashita, A. Tarao, A. Amenomori, Y. Ono, WO99/24432 (**1999**); (b) D. Cavalla, M. Chasin, P. Hofer, WO99/29694 (**1999**); (c) L. J. Landells, M. W. Hensen, D. Spina, D. D. Gale, A. J. Miller, T. Nichols, K. Smith, Y. Rotshteyn, R. M. Burch, C. P. Page, B. J. O'Connor, *Eur. Respir. J.* **1998**, *12* (Suppl 28), Abs P2393; (d) D. D. Gale, L. J. Landells, D. Spina, A. J. Miller, K. Smith, T. Nichols, Y. Rotshteyn, F. Tonelli, P. Lacouture, R. M. Burch, C. P. Page, B. J. O'Connor, *Am. J. Respir. Crit. Care Med.* **1999**, *159*, A108.

82 T. Glaser, J. Traber, *Agents Actions* **1984**, *15*, 341–348.

83 (a) K. Takayama, M. Iwata, Y. Okamoto, M. Aoki, A. Niwa, Y Isomura, *ACS Meeting* (**1997**), *214* (San Francisco): MEDI245; (b) H. Hisamichi, K. Takayama, M. Iwata, H. Kubota, N. Kawano, M. Aoki, Y. Isomura, *ACS Meeting* (**1998**), *215* (Dallas): MEDI049.

84 (a) J. Gracia-Ferrer, M. I. Crespo, A. Vega-Noverola, A. Fernandez-Garcia, WO99/06404 (**1999**).

85 (a) D. J. Pon, M. Plant, J. Tkach, L. Boulet, E. Muise, R. A. Allen, I. W. Rodger, *Cell Biochemistry and Biophysics* **1998**, 29, 159–178; (b) T. Shimamoto, H. Inoue, Y. Hayashi, WO99/07704 (**1999**).

86 (a) Y. Pascal, C. R. Andrianjara, E. Auclair, N. Avenel, B. Bertin, A. Calvet, F. Feru, S. Lardon, I. Moodley, M. Ouagued, A. Payne, M. P. Pruniaux, C. Szilagyi, *Bioorg. Med. Chem. Lett.* **2000**, *10*, 35–38; (b) H. Jacobelli, S. Marc, WO99/50270 (**1999**).

87 C. Burnouf, E. Auclair, N. Avenel, B. Bertin, C. Bigot, A. Calvet, K. Chan, C. Durand, V. Fasquelle, F. Feru, R. Gilbertsen, H. Jacobelli, A. Kebsi, E. Lallier, J. Maignel, B. Martin, S. Milano, M. Ougued, Y. Pascal, M.-P. Pruniaux, J. Puaud, M.-N. Rocher, C. Terrasse, R. Wrigglesworth, A. M. Doherty, *J. Med. Chem.* **2000**, 43, 4850–4867.

88 G. Brenner, J. Göring, E. A. Khan, O. Rothe, M. Tauscher, DE2462367 (**1976**) and DE2402908 (**1975**).

89 (a) J. M. Kaplan, D. J. Herzyk, E. V. Ruggieri, J. O'Leary Bartus,

K. M. Esser, P. J. Bugelski, *Toxicology* **1995**, *95*, 187–196; (b) L. Jensen, D. E. Griswold; H. Aaes, T. Skak-Nielsen, J. R. Hansen, E. Bramm, L. Binderup, *Mediators Inflamm.* **1999**, *8* (Suppl 1), Abs P11-10; (c) D. R. Buckle, J. R. S. Arch, B. J. Connolly, A. E. Fenwick, K. A. Foster, K. J. Murray, S. A. Readshaw, M. Smallridge, D. G. Smith, *J. Med. Chem.* **1994**, *37*, 476–485.

90 A. Vega Noverola, J. M. Prieto Soto, J. Moragues Mauri, R. W. Gristwood, EP435811.

91 G. Sturton, M. Fitzgerald, *Chest* **2002**, *121*, 192S–196S.

92 H. Kuss, N. Höfgen, U. Egerland, *Drugs Future* **2002**, *27*, 111–116.

93 (a) B. Gutterer, H. Amschler, W. R. Ulrich, T. Martin, T. Bär, M. Eltze, R. Beume, D. Häfner, U. Kilian et al., WO98/55481 (**1998**, Byk Gulden); (b) B. Gutterer, H. Amschler, D. Flockerzi, W. R. Ulrich, T. Bär, T. Martin, C. Schudt, A. Hatzelmann, R. Beume, D. Häfner, H. Boss, H. P. Kley, WO99/05113 (**1999**, Byk Gulden); (c) B. Gutterer, H. Amschler, D. Flockerzi, W. R. Ulrich, T. Bär, T. Martin, C. Schudt, A. Hatzelmann, R. Beume, D. Häfner, H. Boss, H. P. Kley, WO99/05112 (**1999**, Byk Gulden); (d) R. Beume, D. Bundschuh, A. Hatzelmann, C. Schudt, C. Weimar, U. Kilian, WO01/13953 (**2001**, Byk Gulden).

94 M. Taira, S. C. Hockman, J. C. Calvo, M. Taira, P. Belfrage, V. C. Manganiello, *J. Biol. Chem.* **1993**, *268*, 18573–18579.

95 V. C. Manganiello, C. J. Smith, E. Degerman, V. Vasta, H. Tornqvist, P. Belfrage, *Adv. Exp. Med. Biol.* **1991**, *293*, 239–248.

96 C. J. Smith, V. Vasta, E. Degerman, P. Belfrage, V. C. Manganiello, *J. Biol. Chem.* **1991**, *266*, 13385–13390.

97 P. Fossa, R. Boggia, L. Mosti, *J. Comp. Mol. Design* **1995**, *9*, 33–43.

98 H. Hidaka, T. Endo, *Adv. Cyclic Nucleotide Protein Phosphoryl. Res.* **1984**, *16*, 245.

99 Y. Abe, R. Ishisu, K. Onishi, K. Sekioka, A. Narimatsu, T. Nakano, *J. Pharmacol. Exp. Ther.* **1996**, *276*, 433–439.

100 T. Bethke, T. Eschenhagen, A. Klimkiewicz, C. Kohl, H. v. d. Leyden, H. Mehl, U. Mende, W. Meyer, J. Neumann, S. Rosswag, W. Schmitz, H. Scholz, J. Starbatty, B. Stein, H. Wenzloff, V. Döring, P. Kalman, A. Haverich, *Arzneim.-Forsch.* **1992**, *42*, 437–445.

101 (a) P. Dorigo, R. M. Gaion, P. Belluco, D. Fracarollo, I. Maragno, G. Bombieri, F. Benetollo, L. Mosti, F. Orsini, *J. Med. Chem.* **1993**, *36*, 2475; (b) P. Fossa, G. Menozzi, P. Dorigo, M. Floreani, L. Mosti, *Bioorg. Med. Chem.* **2003**, *11*, 4749–4759.

102 D. C. Underwood, C. J. Kotzer, S. Bochnowicz, R. R. Osborn, M. A. Luttmann, D. W. Hay, T. J. Torphy, *J. Pharmacol. Exp. Ther.* **1994**, *270*, 250–259.

103 R. F. Kauffman, V. G. Crowe, B. G. Utterback, D. W. Robertson, *Mol. Pharmacol.* **1986**, *30*, 609–616.

104 W. H. Moos, C. C. Humblet, I. Sircar, C. Rithner, R. E. Weishaar, J. A. Bristol, A. T. McPhail, *J. Med. Chem.* **1987**, *30*, 1963–1972.

105 P. Fossa, F. Giodanetto, G. Menozzi, L. Mosti, *Quant. Struct.-Act. Relat.* **2002**, *21*, 267–275.

106 W. Zang, H. Ke, A. P. Tretiakova, B. Jameson, R. W. Colman, *Protein Science* **2001** *10*, 1481–1489.

107 (a) C. Yan et al., *J. Biol. Chem.* **1996**, *272*, 25699–25706; (b) J. Yu et al., *Cell. Signal.* **1997**, *9*, 519–529.

108 V. A. Florio et al., *Biochemistry* **1994**, *33*, 8948–8954.

109 H. S. Ahn, W. Crim, M. Romano, E. Sybertz, B. Pitts, *Biochem. Pharmacol.* **1989**, *38*, 3331.

110 (a) H.-S. Ahn et al., *J. Med. Chem.* **1997**, *40*, 2196–2210; (b) Y. Xia et al., *J. Med. Chem.* **1997**, *40*, 4372–4277; (c) G. D. Ho et al., *Bioorg. Med. Chem. Lett.* **1999**, *9*, 7–12.

111 H. W. Hamilton, EP0201188 (**1986**, Warner Lambert).

112 A. S. Bell, N. K. Terret, EP0911333 (**1999**, Pfizer).

113 D. S. Hlasta, D. C. Bode, J. J. Court, R. C. Desai, E. D. Pagani, P. J. Silver, *Bioorg. Med. Chem. Lett.* **1997**, *7*, 89–94.

114 (a) T. Ota, M. Nakanishi, K. Tomisawa, T. Kobori, Y. Hatanaka, WO99/35142

(**1999**, Taisho); (b) R. Asaki, M. Sodeoka, M. Katoh, JP11199582 (**1999**, Sagami Chemical Research Center).

115 (a) Y. Takase, T. Saeki, JP10175972 (**1998**, Eisai); (b) Y. Takase, N. Watanabe, H. Adachi, K. Kodama, H. Ishihara, T. Saeki, S. Souda, WO95/07267(**1995**, Eisai).

116 R. K. Sharma, J. Kalra, *Biochemistry* **1993**, *32*, 4975–4978.

117 S. A. Martinez, A. Y. Wu, N. A. Glavas, X.-B. Tang, S. Turley, W. G. J. Hol, J. A. Beavo, *Proc. Natl. Acad. Sci. USA* **2002**, *31*, 729–741.

118 G. J. Rosman et al., *Gene* **1997**, *191*, 89–95.

119 (a) H. J. Schaeffer, C. F. Schwender, *J. Med. Chem.* **1974**, *17*, 6–8; (b) T. Podzuweit, P. Nennstiel, A. Müller, *Cell. Signal.* **1995**, *7*, 733–738; (c) P.-F. Méry, C. Pavoine, F. Pecker, R. Fischmeister, in *Phosphodiesterase Inhibitors* (Eds. C. Schudt, G. Dent, K. F. Rabe), Academic Press, London **1996**, pp. 81–88.

120 S. Kozai, T. Maruyama, *Chem. Pharm. Bull.* **1999**, *47*, 574–575.

121 (a) U. Niewöhner, E. Bischoff, J. Hütter, E. Perzborn, H. Schütz, EP0771799 (**1997**, Bayer); (b) H. Haning, U. Niewöhner, U. Rosentreter, T. Schenke, J. Keldenich, E. Bischoff, K.-H. Schlemmer, H. Schütz, G. Thomas, WO98/40384 (**1998**, Bayer); (c) T. Schenke, H. Haning, U. Niewöhner, E. Bischoff, K.-H. Schlemmer, J. Keldenich, WO00/11002 (**2000**, Bayer); (d) H. Haning, U. Niewöhner, U. Rosentreter, T. Schenke, E. Bischoff, K.-H. Schlemmer, WO00/12504 (**2000**, Bayer); (e) U. Niewöhner, D. Schauss, G. König, M. Hendrix, F.-G. Böss, F.-J. van der Staay, R. Schreiber, K.-H. Schlemmer, R. Grosser, WO02/50078, **2002**, Bayer; (f) U. Niewöhner, D. Schauss, M. Hendrix, G. König, F.-G. Böss, F.-J. van der Staay, R. Schreiber, K.-H. Schlemmer, R. Grosser, WO02/68423 (**2002**, Bayer).

122 F.-G. Böss, M. Hendrix, G. König, U. Niewöhner, K.-H. Schlemmer, R. Schreiber, F.-J. van der Staay, D. Schauss, WO02/09713 (**2002**, Bayer).

123 (a) J. Hetman, S. H. Soderling, N. A. Glavas, J. A. Beavo, *Proc. Nat. Acad. Sci. USA* **2000**, *97*, 472–476; (b) Sasaki, T. et al., *Biochem. Biophys. Res. Commun.* **2000**, *271*, 575–583.

124 M. J. Barnes, N. Cooper, R. J. Davenport, H. J. Dyke, F. P. Galleway, F. C. Galvin, L. Gowers, A. F. Haughan, C. Lowe, J. W. Meissner, J. G. Montana, T. Morgan, C. L. Picken, R. J. Watson, *Bioorg. Med. Chem. Lett.* **2001**, *11*, 1081–1083.

125 A. Ohhata, Y. Takaoka, M. Ogawa, H. Nakai, S. Yamamtot, H. Ochiai, WO03/064389 (**2003**).

126 (a) P. Bernadelli, P. Ducrot, E. Lorthiois, F. Vergne, WO02/076953 and WO02/074754; (b) F. Vergne, P. Bernadelli, E. Lorthiois, P. Ducrot, WO03/082277 and WO03/082839 (**2003**); (c) F. Vergne, C. Andrianjara, P. Ducrot, WO02/028847 (**2002**).

127 (a) A. Stolle, D. E. Bierer, Y. Chen, D. F. Can, B. Hart, M. K. Monahan, W. J. Scott, US2003119829 (**2003**); (b) W. J. Scott, D. E. Bierer, A. Stolle, WO03/057149 (**2003**); (c) A. Stolle, D. E. Bierer, Y. Chen, D. F. Can, B. Hart, M. K. Monahan, W. J. Scott, WO02/088138 (**2002**).

128 D. Bundschuh, H.-P. Kley, W. Steinhilber, G. Grundler, B. Gutterer, A. Hatzelmann, J. Stadlwieser, G. J. Sterk, S. Weinbrenner, WO02/040450 and WO02/040449 (**2002**).

129 L. Li, C. Yee, J. A. Beavo, *Science* **1999**, *283*, 848–851.

130 M. Fuhrmann, H.-U. Jahn, J. Seybold, C. Neurohr, P. J. Barnes, S. Hippenstiel, H. J. Krämer, N. Suttorp, *Am. J. Respir. Cell. Mol. Biol.* **1999**, *20*, 292.

131 S. Perez-Torres, R. Cortes, M. Tolnay, A. Probst, J. M. Placios, G. Mengod, *Experimental Neurology* **2003**, *182*, 322–334.

132 D. A. Fisher et al., *Biochem. Biophys. Res. Commun.* **1998**, *246*, 570–577.

133 K. Loughney, *Gene* **1999**, *234*, 109–117.

134 (a) K. Fujishige, J. Kotera, K. Omori, *Eur. J. Biochem.* **1999**, *266*, 1118–1127; (b) T. F. Seeger, B. Bartlett, T. M. Coskran, J. S. Culp, L. C. James, D. Krull, J. Lanfear, A. M. Ryan,

C. J. Schmidt, C. A. Strick, A. H. Varghese, R. D. Williams, P. G. Wylie, F. S. Menniti, *Brain Res.* **2003**, *985*, 113–126.

135 (a) L. Lebel, F. S. Menniti, C. J. Schmidt, US2003/32579 (**2003**, Pfizer); (b) L. Lebel, F. S. Menniti, C. J. Schmidt, EP 1250923 (**2002**, Pfizer).

136 (a) J. K. Ergüden, M. Bauser, N. Burkhardt, D. Flubacher, A. Friedl, V. Hinz, R. Jork, P. Naab, U. Niewöhner, T. O. Repp, K.-H. Schlemmer, J. Stoltefuss, D. Brückner, M. Hendrix, D. Schauss, A. Tersteegen, WO03/00693 (**2003**, Bayer); (b) U. Niewöhner, J.-K. Ergüden, M. Bauser, N. Burkhardt, D. Flubacher, A. Friedl, I. Gerlach, V. Hinz, R. Jork, P. Naab, T.-O. Repp, K.-H. Schlemmer, J. Stoltefuss, WO03/00269 (**2003**, Bayer).

137 (a) U. Niewöhner, M. Bauser, J.-K. Ergüden, D. Flubacher, P. Naab, T.-O. Repp, J. Stoltefuss, N. Burkhardt, A. Sewing, M. Schauer, K.-H. Schlemmer, O. Weber, S. J. Boyer, M. Miglarese, WO02/48144 (**2002**, Bayer); (b) U. Niewöhner, M. Bauser, J.-K. Ergüden, D. Flubacher, P. Naab, T.-O. Repp, J. Stoltefuss, N. Burkhardt, A. Sewing, M. Schauer, K.-H. Schlemmer, O. Weber, S. J. Boyer, M. Miglarese, J. Fan, B. Phillips, B. C. Raudenbusch, Y. Wang, WO03/014116 (**2003**, Bayer); (c) U. Niewöhner, M. Bauser, J.-K. Ergüden, D. Flubacher, P. Naab, T.-O. Repp, J. Stoltefuss, N. Burkhardt, A. Sewing, M. Schauer, K.-H. Schlemmer, O. Weber, S. J. Boyer, M. Miglarese, S. Ying, WO03/14117 (**2003**, Bayer).

138 (a) D. A. Fryburg, E. M. Gibbs, WO03/037432 (**2003**, Pfizer); (b) M. P. Deninno, B. Hughes, M. I. Kemp, M. J. Palmer, A. Wood, WO03/037899 (**2003**, Pfizer).

10
Proteochemometrics: A Tool for Modeling the Molecular Interaction Space

Jarl E. S. Wikberg, Maris Lapinsh, and Peteris Prusis

10.1
Introduction

The function of living matter is determined by the structure, organization, and interaction of its constituents. The composition of living matter is ultimately determined by the genome. The advent of the complete genome sequences of several species has opened completely new vistas to the understanding of the functions of organisms at the molecular level.

However, experimental assignment of functions and mapping interactions over whole genomes and proteomes will involve very large costs. The number of genes in the human genome is estimated to about 30 000–40 000, and the number of proteins in the human proteome may be between 200 000 and 2 million, due to alternative splicing, posttranslational processing, and different subunit assembly. Although genes remain essentially unchanged throughout life, the proteome undergoes constant changes depending on the tissues it resides in, the organism's age, etc., due to different processing in the cells. On top of that, the genomes contain substantial genetic variation, with an estimate of 3 million single nucleotide polymorphisms (SNPs) in the human genome. Considering also the genomes of microorganisms, of which more than 100 bacterial genomes have been cloned with about 4000 genes each, and the genomes of animals and plants of scientific and commercial value, the task is indeed large. If one then also considers mapping the interactions of organic compounds (drugs and metabolites) with the proteome, the task reaches astronomical proportions. The number of organic compounds with molecular weights less that 500 Da is estimated to 10^{200}, of which 10^{60} may be 'drug-like' (i.e., likely to be nontoxic, not containing reactive groups, and showing likelihood of passing biological barriers) [1]. Accordingly, without effective approaches to select the relevant combinations of interacting entities for experimental evaluation, and without the availability of computational methods that correctly predict the functions of the biomolecules and their interactions with the surroundings, mapping of functions over entire genomes may never actually happen.

Chemogenomics in Drug Discovery: A Medicinal Chemistry Perspective.
Edited by Hugo Kubinyi and Gerhard Müller

ISBN: 3-527-30987-X

Current computational methods of assessing protein function rely to a large extent on predictions based on the sequence similarity of proteins with other proteins having known functions. The accuracy of such predictions depends on the ability of the computational methods to extend sequence similarity to functional similarity. Computational approaches to molecular recognition have hitherto essentially required access to protein 3D structures. Computational determination of a 3D structure is resource-demanding and error-prone, and generally requires prior knowledge, such as the 3D structure of a homologous protein. The experimental determination of protein structure is a large bottleneck. For water-soluble proteins the pilot structural genomics project succeeded in obtaining only about 10% of the X-ray structures of the proteins in the trial [2].

Recently, a new bioinformatics approach aiding the mapping of molecular recognition was developed. This technology, which does not necessarily require knowledge of the 3D structure of the biomacromolecules, was termed proteochemometrics (abbreviated PCM) [3]. Proteochemometrics contrasts with most previous bioinformatics approaches in that it starts with information derived from the chemical properties of the biomolecules. This chapter deals with various aspects of proteochemometrics, including its foundation, technical aspects of its use, examples of its application, and its potential for the future.

10.2
Definition and Principles of Proteochemometrics

Proteochemometrics originates in chemometrics. Chemometrics constitutes a collection of technologies that evolved when chemists obtained more and more data characterizing the substances they were working with. It then became necessary to apply sophisticated mathematical and statistical methods to deal with the increasing amounts of information. There is no exact definition of chemometrics, which can be attributed to the fact that it has evolved gradually, but the definition presented at the first International Chemometrics Internet Conference (InCINC'94) gives a good clue: "Chemometrics is the science of relating measurements made on a chemical system or process to the state of the system via application of mathematical or statistical methods." Others have stated that it is a set of "mathematical, statistical, graphical, and symbolic methods to improve the understanding of chemical information" and "the art of extracting chemically relevant information from data produced in chemical experiments." In lieu of a stringent definition, Svante Wold humorously stated that "chemometrics is what chemometricians do," and actually the statement comes very close to the truth, because chemometrics has been a dynamically evolving field since its origin more than 30 years ago. For the sake of the present discussion we can regard chemometrics as a discipline in chemistry that uses mathematical, statistical, and other methods to analyze chemical data [4].

Chemometrics has been most successfully applied in four areas, namely (1) multivariate calibration, (2) quantitative structure–activity relationship (QSAR) studies, (3) pattern recognition, classification, and discriminant analysis, and (4) multivariate modeling and monitoring of processes [5]. The most relevant areas to PCM are QSAR, classification, and discriminant analysis. In QSAR the aim is to correlate chemical data of series of compounds (i.e., compounds contained in a 'chemical space') to a biological activity. 'Biological activity' relates to the strength of interaction of a compound with a target, whatever that target is, e.g., an organism, a cell, or a protein. Thus, in QSAR we deal not only with chemical data, but also with data characterizing the chemical substances' interactions with a target. In PCM, however, we study several targets, such as a series of proteins. In the original setting we simultaneously studied interactions of several targets with several ligands. The benefit of PCM is then that it deals not only with the chemical space of series of compounds, as does QSAR, but also with the 'interaction space' of the targets and the interacting entities. As a result we can extract a much richer set of information about the interactions occurring between targets and the interacting molecules, and the PCM models also become more stable [6]. PCM can accordingly be defined as the area in which we study the molecular interactions between proteins and their interacting entities, using chemometric techniques. Chemometrics can be applied to data characterizing proteins *per se* as well, or for that matter, even to data for any other type of macromolecule and their interactions with their surroundings. For the present discussion we refer to all of this as proteochemometrics.

A central approach used in PCM is schematically outlined in Figure 10.1. PCM is here based on modeling data for sets of biopolymers interacting with sets of

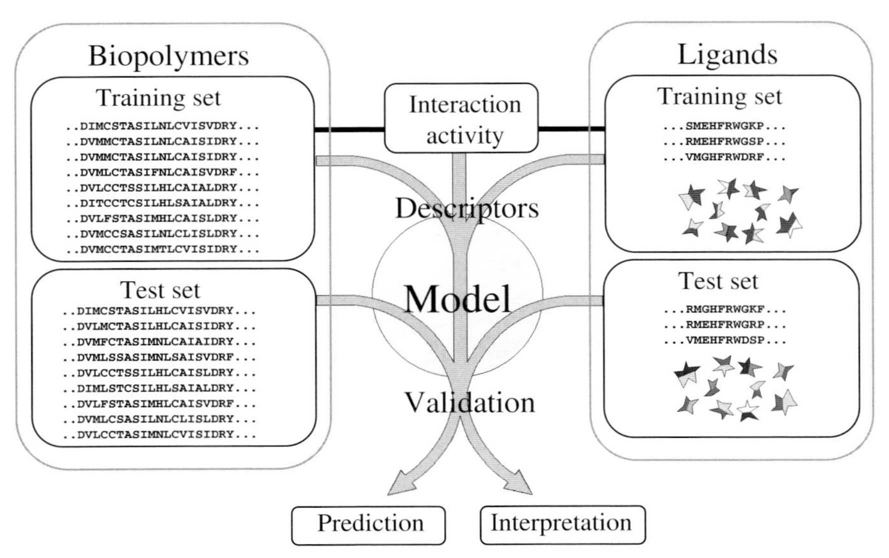

Figure 10.1 Outline of the principles of proteochemometrics. Descriptors of training sets of biopolymers and ligands are correlated with experimentally determined interaction activities, using a suitable mathematical modeling method. The predictive ability of the resulting model is assessed by validation, using test sets of ligands and biopolymers. When its validity has been confirmed, the proteochemometric model can be used for predicting the activity of novel biopolymers and ligands and for interpreting factors that determine biopolymer–ligand interactions.

chemical entities (ligands). To create a PCM model, we need to obtain a set of biopolymers showing sequence variation, as well as a set of ligands showing variation in their chemical structure. These entities are then evaluated for their abilities to interact, which creates training and test sets of data. PCM models are made by correlating suitable descriptors of the biomolecules and the ligands with the experimentally determined interaction activities of the training set. The model thus obtained needs to be validated for correctness (e.g., using the test set) and can thereafter be used for interpretations and predictions (Figure 10.1).

The interaction activities can be measured by any standard interaction assay, such as radioligand binding, plasmon resonance, and enzymatic assays. Other measures, such as of downstream signaling, can also be used for indirectly estimating the interaction activity, but interpretation may then become somewhat different.

10.3 Modeling and Interpretation of Interaction Space

Properties of organic molecules (i.e., ligands) can be described with physicochemical, structural, and binary descriptors, and this is a well studied field in QSAR (Scheme 10.1). In PCM biopolymers (i.e., proteins, nucleic acids, polysaccharides, and assemblies thereof) are described, and the problem is more difficult due to their size and complexity. One option is to consider the polymeric nature of the biomolecules and provide numeric characterizations according to the position and properties of each monomer. However, since any standard description used in QSAR might include from 100 to 20 000 or more descriptors for each monomer, the resulting data table for any normal biopolymer would be extremely large, giving rise to immense problems in subsequent modeling steps. A solution is to consider only so-called principal properties, which are obtained by compressing monomer descriptions by principal component analysis (PCA) [7]. For amino acids such principal property descriptors were developed based on a large number of measured and computed physicochemical properties of the amino acid monomers, resulting in 3–5 principal property scales ('z scales'). Such z scales encapsulate about 70–95% of the physicochemical properties of the amino acids in 3–5 z scales (Scheme 10.1) [8].

With z scales, a whole protein can be directly translated from its amino acid sequence into a vector of numbers. However, sets of proteins of different sequence lengths yield matrices of nonuniform size, which is incompatible with subsequent correlation methods used in PCM modeling. Alignment of sequences is a solution to forming uniform matrices. However, the primary sequences of proteins are seldom conserved to the extent that alignments can be made unambiguously. Wrongly aligned sequences preclude comparisons of the proteins' chemical space. A method called auto cross-covariance (ACC) transforms was developed, which provides a uniform matrix from a set of sequences of unequal lengths by capturing sets of characteristic physicochemical patterns of the sequences. Applying ACC transformations to z-scale coded sequences has proven of value for descriptions from short peptides to proteins [9].

(a) Description of ligands

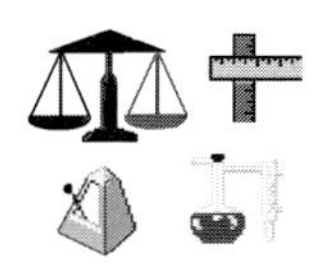

Physico-chemical properties:

Molecular weight, van der Waals volume, molecular surface area, number of hydrogen bond donors and acceptors, polarizability, electronegativity, logP, NMR shifts, chromatographic retention times etc.

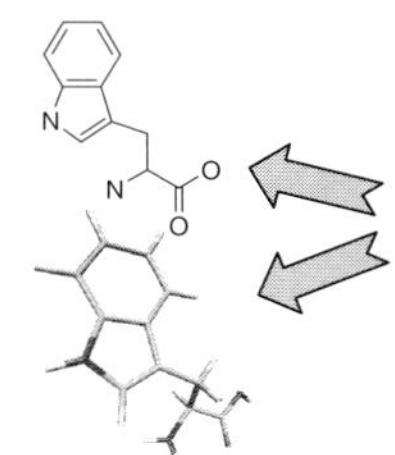

Structural descriptors:

- **One-dimensional (constitutional):** number of atoms, bonds, rings, groups etc.
- **Two-dimensional:** topological (connectivity) indices.
- **Three-dimensional:** molecular modeling + CoMFA (Comparative Molecular Field Analysis) descriptors, CoMSIA (Comparative Molecular Shape Indices Analysis) indices, GRIND (GRind INDependent) descriptors etc.
- **Four and more dimensional:** representation of multiple conformations, orientations, protonation states etc.

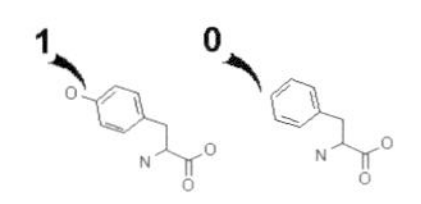

Binary representation: Presence or absence of particular structural features in a molecule can be represented by bitstrings. Binary representation can also be used to characterize values of physico-chemical properties over or below a certain limit.

(b) Description of biopolymers

Principal physico-chemical properties of the monomer sequences

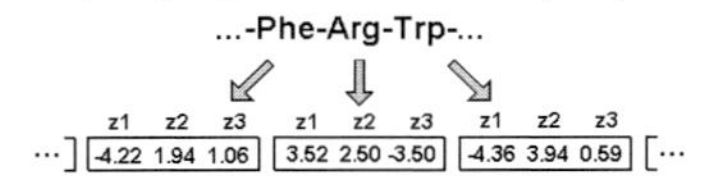

Z-scales are obtained by principal component analysis of physico-chemical properties of monomers. E.g., the first Z-scales of amino acids describe hydrophobicity (z1), steric bulk/polarisability (z2) and polarity (z3) of the amino acids.

Binary representation of chimeric biopolymers

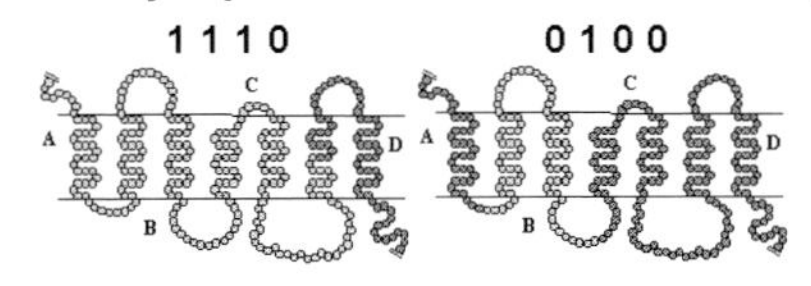

Chimeric biopolymers are formed by interchanging sequence portions of two native biopolymers. Eg. the number **1** can be taken to represent physico-chemical properties of a sequence portion of the first native biopolymer, whereas the number **0** (or **-1**) can be taken to represent the physico-chemical properties of the corresponding sequence portion of the other native biopolymer.

(c) Cross-terms

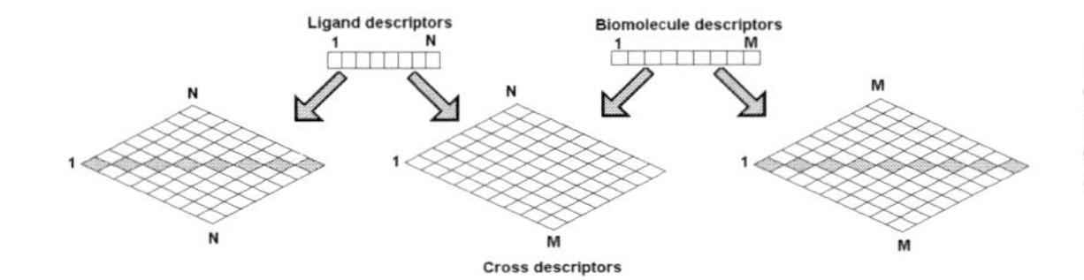

Descriptors of ligands and biopolymers can be used to derive intra-ligand, intra-biopolymer and ligand-biopolymer **cross-terms,** e.g. by multiplication of descriptor-values.

Scheme 10.1 Examples of derivation of descriptors (ordinary terms) for ligands (organic molecules, etc.) and biopolymers and of derivation of cross terms.

If the variation in sequence in a whole biopolymer set is restricted to only two possibilities, binary descriptions can be used. They are useful for describing chimeric proteins containing alternating stretches of amino acids from two different proteins (Scheme 10.1).

PCM modeling aims to find an empirical relation (a PCM equation or model) that describes the interaction activities of the biopolymer–molecule pairs as accurate as possible. To this end, various linear and nonlinear correlation methods can be used. Nonlinear methods have hitherto been used to only a limited extent. The method of prime choice has been partial least-squares projection to latent structures (PLS), which has been found to work very satisfactorily in PCM. PCA is also an important data-preprocessing tool in PCM modeling. Modeling includes statistical model-validation techniques such as cross validation, external prediction, and variable-selection and signal-correction methods to obtain statistically valid models. (For general overviews of modeling methods see [10]).

In PCM we are interested in exploring how molecules interact. When we are using linear modeling, these descriptions reveal the contribution of linear combinations of properties to the interaction. However in reality, complex nonlinear processes govern interactions. Such nonlinearities can be explored by investigating the contribution of nonlinear combinations of descriptors, by forming so-called cross terms. Cross terms are formed from ordinary terms by a suitable mathematical operation, usually multiplication. (Ordinary terms are the descriptors for molecules and biopolymers.) In PCM we can derive three types of cross terms, namely ligand cross terms, ligand–biopolymer cross terms, and biopolymer cross terms (Scheme 10.1). Together with the ordinary terms, these then form five descriptor blocks, which represent a polynomial approximation of the interaction space (Scheme 10.1). Depending on the problem to be solved, one or more of these blocks is sometimes omitted [11].

The key to understanding PCM modeling is to understand the meaning of descriptors. Although the ordinary terms for molecules that correlate with the interaction activity relate to the ability of the underlying property to form interactions with invariant parts of the biopolymers and vice versa, correlated ligand–polymer cross terms reveal points of complementary interactions between varying parts of ligands and varying parts of biopolymers. A large absolute value for a coefficient of a descriptor in the PCM equation indicates a large influence of that descriptor on the interaction activity. With this information, insights into the nature of the molecular interactions can be obtained; it also indicates how we can modify a ligand and/or a biopolymer so as to achieve a desired property for their interaction. Hence, PCM finds application in areas such as functional genomics, drug design, and protein engineering.

From the coefficients of the PCM equation various measures have been derived to aid the interpretation of molecular recognition processes. These include uses of the sum of the absolute values of coefficients for a group of descriptors, computation of contribution estimates, and various measures to reveal the properties underlying ligand selectivity [11].

10.4
Examples of Proteochemometric Modeling

10.4.1
Proteochemometric Modeling of Chimeric MC Receptors Interacting with MSH Peptides

The first successful PCM study that simultaneously used series of proteins and ligands was performed on wild-type and chimeric melanocortin receptors (MCRs) interacting with melanocyte-stimulating hormones (MSHs) and analogs thereof [12a]. The MCRs are G-protein coupled receptors (GPCRs) that contain seven transmembrane regions, TM1–TM7. Their N-terminal portion and three loops (EL1–EL3) are located toward the cell membrane's extracellular face, and three loops (IL1–IL3) and the C-terminal portion are located toward the intracellular face. There are five subtypes of MCRs (MC_{1-5}), but the modeling study included only subtypes MC_1R and MC_3R. Chimeras of the latter had been created by dividing the amino acid sequences into four parts, the first containing the N terminus and TM1; the second IL1, TM2, EL1, and TM3; the third IL2, TM4, EL2, TM5, IL3, and part of TM6; and the fourth the other part of TM6, EL3, TM7, and the C terminus, and then including sequences from the MC_1R and MC_3R subtypes in various combinations [12b]. The eight chimeras thus created, and the wild-type MC_1R and MC_3R subtypes (in total, 10 receptors), were evaluated for their interaction activities with α-MSH, the phage-display-selected MC_1R-selective peptide MS04, and two chimeric variations between α-MSH and MS04 that contained two stretches of three and five amino acids from the original peptides [12c]. Since the chemical variation had been created from a total of four amino acid stretches from two original receptors, and two original initial peptides, the interaction space could be described by six binary descriptors [12a].

Applying PLS modeling by using the six binary descriptors of the peptides and receptors (i.e., using these six binary descriptors as ordinary terms) provided a preliminary model that was substantially improved by including ligand–receptor and receptor–receptor cross terms. The final model had $R^2 = 0.93$ and $Q^2 = 0.75$, which represents a very good model of high validity. (R^2 represents the explained variance of the model and Q^2 is a measure of the predictive ability by cross validation: $Q^2 > 0.4$ is considered acceptable, and $Q^2 > 0.9$ is excellent). The fact that the model was improved very substantially when the cross terms were included showed that a substantial part of the peptide and MCR interactions depend on complimentary interactions between the receptor and the peptides, as well as on interactions within the receptors. Analysis of the model using measures of its PLS regression coefficients showed that the high-binding affinity of MSH peptides was achieved primarily by interactions of the peptides' C-terminal amino acids with TM2 and TM3 of the MCRs and to a lesser extent by the interaction of the N terminus with TM1, TM2, and TM3. Moreover, analysis of the intrareceptor cross terms revealed the existence of strong interaction between TM6/7 and TM2/3 in the receptors.

A method for assessing the interactions responsible for peptide selectivity was also developed, by creating a separate model in which the interaction activity was expressed as the MC_1R selectivity of each peptide, rather than as absolute values for ligand-binding activity. Also here, a very good model was obtained ($R^2 = 0.92$, $Q^2 = 0.72$), which interestingly revealed that the MC_1R selectivity of the peptides were primarily determined by interaction of their N termini with TM2/3 of the MCRs [12a].

The first study was very encouraging, because very robust PCM models could be created, despite the fact that only six binary descriptors were used to describe the underlying dataset. It became of obvious interest to verify that PCM modeling can be applied to other similar datasets and also to broaden the complexity of the problem by using more varied forms for the interacting entities. In a subsequent study, the interactions of a broader set of linear and cyclic MSH peptides were studied, using the wild-type and chimeric MC_1R and MC_3R subtypes mentioned above. Also in this study, binary descriptions were used for the peptides, namely for describing the absence or presence of cyclization in the peptide, the D- or L-conformation of a particular peptide bond, and the presence one of two different amino acid at a certain position in the MSH peptides. Since the receptor set was the same as above, it was described in the same way as before. Also here, the models were improved substantially by adding cross terms, resulting in very good PCM models ($R^2 = 0.97$, $Q^2 = 0.91$ for an affinity model and $R^2 = 0.91$, $Q^2 = 0.83$ for a selectivity model). The validity of the modeling procedure was ascertained by external prediction, by creating a model using the experimental data for only half the observations and using that model to predict the interaction activities for the omitted observations. The results of these external predictions showed that the omitted interaction activities could be 'predicted' with high precision. Interpretation of the models placed the binding pocket for the MSH peptides in the same place as had been deduced from the first study described above. Another interesting result that emerged was the possibility of analyzing cross terms between peptide descriptors, which indicated that PCM modeling can distinguish between differences in the conformational space of peptides that affect the binding affinity and selectivity of the peptides [12d].

10.4.2 Proteochemometric Modeling of α_1 Adrenoceptors Using z Scale Descriptors for Amino Acids

The above studies were conducted on chimeric receptors put together from the sequences of two receptors. Subsequent studies were directed at biopolymer sets containing larger variations. One of these studies utilized a set of chimeric and point-mutated α_1-adrenoceptors (α_1-ARs) that allowed a much richer description compared to the binary descriptions used above. The biopolymer set in the study comprised in total 18 α_1-ARs, namely the wild-type α_{1a}, α_{1b}, and α_{1d}-ARs and 15 mutated α_1-ARs. The latter had been constructed by exchanging TM segments of α_{1a}-AR with the corresponding segments of the α_{1b} or α_{1d}-ARs (i.e., cassette mutations) or by exchanging single amino acids (i.e., point mutations). This resulted in a set of biopolymers that had 11 positions with three amino acids varied and 43 with two amino acids varied. The interaction data comprised the interaction activity of a series of 12 structurally closely related α_1-AR-active 4-piperidyl oxazoles determined in radioligand binding-competition assays. In this study five amino acid z scales were used to describe the receptors. For the positions that differed by only two amino acids, the z scales could be merged into one numeric value by calculating the physicochemical distance between the two amino acids (i.e., the geometric distance between the two property vectors). The quite limited variation among the 4-piperidyl oxazoles, namely changes in substituents at three positions, suggested the use of binary structural descriptors for the organic compounds also [13].

The data for the 3 wild-type, 8 cassette-mutated receptors, and 12 4-piperidyl oxazoles were modeled first. This neatly demonstrated the importance of cross terms. Although neither receptor nor organic compound descriptors alone yielded valid models, it was possible to create a reasonable model based on the ordinary receptor and ligand descriptors ($R^2 = 0.87$, $Q^2 = 0.79$). However, creating a model based on ordinary terms and ligand–receptor cross terms yielded a substantially improved model showing very high validity ($R^2 = 0.96$, $Q^2 = 0.91$). The gain achieved showed that ligand–receptor cross terms explain a substantial part of the interaction activity. Following the success of this modeling, 7 point-mutated receptors were included in the modeling, yielding at total of 18 receptors interacting with 12 4-piperidyl oxazoles. Again, an excellent PCM model was achieved when the appropriate ordinary terms and cross-term blocks were included ($R^2 = 0.94$, $Q^2 = 0.87$).

The models were interpreted by using scaled sums of absolute values of PLS coefficients of ordinary terms (in this study called 'significance of primary variable' or ΣSOP) and sums of absolute values of PLS coefficients of cross terms (in this study called 'significance of a variable cumulated from ligand–receptor cross terms' or ΣSOC). These measures gave different insights into the roles of receptor regions in their interactions with the ligands. ΣSOPs represent the importance of amino acid descriptors for explaining the variation in the affinity of an average ligand for the biopolymers in a dataset, and ΣSOCs describe the joint importance of amino acid and ligand descriptors for creation of ligand selectivity. The ΣSOPs showed that that TM regions two and five have the greater importance for the α_1-ARs' ability to bind 4-piperidyl oxazoles, and the ΣSOCs showed that only TM region two is important for discriminating selective from nonselective 4-piperidyl oxazoles.

An even more detailed picture was obtained by analyzing the model that included both the cassette- and point-mutated receptors. For example, this model showed that the properties of the amino acid at sequence position 86 in TM region two of the α_1-ARs are important for activity. Moreover, this model showed that the single amino acid at position 185 was responsible for almost all of the ΣSOP values, although in the model based only on cassette-mutated receptors high ΣSOP values were assigned to the whole TM region five. These results show that inclusion of more data in a model allows successive refinement of the PCM modeling, aiding the interpretation of the molecular recognition processes for interacting entities [13].

Yet another important observation for the model including both cassette- and point-mutated receptors was a noticeable improvement in its quality upon inclusion of receptor–receptor cross terms – comparable improvement did not occur for the model based only on cassette-mutated receptors. A detailed analysis of the importance of the receptor–receptor cross terms then revealed that important interactions occur mainly between amino acids at positions 85 (TM region two) and 185 (TM region five) and nonmutated parts of TM region five. The results in fact indicated the presence of distant interactions in the receptor, which could affect packing or tilting of the TM regions. Taking the analysis further revealed that it is not just the presence of valine at position 185 of the α_{1A}-ARs that determines their overall high affinities for 4-piperidyl oxazoles, but rather the cooperation of Val185 with other parts of TM region five [13].

10.4.3
Proteochemometric Modeling Using Wild-type Amine GPCRs

The studies reviewed above were performed on chimeric and mutated receptors. The need to use engineered proteins seemed to pose restrictions on the application of PCM on a large scale, in view of the large costs involved in creating such artificially altered proteins. However, the scope and applicability of PCM was recently enlarged significantly by a study that showed that a large set of related wild-type proteins is sufficient to provide highly valid PCM models. The study was performed using 21 GPCRs for amines comprising subtypes of dopamine, serotonin, adrenergic, and histaminergic receptors. The ligands tested on these receptors were a series of 23 organic amines, antipsychotics, and alkaloids, which showed quite varied structures, prompting the need for extensive descriptions of their 3D structures [14a]. To provide a uniform matrix of descriptors from 3D descriptions for use in the subsequent modeling, one option would have been to perform structural alignments of the compounds. However, due to the absence of a common core structure in the compounds, any superimpositions would be at high risk of producing erroneous results if the resulting alignments had no relation to the real orientation of the ligands at the receptors' binding sites. Therefore, an approach to the calculation of alignment-independent 3D descriptors, so-called GRIND descriptors, was used [14b]. Briefly, according to this approach molecular interaction fields (MIFs) are obtained by placing probe 'atoms' on grid points surrounding the molecule and calculating interaction energies at each grid point. The probe atoms used in this study were DRY, O, and N1, representing hydrophobic, H-bond acceptor, and H-bond donor interactions, giving three MIFs. A so-called maximum auto- and cross-correlation (MACC) algorithm was then used to calculate alignment-independent descriptors from the pairs of MIFs. In this way six blocks of descriptors, DRY–DRY, O–O, N1–N1, DRY–O, DRY–N1, and O–N1 were obtained (in all, 230 descriptors) [14].

Amine GPCRs are known to interact with their ligands within the cleft formed between the seven TM regions. Therefore, the receptor descriptors were selected only from the TM segments. This strategy also had the advantage that the sequence homology of the TM segments was high enough to allow their unambiguous alignment. (TM segments were overall 35%–45% homologous to the consensus sequence of amine GPCRs.) In all, 159 amino acids were used and described by 5 z scales each.

Thus, a fair amount of ordinary descriptors were needed, which would give a very large number of cross terms. To reduce the descriptors to a manageable number, PCA was applied separately to the different blocks of ligand descriptors and descriptors for receptor TM regions prior to using them for calculation of cross terms [14a]. It was then shown that using only ordinary descriptors (after PCA) did not allow a valid model to be obtained ($R^2 = 0.41$, $Q^2 = 0.31$). However, after addition of ligand–receptor cross terms and a form of higher-order ligand–receptor 'cross terms' (see [14a] for details) a very good model was obtained ($R^2 = 0.92$ and $Q^2 = 0.75$).

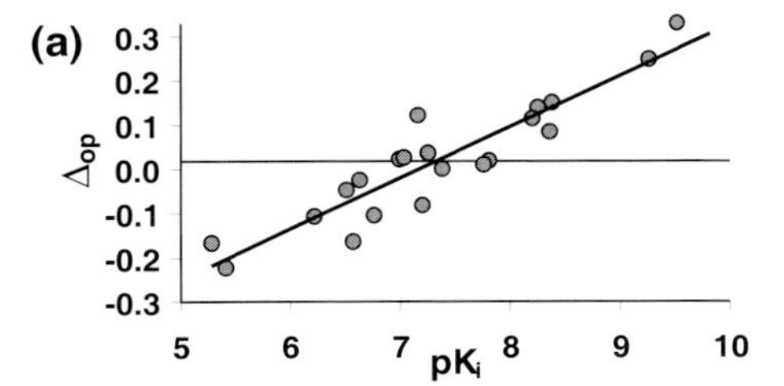

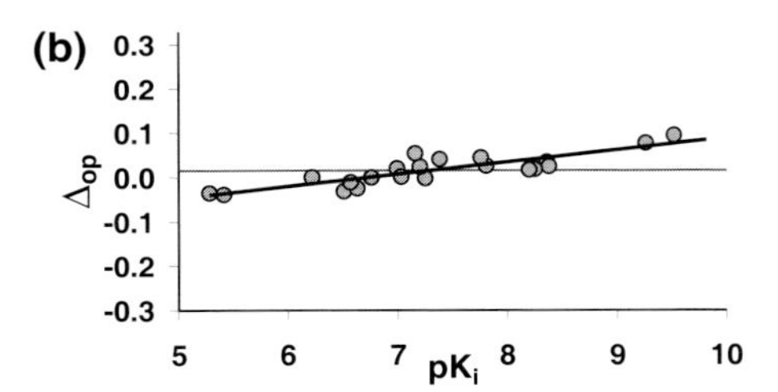

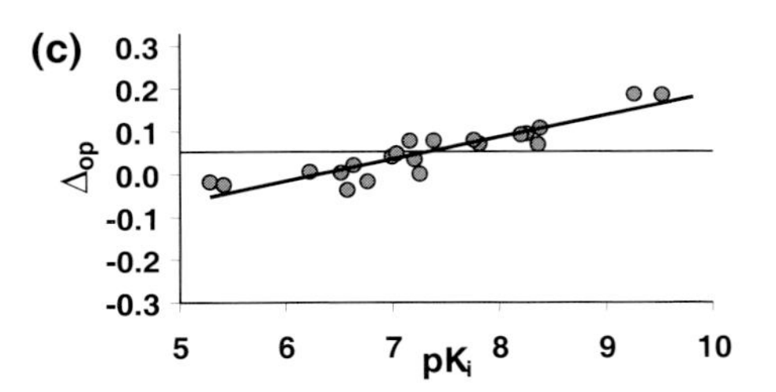

Figure 10.2 The use of contributions of descriptor blocks for evaluating the interactions of organic amines with amine GPCRs. The study used a PCM model based on the interactions of 23 organic amines with 21 amine GPCRs. The *x* axis shows the binding affinity (pK_i) of sertindole for 21 different amine GPCRs, as computed from the model. The *y* axis shows the change in the calculated interaction activity (Δ_{op}) when the values of variables characterizing some receptor property were changed in-silico to the values for a hypothetical amino acid, so that the receptor property's value changes to that of the 'average' property for the descriptor at that amino acid position in the dataset.
(a)–(c) show the Δ_{op} values accumulated from the descriptors of TM region two of the amine GPCRs and the cross terms formed with different molecular interaction fields of the organic compounds (DRY, O, and N1, respectively). Two contribution indices can thereby be obtained for any compound, namely the contribution of any of its described properties to the average affinity (α_{sp}, derived from the horizontal lines) and to the selectivity (σ_{sp}, derived from the slopes of the lines). (Reproduced from *Mol. Pharm.* **2002**, *61*, 1465–1475 by courtesy of the American Society for Pharmacology and Experimental Therapeutics).

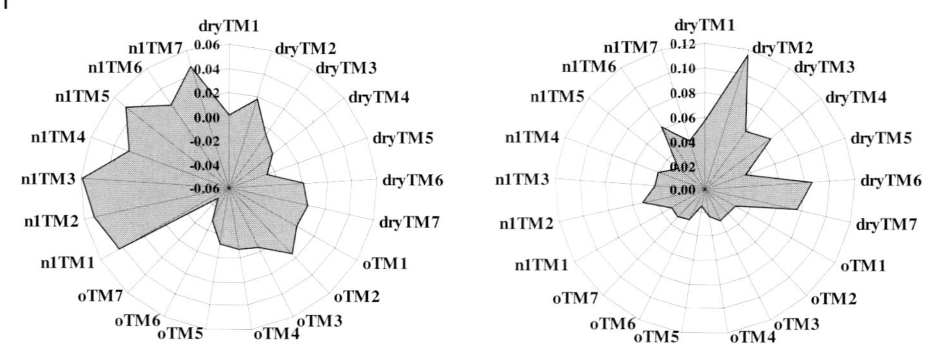

Figure 10.3 Radar plots of α_{sp} (left) and σ_{sp} (right) values for sertindole's DRY, O, and N1 interactions with each of the seven TM regions in amine GPCRs. The plots are based on the same PCM model as in Figure 10.2. As seen, the α_{sp} values (left) do not discriminate very clearly between the different receptor regions. However, the σ_{sp} values (right) reveal that distinct interaction types and TM regions are responsible for selectivity, the DRY–TM2, DRY–TM6, and DRY–TM7 interactions having the largest contributions. (Reproduced from *Mol. Pharm.* **2002**, *61*, 1465–1475 by courtesy of the American Society for Pharmacology and Experimental Therapeutics).

To solve the problems imposed by the large numbers of descriptors and the use of PCA prior to formation of cross terms, as well as the use of higher-order terms, so-called contribution estimates and groupings thereof, were used for model interpretation (see [14a] for details). Thus, contribution values were computed for each descriptor (in the study, called ΔpK_i), and the contributions were summed for regions of interest (in the study, called contribution blocks) (Figures 10.2 to 10.4). Figure 10.2 a–c shows the contributions (called Δ_{op}) to the affinity of one compound (sertindole) of the studied series of (a) hydrophobic, (b) H-bond acceptor, and (c) H-bond donor interactions with one TM region for each amine receptor in the study plotted against the pK_i values (i.e., the negative logarithms of the interaction activity values) of this compound for each receptor. From these plots two further indices can be calculated, namely, the average contribution to the affinity of the TM region of each receptor for a certain type of chemical interaction with a certain compound (in the study called α_{sp}), and the slope of the trend line (in the study called σ_{sp}). The slope indicates how each chemical property contributes to the ability of the selected transmembrane region of the different amine receptors to discriminate the compounds.

The further uses of α_{sp} and σ_{sp}s are illustrated in Figure 10.3, which shows the results for all chemical interaction types for all TM regions binding to one compound. As seen, the set of all regions and chemical interactions reveals widely varied contributions to the average affinity for and selective discrimination of ligands. When these contributions are compared for different compounds, they show major differences, in particular the σ_{sp}, as shown in Figure 10.4 for four different ligands.

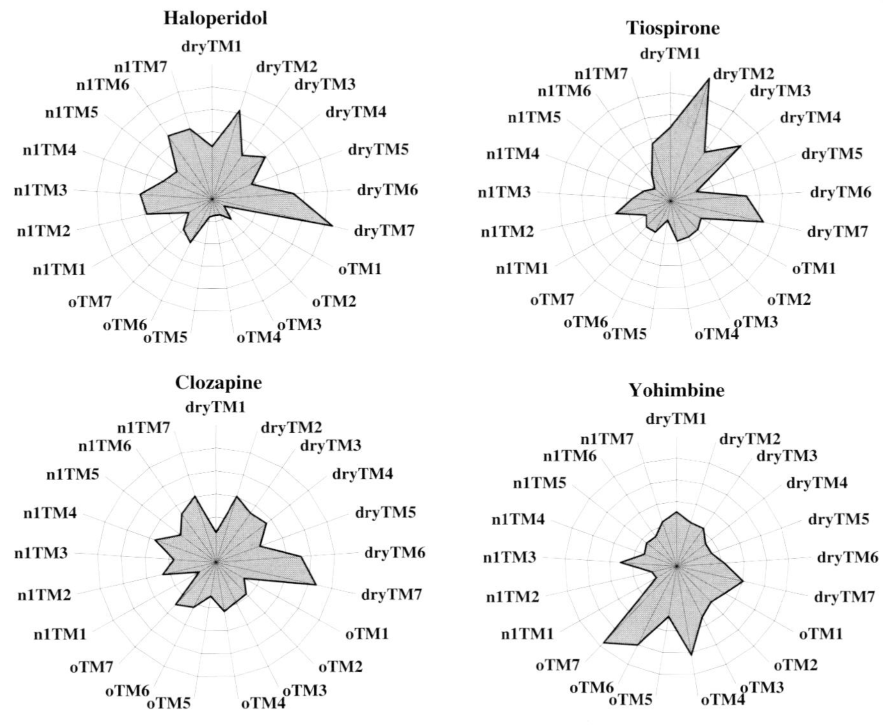

Figure 10.4 Radar plots of σ_{sp} for haloperidol, clozapine, tiospirone, and yohimbine interactions with each of the seven TM regions in amine GPCRs. The plots are based on the same PCM model as in Figure 10.2. Hierarchical clustering of the σ_{sp} data for 23 compounds identified four significant clusters of compounds: the centre compounds – haloperidol, clozapine, tiospirone, and yohimbine – are shown here. As seen, prominent DRY interactions occur between TM6 and TM7 and haloperidol, clozapine, and tiospirone, but not when yohimbine is the ligand. For yohimbine the most prominent feature is instead O-interactions with TM6 and TM7. In addition, tiospirone also shows prominent DRY interactions with TM2 and TM4. The overall profiles of haloperidol and clozapine were similar, although these compounds prefer different receptors (i.e., haloperidol is selective for dopamine receptors, and clozapine shows highest affinity for α_{1A} adrenergic and $5HT_{2B}$ receptors). (Reproduced from *Mol. Pharm.* **2002**, *61*, 1465–1475 by courtesy of the American Society for Pharmacology and Experimental Therapeutics).

These selectivity contributions can be further used to cluster ligands into different categories, indicating that groups of ligands bind to the receptors with distinct binding modes, in turn suggesting the existence of distinct binding pockets for different categories of ligands in the amine GPCRs [14a].

The results of the study also indicate how contribution estimates can be used to map each individual amino acid (and each amino acid physicochemical property) in the receptors with respect to its contribution to the activity and selectivity for every amine ligand [14a].

10.4.4
Interaction of Organic Compounds with Melanocortin Receptor Subtypes

Yet another study using wild-type GPCRs investigated the interactions of a series of 54 organic compounds with the four MCR subtypes $MC_{1,3-5}$ [15]. Since the dataset included only four proteins, it was of interest to find out whether PCM modeling showed any advantage over conventional QSAR modeling. Three modeling approaches were compared, namely, four separate QSAR models (i.e., one for each receptor), multiresponse QSAR, and PCM modeling. The compounds were characterized by structural descriptors, including GRIND descriptors, topological descriptors, and geometrical descriptors.

The low number of receptors suggested that using extensive descriptions of the physicochemical properties of the receptor sequences would yield little benefit in the PCM modeling. The MCRs were therefore described with only four variables, which were based on the receptor sequence identities. For the comparisons with QSAR models, the PCM model was validated so that all four observations of a compound were always included in the same cross-validation group, i.e., when predicting the affinity of a given compound for each of the receptors, no information was present in the model about the binding of this compound to the other receptors. Despite this strict criterion, cross validation gave a high $Q^2 = 0.71$ and $R^2 = 0.80$. Thus, a notable small difference between Q^2 and R^2 was achieved, indicating a high-validity model.

Statistical analysis showed that, although the separate QSAR models and PCM gave similar predictive ability, interpretation of the PCM model was more reliable. After variable selection, each of the four QSAR models retained different sets of descriptors, which made it difficult to find determinants for the receptor-subtype selectivity of the compounds. In contrast, interpretation of the PCM model was straightforward, and the selectivity of the compounds could be explained on the basis of a relatively small number of cross terms [15].

10.4.5
Modeling of Interactions between 'Proprietary Drug-like Compounds' and 'Proprietary Proteins'

A study was recently published [16a], which in essence represents a PCM study. Unfortunately this study was performed on proprietary proteins and proprietary drug-like molecules, so it is a bit difficult to evaluate. The dataset represented the interaction measurements of 576 drug-like compounds with 10 proteins; 3286 binary variables were used to describe topological features of the compounds, and 476 proprietary variables were used to characterize the proteins (no further data was provided). By applying PCA, the number of compound and protein descriptors was reduced to 12 and 5, respectively [16a]. Correlation of these 17 principal components with the affinity data was performed by applying a nonparametric regression algorithm called multivariate adaptive regression splines (MARS) [16b]. Cross validation of the model was performed by leaving out one protein at a time, and the correlation between the observed and predicted values for the four best-predicted proteins was 0.82, 0.66, 0.35, and 0.32. A closer analysis showed that the model was predictive only for proteins that had similar counterparts remaining in the model. The study lead to the conclusion that modeling of a small number of unrelated proteins would likely fail to make predictions for proteins outside the modeled domain. In contrast, rather good predictive ability was observed in cross validation after leaving out 10%–20% of the compounds, correlation between observed and predicted values being 0.78 (corresponding to $Q^2 = 0.6$). The authors also stated that the models were difficult to interpret, in particular with respect to the protein descriptions, something that in fact may be related to the proprietary nature of the modeled problem [16a]. The MARS regression modeling approach itself does not seem to suffer from a lack of interpretability and could be an alternative to calculating all ligand–protein cross terms. However, in a comparative study on some PCM datasets, the predictability, and in particular the robustness, of MARS models appeared inferior to those of PLS models [16c].

10.5 Large-scale Proteochemometrics

The ability to build statistically valid models that can predict the interaction of new macromolecules with new ligand molecules based on descriptors derived from the molecules' primary chemical structure gives great promise in a number of important areas in biology and chemistry. The obvious ones are drug design, protein engineering, and prediction of protein function. PCM can also be used for traditional bioinformatics and functional genomics tasks such as protein annotation, classification, and functional analysis. A great advantage of PCM is that it needs only data that are readily obtainable from the use of simple, well developed assay techniques for measuring the strength of interaction of interacting entities. Interaction assays are routinely performed on recombinant targets expressed in crude preparations on which chemical compounds, often in form of libraries, are tested. Crude preparations from animal tissues may also be useful, if the specificity of the assay method is good enough and so well characterized that the sequence of the interacting biopolymer is known.

Although PCM has until now been applied mainly to modeling the interactions of GPCRs with peptides and organic molecules, for which the interaction strength was quantified at steady state by radioligand binding, the approach is not limited to any particular class of protein or to the particular method used for measuring interactions. It has already been applied in principle to analysis of the kinetics of antibody–antigen interactions and of DNA–protein interactions [17]. PCM can be applied to any protein and fold family. In fact, nothing hinders us from building PCM models covering the interactions of entire proteomes and genomes. In view of the wide applicability of models covering large areas of the proteome and genome, its construction is a matter of high importance. We have elected to call projects aiming at the construction of such models 'large-scale proteochemometrics'

(Figure 10.5). Such projects need not to be limited to any particular species, because data collected for any biopolymer is useful across species. PCM models can be used for predictions regarding any biopolymer and ligand, given that the interaction space of the entities of interest are within the bounds of the model.

Thus, large-scale proteochemometrics would aim to collect interaction data on a broad scale. One possibility is to use existing literature data. A large problem here is that, in the past, interaction data were collected in quite a nonsystematic fashion, with each separate study covering fairly small, nonoverlapping series. This creates problems in finding datasets of high enough quality from the existing literature. Sometimes the existence of overlapping measurements between studies allows the data to be transformed and scaled to make the results of different studies comparable even when different assay methods were used. A very common finding is, however, that the data lack multivariability. Mutational changes were often performed on one or a few single amino acids, which were altered into an unnecessarily large number of different amino acids. Such design of experiments yields datasets that contain limited information and are quite useless for PCM.

In contrast to the well organized information on the structures of biomolecules and synthetic compounds within genome databases and databases of organic compounds, interaction data are still fairly poorly systematized. Open interaction databases include various protein–protein interaction databases such as the Biomolecular Interaction Network Database (BIND), the Database of Interacting Proteins (DIP), the Interact database, the protein–nucleic acid interaction database ProNit, and some protein–organic compounds interaction databases such as Binding DB, the GPCRDB for G-protein coupled receptors, and the PDSP database for GPCRs and other proteins [18]. Although such databases contain information of some usability, there are many deficiencies. For example, protein–protein databases contain information on protein pairs that interact but no information on protein pairs that do not interact. In PCM both types of information are equally important. Many databases also do not contain any information on the interaction strength, which of course is highly desirable information in PCM. Other common problems of protein–ligand databases are the lack of easily accessible structural information on organic compounds (often there is a generic name or a code name with reference to a published article). Assay conditions are also sparsely given (often a reference to an original article in which the desired information was not reported). The situation may be better in the proprietary databases of some large organizations. However, even in these instances one may find that a large number of compounds were tested on a few targets, but data on structurally similar targets are lacking. Hence, the datasets lack multivariability with respect to the interaction space desired for proteochemometric analysis.

It is thus obvious that a systematic approach is needed, to apply PCM on a large scale. Some key factors to success can be identified, namely the use of experimental design to reduce the number of experiments and the development of proper database- and PCM-analysis tools for storage and management of data. Thus, having many thousands of potential macromolecular targets and an essentially unlimited number of interaction partners (i.e., drug-like structures, peptides, proteins, DNA,

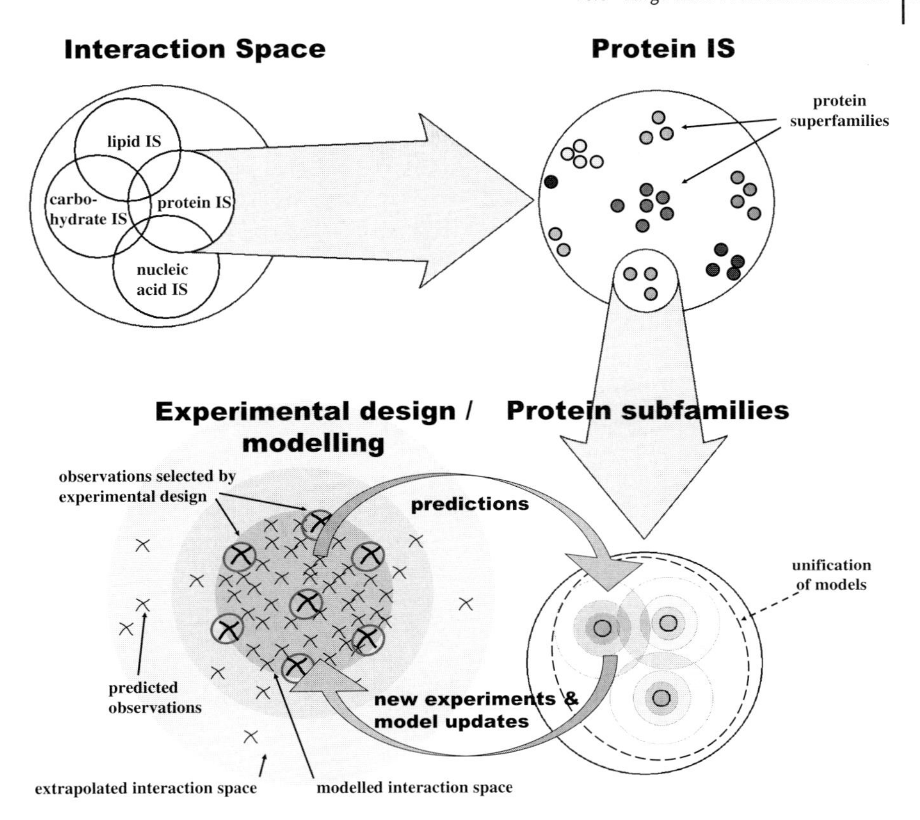

Figure 10.5 Outline of large-scale proteochemometrics. Interaction space (IS) encompasses the noncovalent interactions of all molecules. It contains subsets of interaction spaces, such as the protein interaction space (Protein IS), which includes the noncovalent interactions of all proteins with all molecules (e.g., the interactions of all proteins with all organic compounds, proteins, peptides, lipids, nucleic acids, etc.). The protein IS is then subdivided into protein superfamily interaction spaces. Each protein superfamily IS contains the interaction spaces of protein subfamilies. Experimental design is used to select the optimal observations covering the largest possible area of the interaction space with the fewest number of observations (i.e., selecting the best combinations of proteins and interacting entities to be analyzed experimentally). PCM models created from the data can then be used to predict new experiments, allowing continuous updating and improvement of the models. Models thus created may be used to predict the interaction space within the boundaries of the modeled interaction space, as well as to extrapolate to neighboring interaction spaces (e.g., taking advantage of similarities in domain and fold structures). Unification of the PCM models for several subfamilies will lead to large-scale proteochemometrics models (maps) covering the interaction space of whole protein superfamily classes and beyond.

etc.), only a negligible portion of all possible experiments for finding interaction partners can be performed. Accordingly, experiments must be optimally selected so as to minimize the number of experiments and maximize the information content of the data. Statistical experimental design (SED) has found increasing use in place of testing huge chemical libraries in drug design and could be applied to large-scale proteochemometrics as well. The purpose of SED is to plan and perform experiments in such a way that predictive models can be obtained over the experimental domain with as few experiments as possible. SED should provide estimates as to the influence of each experimental factor (which for PCM includes ligand, protein, and environmental descriptors) on the interaction activity [19].

To reach this goal, multivariate approaches can be used. Thus, PCA can be applied to features so as to obtain a small number of uncorrelated factors. All these factors are then varied simultaneously (rather than one at a time) in a set of experiments, and the data obtained are analyzed by PLS and similar multivariate methods [20].

SED can be divided into screening-phase designs, which include full-factorial, fractional-factorial, Placket–Burman, and D-optimal designs, in which simple models with information about dominating factors are obtained; and optimization or response-surface modeling designs, which include multilevel-factorial, central-composite, and D-optimal designs. In factorial designs, each factor is typically explored on two levels; the required number of observations in full-factorial design being 2^V, where V is the number of factors. In fractional-factorial designs the number of observations most often are between 2^{V-1} and 2^{V-4}. D-optimal design makes use of a computer-selected set of observations that maximizes the volume of factor space covered by the set. It is generally more practically useful than fractional-factorial designs, since D-optimal design is based on real observations, rather than on property combinations that may not be physically available. Detailed reviews of various design types are available [21].

Several studies have demonstrated that using SED enables the structural space of chemical libraries to be spanned by a small number of compounds. For example, the size of a peptoid library was reduced from over 2 billion to only 120 compounds, and in another study 19 peptides were sufficient to create a model for membrane partitioning of 640 000 possible sequences [22]. In a similar manner libraries of biopolymers and interacting entities can be designed, or selected from existing entities, so as to optimize the information content of the interaction space so that a practically manageable number of experiments covering the proteomes and genomes of the organisms can be performed.

In view of its size, the large-scale proteochemometrics project is not particularly suited to any single organization. It is, moreover, not a desired project for a proprietary organization, for many reasons. The pharmaceutical industry has shown a steady decline in productivity over the past decade, the major reasons being nonscientific and related to market considerations (e.g., blockbuster and merger philosophies leading to cancellation of many projects), legal issues (patents and other proprietary considerations precluding the initiation of many promising projects), and the sheer ineffectiveness of large organizations [23]. Rather, building proteochemometric maps on a large scale should be open and provide the infrastructure upon which future developments can be based with equal opportunity. Such maps would be used to increase efficiency in the development of target- (as well as multitarget-) selective pharmaceuticals and to aid in elucidating the functions of the genomes and proteomes. The iterative nature for the creation of these maps is also emphasized. Models are thus used to guide the selection of new experiments, which are then incorporated to improve the models (Figure 10.5). Ultimately, the large-scale proteochemometric map may become the prime source of information in biology and the infrastructure that developments in biomedicine and biotechnology will spring from.

Acknowledgments

Supported by the Swedish VR (04X-05957 and 621-2002-4711).

References

1 International Human Genome Sequencing Consortium, *Nature* **2001**, *409*, 860–921; R. Service, *Science* **2001**, *294*, 2074–2077; G. C. Roberts, C. W. J. Smith, *Curr. Opin. Chem. Biol.* **2002**, *6*, 375–383; P. M. Dean, E. D. Zanders, D. S. Bailey, *Trends Biotechnol.* **2001**, *19*, 288–292.

2 M. L. Lamb, W. L. Jorgensen, *Current Opin Chem.* Biol **1997**, *1*, 449–457; *CASP1–5*, can be found at http://predictioncenter.llnl.gov/; CASP4, *Proteins* **2001**, Suppl 5; N. E. Chayen, *Trends Biotechnol.* **2002**, *20*, 98.

3 P. Prusis, R. Muceniece, P. Andersson, C. Post, T. Lundstedt, J. E. S. Wikberg, *Biochim Biophys Acta* **2001**, *12*, 1544, 350–257; J. E.S Wikberg, F. Mutulis, I. Mutule, S. Veiksina, M. Lapinsh, R. Petrovska, P. Prusis, *Ann NY Acad Sci.* **2003**, *994*, 21–26; M. Lapinsh, P. Prusis, A. Gutcaits, T. Lundstedt, J. E. S. Wikberg, *Biochim Biophys Acta* **2001**, *16*, 180–190.

4 InCINC'94, *The ICS Symbol and Definition of Chemometrics*, can be found at http://www.eigenvector.com/about/definition.html **1994**; B. A. Rock, *An Introduction to Chemometrics: Definitions*, can be found at http://ourworld.compuserve.com/homepages/Catbar/chem_def.htm **1997**; S. Wold, *Chemom. Intell. Lab. Syst.* **1995**, *30*, 109–115; S. Wold, M. Sjöström, *Chemom. Intell. Lab. Syst.* **1998**, *44*, 3–14.

5 S. Wold, M. Sjöström, *Chemom. Intell. Lab. Syst.* **1998**, *44*, 3–14; J. Trygg, *Homepage of Chemometrics*, can be found at http://www.acc.umu.se/~tnkjtg/chemometrics/chemoinfo.html, **2003**.

6 M. Lapinsh, P. Prusis, I Mutule, F. Mutulis, J. E. S. Wikberg, *J. Med. Chem.* **2003**, *46*, 2572–2579.

7 P. Geladi, B. R. Kowalski, *Anal. Chim. Acta* **1986**, *185*, 1–17; L. Eriksson, E. Johansson, *Chemom. Int. Lab. Syst.* **1996**, *34*, 1–19; S. Wold, K. Esbensen, P. Geladi, *Chemom. Int. Lab. Syst.* **1987**, *2*, 37–52.

8 S. Hellberg, M Sjöström, B. Skagerberg, S. Wold, *J. Med. Chem.* **1987**, *30*, 1126–1135; M. Sandberg, L. Eriksson, J. Jonsson, M. Sjöström, S. Wold, *J. Med. Chem.* **1998**, *41*, 2481–2491.

9 S. Wold, J. Jonsson, M. Sjöström, M. Sandberg, S. Rännar, *Anal. Chim. Acta* **1993**, *277*, 239–253; M. Edman, T. Jarhede', M. Sjöström, Å. Wieslander, *Proteins* **1999**, *35*, 195–205; P. M. Andersson, M. Sjöström, T. Lundstedt, *Chemom. Intell. Lab.* **1998**, *42*, 41–50; Å. Nyström, P. M. Andersson, T. Lundstedt, *Quant. Struct.-Act. Relat.* **2000**, *19*, 264–269; M. Sjöström,

S. Rännar, Å. Wieslander, *Chemom. Intell. Lab.* **1995**, *29*, 295–305; M. Lapinsh, A. Gutcaits, P. Prusis, C. Post, T. Lundstedt, J. E. S. Wikberg, *Protein Sci.* **2002**, *11*, 795–805.

10 S. Wold, K. Esbensen, P. Geladi, *Chemom. Intell. Lab. Syst.* **1987**, *2*, 37–52; S. Wold, A. Ruhe, H. Wold, W. J. Dunn, *SIAM J. Sci. Stat. Comput.* **1984**, *5*, 735–743; S. Wold, M. Sjöström, *Chemom. Intell. Lab. Syst.* **1998**, *44*, 3–14; J. Gabrielsson, N.-O. Lindberg, T. Lundstedt, *J. Chemometrics* **2002**, *16*, 141–160; L. Eriksson, E. Johansson, S. Wold, in *Quantitative Structure–Activity Relationships in Environmental Sciences. VII.* (Eds. G. Schuurmann, F. Chen), SETAC, Pensacola FL **1997**, pp. 381–397; I. N. Wakeling, J. J. Morris, *J. Chemometrics* **1993**, *7*, 291–304; B. Efron, *J. Am. Stat. Assoc.* **1987**, *78*, 171–200; A. Golbraikh, A. Tropsha, *J. Mol. Graph. Model.* **2002**, *20*, 269–276; M. Baroni, G. Costadino, G. Cruciani, D. Riganelli, R. Valigi, S. Clementi, *Quant. Struct.-Act. Relat.* **1993**, *12*, 9–20; F. Westad, H. Martens, *J. Near Infrared Spec.* **2000**, *8*, 117–124; H. Martens, M. Martens, *Food Qual. Prefer.* **2000**, *11*, 5–16; A. Yasri, D. Hartsough, *J. Chem. Inf. Comp. Sci.* **2001**, *41*, 1218–1227; L. Xu, W.-J. Zhang, *Anal. Chim. Acta* **2001**, *446*, 477–483; J.-P. Gauchi, P. Chagnon, *Chemom. Int. Lab. Syst.* **2001**, *58*, 171–193; S. Wold, M. Sjöström, L. Eriksson, *Chemom. Int. Lab. Syst.* **2001**, *58*, 109–130; L. Eriksson, E. Johansson, F. Lindgren, M. Sjöström, S. Wold, *J. Comput. Aid. Mol. Des.* **2002**, *16*, 711–726.

11 P. Prusis, R. Muceniece, P. Andersson, C. Post T. Lundstedt J. E. S. Wikberg, *Biochim. Biophys. Acta* **2001**, *1544*, 350–357; M. Lapinsh, P. Prusis, A. Gutcaits, T. Lundstedt, J. E. S. Wikberg, *Biochim. Biophys. Acta* **2001**, 16, 180–190; P. Prusis, T. Lundstedt, J. E. S. Wikberg, *Protein Eng.* **2002**, 15, 305–311; M. Lapinsh, P. Prusis, T. Lundstedt, J. E. Wikberg, *Mol. Pharmacol.* **2002**, 61, 1465–1475.

12 (a) P. Prusis, R. Muceniece, P. Andersson, C. Post T. Lundstedt J. E. S. Wikberg, *Biochim. Biophys. Acta* **2001**, *1544*, 350–357; (b) H.B: Schiöth, P. Yook, R. Muceniece, J. E. S. Wikberg, M Szardenings, *J. Biol. Chem.* **1998**, *54*, 154–161; (c) M. Szardenings, S. Törnroth, R. Muceniece, K. Keinänen, A. Kuusinen, J. E. S. Wikberg, *J. Biol. Chem.* **1997**, *272*, 27943–27948; M. Szardenings, R. Muceniece, I. Mutule, F. Mutulis, J. E. S. Wikberg, *Peptides* **2000**, *21*, 239–243; R. Muceniece, I. Mutule, F. Mutulis, P. Prusis, M. Szardenings, J. E. S. Wikberg, *Biochim. Biophys. Acta* **2001**, *1544*, 278–282; (d) P. Prusis, T. Lundstedt, J. E. S. Wikberg, *Protein Eng.* **2002**, *15*, 305–311.

13 M. Lapinsh, P. Prusis, A. Gutcaits, T. Lundstedt, J. E. S. Wikberg, *Biochim Biophys Acta* **2001**, *16*, 180–190.

14 (a) M. Lapinsh, P. Prusis, T. Lundstedt, J. E. Wikberg, *Mol. Pharmacol.* **2002**, *61*, 1465–1475; (b) M. Pastor, G. Cruciani, I. McLay, S. Pickett, S. Clementi, *J. Med. Chem.* **2000**, *43*, 3233–3243.

15 M. Lapinsh, P. Prusis, I. Mutule, F. Mutulis, J. E. S. Wikberg, *J. Med. Chem.* **2003**, *46*, 2572–2579; F. Mutulis, I. Mutule, M. Lapins, J. E. S. Wikberg, *Bioorg. Med. Chem. Lett.* **2002**, *12*, 1035–1038; F. Mutulis, I. Mutule, J. E. S. Wikberg, *Bioorg. Med. Chem. Lett.* **2002**, *12*, 1039–1042.

16 (a) J. Pittman, J. Sacks, S. S. Young, *J. Chem. Inf. Comput. Sci.* **2002**, *42*, 729–741; (b) J. Friedman, *Ann. Stat.* **1991**, *19*, 1–141; M. Lapinsh, unpublished observations.

17 E. De Genst, D. Areskoug, K. Decanniere, S. Muyldermans, K. Andersson, *J. Biol. Chem.* **2002**, *277*, 29897–29907; K. Andersson, L. Choulier, M. D. Hamalaine, M. H. van Regenmortel, D. Altschuh, M. Malmqvist, *J. Mol. Recognit* **2001**, *14*, 62–71; L. Choulier, K. Andersson, M. D. Hamalainen, M. H. van Regenmortel, M. Malmqvist, D. Altschuh, *Protein Eng.* **2002**, *15*, 373–382; J. Zilliacus, A. P. Wright, U. Norinder, J. A. Gustafsson, J. Carlstedt-Duke, *J. Biol. Chem.* **1992**, *267*, 24941–24947; S. Tomic, L. Nilsson, R. C. Wade, *J. Med. Chem.* **2000**, *43*,

1780–1792; R. Ortiz, M. T. Pisabarro, F. Gago, R. C. Wade, *J. Med. Chem.* **1995**, *38*, 2681–2691.

18 G. D. Bader, D. Betel, C. W. Hogue, *Nucleic Acids Res.* **2003**, *31*, 248–250, database available at http://www.bind.ca/; I. Xenarios, L. Salwinski, X. J. Duan, P. Higney, S. M. Kim, D. Eisenberg, *Nucleic Acids Res.* **2002**, *30*, 303–305; X. J. Duan, I. Xenarios, D, Eisenberg, *Mol Cell Proteomics.* **2002**, *1*, 104–116, database available at http://dip.doe-mbi.ucla.edu/; K. Eilbeck, A. Brass, N. Paton, C. Hodgman, *Proc. Int Conf Intell Syst Mol Biol.* **1999**, 87–94; P. Prabakaran, J. An, M. Gromiha, S. Selvaraj, H. Uedaira, H. Kono, A. Sarai, *Bioinformatics* **2001**, *17*, 1027–1034; A, Sarai, M. M. Gromiha, J. An, P. Prabakaran, S. Selvaraj, H. Kono, M. Oobatake, H. Uedaira, *Biopolymers* **2002**, *61*, 121–126; X. Chen, Y. Lin, M. Liu, M. K. Gilson, *Bioinformatics* **2002**, *18*, 130–139; X. Chen, M. Liu, M. K. Gilson, *J. Combi. Chem. High-Throughput Screen* **2001**, *4*, 719–725, database can be found at http://www.bindingdb.org/; F. Horn, E. Bettler, L. Oliveira, F. Campagne, F. E. Cohen, G. Vriend, *Nucleic Acids Res.* **2003**, *31*, 294–297, database can be found at http://www.gpcr.org/ligand/ligand.html; B. L. Roth, W. K. Kroeze, S. Patel, E. Lopez, *The Neuroscientist* **2000**, *6*, 252–262, database can be found at http://pdsp.cwru.edu/pdsp.asp.

19 E. A. Jamois, *Curr. Opin. Chem. Biol.* **2003**, *7*, 326–330; E. A. Jamois, C. T. Lin, M. Waldman, *J. Mol. Graph. Modell.* **2003**, *22*, 141–149; G. Schneider, *Curr. Med. Chem.* **2002**, 9, 2095–2101; A. Linusson, J. Gottfries, T. Olsson, E. Ornskov, S. Folestad, B. Nordén, S. Wold, *J. Med. Chem.* **2001**, *44*, 3424–3439; P. M. Andersson, M. Sjöström, S. Wold, T. Lundstedt, *J. Chemom.* **2001**, *15*, 353–369; A. Linusson, S. Wold, B. Nordén, *Molecular Diversity* **1998–1999**, *4*, 103–114.

20 S. Wold, *J. Pharm. Biomed. Anal.* **1991**, *9*, 589–596; L. Eriksson, E. Johansson, *Chemom. Intell. Lab.* **1996**, *34*, 1–19.

21 P. F. de Aguiar, B. Bourguignon, M. S. Khots, D. L. Massart, R. Phan-Than-Luu, *Chemom. Intell. Lab.* **1995**, *30*, 199–210; T. Lundstedt, E. Seifert, L. Abramo, B. Thelin, Å. Nyström, J. Pettersen, R. Bergman, *Chemom. Intell. Lab.* **1998**, *42*, 3–40; E. Morgan, in *Chemometrics: Experimental Design*, John Wiley, New York **1991**, p. 275; L. Eriksson, E. Johansson, N. Kettaneh-Wold, C. Wikström, S. Wold, *Design of Experiments: Principles and Application*, Umetrics **2001**, p329.

22 A. Linusson, S. Wold, B. Nordén, *Molecular Diversity* **1998–1999**, *4*, 103–114; L. H. Alifrangis, I. T. Christensen, A. Berglund, M. Sandberg, L. Hovgaard, S. Frokjaer, *J. Med. Chem.* **2000**, *43*, 103–113.

23 T. Peakman, S. Franks, C. White, M. Beggs, *Drug Discovery Today* **2003**, *5*, 203–211; J. Drews, *Drug Discovery Today* **2003**, *8*, 411–420.

III
Chemical Libraries

Chemogenomics in Drug Discovery: A Medicinal Chemistry Perspective.
Edited by Hugo Kubinyi and Gerhard Müller

ISBN: 3-527-30987-X

11
Some Principles Related to Chemogenomics in Compound Library and Template Design for GPCRs

Thomas R. Webb

11.1
Introduction

Several reviews have appeared recently covering various topics that are relevant to the discipline encompassed by the term 'chemogenomics'. Despite this, there currently does not seem to be much agreement as to the exact meaning of this term, although it is generally agreed to be at the interface of cellular/molecular biology and medicinal chemistry [1–8]. This chapter uses a description set forth previously and considers "chemogenomics ... [as] biased towards the rapid identification of novel drugs and drug targets embracing multiple early phase drug discovery technologies ranging from target identification and validation over compound design and chemical synthesis to biological testing and physicochemical profiling" [1]. At the same time, we see that the ultimate goal of chemogenomics is "the discovery and description of all possible drugs to all possible targets" [2]. Clearly, pursuit of this goal will require significant improvements in computational and experimental methods, as well as new insights in data interpretation. A significant part of recent discussion in chemogenomics has focused on important computational methods for deriving information on the relationships between known targets and known ligands, in order to develop the knowledge, and the computational tools, to bridge the small-molecule world to the world of protein targets [3]. This chapter concentrates instead on the development of methods to generate novel templates using information about common features of some native ligands for GPCRs, with the expectation that more such small-molecule template design methods, and the resulting new ligands, will be crucial in developing the data that will help to establish a useful fundamental understanding of the structural relationships between small ligands and large protein families.

Recently, new comprehensive approaches have been developed that use compound libraries for screens in whole living cells using specific phenotype readouts. Some of the applications of this type of approach have been called chemical genetics or chemogenomics [9–27]. In many instances it is probably more accurate to call

Chemogenomics in Drug Discovery: A Medicinal Chemistry Perspective.
Edited by Hugo Kubinyi and Gerhard Müller

ISBN: 3-527-30987-X

these experiments chemical genetics, even though the data that is generated is relevant to chemogenomics. These chemical genetic approaches allow for the investigation of fundamental biology and the discovery of small-molecule tools, or drug leads simultaneously. By the application of specialized cell- or whole-organism-based screens, changes in the phenotypes can be observed to be modulated by specific compounds that are present in diverse compound libraries. With the advent of ultra high-throughput screening (ultra HTS), high-quality compound libraries have become increasingly available, and a significant number of these studies have appeared that rely on commercially available compound libraries such as DiverSet [9–27]. The increasing need for such libraries has led to a new discipline within organic synthesis that we now call 'diversity chemistry'. During the time that diversity chemistry has been developing, great advances have been made in molecular biology and cell-based screening techniques. The combined progress in both fields has led to the successful application of chemical genetics to numerous targets. To pursue these types of projects it is important that researchers have a practical understanding of recent advances in chemical diversity, chemical libraries, and appropriate bioassay technology.

11.2 Diverse Libraries versus Targeted Libraries

It has been reasonably argued that there is a need to refine the early conceptual approaches that were used to design diverse libraries [28]. In library design, if we just try to sample all possible diversity space, the resulting 'virtual chemistry space' (variously measuring up to 10^{60} compounds or more) is beyond all conceivable practical HTS or synthetic reach [29]. Various concepts have been developed to better define 'molecular diversity', including the Tanimoto index, BCUT, and pharmacophore-based diversity [30–32]. These methods attempt to evenly fill and select from 'diversity space', assuming that compounds that are very similar by some relevant measure will give correspondingly similar activity and are therefore redundant for the purposes of screening. Although these relatively early considerations are important, it is also arguable that they are incomplete, since the even filling of diversity space does not optimally match the pharmacophoric preferences of protein targets [33]. We should recognize that the filling of actual diversity space will ultimately be driven by synthetic accessibility and not by pharmacophore and physical property desirability, unless synthetic chemists have understandable practical guidelines for introducing such a bias. Based on these considerations, is it apparent that advanced diversity compound libraries should become more directed toward diverse compounds that are generally 'drug-like'. Initially this has meant compounds that contain the combination of calculated physiochemical or molecular properties (such as cLogP, rotatable bonds, total polar surface area, molecular weight, etc.) that are required of compounds that are expected to penetrate cell membranes and be orally active [34]. These considerations significantly reduce the size of chemical diversity space and are very useful in library design. Truly drug-like compounds should also be further defined as compounds that fill the relevant pharmacophore diversity space.

Other important considerations in library design had been developed from collected compounds and were eventually incorporated into combinatorial library design. Some desirable sample characteristics, such as high purity, lack of any highly reactive functionality, stability during storage, and synthetic scalability, may seem trivial, but are really of critical importance to researchers in chemogenomics, just as they are in high-throughput screening. Accumulated expertise in the design of substructure 'filters' that can be used to remove compound types that are notoriously problematic in bioassays has been used to develop both collected and combinatorial libraries that reduce the presence of false positives in screening. Additional considerations, such as the design of compounds with 'lead-like' characteristics, have also been introduced [35].

11.3
Design of Targeted Libraries via Ligand-based Design

In the area of library design, the concept of 'privileged geometries' that may lie behind some classes of 'privileged structures' has been proposed [36, 37]. I propose that an understanding of these atomic arrangements may allow for the design of novel privileged structures, which can be a source of valuable chemical diversity for the discovery of GPCR molecular tools and drug leads. The concept that implicitly lies behind targeted library design is that there is a relevant pharmacophore diversity space. The patterns of density that fill this space reflect the large, but limited, set of interaction types created by the restrictions imposed by the statistics of the probable (and possible) combination of contact types derived from the functional groups arrayed around the primary, secondary, and tertiary structures that proteins may present to assemble a recognition site for ligands. Furthermore, limitations on the molecular recognition of pharmacophore types by protein families may be imposed by the common features of the ligands that these protein families recognize. Although it is not currently obvious how to take advantage of the former observation to do library design, it is possible to take advantage of the latter fact so as to design libraries that target certain protein families. This type of ligand-based pharmacophore method has in fact been used to design kinase- and GPCR-targeted libraries [32, 38].

11.4
Ligand-based Template Design for GPCR-targeted Libraries

One of the most challenging subclasses in the chemogenomics of GPCRs is the receptors that recognize peptides as ligands [39]. This subclass may also offer some interesting opportunities for chemogenomic approaches via ligand-based methods, since there is sometimes a genetic relationship between the ligand–receptor pairs that may serve as a basis for predicting a ligand's bioactive conformation and, potentially, for predicting the common 'template' core that is recognized [40, 41]. A significant amount of research has been directed toward 'peptide mimics' for these and other targets [42–44]. This section focuses on recent progress and prospects in the design of nonpeptide (non-amino acid-based) templates that mimic the constraints imposed on small peptides by β turns and α helices; it does not focus on templates primarily intended to initiate secondary structures in peptides [45] or other types of protein mimetics [46]. Furthermore, the main topic is mimics of β turns and α helices, since numerous known ligands for GPCRs adopt solution structures containing these constraints [47, 48]. Some previous work on the design of templates mimicking β sheets [33] and β strands [45, 49] is relevant to the chemogenomics of protein families other than GPCRs [50], so it is not discussed here.

Some examples of important peptide ligands for GPCRs, in which β-turn constraints have been implicated in the bioactive conformation or in the native peptide solution structure, are angiotensin [51], bradykinin [52], cholecystokinin [53], melanocyte stimulating factor [54], and somatostatin [55]. Some examples of well known native GPCR ligands that may adopt a solution structure or a known bioactive helical structure are corticotrophin releasing factor [56], parathyroid hormone-related protein [57], neuropeptide Y [58], vasoactive intestinal peptide [59], and growth hormone releasing factor [40].

A useful empirically-based design concept is that of 'preferred structures' for GPCRs and other target families [36, 37]. An important challenge in this approach is the creative construction of new 'preferred structures' based on insights into the origin of the 'preference' [32]. One area that has been explored has been the design and construction of β-turn-mimicking templates [42]. Much of this effort has been guided by the design based on cyclic peptide structural analogs [60]. Such a design method may not be preferred, since such templates are primarily intended to maintain a very specific constraint and are not selected on the basis of desirable 'drug-like' physical properties. It is apparent that synthetic chemists need guiding principles for the design of nonpeptide peptide mimics when designing templates for 'relevant-diversity'-based libraries. At least one example is found in the seminal work of Garland and Dean [61, 62], who had the insight that the geometric

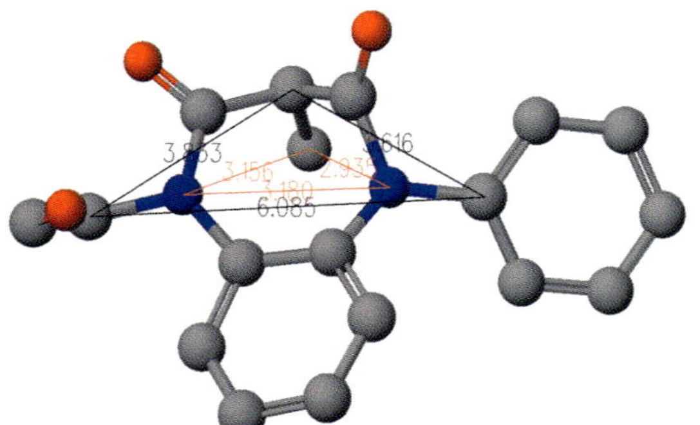

Figure 11.1 An example of one GPCR-preferred structure (the template from GW 5823) [53], showing the rigid conformation that matches two of the three different Garland–Dean constraints.

constraints imposed by peptide ligand conformations can be viewed as three specific C-α atom triangular geometries. They also showed that this geometry is found in certain preferred structures such as the benzodiazepines [61].

Our group has observed that some specific peptide-mimicking GPCR-preferred structures also exhibit the unusual combination of characteristics of matching one or more of the geometric constraints described by Garland and Dean, while at the same time being substituted at the pseudo C-α positions (Figures 11.1 and 11.2) [63]. The example CCK_1 agonist shown in Figure 11.1 has been particularly well studied [53] and is of particular interest in that the vast majority of nonpeptidic CCK_1 receptor agonists are benzodiazepines with substituents in the 1, 3, and 5 positions [53], which are the positions that precisely match the Garland–Dean geometry to create 'pseudo C-α' positions (Figures 11.1 and 11.2). The correct

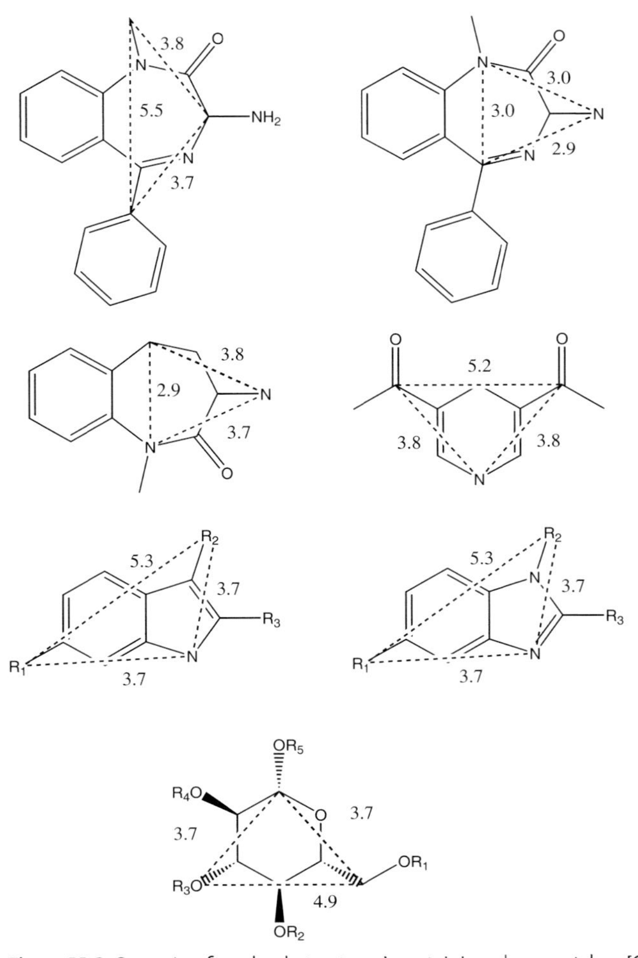

Figure 11.2 Some 'preferred substructures' containing close matches [64] to the Garland–Dean constraints at substituted positions [36, 37, 53, 54].

substitution pattern and precise Garland–Dean match are also seen with 1,4-benzodiazapine derivatives, as well with other 'preferred substructures' [36, 37, 53, 54] found in compounds active against numerous GPCR targets (Figure 11.2) [64]. We recently extended these observations to the design of novel template types that offer desirable physical and synthetic properties [63, 65]. Our design of synthetically tractable templates containing the correct geometry that can be modified to be suitable for substitution with reagents that mimic the sidechains of the natural amino acids is a stepwise process. First we searched > 450 000 compounds in our internal 3D database of library screening structures to find structural starting points, following the published method that was previously applied to a subset of the compounds in the Available Chemical Directory (ACD) [61]. The structures that matched were ranked, giving priority to molecules containing the fewest number of rotatable bonds. The top 50 structures were selected by inspection, for evaluation based on their suitability to redesign. In addition to the general types of structures that were previously reported [61], we also obtained matches with numerous bicyclic (and spirobicyclic) ring systems. The various ring systems that matched were then combined and modified with the introduction of heteroatoms, where needed, until structures that contained multiple Garland–Dean geometries in the same template with no internal rotatable bonds were designed and subjected to retrosynthetic analysis. After several iterations of design, modeling, synthetic analysis, and redesign, we developed several potential molecular targets including the novel general scaffold **1** (Figure 11.3).

Thus, we have developed a method for the design of nonpeptide templates, such as **1**, that can be elaborated to mimic the activity of peptides on a set of closely related receptors [63]. It was shown, by partial exploration of a few of the numerous combinatorial positional and substitution possibilities, that the selectivity and potency to even closely related receptors could be modulated. These examples indicate that this design paradigm, and templates such as **1**, show great potential for the exploration and discovery of new active pharmacophores and corresponding new chemical entities with activity against other protein families that recognize the β-turn motif. Additional exploration and refinement should also lead to more potent and selective compounds active against the somatostatin receptor and potentially to other receptors of interest. Recently, this approach was used to design other new templates, such as **2** (Figure 11.4, see page 320) [65]. Extension of the work to other derivatives of template **1** with the aim of targeting other peptide GPCRs is currently in progress in the laboratories of ChemBridge Corporation and ChemBridge Research Labs., LLC.

As previously mentioned, the design of small-molecule β-turn mimics has received significant attention [42, 44], and recent advancements have been made in evaluating the correct geometry of potential templates [66]. Surprisingly, mimicry of α-helical peptides has received relatively modest consideration from the standpoint of small-molecule or template design [67–69], although some effort has also been directed toward the development of α-helix initiators [70]. The templates that have been reported are highly hydrophobic and have not been designed with 'drug-like' properties in mind. Also so far, no 'stencils' have been elucidated to guide the

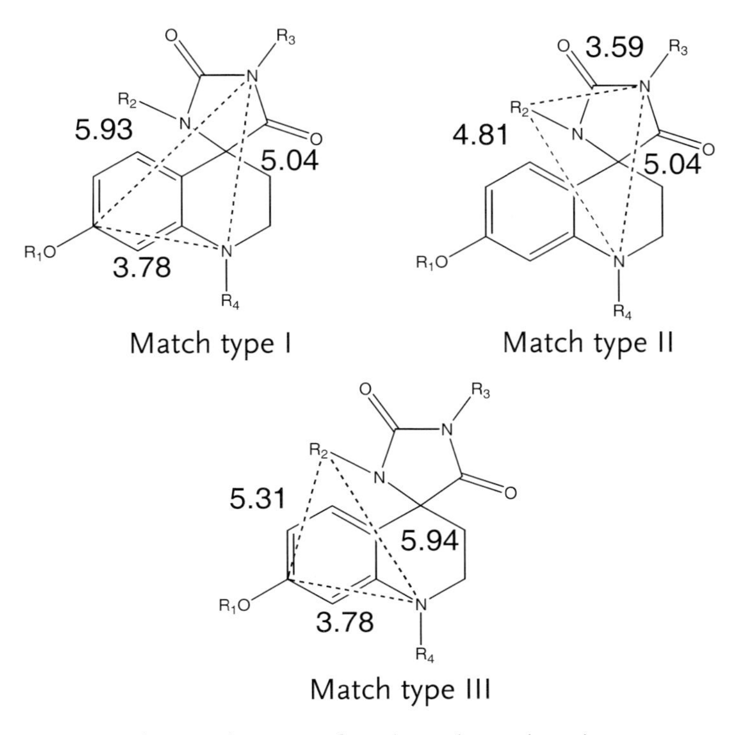

Figure 11.3 The general structure of template **1** along with overlays representing the atoms mimicking the positions of the C-α atoms and the distances in angstroms between these atoms. The Garland–Dean class-3 distances that are the best matches (with standard deviations in parentheses) are 5.42 (± 0.57), 3.82 (± 0.01), 5.44 (± 0.55).

design of new templates that can mimic the constraints imposed by an α helix. as was done for β turns. An analysis that would yield such a prototype would be very beneficial to chemists who are currently looking for principles to direct their synthetic efforts in a meaningful way. It is interesting to note that, in at least certain instances, the important bioactive sidechain interactions of a β turn (RGDF) may be closely mimicked by an α helix (RGYFDV) [71]. This means that the sidechain positioning in some α helices may be mimicked by an appropriately substituted nonpeptide β-turn template.

The need for new 'preferred templates' for GPCRs that offer intellectual property opportunities, as well as desirable physical and synthetic properties (which are all important drivers in the pharmaceutical industry), has generated an interest in developing guiding principles in preferred template design. Garland–Dean geometries have been observed in several 'preferred substructures'. Furthermore, these constraints can be used to design new nonpeptide templates that can be specifically elaborated to show some of the activity associated with the native peptide ligands. If insights similar to that of Garland and Dean can be derived from other

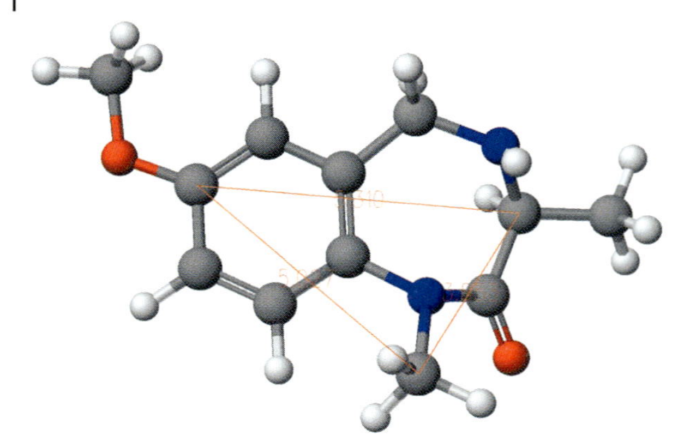

Figure 11.4 Template (2) designed by J.-C. Kim's laboratory [65], showing an overlay of the Garland–Dean coordinates on the C-α atom mimics.

secondary structures, then perhaps other elegant general guiding principles can be elucidated. Secondary structures such as the α helix are obvious candidates for the development of such guiding principles, since many important peptide ligands for GPCRs adopt this conformation in solution. Design work in the area of templates that mimic α helices suggests that the biphenyl group may owe its status as a 'preferred substructure', at least in part, to the fact that, when properly substituted, its derivatives can act as mimics of bioactive α-helical peptides. Clearly, there is a need for simple structural guiding principles to be elucidated and made available to diversity-oriented chemists. Collaborative approaches between computational chemists and synthetic chemists will be one of the keys for the advancement of chemogenomics.

References

1 K. H. Bleicher, Chemogenomics: bridging a drug discovery gap, *Curr. Med. Chem.* **2002**, *9*, 2077–2084.

2 P. R. Caron, M. D. Mullican, R. D. Mashal, K. P. Wilson, M. S. Su, M. A. Murcko, Chemogenomic approaches to drug discovery. *Curr. Opin. Chem. Bio.* **2001**, *5*, 464–470.

3 T. Klabunde, G. Hessler, Drug design strategies for targeting G-protein-coupled receptors, *ChemBioChem* **2002**, *3*, 928–924.

4 A. Schuffenhauer, P. Floersheim, P. Acklin, E. Jocoby, Similarity metrics for ligands reflecting the similarity of the target protein, *J. Chem. Inf. Comp. Sci.* **2003**, *43*, 391–405.

5 L. J. Browne, M. L. Furness, G. Natsoulis, C. Pearson, K. Jamagin, Chemogenomics: pharmacology with genomics tools, *Targets* **2002**, *1*, 59–65.

6 G. H. S. Ashton, J. A. McGrath, A. P. South, Strategies to identify disease genes, *Drugs of Today* **2002**, *38*, 235–244.

7 A. Schuffenhauer, J. Zimmermann, R. Stoop, J.-J. van der Vyver, S. Lecchini, E. Jacoby, An ontology for pharmaceutical ligands and its application for in silico screening and library design, *J. Chem. Inf. Comp. Sci.* **2002**, *42*, 947–955.

8 X. F. S. Zheng, T.-F. Chan, Chemical genomics: a systematic approach in biological research and drug discovery, *Curr. Issues Mol. Bio.* **2002**, *4*, 33–43.

9 Y. Bae, T. G. Lee, J. C. Park, J. H. Hur, Y. Kim, K. Heo, J. Kwak, P. Suh, S. H. Ryu, Identification of a compound that directly stimulates phospholipase C activity. *Molecular Pharmacology* **2003**, *63*, 1043–1050.

10 T. Ma, J. R. Thiagarajah, H. Yang, N. D. Sonawane, C. Folli, L. J. V. Galietta, A. S. Verkman, Thiazolidinone CFTR inhibitor identified by high-throughput screening blocks cholera toxin-induced intestinal fluid secretion, *J. Clin. Invest.* **2002**, *110*, 1651–1658.

11 T. J. F. Nieland, M. Penman, L. Dori, M. Krieger, T. Kirchhausen, Discovery of chemical inhibitors of the selective transfer of lipids mediated by the HDL receptor SR-BI. *Proc. Natl. Acad. Sci. USA* **2002**, *99*, 15422–15427.

12 J. K. Chen, J. Taipale, K. E. Young, T. Maiti, P. A. Beachy, Small molecule modulation of *smoothened* activity. *Proc. Natl. Acad. Sci. USA* **2002**, *99*, 14071–14076.

13 T. Ma, L. Vetrivel, H. Yang, N. Pedemonte, O. Zegarra-Moran, L. J. V. Galietta, A. S. Verkman, High-affinity activators of cystic fibrosis transmembrane conductance regulator (CFTR) chloride conductance identified by high-throughput screening, *J. Biol. Chem.* **2002**, *277*, 37235–37241.

14 T. A. Sohn, R. Bansal, G. H. Su, K. M. Murphy, S. E. Kern, High-throughput measurement of the Tp53 response to anticancer drugs and random compounds using a stably integrated. *Carcinogenesis* **2002**, *23*, 949–957.

15 A. Cheung, J. A. Dantzig, S. Hollingworth, S. M. Baylor, Y. E. Goldman, T. J. Mitchison, A. F. Straight, A small-molecule inhibitor of skeletal muscle myosin II. *Nat. Cell Bio.* **2002**, *4*, 83–84.

16 R. V. Kondratov, P. G. Komarov, Y. Becker, A. Ewenson, A. Gudkov, Small molecules that dramatically alter multidrug resistance phenotype by modulating the substrate specificity of P-glycoprotein. *Proc. Natl. Acad. Sci. USA* **2001**, *98*, 14078–14083.

17 R. T. Peterson, J. D. Mably, J. Chen, M. C. Fishman, Convergence of distinct pathways to heart patterning revealed by the small molecule concentramide and the mutation *heart-and-soul*. *Curr. Biol.* **2001**, *11*, 1481–1491.

18 T. A. Sohn, G. H. Su, B. Ryu, C. J. Yeo, S. E. Kern, High-throughput drug screening of the DPC4 tumor-suppressor pathway in human pancreatic cancer cells. *Annals of Surgery* **2001**, *233*, 696–703.

19 C. M. Grozinger, E. D. Chao, H. E. Blackwell, D. Moazed, S. L. Schreiber, Identification of a class of small molecule inhibitors of the sirtuin family of NAD-dependent deacetylases by phenotypic screening, *J. Biol. Chem.* **2001**, *276*, 38837–38843.

20 A. Degterev, A. Lugovskoy, M. Cardone, B. Mulley, G. Wagner, T. Mitchison, J. Yuan, Identification of small-molecule inhibitors of interaction between the BH3 domain and Bcl-xL. *Nat. Cell Biol.* **2001**, *3*, 173–182.

21 R. Peterson, B. A. Link, J. E. Dowling, S. L. Schreiber, Small molecule developmental screens reveal the logic and timing of vertebrate development. *Proc. Natl. Acad. Sci. USA* **2000**, *97*, 12965–12969.

22 T. Kim, T. Y. Kim, W. G. Lee, J. Yim, T. K. Kim, Signaling pathways to the assembly of an interferon-B enhanceosome, *J. Biol. Chem.* **2000**, *275*, 16910–16917.

23 G. H. Su, A. S. Taylor, R. Byungwoo, S. E. Kern, A novel histone deacetylase inhibitor identified by high-throughput transcriptional screening of a compound library. *Cancer Res.* **2000**, *60*, 3137–3142.

24 T. U. Mayer, T. M. Kapoor, S. J. Haggarty, R. W. King, S. L. Screiber, T. J. Mitchison, Small molecule inhibitor of mitotic spindle bipolarity identified in a phenotype-based screen. *Science* **1999**, *286*, 971–974.

25 P. G. Komarov, E. A. Komarova, R. V. Kondratov, K. Christov-Tselkov, J. S. Coon, M. V. Chernov, A. V. Gudkov, A chemical inhibitor of p53 that protects mice from the side

effects of cancer therapy. *Science* **1999**, *285*, 1733–1337.

26 B. R. Stockwell, J. S. Hardwick, J. K. Tong, S. L. Schreiber, Chemical genetic and genomic approaches reveal a role for copper in specific gene activation, *J. Amer. Chem. Soc.* **1999**, *121*, 10662–10663.

27 A. F. Straight, A. Cheung, J. Limouze, I. Chen, N. J. Westwood, J. R. Sellers, T. J. Mitchison, Dissecting temporal and spatial control of cytokines with a myosin II inhibitor. *Science* **2003**, *299*, 1743–1747.

28 G. Schneider, Trends in virtual combinatorial library design. *Curr. Med. Chem.* **2002**, *9*, 2095–2101.

29 W. P. Walters, M. T. Stahl, M. A. Murcko, Virtual screening – an overview. *Drug Disc. Today* **1998**, *3*, 160–178.

30 V. Gillet, Background theory of molecular diversity *Mol. Diversity Drug Dis.* **1999**, 43–65.

31 R. S. Pearlman, K. M. Smith, Software for chemical diversity in the context of accelerated drug discovery, *Drugs of the Future* **1998**, *23*, 885–895.

32 J. S. Mason, I. Morize, P. R. Menard, D. L. Cheney, C. Hulme, R. F. Labaudiniere, New 4-point pharmacophore method for molecular similarity and diversity applications: overview of the method and applications, including a novel approach to the design of combinatorial libraries containing privileged substructures *J. Med. Chem.* **1999**, *42*, 3251–3264.

33 A. B. Smith III, D. A. Favor, P. A. Sprengeler, M. C. Guzman, P. J. Carroll, G. T. Furst, R. Hirschmann, Molecular modeling, synthesis, and structures of *N*-methylated 3,5-linked pyrrolin-4-ones: toward the creation of a privileged nonpeptide scaffold. *Bioorg. Med. Chem.* **1999**, *7*, 9–22.

34 C. A. Lipinski, F. Lombardo, B. W. Dominy, P. J. Feeney, Experimental and computational approaches to estimate solubility and permeability in drug discovery and development settings. *Adv. Drug Del. Rev.* **1997**, *23*, 3–25.

35 M. M. Hann, A. R. Leach, G. Harper, Molecular complexity and its impact on the probability of finding leads for drug discovery. *J. Chem. Inf. Comput. Sci.* **2001**, *41*, 856–864.

36 D. A. Horton, G. T. Bourne, M. L. Smythe, The combinatorial synthesis of bicyclic privileged structures or privileged substructures. *Chem. Rev.* **2003**, *103*, 893–930.

37 A. A. Patchett, R. P. Nargund, Privileged structures: an update. *Ann. Reports Med. Chem.* **2000**, *35*, 289–298.

38 T. R. Webb, D. Lvovskiy, M. C. Heinrich, J. W. H. Yee, Discovery of novel inhibitors of tyrosine kinases. Abstracts, American Chemical Society (2001), 221st MEDI-228.

39 G. R. Marshall, Peptide interactions with G-protein coupled receptors. *Biopolymers.* **2001**, *60*, 246–277.

40 R. M. Campbell, C. G. Scanes, Evolution of the growth hormone-releasing factor (GRF) family of peptides. *Growth Regul.* **1992**, *2*, 175–191.

41 V. J. Hruby, G. Han, M. E. Hadley, Design and bioactivities of melanotropic peptide agonist and antagonists: design based on a conformationally constrained somatostatin template. *Lett. Pept. Sci.* **1998**, *5*, 117–120.

42 J. B. Ball, P. R. Alewood, Conformational constraints: nonpeptide β-turn mimics, *J. Mol. Rec.* **1990**, *2*, 55–64.

43 G. L. Olson, D. R. Bolin, M. P. Bonner, M. Bos, C. M. Cook, D. Fry, B. J. Graves, M. Hatada, D. E. Hill, M. Kahn, V. S. Madison, V. K. Rusiecki, R. Sarabu, J. Sepinwall, G. P. Vincent, M. E. Voss, Concepts and progress in the development of peptide mimetics, *J. Med. Chem.* **1993**, *21*, 3041–3049.

44 A. E. P. Adang, P. H. H. Hermkens, J. T. M. Linders, H. C. J. Ottenheijm, C. J. van Staveren, Case histories of peptidomimetics: progression from peptides to drugs. *Recl. Trav. Chem. Pays-Bas* **1994**, *113*, 63–78.

45 D. S. Kemp, Peptidomimetics and the template approach to nucleation of beta-sheets and alpha-helices in peptides. *Trends Biotechnol.* **1990**, *8*, 249–255.

46 T. Sasaki, M. Lieberman, Protein mimetics, in *Comprehensive*

Supramolecular Chemistry, Y. Murakami (Ed.), Pergamon: Oxford **1996**, Vol. 4, 193–242.

47 V. J. Hruby, P. M. Balse, Conformational and topographical considerations in designing agonist peptidomimetics from peptide leads. *Curr. Med. Chem.* **2000**, *7*, 945–970.

48 C. S. C. Wu, A. Hachimori, J. T. Yang, Lipid-induced ordered conformation of some peptide hormones and bioactive oligopeptides: predominance of the helix over the beta-form. *Biochemistry* **1982**, *21*, 4556–4562.

49 J. S. Nowick, M. Pairish, I. Q. Lee, D. L. Holmes, J. W. Ziller, An extended β-strand mimic for a larger artificial β-sheet, *J. Am. Chem. Soc.* **1997**, *119*, 5413–5424.

50 C. Nguyen, J. L. Teo, A. Matsuda, M. Eguchi, E. Y. Chi, W. R. Henderson Jr., M. Kahn, Chemogenomic identification of Ref-1/AP-1 as a therapeutic target for asthma. *Proc. Natl. Acad. Sci. USA* **2003**, *100*, 1169–1173.

51 A. G. Tzakos, A. M. Bonvin, A. Troganis, P. Cordopatis, M. L. Amzel, I. P. Gerothanassis, N. A. van Nuland, On the molecular basis of the recognition of angiotensin II (AII): NMR structure of AII in solution compared with the X-ray structure of AII bound to the mAb Fab131, *Eur. J. Biochem.* **2003**, *270*, 849–860.

52 M. Miskolzie, L. Gera, J. M. Stewart, G. Kotovych, The importance of the *N*-terminal beta-turn in bradykinin antagonists, *J. Biomol. Struct. Dyn.* **2000**, *18*, 249–260.

53 C. Giragossian, E. E. Sugg, J. R. Szewczyk, D. F. Mierke, Intermolecular interactions between peptidic and nonpeptidic agonists and the third extracellular loop of the cholecystokinin 1 receptor, *J. Med. Chem.* **2003**, *46*, 3476–3482.

54 S. Z. Li, J. H. Lee, W. Lee, C. J. Yoon, J. H. Baik, S. K. Lim, Type I beta-turn conformation is important for biological activity of the melanocyte-stimulating hormone analogues, *Eur. J. Biochem.* **1999**, *265*, 430–440.

55 L. Abrous, J. Hynes Jr., S. R. Friedrich, A. B. Smith III, R. Hirschmann, Design and synthesis of novel scaffolds for drug discovery: hybrids of β-D-glucose with 1,2,3,4-tetrahydrobenzo[*e*] [1,4]diazepin-5-one, the corresponding 1-oxazepine, and 2-and 4-pyridyldiazepines, *Org. Lett.* **2001**, *3*, 1089–1092.

56 J. Rivier, S. L. Lahrichi, J. Gulyas, J. Erchegyi, S. C. Koerber, A. G. Craig, A. Corrigan, C. Rivier, W. Vale, Minimal-size, constrained corticotropin-releasing factor agonists with $i-(i+3)$ Glu–Lys and Lys–Glu bridges, *J. Med. Chem.* **1998**, *41*, 2614–2620.

57 S. Maretto, S. Mammi, E. Bissacco, E. Peggion, A. Bisello, M. Rosenblatt, M. Chorev, D. F. Mierke, Mono- and bicyclic analogs of parathyroid hormone-related protein. 2. Conformational analysis of antagonists by CD, NMR, and distance geometry calculations, *Biochemistry* **1997**, *36*, 3300–3307.

58 D. A. Kirby, J. H. Boublik, J. E. Rivier, Neuropeptide Y: Y1 and Y2 affinities of the complete series of analogues with single D-residue substitutions, *J. Med. Chem.* **1993**, *36*, 3802–3808.

59 K. Haghjoo, P. W. Cash, R. S. Farid, B. R. Komisaruk, F. Jordan, S. S. Pochapsky, Solution structure of vasoactive intestinal polypeptide (11–28)-NH_2, a fragment with analgesic properties. *Pept. Res.* **1996**, *9*, 327–331.

60 M. Qabar, J. Urban, C. Sia, M. Klein, M. Kahn, Peptide secondary structure mimetics: applications to vaccines and pharmaceuticals. *Farmaco* **1996**, *51*, 87–96.

61 S. L. Garland, P. M. Dean, Design criteria for molecular mimics of fragments of the beta-turn. 1. C alpha atom analysis, *J. Comput.-Aided Mol. Des.* **1999**, *13*, 469–483.

62 S. L. Garland, P. M. Dean, Design criteria for molecular mimics of fragments of the beta-turn. 2. C alpha–C beta bond vector analysis, *J. Comput.-Aided Mol. Des.* **1999**, *13*, 485–498.

63 D. Chianelli, Y.-C. Kim, D. Lvovskiy, T. R. Webb, Application of a novel design paradigm to generate general nonpeptide combinatorial scaffolds mimicking beta turns: synthesis of ligands for somatostatin receptors, *Bioorg. and Med. Chem.* **2003**, *11*, 5059–5068.

64 Measurements on MM2-minimized structures were done with CAChe for Windows and with Chem-3D Pro.

65 I. Im, T. R. Webb, Y. D. Gong, J.-I. Kim, J.-C. Kim, Solid phase synthesis of tetrahydro-1,4-benzodiazepine-2-one derivatives as a beta-turn peptidomimic library, *J. Comb. Chem.* **2004**, *6*, 207–213.

66 G. Müller, G. Hessler, H. Y. Decornez, Are beta-turn mimetics mimics of beta-turns? *Angew. Chem., Int. Ed.* **2000**, *39*, 894–896.

67 D. C. Horwell, W. Howson, W. P. Nolan, G. S. Ratcliffe, D. C. Rees, H. M. G. Willems, The design of dipeptide helical mimetics. I. The synthesis of 1,6-disubstituted indanes, *Tetrahedron* **1995**, *51*, 203–216.

68 D. C. Horwell, W. Howson, G. S. Ratcliffe, H. M. G. Willems, The design of dipeptide helical mimetics: the synthesis, tachykinin receptor affinity and conformational analysis of 1,1,6-trisubstituted indanes, *Bioorg. and Med. Chem.* **1996**, *4*, 33–42.

69 B. P. Orner, J. T. Ernst, A. D. Hamilton, Toward proteomimetics: terphenyl derivatives as structural and functional mimics of extended regions of an alpha-helix, *J. Am. Chem. Soc.* **2001**, *123*, 5382–5383.

70 D. S. Kemp, T. P. Curran, J. G. Boyd, T. J. Allen, Studies of *N*-terminal templates for alpha-helix formation: synthesis and conformational analysis of peptide conjugates of (2*S*,5*S*,8*S*,11*S*)-1-acetyl-1,4-diaza-3-keto-5-carboxy-10-thiatricyclo[2.8.1.04,8]tridecane (Ac-Hel1-OH), *J. Org. Chem.* **1991**, *56*, 6683–6697.

71 G. Mer, E. Kellenberger, J. F. Lefevre, Alpha-helix mimicry of a beta-turn, *J. Mol. Biol.* **1998**, *281*, 235–240.

12
Computational Filters in Lead Generation: Targeting Drug-like Chemotypes

Wolfgang Guba and Olivier Roche

12.1
Introduction

The pharmaceutical industry is struggling with a productivity gap, which is reflected in a reduced number of drug applications to the regulatory authorities, an increase in development times, and sharply rising costs for development [1, 2]. The 'fail early, fail cheap' paradigm requires the early elimination of compounds with pharmacokinetic or toxic liabilities and with a low potential for optimizing potency or selectivity. Knowledge-based decisions about advancing or dropping lead candidates as early as possible in the drug-discovery chain require a systematic approach to gathering and analyzing relevant data for the assessment of compounds.

Therefore, lead generation groups have been established with a mission to discover high-quality leads with a balanced pharmacodynamic and ADME (absorption, distribution, metabolism, absorption) profile [3, 4]. The goal of lead generation is to submit high-content leads to a full-scale lead-optimization program with a high chance of success. However, there will always be a risk of unexpected failures and, therefore, multiple lead series differing in chemotype and pharmacological profiles are required for each target to decrease the risk to the project portfolio.

Computational chemistry has a pivotal role in the interface between chemistry, biology and the corporate data warehouse in extracting relevant information, recognizing correlations between chemical and biological data, and deriving guidelines for knowledge-based decisions. The systematic exploration of similarity relationships between molecular targets and pharmacologically active ligands is in the realm of chemogenomics [5–9]. The efficiency of hit finding is enhanced by organizing targets into protein superfamilies (GPCRs, ion channels, nuclear hormone receptors, proteases, kinases, phosphatases), by identifying the molecular recognition principles, and by matching these generic recognition motifs with chemical master keys (privileged structures) [10, 11]. The common theme for the systematization of chemotype discovery can be summarized as 'similar pharmacophores bind to similar targets'.

Chemogenomics in Drug Discovery: A Medicinal Chemistry Perspective.
Edited by Hugo Kubinyi and Gerhard Müller

ISBN: 3-527-30987-X

Computational strategies for addressing the drug-like character of chemotypes are the topic of this chapter. The emphasis is not on a compilation of available in-silico filters, for which excellent reviews [12–20] are available, but rather, on their pragmatic application within the lead-generation process. Computational filters are categorized into two classes, i.e., 'hard filters', which actually remove compounds from further progression in the lead-generation process, and 'soft filters', which raise alerts for potential liabilities of molecules. The applicability and pitfalls of property-prediction schemes are highlighted, and techniques for using the results of a multivariate in-silico profiling to rank compounds are demonstrated.

12.2 Hard Filters

In the lead-generation process, initial hits are first validated and a limited SAR is generated around each hit class. Compounds meeting these criteria, i.e., qualified hits, are then progressed to lead series, which require further optimization and fine-tuning of pharmacodynamic and pharmacokinetic properties in the lead-optimization process. Lead compounds, therefore, do not necessarily have to display the targeted properties of the final drug, but, nevertheless, a balanced property profile needs to be present to allow for further lead optimization without driving either potency or selectivity and pharmacokinetic properties into an unacceptable range.

Hard filters are the most stringent in-silico filters and are used to shape the property profile of screening or combinatorial libraries and to prune hit lists from primary screening. They are derived from 1D and 2D molecular properties (molecular weight, number of H-bond donors/acceptors, number of rotatable bonds, and so forth) and, as is described below, they are commonly used to reduce the number of false positive hits and to favor lead-like or drug-like chemotypes.

12.2.1 Reducing the Number of False Positive Hits

Reactive, unstable compounds, as well as covalent binders, can be removed from screening collections by substructure searches [21, 22]. At Roche, a global team of experienced medicinal chemists has defined more than 100 functionalities which are reviewed at regular intervals. This list has been augmented by unwanted features (e.g., polyacids, alkyl aldehydes, polyhalogenated phenols, etc.) which are chemically unattractive starting points for a hit-to-lead optimization, because they often result in non-optimizable SAR patterns. These chemotypes have been coded into Markush-type substructures for automated detection and removal of unwanted compounds. However, we need to stress that these filters are fully customizable, and removed chemotypes can be restored if required.

Another source of false positive hits is promiscuous binders, which were shown by McGovern et al. [23–26], via dynamic light scattering, to form aggregates above the critical micelle concentration. These aggregates do not compete with the binding of the natural ligand but prevent access to the binding site by clustering around the target. This is reflected by the lack of a structure–activity relationship (SAR) and by unselective binding to a wide range of targets.

A different approach to detecting promiscuous compounds and to developing an in-silico filter was implemented by Roche et al. [27]. In a first step, frequent hitters were retrieved from the corporate database by identifying molecules that were among the best 1000 hits across 161 high-throughput screening (HTS) assays more than 8 times or that were requested by more than 6 different drug-discovery projects for follow-up testing. A panel of 11 independent teams of medicinal chemists from all major Roche research sites was asked to assess the 'frequent hitter' potential of these compounds, based on their intuition and expert knowledge. Thus, the final votes were based on medicinal chemistry expertise. This approach not only identifies promiscuous compounds based on the formation of aggregates but, in addition, molecules that perturb assays or detection methods are also recognized. The empirically identified 'frequent hitter' compounds are flagged in the corporate database, and this annotation is regularly reviewed and updated.

Based on this in-house dataset, an in-silico prediction model [27] (three-layered neural network, Ghose and Crippen [28, 29] descriptors) was constructed to evaluate the frequent hitter potential before compound libraries are purchased or synthesized. This model was validated with a dataset of the above-mentioned promiscuous ligands published by McGovern et al. [26], in which 25 out of 31 compounds were correctly recognized.

At this point, we need to emphasize that an automated prediction tool for frequent hitters does not necessarily distinguish between an unwanted promiscuous binder and a privileged structure that might be transformed into a selective ligand by further decoration of the 'chemical masterkey' [11]. Therefore, the in-silico tool should be used only for flagging compounds, and the final decision about eliminating a molecule should be left to medicinal chemistry expertise. On the other hand, the strategy outlined above for identifying frequent hitters in the corporate database yields a valid rationale for excluding chemotypes in the screening collection from further follow-up.

12.2.2 Lead-likeness, Drug-likeness

Following a knowledge-driven deductive approach, the molecular structures and properties of marketed oral drugs have been analyzed to derive common structural patterns and molecular property ranges [30–32]. Similarly, drugs were compared with the lead structures from which they originated [33–36]. Analysis of lead–drug pairs yields information about the differences in molecular properties between leads and drugs with important implications for library design.

In two landmark publications, Lipinski [30, 31] related drug-likeness to the balance between potency, permeability, and the required level of solubility for oral absorption. The higher the potency of a compound, the lower is the required solubility for a given permeability, and vice versa. A low permeability needs to be compensated by a correspondingly higher solubility. Lipinski's simple mnemonic 'Rule of 5' (Ro5) predicts that poor absorption or permeation is more likely if two or more of the following criteria are met (exceptions: antibiotics, antifungals, substrates for transporters): > 5 H-bond donors (OH, NH), molecular weight > 500, clogP > 5, > 10 H-bond acceptors (sum of N and O). These rules were derived to encompass 90% of 2245 drugs in the World Drug Index and were recently reevaluated by Wenlock et al. [32], who determined somewhat tighter limits. Oprea [35] and Viswanadhan et al. [37] pointed out that the Ro5 do not discriminate well between drug and nondrug datasets and suggested both additional descriptors (e.g., number of rings, rotatable bonds, polar surface area, etc.) and tighter property ranges. Nevertheless, there is general agreement that high molecular weight and lipophilicity are very rare in oral drugs, due to low solubility and permeability, extensive metabolism, and toxic liabilities via the formation of reactive metabolites. The purpose of the Ro5 and related filters is not to identify a drug candidate but to guide both library design and the selection of chemotypes toward a drug-like physicochemical property profile.

A comparison of the property profiles of lead–drug pairs [32–34, 38] showed that the process of optimizing a lead into a drug results in more complex structures, i.e., higher molecular weight and lipophilicity. To allow for an increase in molecular weight and lipophilicity during lead optimization, Oprea et al. [33, 34, 38] suggested more stringent criteria than contained in the Ro5. An upper limit of 350 for molecular weight, a clogP range between 1 and 3, and the presence of a maximum of one charge (preferably a secondary or tertiary amine) are recommended.

The concept of lead-likeness has been supported by a recent theoretical study by Hann et al. [39], in which increasing complexity of ligands is correlated with a reduced chance for a successful match with the receptor during the molecular recognition event. They concluded that less-complex molecules are more suitable starting points for the discovery of drugs than highly functionalized compounds. As a metric for molecular complexity, the number of bits set in the Daylight [40] fingerprint was suggested.

Summing up, there is no definition for a good lead, but there seems to be a consensus on nonlead-like or nondrug-like properties. The most commonly applied molecular descriptors for assessing drug- or lead-likeness are molecular weight, number of H-bond donors and acceptors, number of rings, lipophilicity, and the number of rotatable bonds. Although these metrics are used as hard filters to eliminate compounds from further progression, the absolute limits have to be set taking into account the number of hits and the required physicochemical property profile for a given target (e.g., protease vs. lipid-binding G-protein coupled receptor). Before actually eliminating molecules, it is absolutely essential to check the list of discarded compounds and to adjust the settings of the filters accordingly. Further criteria for pruning hit lists are patentability and synthetic accessibility, but this assessment requires expert knowledge and presently cannot be automated.

Apart from analyzing molecular property ranges of oral drugs, pattern-recognition methods for predicting drug-likeness have been used to extract features from successful drugs and to embed this information in prediction models. Both Ajay et al. [41] and Sadowski and Kubinyi [42] described the first uses of neural networks for classifying molecules into drug-like and nondrug-like compounds. The frequency of correct classification is about 80%, but these models have a 'black box' character, since the classification result cannot be traced back to the molecular structure. In addition, the definition of a nondrug training dataset is far from trivial, because nondrug-likeness is inferred from the fact that the respective compound does not have a drug history. Thus, prediction of drug-likeness based on the similarity of a molecule to known drugs has to be differentiated from the customizable definition of lead- or drug-like molecular property ranges. The former should be used only as a soft filter, the latter can be used for biasing the design of compound libraries toward a favorable physicochemical property profile.

12.3 Soft Filters

Whereas hard filters can be considered to be knowledge-driven, soft filters are the result of a data-driven approach. A quantitative structure–activity or structure–property relationship (QSAR/QSPR) is established to predict a property from a set of molecular descriptors. Examples are the above-mentioned in-silico prediction tools for frequent hitters [27] and drug-likeness [41, 42]; additional models for ADME properties are described below.

No matter whether linear or nonlinear algorithms are used as the statistical engine in prediction tools, at this point the medicinal chemist can no longer influence the result by adjusting parameters. Compounds are either categorized into classes (e.g., drug-like, nondrug-like) or a property is predicted (e.g., bioavailability). All models contain errors, and valuable chemotypes might be eliminated for the wrong reason. Therefore, soft filters are to be used only as flags to raise alerts, and we will show how to deal with multiple flags for prioritizing compounds. The most common soft filters deal with ADME, toxicity, and physicochemical properties and have been summarized in an excellent review by van de Waterbeemd and Gifford [43]. Thus, only a short overview is given here from our perspective.

12.3.1 Prediction of Physicochemical Properties

Lipophilicity is a key parameter influencing membrane permeability, drug absorption, distribution, and clearance. Several algorithms for logP calculations are available, with each approach showing weaknesses for different structural classes [44]. Thus, multivariate profiling with logP prediction programs is recommended to compensate for errors. The distribution coefficient logD is physiologically more relevant than logP, because the charge state is considered at a given pH (e.g., blood pH 7.4, intestinal pH 6.5). However, the available pKa calculators are not equally well parameterized for all structural fragments, and the error may be unacceptably high. The situation is even worse for solubility prediction: the effects of counter-ions and crystal forms are taken into account only indirectly, by using the experimental melting point as an additional parameter, which very often is not available. At present, there are no methods available to reliably predict the pharmaceutically relevant solubility range of up to 100 μg mL^{-1}.

12.3.2 Prediction of ADME and Toxicity Properties

One of the fundamental assumptions in modern science is the applicability of reductionism to the understanding of complex systems. Complex systems are broken down into a set of simpler components that are modeled separately, and it is assumed that the whole system can be simulated by linking together the less complex parts. Along the lines of reductionist thinking, the complex problem of understanding the molecular basis of pharmacokinetics is approached by reducing it to the sum of simpler components, i.e., absorption, distribution, metabolism, and excretion, which themselves are traced to fundamental physicochemical properties (molecular weight, pKa, solubility, lipophilicity, polar surface area, permeability). Molecular structure descriptors and derived properties are correlated with ADMET (ADME and toxicity) data to generate stable, predictive in-silico models, so as to bias compound libraries towards a favorable pharmacokinetic profile avoiding toxic liabilities. However, we are still far from 'prediction paradise' [43] and, apart from the reductionist approach, which neglects synergistic interactions between system components, the major problems are limitations of the available datasets with respect to the number of data points and the representativeness of molecular structures. Nevertheless, from our perspective, some of the ADMET models have proven to be useful for in-silico profiling within the lead-generation process, and they are listed below.

Both intestinal and brain–blood barrier (BBB) permeation by passive diffusion are predicted reasonably well by the polar surface area (PSA) of a molecule [45–47]. For intestinal absorption the PSA should be < 140 Å^2, and for the tighter BBB the maximum PSA is between 80 and 100 Å^2. Calculation of the PSA requires a 3D structure, but the calculation speed can be greatly increased by the recently published topological PSA (tPSA) approach [48]. The tPSA is calculated from SMILES strings by adding increments for oxygen and nitrogen atoms in their respective topological environments.

A recent study by Kratochwil et al. [49]. critically reviewed the implications of plasma protein binding for the design of lead-like compounds. With the exception of chemotherapeutics (including antibiotic, antiviral, antifungal, and anticancer drugs), in which 77.2% of the compounds display a protein binding below 90%, no clear trend could be observed for other therapeutic classes. The relevance of drug displacement interactions to the clinical efficacy of drugs is often overestimated. However, protein binding values are important for establishing potential safety margins for human exposure (allometric scaling) and for selecting the final dose range for human trials. For this purpose it is essential to differentiate between submicromolar and nanomolar binders, which corresponds to the range between 99.0% and 99.99% protein binding. Therefore, Kratochwil et al. [49] used pharmacophoric similarity descriptors to develop a computational model for the prediction of drug association constants to human serum albumin. The prediction error is comparable to the experimental error, and the calculated binding constants may assist, in combination with other ADME soft flags, in the prioritization of chemotypes.

For modeling pharmacokinetic properties that are closely linked with physicochemical parameters, e.g., oral bioavailability, BBB permeability, and Caco-2 cell permeability, the VolSurf [50] descriptors have been successful in yielding stable, predictive models [51–55]. From GRID [50] molecular interaction fields with the water, the hydrophobic DRY, and optional H-bond donor and -acceptor probes, the size of polar and hydrophobic interaction sites and their spatial distribution is calculated without the need for superposition of 3D molecular structures. Thus, molecular structures are translated into a set of molecular surface descriptors encoding size, shape, lipophilicity, and H-bonding properties. Recently, a chemical space navigation system (ChemGPS, chemical global positioning system) based on VolSurf [50] descriptors was introduced [56–59], which allows for a global ranking of compounds with respect to solubility and permeability. Thus, ChemGPS can be considered as an in-silico analog of the Biopharmaceutics Classification System (BCS), which is recommended by the U. S. Food and Drug Administration (FDA) [60]. The BCS categorizes drugs into four different classes according to combinations of high or low solubility and high or low permeability, and indicates whether the oral bioavailability of a molecule is limited by solubility and/or permeability.

Early assessment of a potential hERG liability is becoming increasingly important in the lead-generation process, since QT prolongation via inhibition of the K^+ hERG channel is associated with cardiac arrhythmia and sudden death [61–63]. The experimental determination of the potency of K^+-channel inhibition via patch–clamp electrophysiology measurements is very time-consuming and has low throughput, therefore, reliable in-silico models are needed. The predictivity of all published hERG models is impaired by limited training datasets [64, 65]. Thus, dynamic refinement with newly generated experimental data is essential to allow for a more universal application of these models. Computational chemists are the driving force behind enlarging the scope of the presently limited in-silico filters and enhancing the efficiency of spotting problematic chemotypes early in the drug-development process.

12.4 Prioritization of Chemotypes Based on Multivariate Profiling

The development of predictive models for drug-likeness, frequent hitters, ADME processes, and toxicological endpoints has so far yielded a great deal of soft filters (see discussion above and the compilation of ADMET computational models by Yu and Adedoyin [66]), and the trend still continues to improve both accuracy and general applicability. Different algorithms and datasets often result in several in-silico models for the same property, e.g., BBB permeation or bioavailability. Since none of these models is a truly global model with equal predictive power for diverse chemotypes, a comparison of the variability of results allows their pragmatic value in drug-discovery projects to be judged.

The medicinal chemist is usually left alone with the task of prioritizing compounds that are often characterized with far more than 10 different computational and experimental parameters representing pharmacodynamic, pharmacokinetic, and physicochemical properties.

The most commonly adopted approach is to rank compounds according to the total number of alerts. If molecules are to be prioritized with respect to properties such as permeability, solubility, and so forth, the goal is to find the best compromise between these continuous parameters. For this purpose, each property is scored by assigning a dimensionless scale d_i between zero ('criterion not met') and one ('target achieved'), with intermediate values being obtained by interpolation between the minimum and maximum for each property. The score profile for each compound is summarized into an overall desirability D by calculating the geometric mean of the individual d_i values [67, 68]:

$$D = (d_1\ d_2\ d_3\ \ldots\ d_n)^{1/n}$$

The desirability D is high if all the individual scores are high as well, and it decreases if individual scores are low. If only one property is in an unacceptable range, the compound gets a total score of zero and is eliminated from a further progression. The final ranking is obtained by sorting according to the desirability D indicating how well molecules meet the targeted profile.

The two methods described above are conceptually simple and easy to implement. However, a great deal of information is lost, because a multivariate profile is condensed into a single number. The analysis of property profiles is analogous to consumer marketing studies in which products are ranked by consumers. A data matrix is generated with one row for each product and one column for each consumer. The matrix elements represent the ranking of a given consumer for the respective product. Preference mapping [69] is used to reveal which products are favored or rejected by the consumers, and whether there are different groups of consumers or individuals with diverse opinions. Within the context of this chapter, 'products' correspond to molecules and 'consumers' to calculated and experimental properties.

Preference mapping can be accomplished with projection techniques such as multidimensional scaling and cluster analysis, but the following discussion focuses on principal components analysis (PCA) [69] because of the interpretability of the results. A PCA represents a multivariate data table, e.g., N rows ('molecules') and K columns ('properties'), as a projection onto a low-dimensional table so that the original information is condensed into usually 2–5 dimensions. The principal components scores are calculated by forming linear combinations of the original variables (i.e., 'properties'). These are the coordinates of the objects ('molecules') in the new low-dimensional model plane (or hyperplane) and reveal groups of similar molecules, trends, and outliers. The scores are sorted in descending importance, i.e., the first principal component t1 explains more variance than does t2, etc. The loadings of each component define how the original variables ('properties') are linearly combined to form the scores, i.e., the projected coordinates of the objects ('molecules'). The loadings indicate the correlation of the original variables ('properties') with the position of the objects ('molecules') in the scores plot. Summing up, a principal components analysis summarizes a data table as the matrix product of scores and loadings ('systematic information') plus a residual matrix ('noise') representing the distance of each object from the model plane (or hyperplane). The original multidimensional space is condensed and simplified into a few components, which are sorted according to their information contents. Molecules are projected into the scores plot, which allows similarity relationships and systematic trends to be visualized, and the loadings plot shows how the original variables contribute to forming patterns between the compounds.

Multivariate preference mapping is illustrated with the following example of a GPCR-targeted library from an external vendor. The initial 3412 compounds were reduced to 2822 by excluding molecules that were either too similar to the Roche compound stock or that contained unwanted structural features. In addition, the following property filters were applied: molecular weight ≤ 500, number of H-bond donors ≤ 5, clogP ≤ 7, number of aromatic rings ≤ 5. After removing molecules with a pairwise Tanimoto distance of ≤ 0.19 (DVS [70], Daylight fingerprints [40]) a final subset of 1064 compounds was obtained for multivariate profiling with the following soft filters: drug-likeness (DND) [42], hERG [65], human serum albumin binding (HSA) [49], frequent hitters (FH) [27], BBB [50], Caco-2 [50], and protein binding (ProtBind) [50]. This profile was complemented with descriptors for size (molecular weight MW), polarity (polar surface area PSA), and lipophilicity (clogP [40], AlogP [28, 29, 71, 72]).

Upon applying PCA to the multivariate property matrix, a 3-component model was obtained. Since the first two components dominated and accounted for 58% of the variation of the data matrix, the analysis here focuses on these components. The scores plot in Figure 12.1 (top) shows an even distribution of compounds without outliers or obvious clusters. From the loadings plot in Figure 12.1 (bottom), we can see that the first component separates lipophilic compounds with high loadings of clogP and AlogP on the right from more-polar compounds on the left (PSA). The second component differentiates between high molecular weight (MW) and highly functionalized (PSA) molecules on the top from smaller molecules on the bottom. Whereas protein binding (ProtBind) calculated by VolSurf [50] is highly collinear with lipophilicity (clogP, AlogP), there is no correlation with the protein binding model (HSA) of Kratochwil et al. [49]. This model predicts HSA association constants based on a pharmacophoric similarity concept and thus has a unique information content within this dataset. Compounds with predicted high BBB and Caco-2 cell permeabilities are located in the lower right region of the scores plot, which also corresponds to both low molecular weight and polar surface area. Neither the predicted hERG liabilities nor the computed HSA association constants contribute to partitioning the compounds along the first two components. The

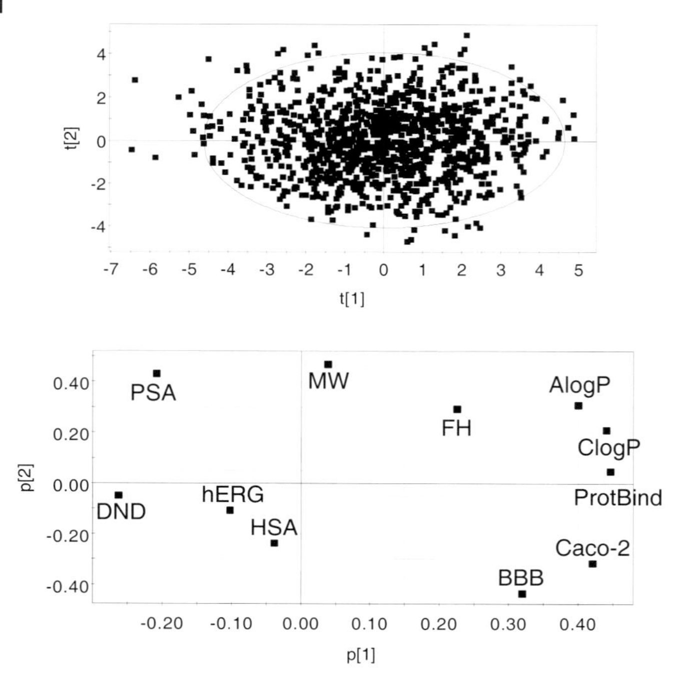

Figure 12.1 Multivariate preference mapping of a set of 1064 compounds that were profiled with 11 descriptors (see text for explanation). The upper scores plot shows the projection of the molecules onto the first two components of the principal components analysis. The lower loadings plot illustrates how the descriptors contribute to the positions of the projected compounds in the scores plot.

predicted drug-likeness DND and frequent hitter score FH occupy unique positions and therefore have to be examined closely if they are to be used for prioritizing molecules. Clustered properties are more reliable as predictors, because possible prediction errors can be compensated for by related soft filters in the same cluster.

Summing up, preference mapping allows one to detect correlations between properties and to spot unique soft filters that should be checked experimentally or be applied cautiously for profiling compounds. Understanding the information contents of soft filters is essential for making knowledge-based decisions about progressing or abandoning chemotypes.

12.5 Concluding Remarks

User-friendly web interfaces allow medicinal chemists to launch a whole battery of in-silico prediction models without the need to know the scope and limitations behind the black box. Only a combination of 'in silico' with 'in cerebro' will avoid frustration and provide for judicious and productive application of these models [73].

Therefore, the prime question should always be: What is the predictivity range and the confidence interval of the model? QSAR/QSPR models allow only for an interpolation but not for an extrapolation to novel chemotypes. Good in-silico models check whether the prediction dataset is contained within the property space of the training dataset, otherwise a warning is issued. If a model is based on a small number of molecules, one cannot assume that predictions will be reliable for compounds that are structurally diverse from the training set. This is illustrated in Figure 12.2 [74], where the training sets of the hERG and BBB in-silico filters are compared via PCA. Both sets were characterized with Ghose and Crippen [28, 29] descriptors, and after autoscaling, the molecules were projected onto the scores plot of the first two principal components. It is obvious that the two sets occupy different regions of the plot with only minor overlap. This is because the structural series are constrained to different locations in chemistry space, and the prediction of an hERG liability is therefore risky for those compounds of the BBB set that are distant from the hERG cluster.

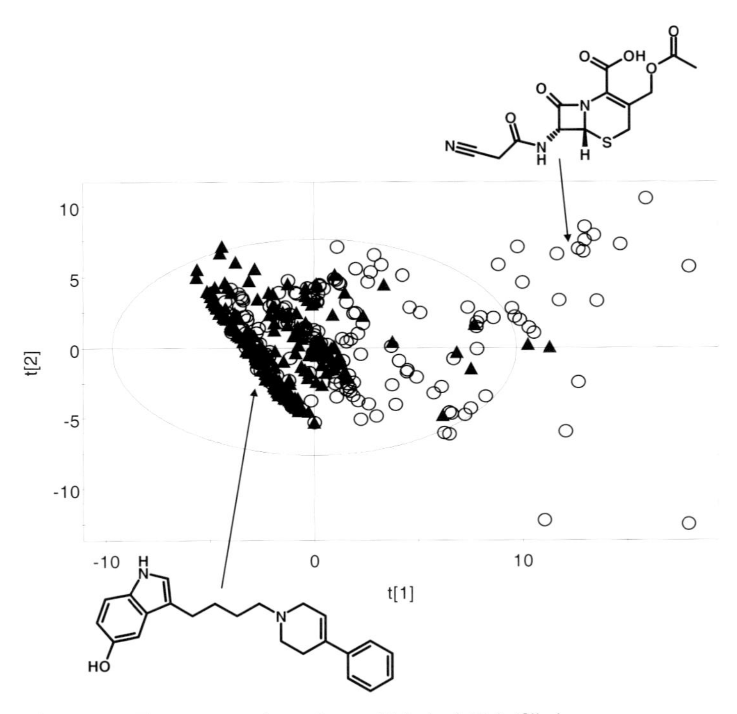

Figure 12.2 The compound sets from which the hERG (filled triangles) and the BBB (open circles) in-silico filters were derived are compared by principal components analysis, and one structure of each set is depicted. Ghose and Crippen descriptors were calculated for all the molecules, and after autoscaling, the compounds were projected onto the scores plot of the first two components [74].

Furthermore, in-silico models need to be challenged with new data, since models derived from small training datasets may become unstable upon the addition of new compounds. Finally, the prediction of physicochemical properties should be checked against experimental data for representative members of a lead series.

The predictive power of in-silico filters depends on both the accuracy of identifying true liabilities ('true positive') and the probability for raising false alerts ('false positive'). The prime question for ranking compounds is: What is the probability that a flagged molecule actually has the indicated liability? Considering as an example the 'frequent hitters' filter by Roche et al. [27], this question is discussed by presenting the statistical information in natural frequencies [75–80] instead of by using Bayes' theorem (Figure 12.3). The total dataset contained 902 structures with 479 frequent hitters and 423 nonfrequent hitters; 460 out of 479 molecules (96%) were correctly classified as frequent hitters, and 17 of the 423 nonfrequent hitters were predicted to be false positives (i.e., 4% false positives among the nonfrequent hitters). Based on the total dataset of 902 compounds, the probability that a raised alert correctly identifies a frequent hitter ('true positive') equals 460/(460 + 17), i.e., an excellent 96%. However, if the same probabilities for true and false positives are applied to a dataset that contains, not 53%, but only 5% frequent hitters, the results are considerably different (Figure 12.3): the probability

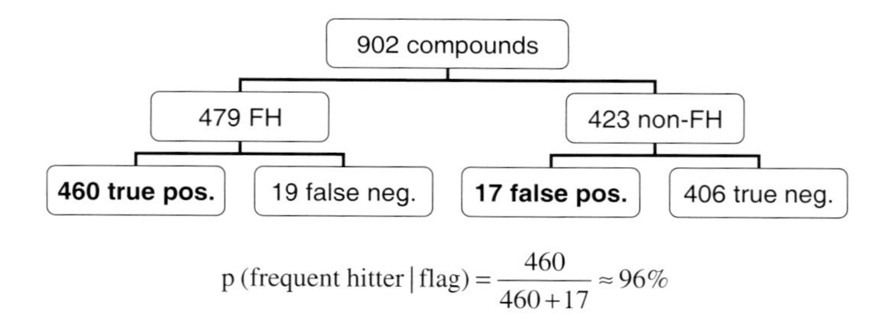

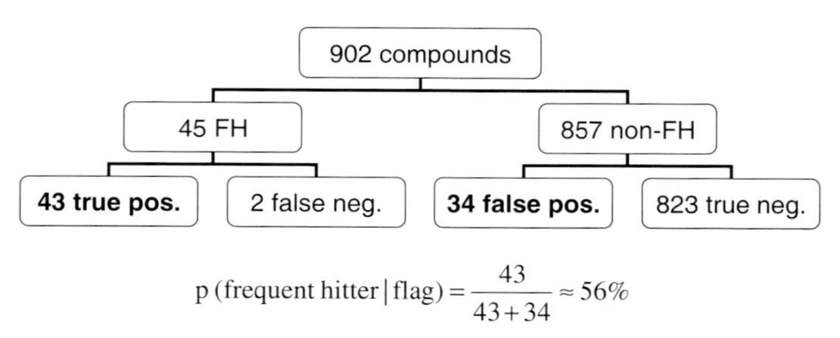

Figure 12.3 Using the example of the 'frequent hitters' prediction model of Roche et al. [27], the probability that a flagged molecule actually has the indicated liability is calculated for a dataset with an even distribution between frequent and nonfrequent hitters (top) and for one with only 5% frequent hitters (bottom). In both scenarios it is assumed that frequent hitters are correctly classified with 96% and that the false positive rate equals 4%.

that a flagged compound is a true positive has now dropped to 56% (43 true positives/(43 true positives + 34 false positives)). Flipping a coin would yield a 5% probability of correctly diagnosing a frequent hitter liability in this dataset.

Summing up, the accuracy with which true positives are identified is not sufficient for correctly judging the performance of a prediction model. The critical factor is the number of false positives, which in turn depends on the enrichment of the dataset with compounds raising alerts. The fewer true positives are contained within a dataset, the higher will be the number of false positives flagged by a soft filter. Assuming that in typical applications the proportion of molecules with liabilities is well below 50%, this example shows that in-silico prediction tools work reasonably well for prioritizing large compound sets but are not suitable for ranking single molecules. In addition, multivariate profiling will enable more reliable identification of liabilities than ranking based on single soft filters.

In conclusion, the quest for novel chemotypes is a long, expensive process. Therefore, the statistical relevance, predictivity range, and stability of computational filters need to be scrutinized in order to prevent a potential lead series from being rejected for the wrong reason.

Acknowledgments

We thank our coworkers in Molecular Modeling/Cheminformatics at Roche, Basel, for scientifically stimulating discussions and valuable contributions. The fruitful collaboration with colleagues from the Lead Generation, Medicinal Chemistry, Roche Compound Depository, Assay Development & HTS departments and the Molecular Properties Group is gratefully acknowledged.

References

1 S. Frantz, A. Smith, *Nat. Rev. Drug Discov.* **2003**, *2*, 95–96.
2 J. Drews, *Drug Discov. Today* **2003**, *8*, 411–420.
3 A. Alanine, M. Nettekoven, E. Roberts, A. W. Thomas, *Comb. Chem. High Throughput Screen.* **2003**, *6*, 51–66.
4 K. H. Bleicher, H. J. Böhm, K. Müller, A. Alanine, *Nat. Rev. Drug Discov.* **2003**, *2*, 369–378.
5 E. Jacoby, *Quant. Struct.-Act. Relat.* **2001**, *20*, 115–123.
6 E. Jacoby, A. Schuffenhauer, P. Floersheim, *Drug News & Perspectives* **2003**, *16*, 93–102.
7 T. Klabunde, G. Hessler, *ChemBioChem* **2002**, *3*, 928–944.
8 A. Schuffenhauer, J. Zimmermann, R. Stoop, J.-J. van der Vyver, S. Lecchini, E. Jacoby, *J. Chem. Inf. Comput. Sci.* **2002**, *42*, 947–955.
9 A. Schuffenhauer, P. Floersheim, P. Acklin, E. Jacoby, *J. Chem. Inf. Comput. Sci.* **2003**, *43*, 391–405.
10 S. V. Frye, *Chemistry & Biology* **1999**, *6*, R3–R7.
11 G. Müller, *Drug Discov. Today* **2003**, *8*, 681–691.
12 D. E. Clark, P. D. J. Grootenhuis, *Curr. Opin. Drug Discov. Dev.* **2002**, *5*, 382–390.
13 D. E. Clark, S. D. Pickett, *Drug Discov. Today* **2000**, *5*, 49–58.
14 W. J. Egan, W. P. Walters, M. A. Murcko, *Curr. Opin. Drug Discov. Dev.* **2002**, *5*, 540–549.

15 A. K. Ghose, V. N. Viswanadhan, J. J. Wendoloski, *J. Comb. Chem.* **1999**, *1*, 55–68.
16 H. Matter, K.-H. Baringhaus, T. Naumann, T. Klabunde, B. Pirard, *Comb. Chem. High Throughput Screen.* **2001**, *4*, 453–475.
17 T. Mitchell, G. A. Showell, *Curr. Opin. Drug Discov. Dev.* **2001**, *4*, 314–318.
18 W. P. Walters, A. Murcko, M. A. Murcko, *Curr. Opin. Chem. Biol.* **1999**, *3*, 384–387.
19 W. P. Walters, M. A. Murcko, *Adv. Drug Delivery Rev.* **2002**, *54*, 255–271.
20 S. Ekins, C. L. Waller, P. W. Swaan, G. Cruciani, S. A. Wrighton, J. H. Wikel, *J. Pharmacol. Toxicol. Methods* **2000**, *44*, 251–272.
21 G. M. Rishton, *Drug Discov. Today* **1997**, *2*, 382–384.
22 G. M. Rishton, *Drug Discov. Today* **2002**, *8*, 86–96.
23 J. Seidler, S. L. McGovern, T. N. Doman, B. K. Shoichet, *J. Med. Chem.* **2003**, *46*, 4477–4486.
24 S. L. McGovern, B. T. Helfand, B. Feng, B. K. Shoichet, *J. Med. Chem.* **2003**, *46*, 4265–4272.
25 S. L. McGovern, B. K. Shoichet, *J. Med. Chem.* **2003**, *46*, 1478–1483.
26 S. L. McGovern, E. Caselli, N. Grigorieff, B. K. Shoichet, *J. Med. Chem.* **2002**, *45*, 1712–1722.
27 O. Roche, P. Schneider, J. Zuegge, W. Guba, M. Kansy, A. Alanine, K. Bleicher, F. Danel, E. M. Gutknecht, M. Rogers-Evans, W. Neidhart, H. Stalder, M. Dillon, E. Sjogren, N. Fotouhi, P. Gillespie, R. Goodnow, W. Harris, P. Jones, M. Taniguchi, S. Tsujii, W. von der Saal, G. Zimmermann, G. Schneider, *J. Med. Chem.* **2002**, *45*, 137–142.
28 V. N. Viswanadhan, A. K. Ghose, G. R. Revankar, R. K. Robins, *J. Chem. Inf. Comput. Sci.* **1989**, *29*, 163–172.
29 A. K. Ghose, A. Pritchett, G. M. Crippen, *J. Comput. Chem.* **1988**, *9*, 80–90.
30 C. A. Lipinski, F. Lombardo, B. W. Dominy, P. J. Feeney, *Adv. Drug Delivery Rev.* **1997**, *23*, 3–25.
31 C. A. Lipinski, *J. Pharmacol. Toxicol. Methods* **2000**, *44*, 235–249.
32 M. C. Wenlock, R. P. Austin, P. Barton, A. M. Davis, P. D. Leeson, *J. Med. Chem.* **2003**, *46*, 1250–1256.
33 T. I. Oprea, *J Comp.-Aided Mol. Des.* **2002**, *16*, 325–334.
34 T. I. Oprea, A. M. Davis, S. J. Teague, P. D. Leeson, *J. Chem. Inf. Comput. Sci.* **2001**, *41*, 1308–1315.
35 T. I. Oprea, *J Comp.-Aided Mol. Des.* **2000**, *14*, 251–264.
36 J. R. Proudfoot, *Bioorg. Med. Chem. Lett.* **2002**, *12*, 1647–1650.
37 V. N. Viswanadhan, C. Balan, C. Hulme, J. C. Cheetham, Y. Sun, *Curr. Opin. Drug Discov. Dev.* **2002**, *5*, 400–406.
38 S. J. Teague, A. M. Davis, P. D. Leeson, T. Oprea, *Angew. Chem., Int. Ed.* **1999**, *38*, 3743–3748.
39 M. M. Hann, A. R. Leach, G. Harper, *J. Chem. Inf. Comput. Sci.* **2001**, *41*, 856–864.
40 Daylight Chemical Information Systems, Inc. **2002**, http://www.daylight.com/.
41 Ajay, W. P. Walters, M. A. Murcko, *J. Med. Chem.* **1998**, *41*, 3314–3324.
42 J. Sadowski, H. Kubinyi, *J. Med. Chem.* **1998**, *41*, 3325–3329.
43 H. van de Waterbeemd, E. Gifford, *Nature Rev. Drug Discov.* **2003**, *2*, 192–204.
44 R. Mannhold, H. Van de Waterbeemd, *J. Comp.-Aided Mol. Des.* **2001**, *15*, 337–354.
45 D. E. Clark, *J. Pharm. Sci.* **1999**, *88*, 807–814.
46 D. E. Clark, *J. Pharm. Sci.* **1999**, *88*, 815–821.
47 D. E. Clark, *Drug Discov. Today* **2003**, *8*, 927–933.
48 P. Ertl, B. Rohde, P. Selzer, *Rational Approaches to Drug Design*, Proceedings of the European Symposium on Quantitative Structure–Activity Relationships, 13th, Duesseldorf, Germany, Aug. 27–Sept. 1, 2000. **2001**, 451–455.
49 N. A. Kratochwil, W. Huber, F. Muller, M. Kansy, P. R. Gerber, *Biochem. Pharmacol.* **2002**, *64*, 1355–1374.
50 http://www.moldiscovery.com/
51 P. Crivori, G. Cruciani, P.-A. Carrupt, B. Testa, *J. Med. Chem.* **2000**, *43*, 2204–2216.
52 G. Cruciani, M. Pastor, W. Guba, *Eur. J. Pharmaceutical Sci.* **2000**, *11*, S29–S39.
53 G. Cruciani, P. Crivori, P. A. Carrupt, B. Testa, *THEOCHEM* **2000**, *503*, 17–30.

54 G. Cruciani, M. Pastor, R. Mannhold, *J. Med. Chem.* **2002**, *45*, 2685–2694.

55 W. Guba, G. Cruciani, *Molecular Modeling and Prediction of Bioactivity*, Proceedings of the European Symposium on Quantitative Structure–Activity Relationships: Molecular Modeling and Prediction of Bioactivity, 12th, Copenhagen, Denmark, Aug. 23–28, 1998. **2000**, 89–94.

56 T. I. Oprea, J. Gottfries, *Rational Approaches to Drug Design*, Proceedings of the European Symposium on Quantitative Structure–Activity Relationships, 13th, Duesseldorf, Germany, Aug. 27–Sept. 1, 2000. **2001**, 437–446.

57 T. I. Oprea, J. Gottfries, *J. Comb. Chem.* **2001**, *3*, 157–166.

58 T. I. Oprea, I. Zamora, A.-L. Ungell, *J. Comb. Chem.* **2002**, *4*, 258–266.

59 T. I. Oprea, *Curr. Opin. Chem. Biol.* **2002**, *6*, 384–389.

60 http://www.fda.gov/cder/OPS/BCS_guidance.htm.

61 B. Fermini, A. A. Fossa, *Nature Rev. Drug Discov.* **2003**, *2*, 439–447.

62 R. Webster, D. Leishman, D. Walker, *Curr. Opin. Drug Discov. Dev.* **2002**, *5*, 116–126.

63 I. Cavero, M. Mestre, J.-M. Guillon, W. Crumb, *Exp. Opin. Pharmacother.* **2000**, *1*, 947–973.

64 G. M. Keseru, *Bioorg. Med. Chem. Lett.* **2003**, *13*, 2773–2775.

65 O. Roche, G. Trube, J. Zuegge, P. Pflimlin, A. Alanine, G. Schneider, *ChemBioChem* **2002**, *3*, 455–459.

66 H. Yu, A. Adedoyin, *Drug Discov. Today* **2003**, *8*, 852–861.

67 M. Bertuccioli, S. Clementi, G. Cruciani, G. Giulietti, I. Rosi, *Food Qual. Preference* **1990**, *2*, 1–12.

68 V. Cecchetti, E. Filipponi, A. Fravolini, O. Tabarrini, D. Bonelli, M. Clementi, G. Cruciani, S. Clementi, *J. Med. Chem.* **1997**, *40*, 1698–1706.

69 L. Eriksson, E. Johansson, N. Kettaneh-Wold, S. Wold, *Multi- and Megavariate Data Analysis: Principles and Applications*, Umetrics AB, Umeå **2001**.

70 DiverseSolutions version 4.0.9, http://www.tripos.com/

71 A. K. Ghose, G. M. Crippen, *J. Comput. Chem.* **1986**, *7*, 565–577.

72 A. K. Ghose, G. M. Crippen, *J. Chem. Inf. Comput. Sci.* **1987**, *27*, 21–35.

73 H. Kubinyi, *Nature Rev. Drug Discov.* **2003**, *2*, 665–668.

74 R. A. Goodnow Jr., W. Guba, W. Haap, *Comb. Chem. High Throughput Screen.* **2003**, *6*, 649–660.

75 U. Hoffrage, S. Lindsey, R. Hertwig, G. Gigerenzer, *Science* **2001**, *292*, 854–855.

76 S. Kurzenhauser, U. Hoffrage, *Med. Teach.* **2002**, *24*, 516–521.

77 U. Hoffrage, G. Gigerenzer, S. Krauss, L. Martignon, *Cognition* **2002**, *84*, 343–352.

78 U. Hoffrage, S. Lindsey, R. Hertwig, G. Gigerenzer, *Science* **2000**, *290*, 2261–2262.

79 G. Gigerenzer, U. Hoffrage, *Psychol. Rev.* **1999**, *106*, 425–430.

80 G. Gigerenzer, U. Hoffrage, *Psychol.Rev.* **1995**, *102*, 684–704.

13
Navigation in Chemical Space: Ligand-based Design of Focused Compound Libraries

Gisbert Schneider and Petra Schneider

An aim of research in the field of chemogenomics and chemical biology is to a provoke defined biological response by interaction of sets of chemical agents with macromolecular targets [1]. A substance does not necessarily interact specifically with only one receptor; a compound library-based approach rather helps to understand the relationship between sets of small molecules and their biological activity profiles. Understanding the connections between target space and ligand space will provide us with a chemogenomic foundation for the rapid identification of hits and lead compounds by making possible heuristic predictions of desired and potential undesired activities in early-phase virtual screening and thereby facilitating the selection of appropriate leads in the drug-discovery process [2]. Recently, focus on such activities has increased, due to high clinical failure rates and constantly increasing development costs, and as a consequence "the vast emerging opportunities from efforts in functional genomics and proteomics demands a departure from the linear process of identification, evaluation and refinement activities towards a more integrated parallel process", as stated by Alanine and coworkers [3]. Early awareness of the expected activity profile is required to make an informed selection and prioritization of candidate leads with reduced attrition liability. The library-based approach is well suited for following several search tracks in chemical space ('backup' compounds) in parallel and for assisting the medicinal chemist in the task of scaffold hopping, i.e., finding an isofunctional but different molecular architecture to an already known hit or lead structure [4, 5]. The general idea is to define similarity between molecules on a level that permits escaping a local optimum in search space without losing the desired activity. A well known example of such a scaffold hop is given by structures **1** and **2** (Scheme 13.1). Using a topological pharmacophore descriptor of molecule **1** (Mibefradil), structure **2** was predicted to be isofunctional and was indeed shown to exhibit the same target function, namely blocking the human T-type calcium channel [4]. This chapter provides an overview of a straightforward chemogenomics approach to ligand-based library design, highlighting selected emerging computer-based techniques and their application.

We should stress that modern combinatorial chemistry approaches span a chemistry space that easily contains 10^{60} drug-like molecules [6]. No computational

Chemogenomics in Drug Discovery: A Medicinal Chemistry Perspective.
Edited by Hugo Kubinyi and Gerhard Müller

ISBN: 3-527-30987-X

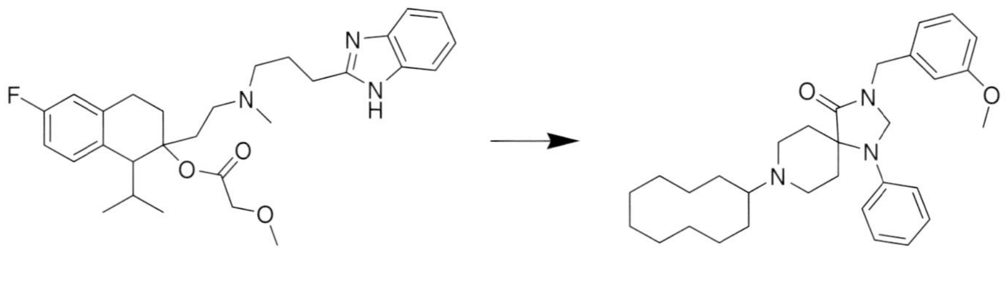

1

2

Scheme 13.1

technique is able to deal directly with such large quantities of data, and therefore the full enumeration of chemistry space is impossible. (A hypothetical enumerator that began to work at the time of the Big Bang some 13 billion years ago would have needed to produce about 10^{43} structures per second to have completed the job by today.) Even when the synthesis strategies used are restricted to feasible chemical reactions, the combinatorial explosion still leads to huge chemistry spaces, typically between 10^{6} and 10^{20} virtual compounds. Although for most of these still large (but manageable) virtual libraries a full enumeration might be doable, it would not be time- or cost-effective. Therefore, other strategies, like virtual de-novo synthesis and adaptive optimization methods, are necessary for dealing with large combinatorial spaces [7]. The basic problem can be formulated as the selection of the most promising candidates for real synthesis from a large virtual combinatorial library. Promising candidates are those compounds that are likely to meet certain selection criteria, most importantly drug- or lead-likeness, including sufficient aqueous solubility, and a desired activity profile. Here we present some product-based selection methods, but the discussion of de novo techniques is beyond the focus of this chapter.

13.1
Defining Reference and Target

Ligand-based design of focused drug-like compound libraries requires known molecules exhibiting the desired properties or pharmacological activity as a starting point. These molecules are often called seed structures or reference compounds. Novel structures having identical or very similar activity can be designed in silico or picked from collections of physically available compounds so that they exhibit similarity to the reference set. Such focused compound libraries can then be tested for activity in vitro. The choice of an appropriate similarity metric depends on the drug-discovery project and is context-dependent (see below) [8]. Two complementary compound sources are accessible for virtual screening: databases of physically available structures and virtual libraries including enumerated combinatorial libraries. Several commercially available databases are commonly relied on as sources of reference compounds for library design purposes. The most popular are (1) the World Drug Index (WDI; Derwent Information London, UK), containing over 60 000 pharmaceutical compounds in all stages of development; (2) the Comprehensive Medicinal Chemistry database (CMC; MDL Information Systems, Inc., San Leandro, CA, USA), with over 7500 structures and properties of drugs; and (3) MDDR (Drug Data Report; MDL Information Systems), representing a compilation of more than 100 000 structures and activity data of compounds in the early stages of drug development. In addition, several companies offer large libraries of both combinatorial and historical collections on a commercial basis. Usually the combinatorial collections contain 10 000–100 000 structures, and commercially available historical collections rarely exceed 200 000 compounds. Most of the major pharmaceutical companies have compound collection in the range ≥ 700 000. We should mention that combinatorial synthesis methods have significantly advanced from straightforward approaches, such as mixtures of Ugi reaction products, to fully automated parallel synthesis of well characterized compounds of high purity. Once these samples are fully characterized – e.g., by HPLC and mass spectroscopy – the data are of interest for structure–activity purposes. In most companies, these combinatorial products are also present in the 'historical' collection of compounds, generally derived from classical medicinal chemistry programs, and most of them have very well defined chemical characteristics. Commercial compound collections can also be purchased that fall in between these two extremes. Collectively, therefore, the information used to relate biological activity and chemical structure must clearly integrate all of these types of compounds. Combinatorial chemistry, high-throughput screening (HTS), and the availability of large compound selections have put us in the comfortable position of having a large number of hits to choose from for lead optimization – at least for certain classes of drug targets.

Although these data collections provide a large number of pharmacologically active compounds with some activity information, their usefulness for compilation of reference sets can be limited. Detailed information is necessary but not always provided or cannot be easily extracted from the databases. On a minimalist basis, reference compounds for ligand-based library design should be annotated according to target receptor class and receptor subtype, including quantitative activity values, e.g., IC_{50} or K_i (the IC_{50} and K_i of a compound can vary significantly). Ideally, complementary information about assay conditions and specificity data should be accessible. Such a body of information facilitates reasonable grouping of molecules according to meaningful attributes (features), compilation of conclusive reference data, and implicit or explicit formulation of quantitative structure–activity relationship (QSAR) models. As a step toward such a reference collection that can be used for chemogenomics studies, we have compiled approximately 6000 pharmacologically active molecules from the recent scientific literature, including receptor information and activity data (COBRA, Collection of Bioactive Reference Analogues) (Figure 13.1) [9].

The GPCR group consists of compounds binding to mainly class A rhodopsin-like GPCR; olfactory GPCR ligands are excluded. 'Hormone' represents the class of nuclear hormone receptor ligands. 'Enzyme' contains those enzyme targets that

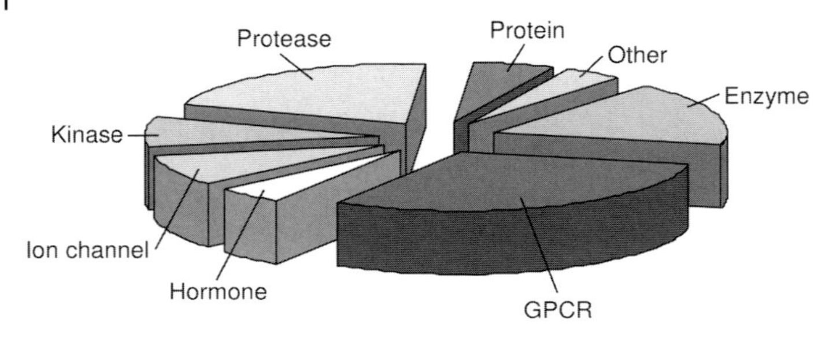

Figure 13.1 Receptor-class distribution of 4825 ligands of the COBRA collection.

are not contained in the Protease or Kinase sets. The Proteins group consists of ligands to various protein targets, including integrins, transfer proteins, transporters and symporters, viral envelope proteins, and others. Finally, the 'Other' set is used for a set of compounds with unidentified targets or additional targets, e.g., DNA. Antiinfective agents are not included here. The relative distribution of ligands reflects the fact that enzymes, in particular proteases, and membrane-bound receptors, specifically GPCR, represent the dominant target classes. Therapeutic target classes have been subdivided into seven main categories by Drews [10]: receptors 45%, enzymes 28%, hormones and factors 11%, ion channels 5%, nuclear receptors 2%, nucleic acids 2%, and unknown 7%. Constructing libraries targeted to one of these classes is an ongoing effort of common interest. For example, neural network techniques have been used for designing GPCR-targeted libraries [11] and serine protease-targeted libraries [12] using the Ensemble Database (Prous Science 2002) as a training set. It is evident that a target family can be defined on various levels of abstraction. For example, 'GPCR ligand' describes a superfamily that can be subdivided into further categories down to the level of an individual receptor subtype, e.g., 5HT-2A. Although this seems trivial, it is important to keep this in mind for application of library-filtering criteria.

The molecular weights (MW) and calculated octanol/water partition coefficients (clogP) obtained from all COBRA compounds have average values of $\langle MW \rangle = 412$ and $\langle clogP \rangle = 3.2$. These values are within both the limits proposed by Lipinski's 'rule of five' ($MW < 500$; $clogP < 5$) [13] and the quantifying ranges suggested by Ghose and coworkers ($MW < 480$; $clogP < 5.6$) [14]. This means that published bioactive structures generally 'obey' the rules for recommended maximal values of MW and clogP. With the slight exception of the protease set ($\langle MW \rangle = 495$), none of the average values calculated for the individual sets of molecules is above the recommended maximum values, but some general trends become visible. Most strikingly, the COBRA molecules tend to have a higher molecular weight and higher clogP than reported earlier for other collections of drugs or drug-like molecules [15–17]. For example, Ghose and coworkers used the CMC database as a compound resource (6304 molecules) and found $\langle MW \rangle = 360$ and $\langle clogP \rangle = 2.3$ [14]. Considering that most compounds in COBRA are recently developed molecules and that the CMC database consists of comparably 'older' compounds, we conclude that modern

Table 13.1 Molecular weight (MW) and lipophilicity (clogP) of sets of ligands taken from the COBRA compound collection [9]. Average, maximum and minimum values, and standard deviations (σ) are given. Properties were calculated with the software suite ChemOffice (CambridgeSoft Corp., Cambridge MA).

Receptor class (number of compounds)	***MW***				***clogP***			
	Avg.	***Min.***	***Max.***	σ	***Avg.***	***Min.***	***Max.***	σ
GPCR ($N = 1467$)	406	121	993	134	3.6	−11.1	13.7	2.6
Protease ($N = 1015$)	495	136	945	114	2.9	−8.2	10.3	2.4
Kinase ($N = 387$)	395	74	717	104	3.1	−5.6	8.7	2.2
Enzyme ($N = 839$)	364	68	849	119	2.6	−5.2	10.9	2.5
Hormone ($N = 227$)	336	142	949	115	4.0	−5.2	10.1	2.9
Ion channel ($N = 412$)	375	208	969	106	3.1	−11.1	10.8	2.7

drug discovery tends toward the development of larger, more lipophilic molecules. A similar general observation was recently made by Lipinski [18], and a shift to higher molecular weight and lipophilicity is also apparent for the development process from leads to drugs [19].

Relying on a definition of a target family and a collection of corresponding reference compounds, we can begin to systematically design target-family-specific libraries [20]. A straightforward method is to start with the analysis of general ligand properties. Table 13.1 gives an overview of the characteristic MW and clogP distributions for the target families mentioned above. It is apparent that, e.g., protease inhibitors tend to have a comparably high molecular weight and a low clogP, which means that these compounds tend to be large and hydrophilic In contrast, nuclear hormone receptor ligands (mostly steroids) are relatively smaller and significantly more lipophilic. Careful use of such thresholds is necessary to find a balance between a stringent focus on the one hand and exploration of a compound library on the other. The latter is critically important for allowing scaffold hops to occur. For example, applying a strict cutoff at MW = 500 would lead to a dramatic loss of potential protease inhibitors. The rule-of-5 does indeed indicate a potential problem with oral bioavailability for molecules with molecular weights greater than approximately 500, but we should stress that this rule's recommended parameter values were derived from all kinds of drugs, not just proteases [18]. If one intends to specify an unrefined focal area for library design by using property thresholds, an analysis of carefully selected, well defined reference compounds can be useful. Note that the minimum and maximum values and the standard deviations in Table 13.1 indicate a wide distribution of property values, which means that there will always be a significant proportion of exceptions to the rule.

An example of library design by multiple property optimization is shown in Figure 13.2. A pool of 83 400 molecules was subjected to property analysis, and a subset was cherry-picked to form the selected library containing 7350 members.

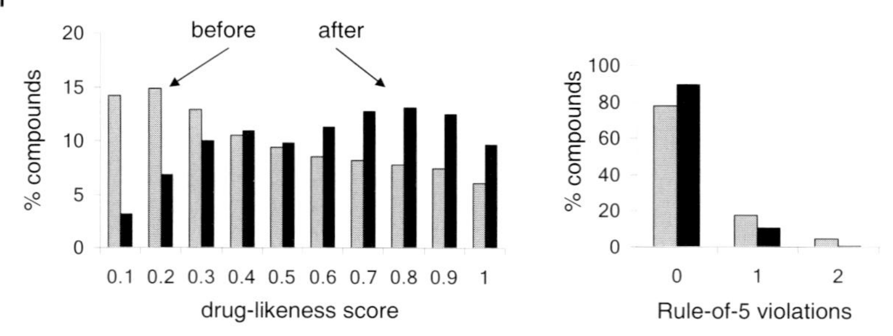

Figure 13.2 Multiple property-based library shaping. Property distributions are shown for the raw collection (gray bars) and for the selected library members (black bars). The drug-likeness score corresponds to the output value of an artificial neural network, where a value of 1 indicates maximal drug-likeness (for details, see the text). The rule-of-5 violations are counted per molecule.

The particular method used in this sample application for selection of compounds is detailed later in this chapter. However, note that several properties were considered in parallel and no cutoff or threshold values were applied to any of the properties. This can be seen from the rule-of-5 violations in Figure 13.2: one might reject compounds having two or more violations, but the 'soft' selection method does not discard candidates simply because of this; rather, a multidimensional ranking of molecules is performed, and the high-ranking candidates become members of the library subset. Further examples and related methods for library design by optimization of property distributions can be found elsewhere, e.g., in a special volume on combinatorial library design and evaluation edited by Ghose and Viswanadhan [21].

13.2
A Straightforward Approach: Similarity Searching

To reliably predict molecular properties, the molecules under investigation must be represented in a suitable fashion. In other words, the appropriate level of abstraction must be defined to perform rational virtual screening. A convenient way to do this is to employ molecular descriptors, which can be used to generate molecular encoding schemes reaching from general properties (e.g., lipophilicity, molecular weight, total charge, volume in solution, etc.) to very specific structural and pharmacophoric attributes (e.g., multipoint pharmacophores, field-based descriptors) [22]. Filtering tools can be constructed using a simplistic model relating the descriptors to some kind of bioactivity or molecular property. However, the selection of appropriate descriptors for a given task is not trivial, and careful statistical analysis is required. Besides an appropriate representation of the molecules under investigation, any useful feature-extraction system must be structured in such a way that meaningful analysis and pattern recognition are possible. Technical systems for information processing are intuitively considered to mimic some aspects of human capabilities in the fields of perception and cognition. Despite great achievements in artificial intelligence research during recent decades and an increasing application of machine learning methods in virtual screening, we are still far from understanding complex biological information-processing systems in detail. This means that a feature-extraction task that looks very simple to a human expert can be extremely hard or even impossible for a technical system, e.g., certain virtual screening software. As we have learned from many years of artificial intelligence research, it is extremely difficult (if not impossible) to develop virtual screening algorithms that mimic the medicinal chemists' intuition. Furthermore, there is no common 'gut feeling', because different medicinal chemists have different educational backgrounds, skills, and experience. Despite such limitations there is, however, substantial evidence that it is possible to support drug discovery in various ways with the help of computer-assisted library design and selection strategies. Two specific properties of computer-based approaches make virtual screening very attractive for library design:

- Speed and throughput of virtual testing can be much better than what is possible by means of wet-bench experimental systems.
- Virtual screening of virtual libraries, or virtual library construction, enables hitherto unknown parts of chemical space to be explored.

Due to its ease of implementation and speed of execution, chemical similarity searching has a long tradition in this area, and many different similarity metrics have been proposed for rapidly analyzing very large virtual libraries [23]. The general idea is to define a query, e.g., a single reference molecule, a set of molecules, or a pharmacophore model, and then to rank the members of the virtual library so that the compounds that are most similar appear at the top of the ranked list. The aim of a similarity searching can be characterized in one of the following two ways. First, it can be used with a set of n known active molecules. Then one can evaluate the parameters used (query structures, descriptor, distance metric) by means of the enrichment factor (Eq. 13.2) [24]. This application of a similarity search is called retrospective screening. In contrast, prospective screening can be performed to find molecules that potentially exhibit activity for the same target as the query structure. The decision as to which specific parameters should be employed for a prospective screen has to be based on prior experience, and retrospective screening provides a useful means for this purpose.

The enrichment factor *ef* provides a way to rate a similarity search. Given a database containing D_{all} compounds, of which D_{act} have known biological activity against a desired target, a certain fraction F, e.g., the top 5%, is taken from a similarity ranked list. The fraction contains F_{all} compounds, of which F_{act} are indeed experimentally validated actives. Provided that D_{act} is randomly distributed among D_{all}, the expected number of active molecules among F_{all} is

$$F_{act,expected} = F_{all} \frac{D_{act}}{D_{all}} \tag{13.1}$$

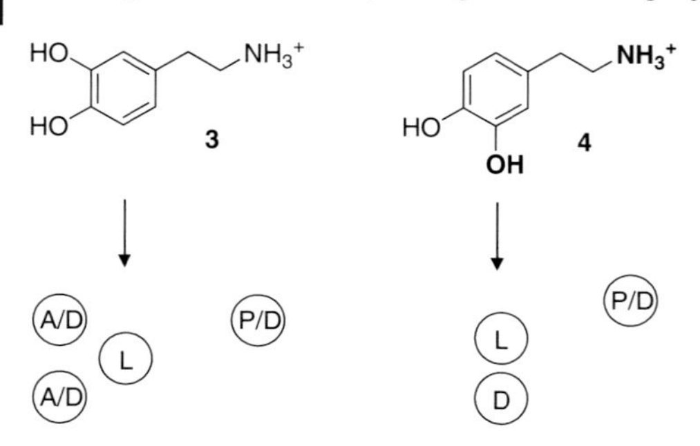

Figure 13.3 Two rotamers of dopamine and derived pharmacophore models (adapted from [25]). A: H-bond acceptor, D: H-bond donor, P: positively charged, L: lipophilic. The pharmacophore model on the right resulted from inspection of several other dopamine receptor ligands, which led to a single remaining donor site at one of the hydroxyl groups.

Thus, a similarity search can be qualified by calculating the enrichment of active molecules within F_{all} over a random distribution of the active molecules:

$$\mathrm{ef} = \left(\frac{F_{act}}{F_{all}}\right) \Big/ \left(\frac{D_{act}}{D_{all}}\right) \tag{13.2}$$

where *ef* is called the 'enrichment factor'. An enrichment factor greater than 1 is returned by a method that is superior to random selection of compounds within F_{all}. Typically, the enrichment factor is obtained by retrospective analysis of reference data. The trick for scaffold hopping is first to find an appropriate set of descriptors and a suitable similarity metric by retrospective screening, and then to make predictions for those molecules that appear in between the top-ranking known actives.

In the following sections, we discuss the use of retrospective screening and the meaning of 'similarity' and 'enrichment of actives' in more detail, beginning with a classical example of similarity searching. The neurotransmitter dopamine and two derived pharmacophore models are shown in Figure 13.3 [25]. Depending on the dopamine rotamer considered (**3** or **4**), different pharmacophore models appear to be equally plausible. When additional reference compounds are available for mutual alignment of compounds, rotamer **4** and the corresponding pharmacophore can be assigned an agonist motif, under the assumption of similar binding modes. If only a single reference compound is used for similarity searching, both pharmacophore models in Figure 13.3 are equally valid, since there is no additional information that might be of use for weighting the models. We must be aware of this kind of situation whenever a similarity search is performed that relies on a single query molecule.

An example may help to clarify this statement. Figure 13.4 shows the 10 highest-ranking compounds that were retrieved from the COBRA database by a topological pharmacophore similarity search (CATS method, see below). The query structure was Haloperidol, a dopamine (D2) receptor antagonist. Not surprisingly, classic variations of the query structure are found in ranks 1 and 2. These are not very

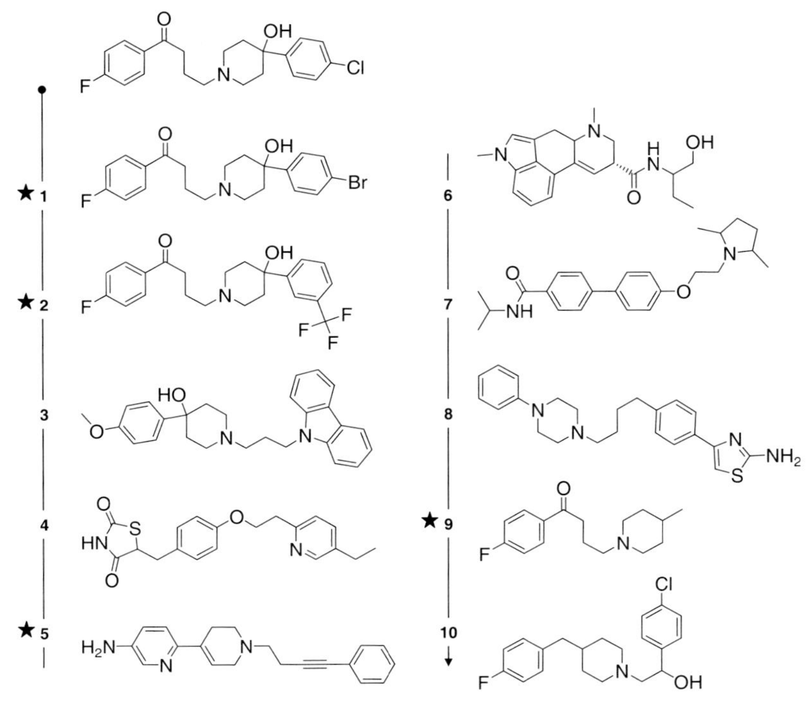

Figure 13.4 Results of a CATS similarity search. Similarity between the query structure (Haloperidol, a D2 antagonist; upper left) and database compounds was defined in terms of a topological pharmacophore descriptor. The top 10 most-similar molecules found are shown. Stars indicate the ranks occupied by known D2-receptor ligands.

interesting from the library design or scaffold-hopping points of view. At ranks 5 and 9, two additional D2 receptor ligands are found, one of which, surprisingly, is an agonist (rank 5), and the other is the well known Melperone structure (rank 9), which represents a substructure of Haloperidol. Retrieval of the rank-5 molecule could already be regarded as a scaffold hop, because the compound has a different structure than the query. This molecule is a D2 ligand and can be regarded as isofunctional on this description level of bioactivity, but it does not necessarily exhibit the same kind of functional activity (the compound at rank 5 is an agonist, not an antagonist, as is the query molecule). Looking at the first molecules ranked between known D2 ligands, we find an annotated ion-channel blocker (GABA transporter type I, GAT1) at rank 3 and an antiinflammatory PPAR-γ agonist (Pioglitazone) at rank 4. Based on the similarity ranking, it would now be worthwhile to test these molecules in a dopamine receptor binding assay. Indeed, coinhibition of dopamine transporter and GAT1 has been reported for Orphanin FQ, an endogenous antagonist of the dopamine transporter [26], and Pioglitazone has been found to prevent dopaminergic cell loss [27]. These are first indications that the similarity search might have produced useful results. Still, only a biochemical test can validate the results. An argument against the compound at rank 4 might be the lack of a basic amine function. Looking at the lower-ranking structures, we find a serotonin receptor 5-HT2C antagonist (rank 6), a histamine receptor H3 antagonist (rank 7), a TNF-α inhibitor (rank 8), and another ion-channel blocker (Eliprodil, rank 10).

To assess these findings, it is important to learn about other activities of the query structure: K_i (D1) = 270 nM, K_i (D3) = 21 nM, K_i (D4.2) = 11 nM, K_i (5-HT2A) = 25 nM, K_i (α1) = 19 nM, K_i (H1) = 730 nM. This means that Haloperidol exhibits binding activity against a whole family of targets and is not specific for the D2 receptor. Therefore, retrieving an H3 ligand at rank 7 can be considered a success if we keep in mind that the query has significant binding potential at the H1 receptor. This brief example of a pharmacophore-based similarity search demonstrates that one has to be very careful when analyzing a ranked list, and a seeming contradiction with what was expected as an outcome of the experiment might be resolved by considering multiple activity of the query structure.

Similarity searching can be successful only when molecules are represented by a suitable description of the chemical space. The definition of 'important' attributes depends heavily on the query structure and therefore on its associated binding partner. Descriptors of chemical space can be categorized, e.g., according to their data representation and according to the dimensionality of molecular attributes (1D, 2D, or 3D) they describe. Binary fingerprints are a typical data representation for similarity searching [28]. They describe the presence or absence of a feature, e.g., a substructure [29], or a certain pharmacophore [30, 31], in a linear bit-string format. Fingerprints vary in length from 57 bits for mini-fingerprints (a collection of 1D and 2D molecular descriptions) [32] up to millions of bits for 4D-pharmacophore fingerprints (all combinations of four-point pharmacophores) [31]. For an extensive review of issues related to conformer generation in the process of property calculations, see [33].

Pharmacophore models seem to be specifically suited for scaffold hopping and respective library design [34]. If we want to pick members of a compound library from a very large virtual chemistry space, calculation of 3D conformers and subsequent structural or potential pharmacophore point-based alignment of molecules can be a limiting factor. Therefore, alignment-free models have value, particularly during the early phases of library design [35]. To demonstrate the idea of retrospective screening and its use for library-design purposes, one representative of these methods – correlation vector representations (CVR) – is discussed in more detail.

The correlation vector approach was introduced to the field of cheminformatics by Broto and Moreau in 1984 [36] and was brought to wider attention through studies by Gasteiger and coworkers [37]. The basic idea of CVR is to map molecular features, e.g., pharmacophore points or properties, to a numerical vector of fixed length. As a consequence, each molecule is encoded by such a vector of a given dimension, and pairwise comparison of vectors (similarity calculation) can be executed very quickly without having to explicitly align the molecular structures. CVR belongs to the class of alignment-free descriptors. Several applications of CVR to similarity searching have been reported by our group and others previously, exploiting the possibility for very fast virtual screening of large compound collections and their use in de-novo design [2, 4, 37, 38].

The CATS descriptor follows the CVR idea and was introduced to provide a concept for scaffold hopping [4]. Its original implementation was based on the 2D structure of a molecule and therefore avoids the issue of conformational flexibility. It belongs to the category of atom-pair descriptors and encodes topological information of a molecule [39]. The centers of the atom pairs are not characterized by their chemical element type, but by their membership in a potential pharmacophore point (PPP) group, i.e., generalized atom types. Typically, five PPP groups are considered: H-bond donor (D), H-bond acceptor (A), positively charged or ionizable (P), negatively charged or ionizable (N), and lipophilic (L). These five groups are assumed to represent potential pharmacophore points of a molecular structure. If an atom does not belong to one of the five PPP types it is not considered; i.e., in the standard implementation of CATS the atom type 'null' does not exist. However, it would be a worthwhile exercise to quantify a null atom's influence on similarity searching results. A straightforward definition of atom types is given by the following assignment [40]: lipophilic: {C(C)(C)(C)(C), Cl}; positive: {[+], NH2}; negative: {[–], COOH, SOOH, POOH}; H-bond donor: {OH, NH, NH2}; H-bond acceptor: {O, N[!H]}. Thus, every atom of a molecule is assigned to none, one, or two PPP types. Of course, many other sets of generalized atom types have been described and may be used instead [41].
The occurrences of all 15 possible pairs of PPP types (DD, DA, DP, DN, DL, AA, AP, AN, AL, PP, PN, PL, NN, NL, LL) are then counted, and the resulting histograms are divided by the number of nonhydrogen atoms in the molecule to obtain a scaled vector. Other scaling methods are applicable also, e.g., scaling by the expected background frequency of an atom type. After scaling, all 15 possible pairs of CATS types are associated with the number of intervening bonds between the two corresponding atoms, using the shortest path length. The minimum distance between a pair of CATS types is zero bonds, and the maximum distance is typically chosen to be nine bonds. Thus, the result of the calculation of the CATS descriptor is a 150-dimensional correlation vector representation. The general procedure to obtain the CATS 2D pharmacophore model is

1. Extract the unweighted, hydrogen-depleted molecular graph.
2. Assign PPP atom types to the nodes of the molecular graph.
3. Calculate the distance matrix.
4. Calculate the correlation vector representation CVR:

$$\mathrm{CVR}_d^T = \frac{1}{A}\sum_{i=1}^{A}\sum_{j=1}^{A}\delta_{ij,d}^T \tag{13.3}$$

where d is the path length, T is the atom type pair, A is the number of nonhydrogen atoms, and δ^{T} is the Kronecker delta that equals 1 if a pair T exists and 0 otherwise. This descriptor can be easily extended to the 3D situation, in which distances

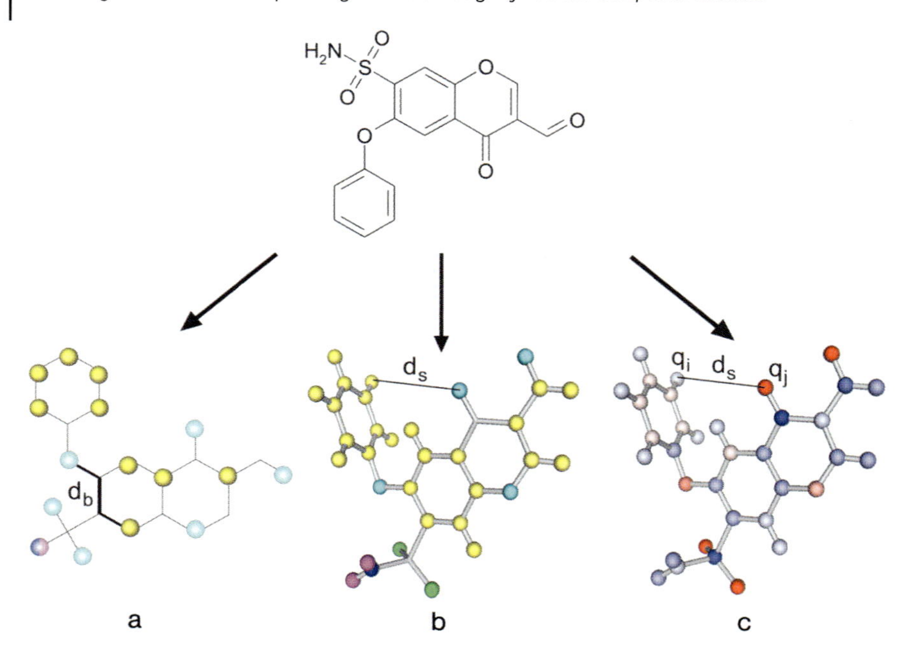

Figure 13.5 The principle of correlation vector representation (CVR).
(a) During calculation of the topological (2D) CVR descriptor, the 2D structure of the molecule is converted to its molecular graph representation and generic atom types are assigned to the vertices of this graph. Then all possible pairs of atom types are counted; only the shortest path (number of bonds) between two vertices is taken into account. Yellow balls represent lipophilic centers, cyan balls H acceptors, blue balls positive centers, and magenta balls H donor. An example of an acceptor–lipophilic pair spaced three bonds apart is depicted by bold edges, where d_b is the distance in number of bonds.
(b) For calculation of the 3D CVR descriptor, the explicit 3D conformation of a molecule is converted into a 3D distribution of potential pharmacophore points. The descriptor encodes the number of pairs of generalized atom types that fall into predefined distance bins. The color scheme of the atom types is consistent with (a). Additionally, green balls represent polar atoms. An example of an acceptor–lipophilic pair is depicted with a black line; d_s indicates the spatial distance between the two atoms.
(c) The atom charge-based CVR descriptor maps the partial atom charges of a molecule into predefined distance bins. Blue represents positive charge, red represents negative charge; color intensity indicates the charge value. The black line shows the spatial distance d_s between a partially positive atom with charge q_i and a partially negative atom with charge q_j (adapted from [40]).

measured in 3D space are used instead of topological distances. The original application of the autocorrelation idea was grounded on an atom charge model instead of on generalized PPP types (Figure 13.5) [37].

Once descriptors have been calculated for all compounds, the question arises as to which similarity metric is to be used to obtain maximal enrichment of actives in the library subset [42]. Several comparative studies have revealed that there is no single best similarity metric that outperforms all others [37, 40, 42–44]. Nevertheless, there seem to exist certain application domains for individual similarity metrics, i.e., in combination with a particular descriptor and target. This is shown in Figure 13.6, which demonstrates the application of the three CVR methods outlined above. Known ligands of three selected protein targets, cyclooxygenase-2 (COX-2), human immunodeficiency virus (HIV) protease, and matrix metalloproteinases (MMP), were used in a retrospective virtual screening study [40]. A comparison of the three descriptors revealed that none is generally superior for all datasets, but often one descriptor is preferred for a given dataset. This reflects the suitability of a specific descriptor for a specific ligand–receptor interaction pattern.

However, the information induced in each descriptor seems to differ, as illustrated in Figure 13.6 by the number of compounds that were exclusively identified by a

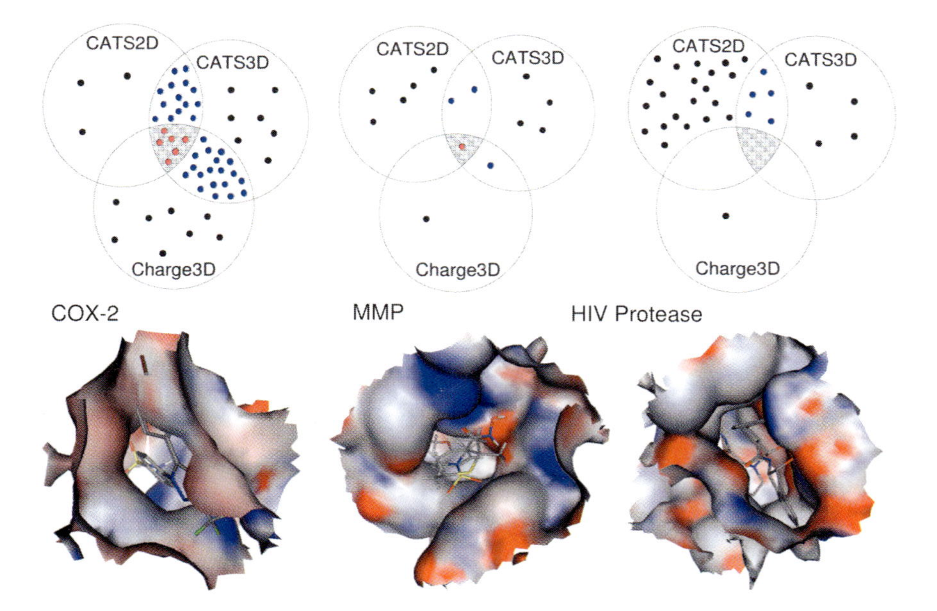

Figure 13.6 Results of retrospective screening with CVR methods. *Top:* Elements of the Euler–Venn diagrams represent compounds that can be found among the first 5% of the ranked lists resulting from similarity searching with COX-2, MMP, and HIV protease subsets of the COBRA database. The Manhattan distance was used as a distance metric. Membership in a set indicates that the respective compound was retrieved by retrospective screening with the corresponding descriptor. The diagrams reveal that the three descriptors complement one another to different extents, depending on the underlying dataset (adapted from [40]). *Bottom:* Models of the binding pockets containing bound ligands of COX-2 (PDB identifier: 1CX2), MMP-3 (PDB identifier: 1D5J), and HIV protease (PDB identifier: 1HSG). Molecular surfaces are colored according to partial charges as implemented in the MOE modeling software (Chemical Computing Group Inc., Montreal).

single descriptor. The sizes of the intersection (shaded area) are six, one, and zero compounds for the COX-2, MMP, and HIV protease subsets, respectively. The small intersection sizes suggest that the information contents of the individual descriptors complement one another. Exclusive selection of compounds that are recognized as potential candidates by all three descriptors would result in a significant loss of actives. This finding implies that a single descriptor alone does not cover all aspects of molecular features that are necessary for a ligand to bind successfully. Thus, it seems appropriate to combine the most promising candidates from each ranked list of compounds that was produced by individual similarity searches of a compound library [45]. This strategy might be particularly useful for combinatorial library design and building-block selection aiming at activity-enriched subsets, complementing computationally more demanding techniques for target-family biased library design [46, 47].

Figure 13.6 also shows that there is a possible structural explanation for the different performance of descriptors. The binding pocket of COX-2 is buried and narrow, and almost all portions of the small ligands participate in binding. This is reflected by the considerable performance of all three correlation vector descriptors for the COX-2 dataset. The binding pocket of MMP-3 is rather shallow, and a great portion of the ligand surface is accessible to the solvent. Since the CVR descriptors described above take into account PPP types and partial charge information for the entire molecule, much of the encoded information is worthless for the binding pattern of the MMP ligands. This may contribute to noise in the descriptor data. Finally, the HIV protease binding pocket is deep, long, and tunnel-like. Here, the topological correlation vector descriptor achieved better overall results than the two descriptors that are based on 3D distances. A closer look at the HIV protease ligands revealed that they contain more rotatable bonds (on average 19 compared to 6 and 15 for COX-2 and MMP ligands, respectively) and have a higher molecular weight (on average 607 compared to 360 and 439 for COX-2 and MMP ligands, respectively) than the COX-2 and MMP ligands. In consequence, it is much more difficult to generate conformations resembling the receptor-bound ligand conformation which might explain the performance of the two 3D correlation vector descriptors with the HIV protease test data.

Despite several limitations, correlation vector methods were shown to be suited for ligand-based similarity searching, i.e., considerable enrichment of actives was obtained by retrospective analysis [40]. These alignment-free descriptors seem to be applicable to early-phase virtual screening campaigns, where a course-grained filtering of datasets is required in combination with a high execution speed. If a single similarity metric is to be used, the Manhattan distance seems to be particularly applicable because of its computational simplicity and ability to produce significant enrichment in actives. Again, this statement does not imply that this is the metric of choice for any similarity search. Appropriate tuning of method parameters is essential, e.g., binning options or the definition of pharmacophore types, as well as meaningful data preprocessing, e.g., descriptor scaling. This aspect is of utmost importance for application of similarity concepts in chemogenomics and large-scale HTS data analysis.

13.3 Fuzzy Pharmacophore Models

Pharmacophore models represent the location of generalized interaction sites in 3D space which are considered to be related to biological activity. Two basically different approaches for representing the pharmacophore can be distinguished, traditional 3D pharmacophore models and pharmacophore fingerprints [41]. The traditional approach, implemented in program packages like Catalyst [48], DISCO [49], GASP [50], and MOE [51], usually determines the most-conserved features of a set of structurally aligned known active ligands. The spatial configuration of generalized interaction sites is the basis for screening new molecules with the same biological activity. Molecules that have all or a user-defined minimum number of the features are presumed to be active. One significant drawback of this approach is the necessity to align a molecule to the pharmacophore query before it can be classified as potentially active or inactive. This step can be very time-consuming and thus prevent the screening of very large databases. Another drawback is the lack of information about less-conserved regions for virtual screening. It is not easy to take such regions into consideration, since a minimum number of features have to be satisfied by a molecule for it to be classified as active. Excluded volumes can sometimes compensate for part of this problem by preventing the selection of molecules that are too large for the binding pocket [41].

Pharmacophore fingerprints describe the spatial arrangement of pharmacophoric features as a bitstring in which each bit corresponds to a certain feature, or in the form of a CVR (see above) [40, 52, 53]. The latter has the advantage that no alignment is needed to estimate the activity of a molecule under consideration, which allows for rapid screening of large compound databases and makes the method prone to alignment errors. Pharmacophore fingerprints contain information about the number or the presence or absence of the spatial arrangement of defined multiplets of generalized pharmacophoric features. Although information is retained about all PPP features present in a molecule, a drawback of these methods is usually that no information is available about the conservation and tolerance of the features. Successful attempts have been made to include conservation information extracted from conserved positions in the CVRs of sets of known active molecules [54]. However, information about conserved features extracted from molecular alignments needs to be more reliable. Initial studies on introducing fuzziness into CVR-based retrospective screening showed no significant improvement [55]. A shortcoming of the initial approach might have been that a uniform degree of fuzziness was used for all pairs of features of the molecules. Another property of pharmacophore fingerprints – which can either be considered a benefit or a drawback – is that the descriptor of a molecule is not necessarily unique to that molecule. On the one hand this can lead to molecules that are different overall from the query molecule and do not have any of the desired biological activity. On the other hand this provides an opportunity to perform scaffold hopping.

An example of fuzzy pharmacophores is shown in Figure 13.7 (S. Renner, G. Schneider, unpublished results) [56]. Three known COX-2 inhibitors, **5** (SC-558),

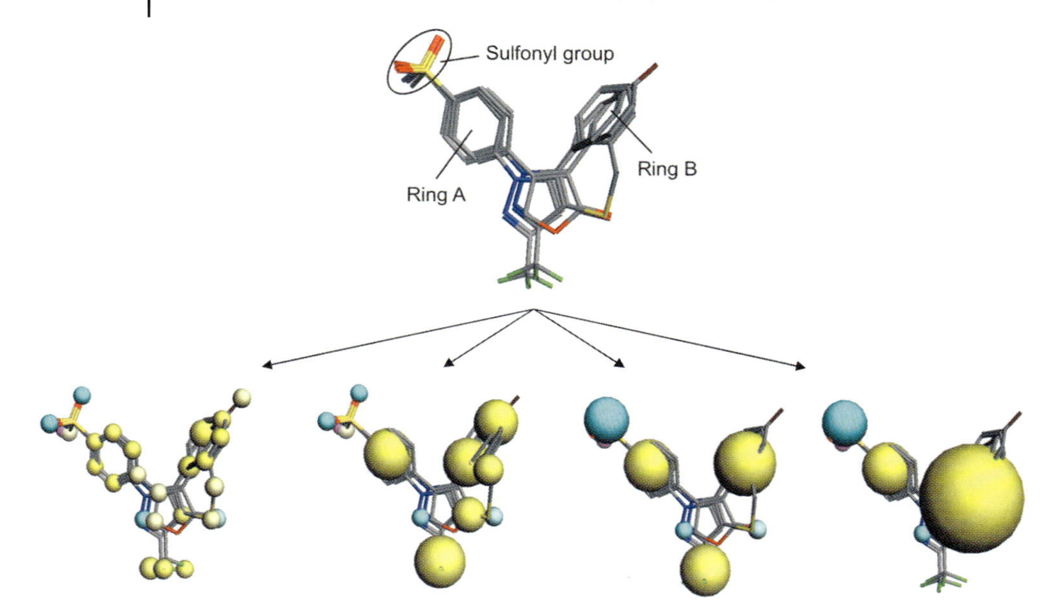

Figure 13.7
Top: 3D alignment of COX-2 inhibitors **5**, **6**, and **7**. Rofecoxib and M5 were aligned to the crystal structure conformation of SC-558 bound to COX-2. *Bottom:* Pharmacophore models calculated with different cluster radii from the aligned COX-2 inhibitors. The cluster radii of the four shown models are 1.0 Å, 1.5 Å, 3.0 Å, and 3.5 Å. The colored spheres represent potential pharmacophore points (PPP) of the models. The radii of the spheres denote the standard deviations of the spatial distributions of the atoms contributing to each PPP. Yellow: hydrophobic interactions, magenta: H-bond donors, cyan: H-bond acceptors. Color intensity denotes the extent of conservation of the PPP among the aligned molecules (courtesy S. Renner).

6 (M5), and **7** (Rofecoxib), were used as reference structures for calculation of a 3D pharmacophore model (Scheme 13.2). 3D conformers and pharmacophore alignments were generated with the MOE software suite [51]. Figure 13.7 shows the resulting superpositioning, with crucial pharmacophoric points indicated. According to Palomer and coworkers, essential interactions for specific COX-2 inhibition are mediated by the aromatic rings A and B and the sulfonyl group [57]. Four resulting pharmacophore models with different degrees of fuzziness are illustrated below the superposed molecules. Each of these models can serve as a query for similarity searching. The important difference from standard similarity searching is that ensemble information from several reference structures is used to define important PPP sites. Increased fuzziness can be introduced by larger clustering radii for determination of pharmacophore points. The pharmacophore models consist of only three generalized interaction types: H-bond donors, H-bond acceptors, and hydrophobic interactions. The model resulting from a 1-Å cluster radius is the most detailed model. In most instances, only close-by atoms are combined to PPPs,

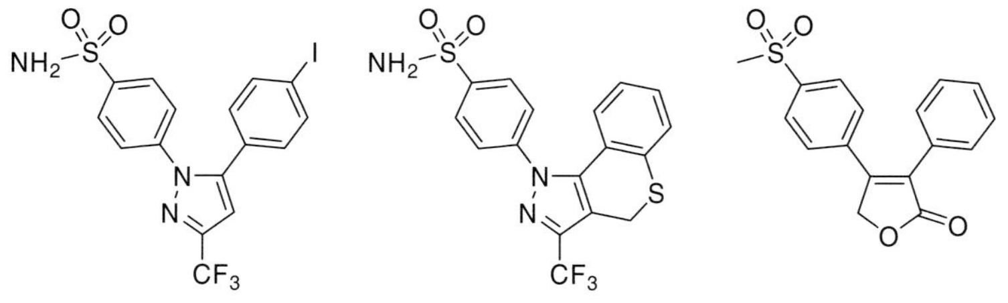

5 **6** **7**

Scheme 13.2

which results in a low degree of abstraction from the scaffold of the molecules in the alignment. Note that, in contrast to the other models shown here, the preferred angles between the two aromatic rings A and B are preserved in this model. The models resulting from 1.5-Å and 3.0-Å cluster radii exhibit a higher degree of generalization from the molecular alignment. Many atoms, especially in the regions of the aromatic rings A and B, are combined into large PPPs, covering several atoms in each of the molecules. Finally, in the model resulting from 3.5-Å cluster radius, the shape of the underlying alignment is only marginally visible, thus this model represents the highest degree of abstraction from the chemical entities **5**, **6**, and **7**.

So, which of the models should be used for library design? One way of finding out is to again conduct a retrospective screening study. The different models will lead to different enrichment factors. In the present example, the 1.5-Å model performed best [56]. The appropriate degree of PPP fuzziness again depends on the particular project and must be determined separately for each alignment of reference molecules. A challenge for any pharmacophore model is to assess its

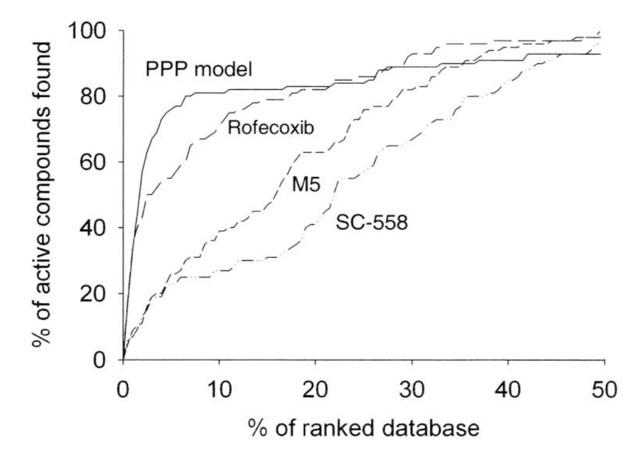

Figure 13.8 Comparison of the enrichment obtained with the combined pharmacophore model (PPP model) resulting from a 1.5-Å PPP cluster radius with the results of retrospective screening using the COX-2 inhibitors Rofecoxib, M5, and SC-558 as seed structures (courtesy S. Renner).

performance in comparison to straightforward similarity searching taking the individual reference molecules as queries. Figure 13.8 shows the results of such a comparison obtained for the COX-2 example. Generally, a significant enrichment of actives among the top-ranking library members was observed. In addition, Rofecoxib performed much better than the other two COX-2 inhibitors. Both pharmacophore models performed similar to Rofecoxib for the first 1% of the database. Up to the first 20% of the database, the pharmacophore model clearly outperformed Rofecoxib. The pharmacophore model retrieved 80% of the known active COX-2 inhibitors into the first 8% of the ranked database, yielding an enrichment factor of 35 for the 8% subset. In comparison, Rofecoxib ranked 80% of the actives into the first 18% of the library. This means that a COX-2-focused library could be compiled from the first 8% of the ranked database and that the inherent fuzzy description of the molecules should support the goal of scaffold hopping, especially with higher degrees of fuzziness.

In the past, assay data were analyzed primarily by medicinal chemists, by looking at the active compounds and then deciding on which hits to focus their efforts. With the increase in the number of experimentally determined actives, this approach becomes increasingly ineffective, and computational techniques are increasingly used to classify the hits and derive hypotheses. One should keep in mind that it is basically impossible for a human being to also take into account the large number of inactive compounds. The development of a pharmacophore hypothesis, for example, ideally incorporates additional information on inactive compounds. As we have seen, sets of candidate structures can be rapidly compiled from databases or virtual chemical libraries by similarity searching. Practical experience shows that such hypotheses are often weak, and there clearly is no cure-all recipe or generally valid hypothesis leading to success in chemical similarity searching. Nevertheless, similarity searching methods provide a useful concept for rapid first-pass virtual screening and focused library design.

13.4 Fast Binary Classifiers for Library Shaping

Automated compound classification is a further approach for rapid library design. It can be divided into three major steps:

1. Molecule encoding. Several molecular attributes, such as predicted and measured properties and structural descriptors, span a high-dimensional feature space. The selection of descriptors is mainly driven by the experience of the scientists, project-specific considerations, and existing knowledge about putative structure–activity relationships.
2. Class assignment. If activity values are available at this stage, the tested molecules can be assigned class labels representing activity classes (e.g., low, medium, high). Otherwise, classes can be automatically assigned by analysis of the data distribution, e.g., by cluster analysis.

3. Feature extraction. A small set of class-discriminating features is selected (extracted) from the descriptor space, which provide the basis for activity (class) predictions. Traditional feature-extraction methods are based on factor analysis and projection methods [58].

A common theme in molecular feature extraction is the transformation of raw data into a new coordinate system, in which the axes of the new space represent 'factors' or 'latent variables' – features that might help to explain the shape of the original distribution. The most widely applied statistical feature-extraction method in drug design belongs to the class of factorial methods: principal component analysis (PCA). PCA performs a linear projection of data points from the high-dimensional space onto a low-dimensional space. In addition to PCA, nonlinear projection methods like the self-organizing map (SOM) and various types of encoder networks have been employed in drug design projects to address the fact that most structure–activity relationships are inherently nonlinear [7, 59, 60]. Since none of these methods require a priori knowledge of target values (e.g., inhibition constants, properties) or active/inactive assignments, they are considered 'unsupervised'. Unsupervised procedures can be used to perform a first data analysis step, complemented by supervised methods later during the molecular design process (see below).

A straightforward approach to library shaping employs filtering routines that are based on binary classifiers solving two-class problems and are meant to either eliminate potentially unwanted molecules from a compound library or enrich a library with molecules predicted to reveal some kind of desired activity [61, 62]. The basic idea is to define two classes of compounds, one sharing a desired property (the positive set), and another lacking this property (the negative set). Then a binary classifier, e.g., a separating hyperplane, is developed, which can be applied in early-phase virtual screening and compound library shaping. Independent of the particular classification method used and the particular project under investigation, appropriate preprocessing of data is essential for successful feature extraction. Currently, two classifier systems are most often used in these applications: PLS-based classifiers [63, 64] and various types of artificial neural networks (ANN) [65–69]. Typically, these systems follow the 'likeness concept' in virtual screening [66, 67]. A recent addition to this set of methods is the support vector machine (SVM) approach, which was first introduced by Vapnik as a potential alternative to conventional artificial neural networks [70, 71]. Its popularity has grown ever since in various areas of research, and its first applications in molecular informatics and pharmaceutical research have been described [72].

The basic idea of an ANN is to find a typically nonlinear classifier directly in the n-dimensional descriptor space, whereas SVM relies on preprocessing the original n-dimensional data to represent patterns in a much higher-dimensional space. With an appropriate nonlinear mapping to a sufficiently high dimension, two data classes can always be separated by a hyperplane [73, 74]. The classifier found in the very high-dimensional space can be applied to the original n-dimensional data by an elegant method using so-called kernel functions [75, 76].

13.4.1 Artificial Neural Networks

Conventional two-layered neural networks with a single output neuron are the most frequently employed class of ANN models for virtual screening and library design purposes (Figure 13.9). As a result of network training, a decision function is chosen from the family of functions represented by the network architecture. This function family is defined by the complexity of the neural network, i.e., the number of hidden layers, number of neurons in these layers, and topology of the network. The decision function is determined by choosing appropriate weights for the neural network. Optimal weights usually minimize an error function for the particular network architecture. The error function describes the deviation of predicted target values from observed or desired values. For a class/nonclass classification problem, the target values could be 1 for 'class' (e.g., GPCR modulators) and –1 for 'nonclass' (e.g., molecules which are known or expected not to be GPCR modulators). Standard two-layered neural network with a single output neuron can be represented by the following equation:

$$y = \tilde{g}\left[\sum_{j=1}^{M} w_j^{(2)} \cdot g\left(\sum_{i=1}^{d} w_{ji}^{(1)} \cdot x_i + w_{j0}^{(1)}\right) + w_0^{(2)}\right] \tag{13.4}$$

In most of the current ANN applications in virtual screening, $\tilde{g}$ and g are sigmoidal transfer functions of the form $g(in) = 1 / (1 + \exp(-in))$.

A neural network is typically trained by variations of gradient descent-based algorithms, trying to minimize an error function [77]. It is important that additional validation data be left untouched during ANN training, so as to have an objective measure of the model's generalization ability [78].

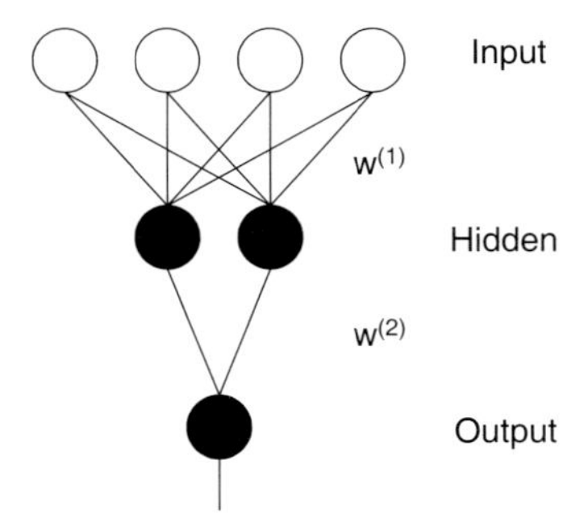

Figure 13.9 Architecture of a fully connected feed-forward network. Formal neurons are drawn as circles, and weights are represented by lines connecting the neuron layers. 'Fan-out' neurons are drawn in white, sigmoidal neurons in black.

13.4.2
Support Vector Machines

SVM classifiers are generated in a two-step procedure. First, the sample data vectors are mapped (projected) to a very high-dimensional space. The dimension of this space is significantly greater than the dimension of the original data space. Then the algorithm finds a hyperplane in this space that separates classes of data by the largest amount (Figure 13.10). Classification accuracy usually depends only weakly on the specific projection, provided that the target space is sufficiently high-dimensional [70]. The decision function represented by an SVM can be expressed as follows (Figure 13.10):

$$D(\mathbf{x}) = \mathrm{sign}\left[\sum_i \alpha_i\ (\mathbf{x}_i^{\mathrm{SV}}, \mathbf{x}) + b\right] \tag{13.5}$$

where α_i are Lagrange multipliers determined during training of SVM. The sum is only over the support vectors $\mathbf{x}^{\mathrm{SV}}$, i.e., those molecules that determine the separating hyperplane. Parameter b determines the shift of the hyperplane, which is also determined during SVM training. For further details see references [70, 71].

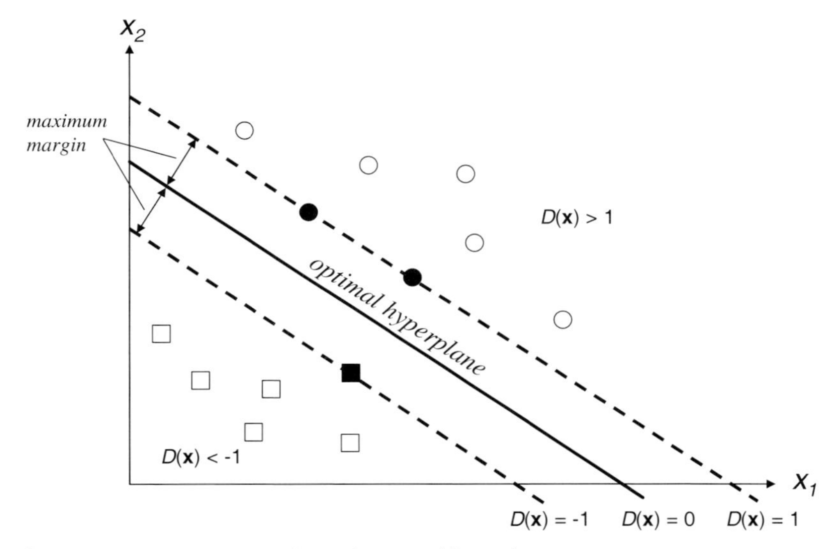

Figure 13.10 SVM training results in the optimal hyperplane separating classes of data. The optimal hyperplane is the one with the maximum distance from the nearest training patterns (support vectors). The three support vectors defining the hyperplane are shown as solid symbols. $D(\mathbf{x})$ is the SVM decision function (classifier function).

13.4.3
An Important Step: Data Scaling

Independent of the particular classification method used and the particular project under investigation, appropriate preprocessing of data is essential for successful feature extraction. We bring this fact to attention here, because appropriate data scaling can be a key to successful library design. In our experience, four scaling methods have been proven useful for meaningful feature extraction: autoscaling, logistic scaling, histogram equalization, and block scaling.

Autoscaling results in data with zero mean and unit variance. Logistic scaling, which is also called 'softmax' transformation, first performs autoscaling and then transforms the data with a logistic function. Histogram equalization transforms the descriptor vectors in three steps: (1) ascending ordering of the values of the descriptor vector, (2) replacing the values with their ordinal numbers, and (3) scaling the ordinal numbers to the interval [0,1]. Block scaling is based on the idea that parts of the descriptor vector having a similar meaning, e.g., topological features vs. charge-based features, are independently scaled (as separate blocks). This can facilitate identification of feature sets that are relevant for a certain structure–activity relationship.

The choice of an appropriate scaling method critically depends on the dataset, on the set of descriptors, and on the classification goal. To demonstrate the influence of scaling on a classification experiment, let us consider drug/nondrug classification as an example. Supervised neural networks were developed with raw and with scaled data (A. Givehchi, G. Schneider, unpublished). The networks contained a single hidden layer containing six hidden units (Figure 13.9). The mean-square-error (mse) values obtained were mse = 0.13 for logistic scaling, mse = 0.14 for histogram scaling, and mse = 0.41 without scaling. With logistic scaling the network model produced the most accurate prediction. To see whether scaling would deliver lower error and better prediction with other update algorithms, various network training methods were used and additional network architectures were tested. It turned out that logistic scaling and histogram equalization generally resulted in predictions with lower error than no scaling. In some combinations of network size and training methods logistic scaling was favorable, and in others histogram equalization yielded better predictions. But the highest mse values were always obtained without scaling, irrespective of the training method and the size of the neural network. These studies show that appropriate data scaling can lead to significantly improved results.

13.4.4
Application to Library Design

How can we exploit such fast filtering systems for library design? One possibility is to use predicted values like drug-likeness, cytotoxicity, 'frequent-hitter-likeness', or aqueous solubility and other generic properties in combination with target-class specific predictions of 'GPCR-ligand-likeness', 'kinase-inhibitor-likeness', and similar properties to define a fitness or quality space (Figure 13.11). SVM and ANN are methods that can be used for this purpose. As a result, each compound represents a point in a prediction space. To eliminate correlation between the individual axes, principal component analysis (PCA) can be performed, and the coordinates of the original quality space are thus represented by orthogonal principal components, i.e., score values of the PCA:

$$\mathbf{X} = \mathbf{S}\,\mathbf{L}^{\mathrm{T}} \tag{13.6}$$

where **X** is the original data matrix (quality space) consisting of n rows (number of compounds in a library) and p columns (prediction values), **S** is the scores matrix with n rows and d columns (number of principal components considered; note that p principal components exist, but usually only the first d components having an eigenvalue > 1 are used), and **L** is the loadings matrix with d columns and p rows representing the correlation between the principal components and the axes of the original data space. T denotes the transpose of the matrix. The new coordinates are linear combinations of the original variables, and the principal components can be considered projections of the original data matrix **X** onto the scores matrix **S**:

$$\mathbf{S} = \mathbf{X}\,\mathbf{L} \tag{13.7}$$

A straightforward method for PCA is the NIPALS (nonlinear iterative partial least squares) algorithm; it is quickly implemented and can be applied to large datasets [58, 79].

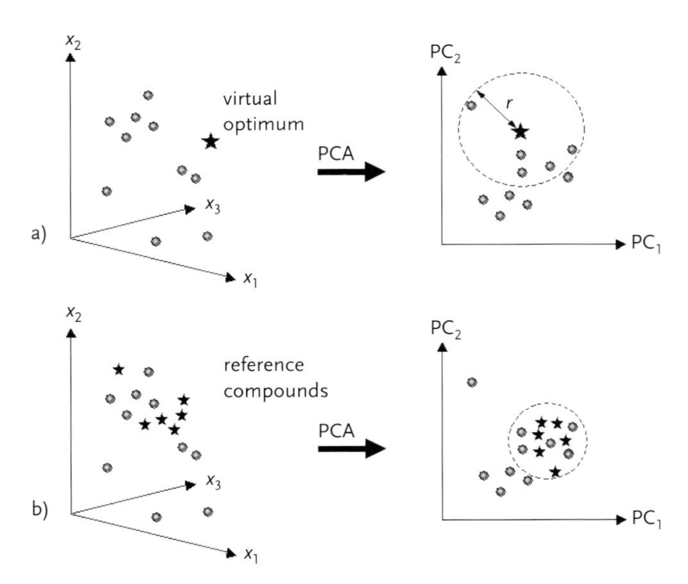

Figure 13.11 Principle of PCA-based library design. The original data space **X** is spiked with a virtual optimum (denoted by a star in a); whereas in b) a set of reference compounds defines the optimum), thereby defining a target area for compound selection. After PCA, the relevant principal components (PC) span an orthogonal space, and distance criteria are used for picking compounds that are closest to the a) virtual optimum, or b) reference compounds.

Now the whole trick of this procedure is the following. Prior to PCA, artificial compounds (or a real set of reference compounds) are added to the data matrix **X**. These additional data points (library 'spikes') have coordinates that represent idealized properties of the library. For example, if the aim is to generate a library for the cannabinoid receptor family, possible coordinates of the idealized artificial compound might be 1 for drug-likeness, 1 for GPCR-ligand-likeness, 1 for cannabinoid-likeness, 0 for dopamine-likeness, 0 for kinase-inhibitor-likeness, and so forth. In this example, the value '1' indicates the maximum value of a property (presence of a feature), '0' indicates minimum values (absence of a feature). Of course, appropriate prediction models must be at hand.

Then PCA is performed, and the compounds of a virtual library are ranked according to their distance from the optimum in principal component space. A focused library can then be compiled by picking those compounds that are closest to the optimum. Of course, many variations of this scheme are feasible, e.g., by application of experimental design techniques. Also, instead of an artificial optimum defined by extreme values, experimentally determined values of reference compounds can be used. For example, if an activity profile of a set of reference molecules is available for a panel of receptor classes, it might be wise to define the relative activities of reference compounds against different target classes as optimal for library design. This might help in the identification of truly isofunctional molecules with a comparable activity panel, yet with a lower risk of generating artifact designs, since constellations of idealized compound features do exist that represent an ill-posed problem, i.e., a contradiction in itself. We should stress that this library-design tactic can only help identify trends and generate an enrichment of activity for a set of molecules (a library). It is not recommended for application to the design of single molecules, simply because all prediction systems are faulty and the definition of optimal properties is somewhat arbitrary. Despite this obvious limitation, the method offers the following advantages:

- Straightforward implementation.
- Speed of execution; large virtual or physically existing libraries can be analyzed.
- Definition of user-, project-, or company-specific filtering criteria and optimal compound properties.
- Possibility to include experimental observations made for reference compounds, e.g., activity profiles, solubility issues, etc.

PCA is not the only projection method that can be used. Various types of nonlinear projections have been employed, e.g., Sammon mapping and nonlinear PCA [80], and several software packages can be used to graphically visualize library distributions and aid compound selection [81].

Fast binary filtering methods can also be used for scaffold ranking, i.e., the prioritization of combinatorial scaffolds based on predicted properties. 'Privileged' scaffolds were selected to demonstrate this idea [82]. Piperazines **S1**, benzodiazepines **S2**, and spiroindolines **S3** have been described as GPCR-privileged scaffolds [83]. Scaffold **S4** represents a SPIKET motif for tubulin binding which is effective for inhibiting cellular proliferation [84]. Dysidiolide-derived compounds

Table 13.2 Rank order of virtual combinatorial libraries based on predicted drug-likeness, cytotoxicity, GPCR-ligand likeness, and kinase-ligand likeness. ++ indicates pronounced positive prediction, + mediocre to slightly positive prediction, – negative prediction (absence of the property).

Score	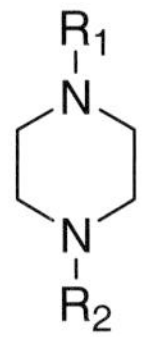 ***S1***	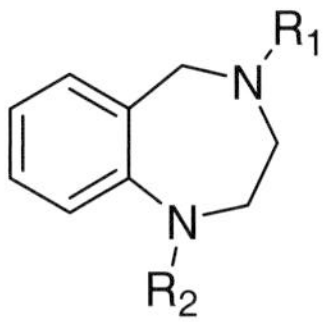 ***S2***	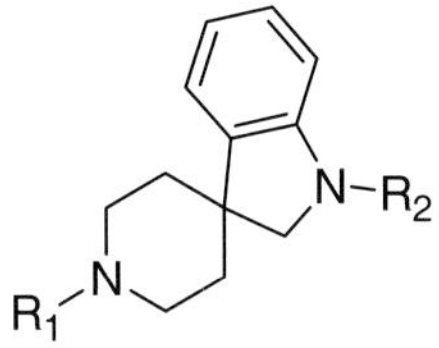 ***S3***	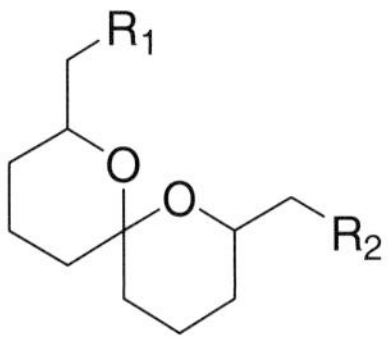 ***S4***	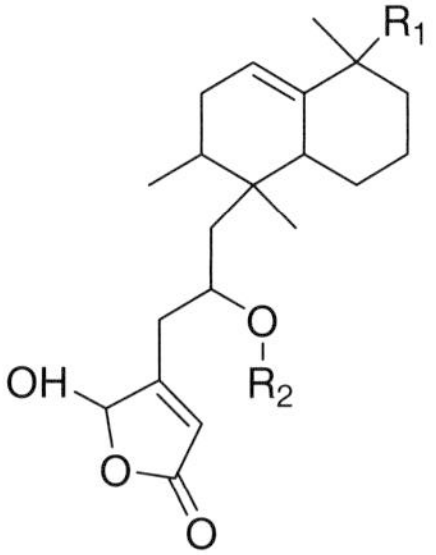 ***S5***	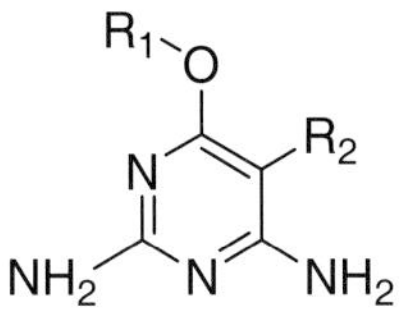 ***S6***
Drug-likeness	++	++	++	++	++	++
Cytotoxicity	+	+	+	++	++	–
GPCR-ligand likeness	++	++	++	–	+	+
Kinase-ligand likeness	–	–	–	–	–	+

S5 exhibit antiproliferative properties [85]. Finally, scaffold **S6** represents a kinase-privileged structure. Derivatives inhibit cyclin-dependent kinases (CDK) 1 and 2 and prevent loss of cell cycle control and uncontrolled proliferation of cancer cells [86]. Each of these scaffolds contains two attachment sites, R_1 and R_2. Virtual libraries were constructed from a set of 60 generic building blocks which were linked to the attachment sites, leading to a maximum of $60 \times 60 = 3600$ virtual products per scaffold [87]. Then the virtual products were encoded by the topological CATS descriptor and subjected to library analysis by various prediction systems. The results are summarized in Table 13.2. Generally, the predictions met the expectations, i.e., target-family preferences become apparent when library averages are considered. It is noteworthy that all six virtual libraries received a high drug-likeness score (indicated by ++), although scaffolds **S4** and **S5** tend to induce a library bias toward antiproliferative properties. The comparison of virtual combinatorial libraries can be further extended, from such viable libraries that are relatively small in size and are assembled from chemical building blocks that have been filtered by medicinal chemists, to massive virtual libraries. As briefly outlined in the introduction to this chapter, such virtual libraries can never be physically synthesized in their entirety; hence, novel methods for in-silico screening must be developed that can cope with this problem. Several such systems have already been conceived, and we expect this area to deliver valuable novel tools soon [7, 88].

13.5 Mapping Chemical Space by Self-organizing Maps: A Pharmacophore Road Map

The introduction of combinatorial chemistry, HTS, and the presence of large compound selections have put us in the comfortable position of having a large number of hits to choose from for lead optimization – at least for certain classes of drug targets. We anticipate that, although the size of the compound libraries and the number of HTS hits will continue to increase, leading to a larger number of hits, the number of leads actually being followed up per project will remain roughly the same. The challenge is to select the most promising candidates for further exploration, and computational techniques will play a very important role in this process. Assuming a hit rate of 0.1%–1% and a compound collection size of 10^6 compounds, we have (or will have) about 1000–10 000 hits that are potential starting points for further work. It is important to realize that, although screening throughput has increased significantly, the throughput of a traditional chemistry laboratory has not. While it is true that automated and/or parallel chemistry is now routinely used, there are still many molecules that are not amenable to these more automated and high-throughput approaches. Thus, the question to be answered is: How can we select the most promising compounds for library design and subsequent optimization? Various computational approaches toward defining pharmacophore road maps that reflect findings in chemogenomics might play a practical role here. The self-organizing map (SOM) has proven its usefulness for drug discovery, in particular, for the tasks of data classification, feature extraction, and visualization

[37, 89–91]. Therefore this method is described in some more detail as one possible way of dealing with the task of similarity-based design of target-family-focused libraries.

The SOM belongs to the class of unsupervised neural networks and was pioneered by Kohonen in the early 1980s [92]. Among other applications, it can be used to generate low-dimensional, topology-preserving projections of high-dimensional data. In contrast to the supervised, multilayered ANN discussed in the previous section of this chapter, the neurons of a SOM adopt either an active or an inactive state. For data processing, the input pattern (a molecular descriptor vector) is compared to all neurons of the SOM, and the one neuron that is most similar to the input pattern – the so-called winner neuron – fires a signal, i.e., it is active. All other neurons are inactive. In this way, each input pattern is assigned to exactly one neuron. The data patterns belonging to a neuron form a cluster, since they are more similar to their neuron than to any other neuron of the SOM. During the SOM training process – an optimization procedure following the principles of unsupervised Hebbian learning [93] – the original high-dimensional space is tessellated, resulting in a certain number of data clusters. As many clusters are formed as there are neurons in the SOM. The neurons represent prototype vectors of each cluster, and the resulting prototype vectors capture features of the input space that are unique to each data cluster. Feature analysis can be done, e.g., by comparing adjacent neurons. Kohonen's SOM algorithm represents a strikingly efficient way of mapping similar patterns, given as vectors close to each other in input space, onto contiguous locations in the output space (the so-called map). For an introductory overview to SOM and further details, see references [7, 37, 92].

The SOM approach can be applied to visualizing a chemical space, e.g., the distribution of reference compounds in a pharmacophore space. Figure 13.12 displays the areas populated by six different ligand classes. It is evident that the particular molecular descriptor used for this purpose (the 2D CATS descriptor) provides a basis for rough discrimination of the classes. Although the separation is not perfect and the descriptor may not be suited for all ligand classes (see. e.g., the scattering of kinase inhibitors), several activity islands are visible, i.e., clusters of neurons containing a significant fraction of ligands of one class. This visualization serves two purposes: (1) to assess the suitability of a molecular representation and its discrimination power; and (2) to use the map for library design. The latter has been exemplified for the task of identifying members of a combinatorial library that specifically bind to the purinergic receptor subtype A_{2A} [89]. Scaffold **8** provided the basis for virtual library enumeration and projection onto a SOM that was developed by using a reference set of known purinergic receptor ligands. Compound **9** was picked from the most promising activity island, synthesized, and tested; it had a $K_i = 2.4$ nM and 120-fold selectivity over the A_1 receptor subtype. Overall, the focused library had binding affinity three times that of the reference set.

A list of privileged scaffolds – several of which are natural-product derived – for target-family-biased combinatorial libraries was recently presented by Müller [94]. These scaffolds were proven to produce biologically active compounds for more than one member of a given target family. A rough-and-ready in-silico evaluation

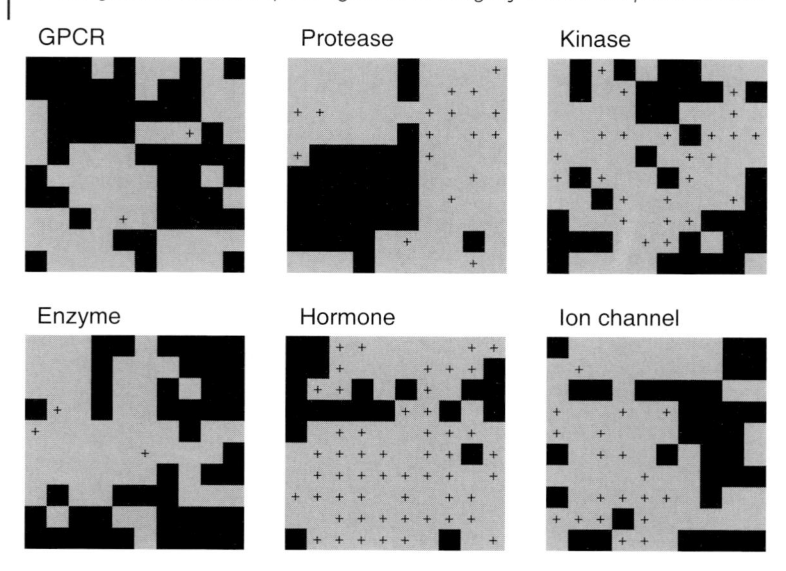

Figure 13.12 A SOM-based pharmacophore road map. Different sets of ligands were projected onto a SOM that was generated by using the complete COBRA library. Black areas indicate the characteristic distributions of the compounds. Crosses indicate empty neurons in the map, i.e., areas of pharmacophore space that are not populated by the respective compound class. All molecules were encoded by a topological pharmacophore descriptor (CATS) [4]. Note that each map forms a torus.

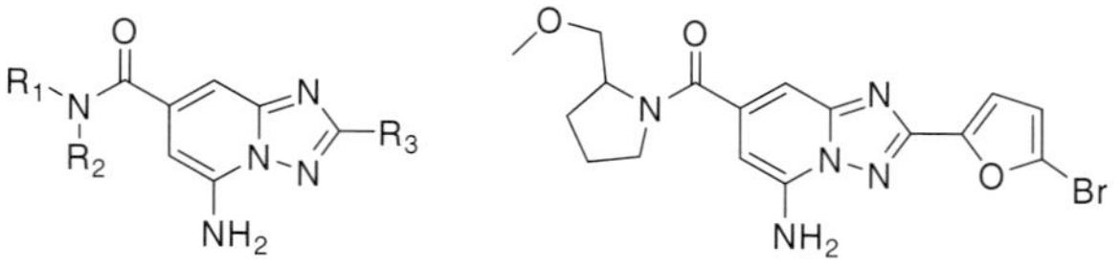

8 **9**

Scheme 13.3

of the preferred target families of such structures can be made by projecting the appropriate virtual combinatorial libraries onto a SOM that was developed using a representative set of drugs and drug-like compounds, e.g., the COBRA reference structures. Figure 13.13 gives an impression of the distribution of a privileged library: 20 generic building blocks were used for scaffold decoration, obtaining a virtual library containing $20 \times 20 \times 20 = 8000$ compounds. Although the building blocks were not specifically selected, an apparent overlap of the densely populated areas (shown in red) of both maps can be observed. This straightforward analysis demonstrates how the approach can help qualify combinatorial libraries. The study was performed in product space (rather than educt space) and complements established techniques for building-block selection, such as statistical design or

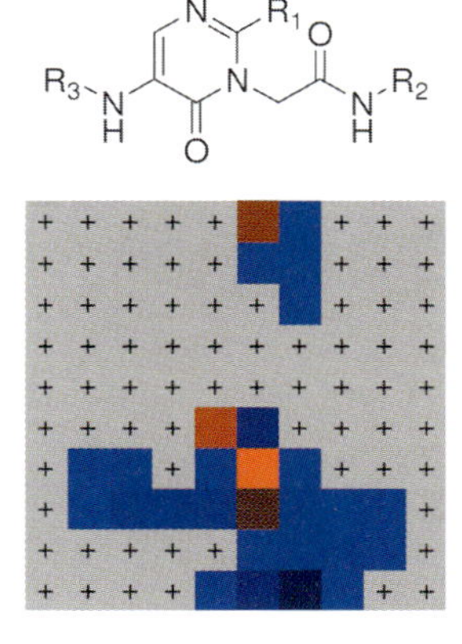

Figure 13.13 SOM showing the distribution of known serine protease inhibitors (left), and a virtual combinatorial library that was constructed around a serine protease-privileged scaffold [94]. Red areas indicate high compound density. Note that each map forms a torus.

heuristic sampling procedures [21, 95]. Of course, it accordingly is limited to manageable library sizes. A related SOM approach was also successfully employed for identification of novel natural-product-derived scaffolds, following the strategy outlined here [90].

Although the sample SOMs shown in this chapter are relatively coarse, they already point to the matter of overlapping activity areas, i.e., potential promiscuous binding behavior of compounds. An example is provided by the well known unspecific binding behavior of many serotonin receptor ligands (Figure 13.14). The areas populated by known dopamine and serotonin receptor ligands are in fact very similar and cannot be distinguished, considering the resolution and inherent noise of the SOM. Two promiscuous binders, Sertindole (primary target: 5-HT2a) and Clozapine (primary target: D4.2), were projected onto the COBRA SOM. Based on their location on the SOM, they would actually be predicted to exhibit binding activity to various dopamine and serotonin receptor subtypes and some related receptors. A literature study of known activities reported for the two compounds confirmed the SOM prediction. We want to again stress that, due to the coarseness of the approach, such predictions should be made only for whole libraries, and this example is just meant to demonstrate the concept. Despite the appeal of the SOM technique, we should point out that several alternatives exist, and the SOM concept might not even be the wisest option for focused library design. SOMs can even be misleading, e.g., due to mapping errors, the usual high-dimensionality of the data, premature end of training, the problem of local optima, and other issues. Additional information and different approaches to the nonlinear mapping task can be found elsewhere [96]. An extension of the SOM, the visualization-induced SOM (ViSOM), was presented recently to overcome some of the limitations of manifold mapping by conventional SOMs. In particular, ViSOM directly preserves distance information on the map, along with the topology [97].

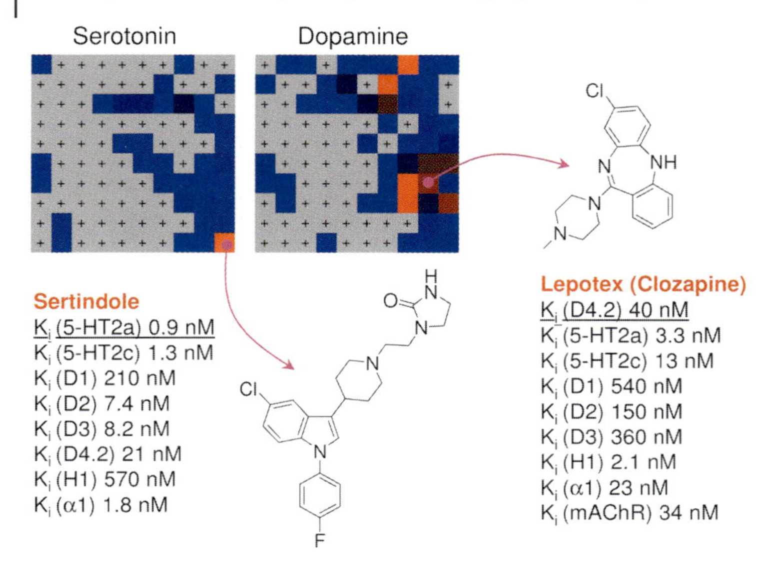

Figure 13.14 Projection of two promiscuous binders, Sertindole (primary target: 5-HT2a, left) and Clozapine (primary target: D4.2, right), onto the COBRA SOM. The distribution of known serotonin receptor ligands is shown in the left map, the distribution of known dopamine receptor ligands is shown in the right map. The colored areas of the two maps overlap, indicating similar activity of the compounds. Blue: few compounds; red: many compounds; crosses indicate unpopulated areas of pharmacophore space. Colors were scaled separately for each map.

Although this method has not yet been applied to library design, it represents a promising approach that might help improve the SOM-based pharmacophore road map.

The observation of promiscuous binding behavior has also been described for other ligand classes, e.g., steroids and kinase inhibitors [94, 98]. We will certainly see many more such surprises in the future as chemogenomics techniques are increasingly applied. A pharmacophore road map can help unearth such candidates. A related large-scale analysis of compound activity was carried out by Covell and coworkers [99], who performed a SOM cluster analysis using the National Cancer Institute's tumor-screening database and found indications of compound selectivity between various types of cellular activity. This study supports the idea of SOM-based compound clustering to identify receptor-family-specific pharmacophore patterns.

13.6 Concluding Remarks

The issue of designing compound libraries that are focused toward a target family has been approached by many research teams in the pharmaceutical industry and in academia, and various strategies are emerging within the teams, relying on similarity searching and virtual screening, structure-based design, and high-throughput analytical methods [100]. Despite the many studies that have already been performed, the availability of chemogenomics data will enable a more thorough investigation of neighborhood behavior of virtual chemistry spaces with respect to biological activity and receptor spaces. In this chapter, we presented only a small fraction of what is already possible from a ligand-based virtual screening perspective. Still, this process of connecting target with ligand space has only just begun, and there are several crucial questions to be answered – independent of the particular computational concept chosen, e.g., How can the large body of available HTS data be fully exploited for target-family-specific SAR modeling [101]? Which level of fuzziness of pharmacophoric descriptors is appropriate for a particular target family and allows for scaffold hopping within the respective ligand family [102]? How can multidimensional optimization be implemented so that pharmacokinetic and pharmacodynamic profiling of focused compound libraries can be addressed during the early library design phase [103]? How can promiscuous binders, frequent hitters, and target-family-specific preferred scaffolds be differentiated and systematically identified [16, 94, 104]? The concept of chemogenomics might provide an approach to solving some of these questions, and molecular informatics represents a key discipline for coming up with practical solutions. Fragment-based library assembly, e.g., by combinatorial design, has proven to be suitable for rapid lead identification [105]. A challenging task in this respect is to refine rules for lead-likeness (in contrast to generic drug-likeness) so that they are amenable to virtual screening of virtual lead-like libraries [106], and it remains to be seen whether generic definitions of lead-likeness will have to be adapted to meet the requirements of target-family-specific ligand design. Chemogenomics studies will undoubtedly push virtual screening concepts and methods to a higher level, as we will once more have to face the fact that both the drug-design process and the target organisms represent complex systems, and adaptation of project strategies and technologies are key to future success in this game [107].

Acknowledgements

We are most grateful to Evgeny Byvatov, Norbert Dichter, Uli Fechner, Alireza Givehchi, Lutz Franke, Steffen Renner, and Jochen Zuegge for valuable contributions to this chapter and many stimulating discussions. This work was supported by the Beilstein-Institut zur Förderung der Chemischen Wissenschaften.

References

1 (a) E. Jacoby, A. Schuffenhauer, P. Floersheim, *Drug News Perspect.* **2003**, *16*, 93–102; (b) D. K. Agrafiotis, V. S. Lobanov, F. R. Salemme, *Nat. Rev. Drug Discov.* **2002**, *1*, 337–346; (c) K. H. Bleicher, *Curr. Med. Chem.* **2002**, *9*, 2077–2084; (d) X. F. Zheng, T. F. Chan, *Curr. Issues Mol. Biol.* **2002**, *4*, 33–43; (e) C. Nislow, G. Giaever, *Pharmacogenomics* **2003**, *4*, 15–18.

2 (a) A. Schuffenhauer, P. Floersheim, P. Acklin, E. Jacoby, *J. Chem. Inf. Comput. Sci.* **2003**, *43*, 391–405; (b) T. Naumann, H. Matter, *J. Med. Chem.* **2002**, *45*, 2366–2378.

3 K. H. Bleicher, H.-J. Böhm, K. Müller, A. I. Alanine, *Nat. Rev. Drug Discov.* **2003**, *2*, 369–378.

4 G. Schneider, W. Neidhart, T. Giller, G. Schmid, *Angew. Chem., Int. Ed. Engl.* **1999**, *38*, 2894–2896.

5 D. V. Green, *Prog. Med. Chem.* **2003**, *41*, 61–97.

6 R. S. Bohacek, C. McMartin, W. C. Guida, *Med. Res. Rev.* **1996**, *16*, 3–50.

7 G. Schneider, S.-S. So, *Adaptive Systems in Drug Design*, Landes Bioscience, Georgetown TX **2003**.

8 P. Willett, in *Handbook of Chemoinformatics* (Ed. J. Gasteiger), Wiley-VCH, Weinheim **2003**, pp. 904–915.

9 P. Schneider, G. Schneider, *QSAR Comb. Sci.* **2003**, *22*, 713–718.

10 J. Drews, *Science* **2000**, *287*, 1960–1964.

11 (a) K. V. Balakin, S. A. Lang, A. V. Skorenko, S. E. Tkachenko, A. A. Ivashchenko, N. P. Savchuk, *J. Chem. Inf. Comput. Sci.* **2003**, *43*, 1533–1562; (b) K. V. Balakin, S. E. Tkachenko, S. A. Lang, I. Okun, A. A. Ivashchenko, N. P. Savchuk, *J. Chem. Inf. Comput. Sci.* **2002**, *42*, 1332–1342.

12 S. A. Lang, A. V. Kozyukov, K. V. Balakin, A. V. Skorenko, A. A. Ivashchenko, N. P. Savchuk, *J. Comput. Aided Mol. Des* **2002**, *16*, 803–807.

13 C. A. Lipinski, F. Lombardo, B. W. Dominy, P. J. Feeney, *Adv. Drug Delivery Rev.* **1997**, *23*, 3–25.

14 A. K. Ghose, V. N. Viswanadhan, J. J. Wendoloski, *J. Comb. Chem.* **1999**, *1*, 55–68.

15 (a) Ajay, *Curr. Top. Med. Chem.* **2002**, *2*, 1273–1286; (b) W. P. Walters, M. A. Murcko, *Adv. Drug Deliv. Rev.* **2002**, *54*, 255–271.

16 O. Roche, P. Schneider, J. Zuegge, W. Guba, M. Kansy, A. Alanine, K. Bleicher, F. Danel, E.-M. Gutknecht, M. Rogers-Evans, W. Neidhart, H. Stalder, M. Dillon, E. Sjogren, N. Fotouhi, P. Gillespie, R. Goodnow, W. Harris, P. Jones, M. Taniguchi, S. Tsujii, W. von der Saal, G. Zimmermann, G. Schneider, *J. Med. Chem.* **2002**, *45*, 137–142.

17 W. J. Egan, W. P. Walters, M. A. Murcko, *Curr. Opin. Drug Discov. Devel.* **2002**, *5*, 540–549.

18 (a) C. A. Lipinski, *Pharmacol. Toxicol. Meth.* **2000**, *44*, 235–249; (b) see also http://www.iainm.demon.co.uk/spring99/lipins.pdf.

19 T. I. Oprea, *J. Comput. Aided Mol. Des.* **2002**, *16*, 325–334.

20 (a) T. Klabunde, G. Hessler, *ChemBioChem* **2002**, *3*, 928–944; (b) V. J. Gillett, P. Willett, P. J. Fleming, D. V. Green, *J. Mol. Graph. Model.* **2002**, *20*, 491–498; (c) D. K. Agrafiotis, *Mol. Divers.* **2002**, *5*, 209–230.

21 A. K. Ghose, V. N. Viswanadhan (Eds.), *Combinatorial Library Design and Evaluation*, Marcel Dekker, New York **2001**.

22 R. Todeschini, V. Consonni, *Handbook of Molecular Descriptors*, Wiley-VCH, Weinheim **2000**.

23 (a) P. Willett, *Curr. Opin. Biotechnol.* **2000**, *11*, 85–88; (b) P. M. Dean, R. A. Lewis (Eds.), *Molecular Diversity in Drug Design*, Kluwer Academic, Dordrecht **1999**; (c) M. Rarey, M. Stahl, *J. Comput. Aided Mol. Des.* **2001**, *15*, 497–520; (d) J. W. Raymond, P. Willett, *J. Comput. Aided Mol. Des.* **2002**, *16*, 59–71.

24 H. Xu, D. K. Agrafiotis, *Curr. Top. Med. Chem.* **2002**, *2*, 1305–1320.

25 H. Kubinyi, in *Handbook of Chemoinformatics* (Ed. J. Gasteiger), Wiley-VCH, Weinheim **2003**, pp. 1555–1574.

26 Z. Liu, Y. Wang, J. Zhang, J. Ding, L. Guo, D. Cui, J. Fei, *NeuroReport* **2001**, *12*, 699–702.

27 T. Breidert, J. Callebert, M. T. Heneka, G. Landreth, J. M. Launay, E. C. Hirsch, *J. Neurochem.* **2002**, 82, 615–624.

28 L. Xue, J. W. Godden, F. L. Stahura, J. Bajorath, *J. Chem. Inf. Comput. Sci.* **2003**, *43*, 1151–1157.

29 (a) MACCS keys. MDL Information Systems, Inc., San Leandro, CA; (b) C. A. James, D. Weininger, *Daylight Theory Manual*. Daylight Chemical Information Systems, Inc., Irvine, CA **2003**; (c) UNITY. Chemical Information Software, Tripos, Inc., St. Louis, MO.

30 M. J. McGregor, S. M., *J. Chem. Inf. Comput. Sci.* **1999**, *39*, 569–574.

31 J. S. Mason, D. L. Cheney, *Pac. Symp. Biocomput.* **1999**, *4*, 456–467.

32 L. Xue, J. Godden, H. Gao, J. Bajorath, *J. Chem. Inf. Comput. Sci.* **1999**, *39*, 699–704.

33 D. J. Livingstone, *J. Chem. Inf. Comput. Sci.* **2000**, *40*, 195–209.

34 A. K. Ghose, V. N. Viswanadhan, J. J. Wendoloski, in *Combinatorial Library Design and Evaluation* (Eds. A. K. Ghose, V. N. Viswanadhan), Marcel Dekker, New York **2001**, pp. 51–71.

35 (a) J. Schnitker, R. Gopalaswamy, G. M. Crippen, *J. Comput. Aided Mol. Des.* **1997**, *11*, 93–110; (b) S. Cui, X. Wang, S. Liu, L. Wang, *SAR QSAR Env. Res.* **2003**, *14*, 223–231; (c) N. E. Jewell, D. B. Turner, P. Willett, G. J. Sexton, *J. Mol. Graph. Model.* **2001**, *20*, 111–121.

36 P. Broto, G. Moreau, C. Vandyke, *Eur. J. Med. Chem.* **1984**, *19*, 66–70.

37 (a) H. Bauknecht, A. Zell, H. Bayer, P. Levi, M. Wagener, J. Sadowski, J. Gasteiger, *J. Chem. Inf. Comput. Sci.* **1996**, *36*, 1205–1213; (b) S. Anzali, G. Barnickel, M. Krug, J. Sadowski, M. Wagener, J. Gasteiger, J. Polanski, *J. Comput. Aided Mol. Des.* **1996**, *10*, 521–534; (c) J. Zupan, J. Gasteiger, *Neural Networks in Chemistry and Drug Design*, Wiley-VCH, Weinheim **1999**.

38 G. Schneider, O. Chomienne-Clement, L. Hilfiger, S. Kirsch, H.-J. Böhm, P. Schneider, W. Neidhart, *Angew. Chem. Int. Ed.* **2000**, *39*, 4130–4133.

39 R. E. Carhart, D. H. Smith, R. Venkataraghavan, *J. Chem. Inf. Comput. Sci.* **1985**, *25*, 64–73.

40 (a) U. Fechner, L. Franke, S. Renner, P. Schneider, G. Schneider, *J. Comput. Aided Mol. Des.* **2003**, in press; (b) U. Fechner, G. Schneider, *ChemBioChem* **2004**, in press.

41 (a) O. Guner (Ed.), *Pharmacophore Perception, Development and Use in Drug Design*, International University Line, La Jolla CA **2000**; (b) S. Pickett, in *Protein–Ligand Interactions* (Eds. H.-J. Böhm, G. Schneider), Wiley-VCH, Weinheim **2003**, pp. 73–105; (c) T. Langer, E. M. Krovat, *Curr. Opin. Drug Discov. Devel.* **2003**, *2*, 481–485.

42 J. M. Barnard, G. M. Downs, P. Willett, in *Virtual Screening of Bioactive Molecules* (Eds. H.-J. Böhm, G. Schneider), Wiley-VCH, Weinheim **2003**, pp. 59–80.

43 M. Stahl, M. Rarey, G. Klebe, in *Bioinformatics: From Genomes to Drugs, Vol. 2* (Ed. T. Lengauer) Wiley-VCH, Weinheim **2001**, pp. 137–170.

44 X. Chen, C. H. Reynolds, *J. Chem. Inf. Comput. Sci.* **2002**, *42*, 1407–1414.

45 (a) R. D. Brown, Y. C. Martin, *J. Chem. Inf. Comput. Sci.* **1996**, *36*, 572–584; (b) R. D. Brown, Y. C. Martin, *J. Chem. Inf. Comput. Sci.* **1997**, *37*, 1–9.

46 G. Schneider, *Curr. Med. Chem.* **2002**, *9*, 2095–2101.

47 A. Tropsha, W. Zheng, *Comb. Chem. High Throughput Screen.* **2002**, *5*, 111–123.

48 J. Greene, S. Kahn, H. Savoj, P. Sprague, S. Teig, *J. Chem. Inf. Comput. Sci.* **1994**, *34*, 1297–1308.

49 Y. C. Martin, M. G. Bures, E. A. Danaher, J. DeLazzer, I. Lico, P. A. Pavlik, *J. Comput. Aided Mol. Des.* **1993**, *7*, 83–102.

50 G. Jones, P. Willett, R. C. Glen, *J. Comput. Aided Mol. Des.* **1995**, *9*, 532–549.

51 MOE, Molecular Operating Environment. Distributor: Chemical Computing Group, Montreal, Canada.

52 A. C. Good, I. D. Kuntz, *J. Comput. Aided Mol. Des.* **1995**, *9*, 373–379.

53 J. S. Mason, I. Morize, P. R. Menard, D. L. Cheney, C. Hulme, R. F. Labaudiniere, *J. Med. Chem.* **1999**, *42*, 3251–3264.

54 L. Xue, F. L. Stahura, J. W. Godden, J. Bajorath, *J. Chem. Inf. Comput. Sci.* **2001**, *41*, 746–753.

55 D. Horvath, B. Mao, *QSAR Comb. Sci.* **2003**, *22*, 498–509.

56 S. Renner, G. Schneider **2004**, unpublished.

57 A. Palomer, F. Cabre, J. Pascual, J. Campos, M. A. Trujillo, A. Entrena, M. A. Gallo, L. Garcia, D. Mauleon, A. Espinosa, *J. Med. Chem.* **2002**, *45*, 1402–1411.

58 (a) D. N. Lawley, A. E. Maxwell, *Factor Analysis as a Statistical Method*, Butterworth, London **1971**; (b) L. Eriksson, E. Johansson, N. Kettaneh-Wold, S. Wold, *Introduction to Multi- and Megavariate Data Analysis Using Projection Methods (PCA & PLS)*, Umetrics, Umeå **1999**.

59 H. Kubinyi, in *Computer-Assisted Lead Finding and Optimization: Current Tools for Medicinal Chemistry* (Eds. H. van de Waterbeemt, B. Testa, G. Folkers), Wiley-VCH, Weinheim **1997**, pp. 9–28.

60 D. K. Agrafiotis, V. S. Lobanov, *J. Chem. Inf. Comput. Sci.* **2000**, *40*, 1356–1362.

61 D. E. Clark, S. D. Pickett, *Drug Discov. Today* **2000**, *5*, 49–58.

62 G. Schneider, H.-J. Böhm, *Drug Discov. Today* **2002**, *7*, 64–70.

63 S. Wold, *Chemom. Intell. Lab. Syst.* **1994**, *23*, 149–161.

64 M. Forina, M. C. Casolino, C. de la Pezuela Martinez, *J. Pharm. Biomed. Anal.* **1998**, *18*, 21–33.

65 J. Devillers (Ed.), *Neural Networks in QSAR and Drug Design*, Academic Press, London **1996**.

66 Ajay, W. P. Walters, M. A. Murcko, *J. Med. Chem.* **1998**, *41*, 3314–3324.

67 J. Sadowski, H. Kubinyi, *J. Med. Chem.* **1998**, *41*, 3325–3329.

68 J. Sadowski, *Curr. Opin. Chem. Biol.* **2000**, *4*, 280–282.

69 G. Schneider, *Neural Networks* **2000**, *13*, 15–16.

70 C. Cortes, V. Vapnik, *Machine Learning* **1995**, *20*, 273–297.

71 V. Vapnik, *The Nature of Statistical Learning Theory*, Springer, Berlin **1995**.

72 E. Byvatov, G. Schneider, *Appl. Bioinf.* **2003**, *2*, 67–77.

73 T. F. Coleman, Y. Li, *SIAM J. Optimization* **1996**, *6*, 1040–1058.

74 T. Joachims, in *Advances in Kernel Methods: Support Vector Learning* (Eds. B. Schölkopf, C. Burges, A. Smola), MIT Press, Cambridge MA **1999**, pp. 41–56.

75 N. Cristianini, J. Shawe-Taylor, *An Introduction to Support Vector Machines and Other Kernel-based Learning Methods*, Cambridge University Press, Cambridge **2000**.

76 C. J. C Burges, *Data Mining and Knowledge Discovery* **1998**, *2*, 121–167.

77 (a) D. E. Rumelhart, J. L. McClelland, The PDB Research Group, *Parallel Distributed Processing*, MIT Press, Cambridge MA **1986**; (b) M. T. Hagan, H. B. Demuth, M. H. Beale, *Neural Network Design*, PWS Publishing, Boston **1996**.

78 R. O. Duda, P. E. Hart, D. G. Stork, *Pattern Classification*, Wiley Interscience, New York **2000**.

79 M. Otto, *Chemometrics*, Wiley-VCH, Weinheim **1999**.

80 (a) D. K. Agrafiotis, V. S. Lobanov, *J. Chem. Inf. Comput. Sci.* **2000**, *40*, 1356–1362; (b) D. K. Agrafiotis, *J. Comput. Chem.* **2003**, *24*, 1215–1221; (c) B. Bienfait, J. Gasteiger, *J. Mol. Graph. Model.* **1997**, *15*, 203–215.

81 For example: a) A. R. Leach, J. Bradshaw, D. V. Green, M. M. Hann, J. J. Delany 3rd, *J. Chem. Inf. Comput. Sci.* **1999**, *39*, 1161–1172; (b) J. S. Mason, B. R. Beno, *J. Mol. Graph. Model.* **2000**, *18*, 438–451; (c) A. Givehchi, A. Dietrich, P. Wrede, G. Schneider, *QSAR Comb. Sci.* **2003**, *22*, 549–559; (d) R. S. Pearlman, K. M. Smith, *Persp. Drug Discov. Des.* **1998**, *9–11*, 339–353.

82 D. J. Triggle, *Cell. Mol. Neurobiol.* **2003**, *23*, 293–303.

83 K. H. Bleicher, *Curr. Med. Chem.* **2002**, *9*, 2077–2084.

84 M. Faith, PCT Int. Appl. WO 00/00514, **2000**.

85 D. Brohm, N. Philippe, S. Metzger, A. Bhargava, O. Mueller, F. Lieb,

H. Waldmann, *J. Am. Chem. Soc.* **2002**, *124*, 13171–13178.

86 V. Mesguiche, R. J. Parsons, C. E. Arris, J. Bentley, F. T. Boyle, N. J. Curtin, T. G. Davies, J. A. Endicott, A. E. Gibson, B. T. Golding, R. J. Griffin, P. Jewsbury, L. N. Johnson, D. R. Newell, M. E. Noble, L. Z. Wang, I. R. Hardcastle, *Bioorg. Med. Chem. Lett.* **2003**, *13*, 217–222.

87 A. Schüller, G. Schneider, E. Byvatov, *QSAR Comb. Sci.* **2003**, *22*, 719–721.

88 V. S. Lobanov, D. K. Agrafiotis, *Comb. Chem. High Throughput Screen.* **2002**, *5*, 167–178.

89 G. Schneider, M. Nettekoven, *J. Comb. Chem.* **2003**, *5*, 233–237.

90 M.-L. Lee, G. Schneider, *J. Comb. Chem.* **2001**, *3*, 284–289.

91 D. Korolev, K. V. Balakin, Y. Nikolsky, E. Kirillov, Y. A. Ivanenkov, N. P. Savchuk, A. A. Ivashchenko, T. Nikolskaya, *J. Med. Chem.* **2003**, *46*, 3631–3643.

92 T. Kohonen, *Self-Organization and Associative Memory*. Springer, Heidelberg **1984**.

93 (a) D. O. Hebb, *The Organization of Behaviour*, Wiley, New York **1949**; (b) C. M. Bishop, *Neural Networks for Pattern Recognition*, Clarendon Press, Oxford.

94 G. Müller, *Drug Discov. Today* **2003**, *8*, 681–691.

95 (a) D. K. Agrafiotis, *Mol. Divers.* **2002**, *5*, 209–230; (b) A. Linusson, J. Gottfries, F. Lindgren, S. Wold, *J. Med. Chem.* **2000**, *43*, 1320–1328; (c) R. P. Sheridan, S. G. SanFeliciano, S. K. Kearsley, *J. Mol. Graph. Model.* **2000**, *18*, 320–334; (d) R. D. Brown, M. Hassan, M. Waldmann, *J. Mol. Graph. Model.* **2000**, *18*, 427–437; (e) S. Shi, Z. Peng, J. Kostrowicki, G. Paderes, A. Kuki, *J. Mol. Graph. Model.* **2000**, *18*, 478–496; (f) G. Schneider, *Curr. Med. Chem.* **2002**, *9*, 2095–2101.

96 (a) C. L. Chang, R. C. T. Lee, *IEEE Trans. Syst. Man. Cybern.* **1973**, *3*, 197–200; (b) R. C. T. Lee, J. R. Slagle, H. Blum, *IEEE Trans. Comput.* **1977**, *C-27*, 288–292; (c) G. Biswas, A. K. Jain, R. C. Dubes, *IEEE Trans. Pattern Anal. Machine Intell.* **1981**, *3*, 701–708; (d) J. Mao, A. K. Jain, *IEEE Trans. Neural Networks* **1995**, *6*, 296–317; (e) D. J. Livingstone, in *Neural Networks in QSAR and Drug Design* (Ed. J. Devillers), Academic Press, London **1996**, pp 157–176.

97 H. Yin, *Neural Networks* **2002**, *15*, 1005–1016.

98 S. L. McGovern, B. K. Shoichet, *J. Med. Chem.* **2003**, *46*, 1478–1483.

99 A. A. Rabow, R. H. Shoemaker, E. A. Sausville, D. G. Covell, *J. Med. Chem.* **2002**, *45*, 818–840.

100 (a) D. F. Wyss, *Drug Discov. Today* **2003**, *8*, 924–926; (b) G. Wess, M. Urmann, B. Sickenberger, *Angew. Chem., Int. Ed. Engl.* **2001**, *40*, 3341–3350; (c) J. D. Holliday, N. Salim, M. Whittle, P. Willett, *J. Chem. Inf. Comput. Sci.* **2003**, *43*, 819–828; (d) X. Chen, C. H. Reynolds, *J. Chem. Inf. Comput. Sci.* **2003**, *43*, 1407–1414; (e) R. D. Brown, Y. C. Martin, *SAR QSAR Environ. Res.* **1998**, *8*, 23–39; (f) R. D. Brown, Y. C. Martin, *J. Chem. Inf. Comput. Sci.* **1996**, *36*, 572–584; (g) C. M. R. Ginn, P. Willett, J. Bradshaw, *Perspec. Drug. Disc. Des.* **2000**, *20*, 1–16.

101 E. H. Kerns, L. Di, *Drug Discov. Today* **2003**, *8*, 316–323.

102 (a) C. Merlot, D. Domine, C. Cleva, D. J. Church, *Drug Discov. Today* **2003**, *8*, 594–602; (b) D. Horvath, C. Jeandenans, *J. Chem. Inf. Comput. Sci.* **2003**, *43*, 680–690.

103 (a) F. Darvas, G. Keseru, A. Papp, G. Dorman, L. Urge, P. Krajcsi, *Curr. Top. Med. Chem.* **2002**, *2*, 1287–1304; (b) A. Alanine, M. Nettekoven, E. Roberts, A. W. Thomas, *Comb. Chem. High Throughput Screen.* **2003**, *6*, 51–66; (c) O. Schwardt, H. Kolb, B. Ernst, *Curr. Top. Med. Chem.* **2003**, *3*, 1–9; (d) O. Roche, G. Trube, J. Zuegge, P. Pflimlin, A. Alanine, G. Schneider, *ChemBioChem* **2002**, 3, 455–459.

104 S. L. McGovern, E. Caselli, N. Grigorieff, B. K. Shoichet, *J. Med. Chem.* **2002**, *45*, 1712–1722.

105 (a) C. A. Lepre, J. Peng, J. Fejzo, N. Abdul-Manan, J. Pocas, M. Jacobs, X. Xie, J. M. Moore, *Comb. Chem. High Throughput Screen.* **2002**, *5*, 583–590;

(b) A. R. Leach, M. M. Hann, *Drug Discov. Today* **2000**, *5*, 326–336; (c) H. M. Vinkers, M. R. de Jonge, F. F. Daeyaert, J. Heeres, L. M. Koymans, J. H. van Lenthe, P. J. Lewi, H. Timmerman, K. Van Aken, P. A. J. Janssen, *J. Med. Chem.* **2003**, *46*, 2765–2773.

106 (a) M. Congreve, R. Carr, C. Murray, H. Jhoti, *Drug Discov. Today* **2003**, *8*, 876–877; (b) J. R. Proudfoot, *Bioorg. Med. Chem. Lett.* **2002**, *12*, 1647–1650; (c) S. J. Teague, A. M. Davis, P. D. Leeson, T. Oprea, *Angew. Chem. Int. Ed.* **1999**, *38*, 3743–3748; (d) M. Hann, A. R. Leach, G. Harper, *J. Chem. Inf. Comput. Sci.* **2001**, *41*, 856–864; (e) T. I. Oprea, *J. Chem. Inf. Comput. Sci.* **2001**, *41*, 1308–1315.

107 (a) J. Drews, S. Ryser, *Nat. Biotechnol.* **1997**, *15*, 1318–1319; (b) J. Drews, *Drug Discov. Today* **2003**, *8*, 411–420.

14
Natural Product-derived Compound Libraries and Protein Structure Similarity as Guiding Principles for the Discovery of Drug Candidates

Marcus A. Koch and Herbert Waldmann

14.1
Introduction

The last decade brought tremendous gains in biological information through large-scale and global approaches addressing the aspects of DNA sequence (genomics), protein structure (structural genomics), and protein expression and interactions (proteomics). Bioinformatics tools help to convert this vast amount of basic data into actual knowledge exploitable for the benefit of mankind, in particular for the development of new therapies for diseases. Of preeminent interest is the relationship between protein structure and function, as its understanding will help to find small molecules that alter protein function by selective inhibition or activation [1, 2]. On the other hand, tight-binding, target-specific small molecule tool-compounds can be used to study the biological functions of a known target protein or to validate it as a drug target. This chemobiological strategy, commonly subsumed under the terms 'chemical genetics' and 'chemical genomics', implicates the need for cell-permeable chemical ligands for any interesting target protein that allow modulation of the protein's activity at low concentrations as selectively as possible. Compound development via combinatorial chemistry techniques will become the method of choice in undertaking this herculean task. But since the universe of thinkable chemical compounds is almost infinite [3], one important question arises: Where in chemical structural space are compounds with the desired biological properties to be found?

The original expectation that the synthesis of vast random compound libraries will produce as many or even more drug candidates as have historical libraries of pharmaceutical companies and that such libraries will overcome the problem of efficient hit and lead finding was not fulfilled. It was soon recognized that it is not numbers that determine the quality of a library, but its diversity [4–6], its drug-likeness [7–11], and its biological relevance [12, 13]. A central and crucial task is the identification of compound classes that represent already biologically validated starting points in structural space, to find a synthetic access to them that is amenable to combinatorial variation, and to design and synthesize combinatorial libraries centered on the identified underlying structural frameworks of these compound classes.

Chemogenomics in Drug Discovery: A Medicinal Chemistry Perspective.
Edited by Hugo Kubinyi and Gerhard Müller

ISBN: 3-527-30987-X

Biologically active natural products, usually low molecular weight chemical compounds, are synthesized by biological organisms as secondary metabolites and endow their producers with a survival advantage, for example, plants that synthesize metabolites that act as deterrents against herbivores. Natural products are selected by Nature in the process of evolution along the parameter of interaction capability with biomolecules, usually proteins. The necessity of gaining fitness through potent biologically active compounds and the strict structural requirements for tight binding to the respective target protein have acted as important evolutionary constraints on the producing organisms. Using these naturally preselected molecules, with their unique diversity as examples, provides biologically validated and thus relevant starting points for library design. Natural product-based libraries permit finding of hit or lead compounds with enhanced probability and quality if they are included in high-throughput screening programs [12–14]. Scaffolds of certain natural products and nonnatural compounds embody so-called privileged structures. This term was originally coined by B. E. Evans and coworkers at Merck, who recognized in their pharmacological studies of benzodiazepines that derivatives within this compound class bind not only to benzodiazepine receptors of the central nervous system, but also to cholecystokinin receptors and to the unrelated class of peripheral benzodiazepine receptors [15]. Being peptidomimetics, benzodiazepines can be assumed to have intrinsically good binding affinity to various proteins that bind similar regions of peptides or other proteins. According to Evans' definition, privileged structures constitute a class of structural frameworks that can bind to various proteinaceous receptor surfaces, implying that they can be reused as common 'shape themes' in widely divergent drug-design situations [16]. The biological relevance of natural products and privileged structures can be understood in the light of the structural and/or functional relationships of proteins.

14.2
Protein Folds and Protein Function

Proteins can be regarded as modularly built biomolecules assembled from individual building blocks. These building blocks are called domains – discrete parts of the proteins with their own functions that fold independently from the rest of the structure into a compact arrangement of secondary structural elements interconnected via more or less complex linker peptides. The term domain family, as it is used here, refers to a family of related sequences that have an ancient common ancestor, which means that they have developed via divergent evolution. Different sequence families (domains) can adopt the same fold. This can be regarded either as convergence due to functional and physical constraints, because of the limited number of acceptable spatial arrangements of secondary structural elements, or as a result of divergent evolution to such an extent that the sequence relationship is no longer recognizable [17, 18]. Protein domains can be regarded as structurally conserved yet genetically mobile units [19].

Although the estimate for the number of different proteins in humans ranges from 100 000 to 450 000, there is common agreement that the number of domain families and – even more – of topologically distinct folds is much smaller. At present, approximately 600 folds are known, derived by classifying all structurally characterized proteins according to their 3D structures [18, 20–22]. Data from the ongoing genome sequencing projects allow the number of existing folds and families in Nature to be estimated. Current estimates vary between 600 and 8000 distinct folds and between 4000 and 60 000 sequence families [22–26]. In this context we must mention that fold definition often remains an empirical approximation, and even experts disagree on fold assignments for many proteins. This is mainly due to the fact that the criteria used are often rather loose. Frequently, not only structural data but also evolutionary and functional considerations are taken into account. Instead, categorization along exclusively structural aspects would be more appropriate, because proteins having the same fold do not necessarily share a common ancestor or play similar physiological roles.

There is an ongoing effort to reveal the correlation patterns of protein functions and sequences [27–29]. Although we are still far away from a deep and consistent understanding, and analysis is hampered by the small number of available X-ray structures of proteins with bound small-molecule ligands, some interesting observations relevant to the topic of the theme discussed in this chapter have been made about the diversity and evolutionary relationships of ligand binding sites in proteins [30–32].

14.3 Implications for Library Design: Nature's Structural Conservatism and Diversity

In a classical chemical genomics approach, potential targets are clustered into target families on the basis of functional relatedness and amino acid sequence homology alignments reflecting their genetic relationship. This categorization is then used to pool known ligands of a target family and to take them as starting points for combinatorial library design. This strategy constitutes a rationale that allows direct conversion of genetic information and relatedness into actual chemical ligand design. A further, analogous principle was outlined as the structure–activity relationship homology concept. Potential drug discovery targets are grouped into families based on the relatedness of the structure–activity relationships of their ligands [33]. It is assumed that the conservation of binding site architectures and thus the relatedness of molecular recognition specificities within a target family or one of its subfamilies translate into a conservation of ligand scaffolds that bind to these targets [34]. The major limitation of these concepts is that usually only close sequence homologs are considered, because target proteins and their ligands are predominantly categorized on the basis of function and sequence similarity. Family assignment derived from sequence information alone in the absence of structural information usually requires sequence identities greater than 30% [31].

An approach based on both sequence and fold analysis promises that in the long run the process from gene identification to lead discovery may be shortened and accelerated significantly. For instance, the analysis of a newly discovered gene with bioinformatics tools may suggest that the corresponding protein will be a multi-domain protein composed of already-known structural domains [35]. Subsequent comparison of domain architectures can reveal highly diverged homologs even if they have completely different biological functions, for example, catalyze different reactions. This requires in the end that, for every unique fold, at least one 3D structure must be solved so that protein sequences of unknown 3D structures can be modeled by structure prediction without the immediate need for experimental verification by X-ray or NMR techniques. Very often, Nature's structural conservatism is confined to the domain's overall architecture, whereas the binding sites for ligands are structurally diverse yet often topologically similarly located. This observation has led to the introduction of the term supersites, for example, concerning the ferredoxin-like fold [36]. A potential supersite can also be observed in the cystatin-like fold. These similarities are thought to result from divergent evolution [37]. Sometimes conserved sequence elements that are required for the recognition of certain ligand partial structures or for the catalytic mechanism may remain. If small-molecule binders to these domains are already known – for example, natural products whose binding capabilities for a certain protein domain were selected during a long-term evolutionary process – then these can serve as starting points for the design and synthesis of libraries targeting a structurally

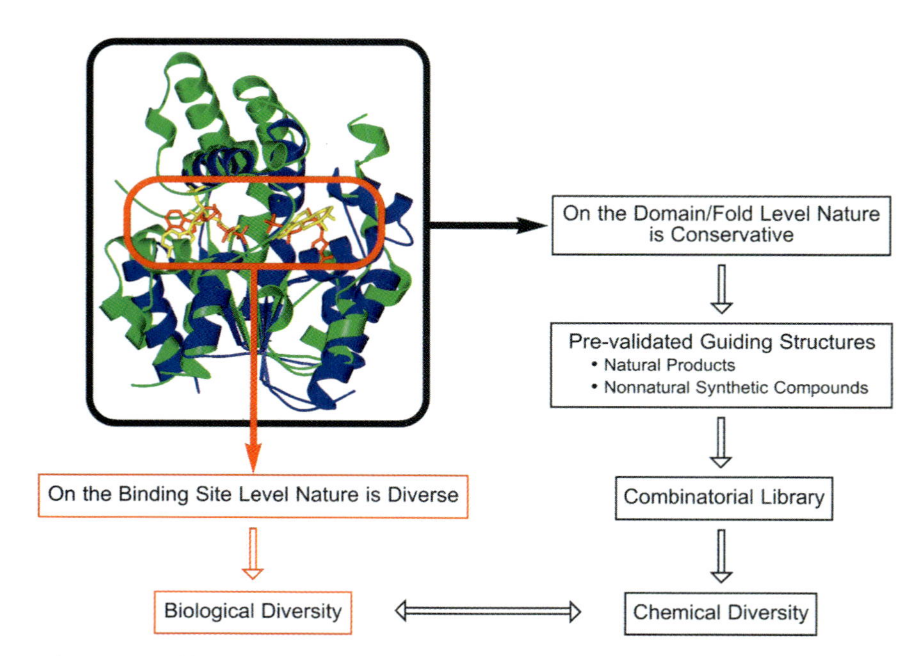

Scheme 14.1 Fold similarity and binding site diversity and their implications for combinatorial library development.

related protein. Despite their particular biological prevalidation, not only natural products can serve as guiding structures for combinatorial libraries. Of course, nonnatural synthetic compounds may also meet this selection criterion. The chemical diversity subsequently generated by combinatorial elaboration of a thus-identified core structure has to match the biological diversity occurring in the ligand-binding cavities of the template domain and the respective target domain, thus evolving optimal binding properties.

The conserved overall architecture of protein domains is used as an abstracting guiding principle leading to the core structure around which chemical diversity is generated (Scheme 14.1). This development of small-molecule binders, in principle (although it is not necessarily desirable), could be initiated without further knowledge of the target's biological functions, binding partners, and so forth – information that is usually obtained by laborious biochemical and cell biological techniques. In fact, the evolved ligands can be used for further characterization of the physiological role of the target protein, which is of outstanding importance in the target-validation process.

14.4 Development of Natural Product-based Inhibitors for Enzymes Belonging to the Same Family

14.4.1 Nakijiquinone Derivatives as Selective Receptor Tyrosine Kinase Inhibitors

The value of using a natural product as a guiding structure, elaborating it into a combinatorial compound library, and screening the library against a set of proteins exhibiting the same fold was demonstrated by Waldmann and Giannis and co-workers [38, 39]. Receptor tyrosine kinases (RTKs) represent a family of closely related homologs. The challenge here is to generate selectivity for one or a distinct group of this protein family, resulting in a biological effect that is exploitable for therapeutic intervention. Nakijiquinones **1a–d** (Scheme 14.2) are naturally occurring inhibitors of human epidermal growth factor receptor 2 (HER-2/neu), which is over-expressed in about 30% of primary breast, ovarian, and gastric carcinomas. In light of the concept described above, a focused compound library of 56 analogs of this lead structure was synthesized and screened for its inhibitory activity toward other RTKs involved in cell signaling and proliferation, such as the vascular endothelial growth factor receptors (VEGFR-2 and VEGFR-3), the Tie-2-receptor, the insulin-like growth factor 1 receptor (IGF1R), and the epidermal growth factor receptor (EGFR) (Scheme 14.2, Figure 14.1). The nakijiquinone-based library was designed on the basis of the modular structure of the natural products. The nakijiquinones consist of a hydrophobic diterpene unit, which may interact with a hydrophobic pocket close to the ATP binding site, a quinone-type building block, and an amino acid. The quinone group and the amino acid may form H bonds to the ATP binding site of kinases. Consequently, the diterpene part was replaced with simple hydrophobic

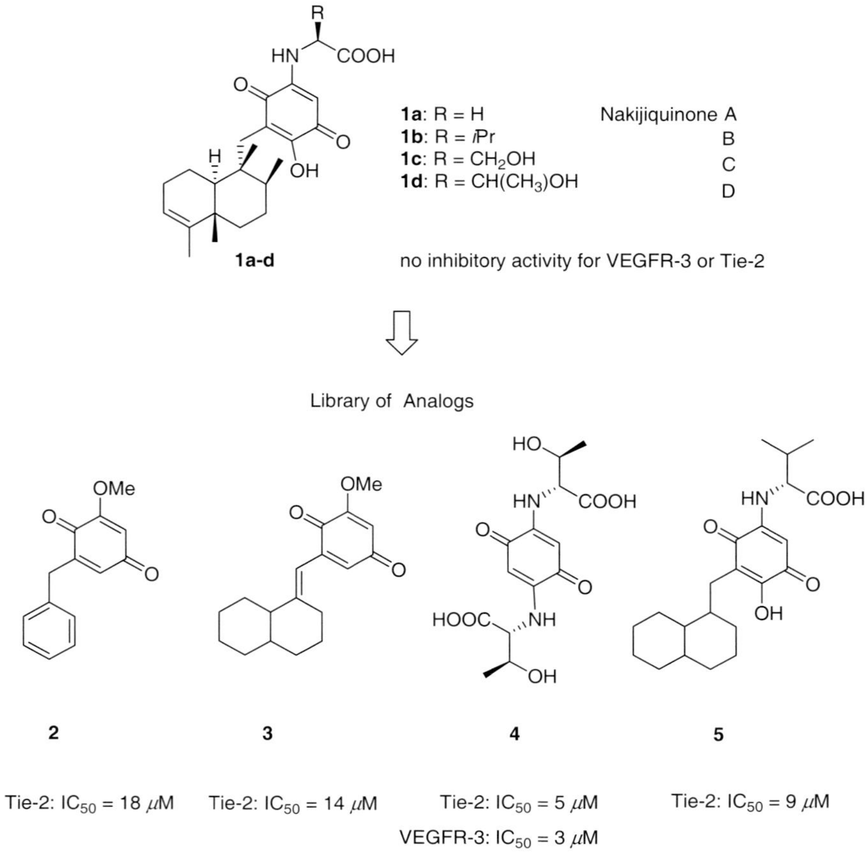

1a-d

1a: R = H — Nakijiquinone A
1b: R = *i*Pr — B
1c: R = CH_2OH — C
1d: R = $CH(CH_3)OH$ — D

no inhibitory activity for VEGFR-3 or Tie-2

Library of Analogs

2 Tie-2: IC_{50} = 18 μM

3 Tie-2: IC_{50} = 14 μM

4 Tie-2: IC_{50} = 5 μM; VEGFR-3: IC_{50} = 3 μM

5 Tie-2: IC_{50} = 9 μM

Scheme 14.2 Representatives of a 56-member library of nakijiquinone analogs leading to a structural pattern for the targeting of Tie-2 RTK and VEGFR-3.

structures (**2**, **3** and **5**, Scheme 14.2). The hydrophilic amino acids serine and threonine (**4**, Scheme 14.2) and the hydrophobic amino acids valine (**5**, Scheme 14.2) and glycine were chosen, and the stereochemistry was also varied. To modify the type and number of putative H-bond donors and acceptors, either one or two amino acids, an amino acid and an OH group, or only one amino acid were introduced.

Although none of the natural nakijiquinones exhibited significant inhibitory activity against the new set of RTKs, six members of the library of analogs were identified as kinase inhibitors in the low micromolar range. In particular, a structural pattern emerged that may allow selective targeting of Tie-2 RTK, which is of paramount importance in the regulation of angiogenesis, that is, the formation of blood vessels from preexisting vessels and, thereby, in cancer development. This result stresses the importance of compound libraries based on natural products, in contrast to using only the natural substances themselves. In a screen with the natural products alone, these inhibitors would have been missed.

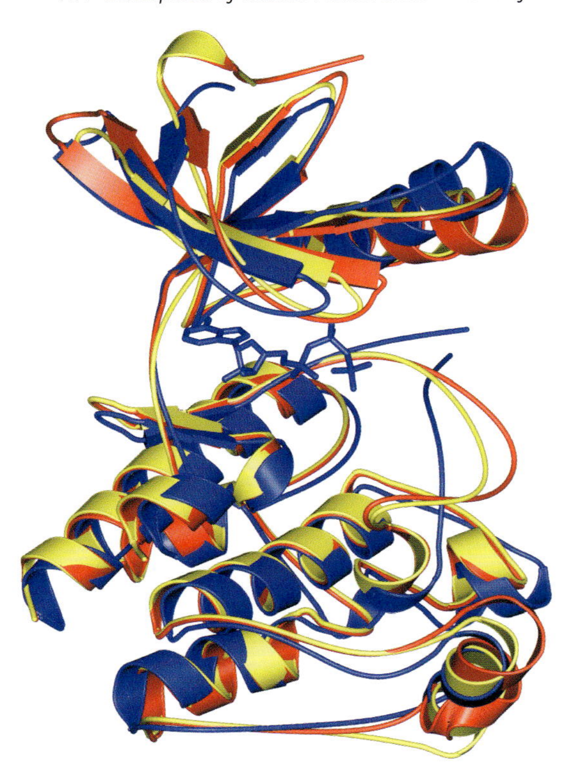

Figure 14.1 Superposed receptor tyrosine kinases inhibited by members of the nakijiquinone-based library. Yellow: Tie-2 RTK (homology model); red: VEGFR-3 (homology model); blue: IGF1R with bound ATP analog (X-ray structure).

14.4.2
Dysidiolide Derivatives as Cdc25 Phosphatase Inhibitors

An interesting example of a natural product-based focused library of inhibitors targeting the Cdc25 family was elaborated by Waldmann and coworkers [40, 41]. The natural product dysidiolide (**6**, Scheme 14.3) was found to inhibit the dual-specificity Cdc25 protein phosphatase family and was used as a guiding principle. Dysidiolide was considered particularly promising, since it inhibits Cdc25A with an IC_{50} value of 9.4 μM but does not inhibit the phosphatases calcineurin, CD45, and LAR [42]. In addition, dysidiolide induces growth arrest of various cancer cell lines and arrest in the G_1 phase of the cell cycle or apoptosis [42, 43].

The 6-*epi* diastereomer of dysidiolide (**7**, Scheme 14.3) and seven analogs of it were synthesized using a solid-phase approach. A notable feature of the multistep reaction sequence on solid phase is that a wide range of transformations with vastly differing requirements could be successfully developed. Key transformations of the synthesis include an asymmetric Diels–Alder reaction with the chiral dienophile

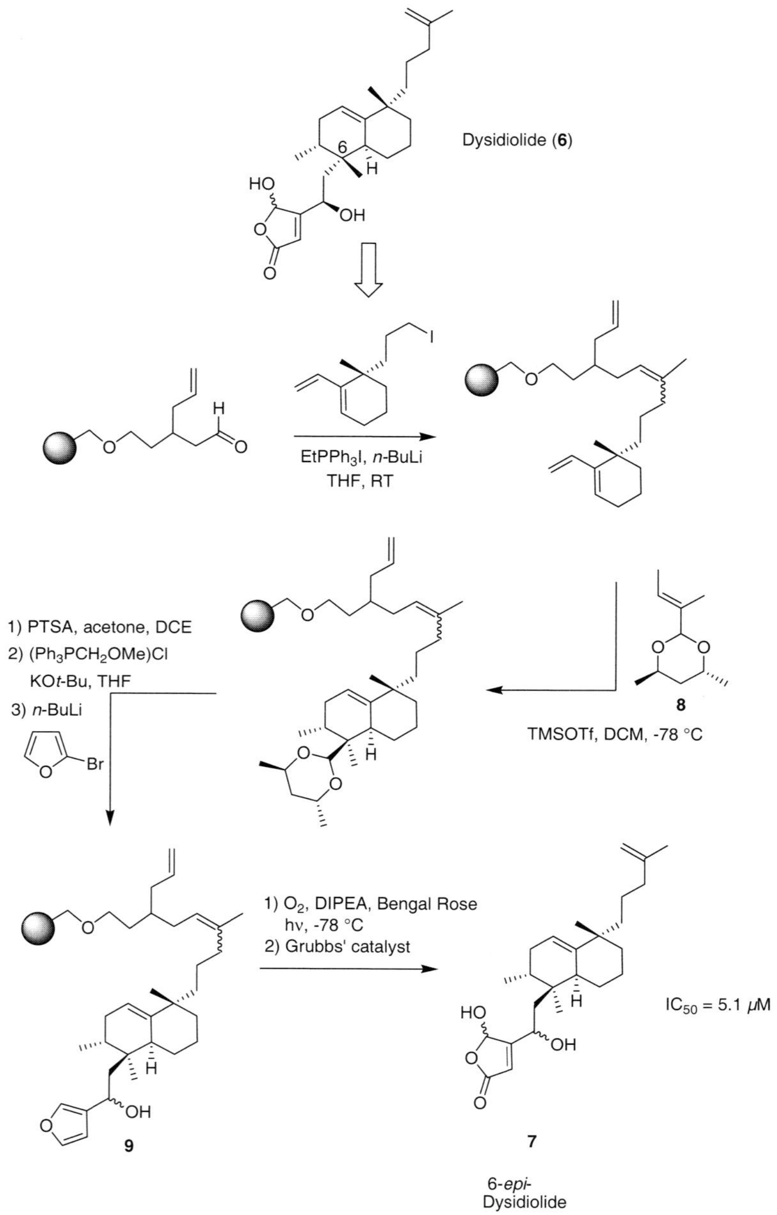

Scheme 14.3 Solid-phase synthesis of 6-*epi*-dysidiolide. Dysidiolide is a naturally occurring inhibitor of Cdc25A.

8 and an oxidative elaboration of the furan **9** with singlet oxygen on solid phase, as well as traceless cleavage of the products, via olefin metathesis, from the support (Scheme 14.3).

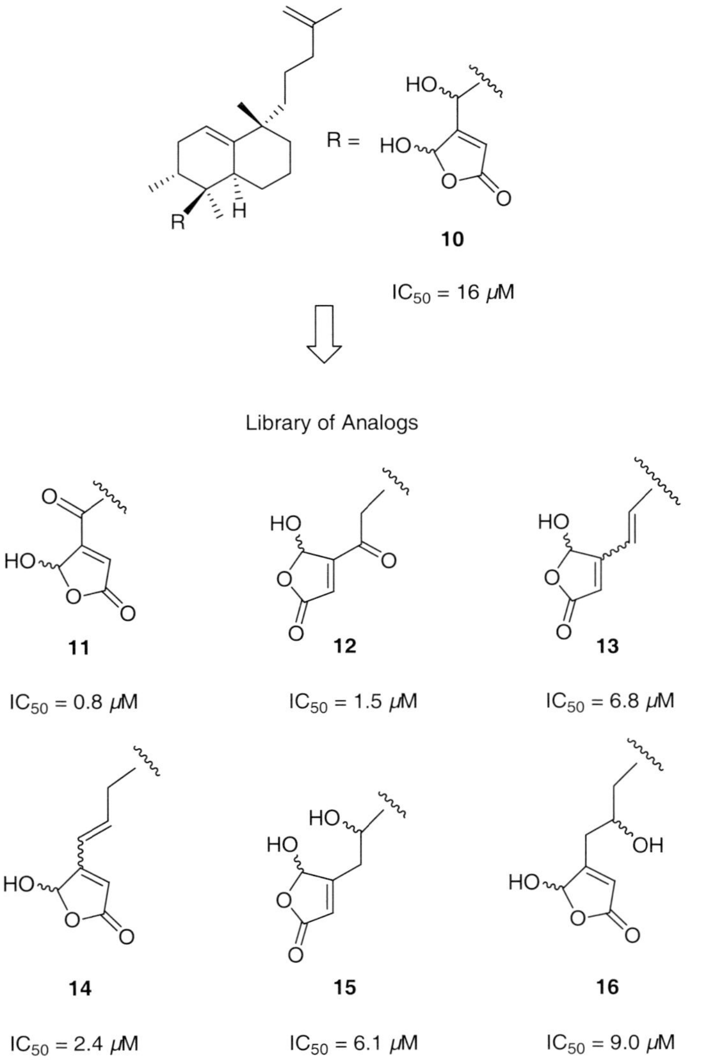

R =

10

IC_{50} = 16 μM

Library of Analogs

11 IC_{50} = 0.8 μM

12 IC_{50} = 1.5 μM

13 IC_{50} = 6.8 μM

14 IC_{50} = 2.4 μM

15 IC_{50} = 6.1 μM

16 IC_{50} = 9.0 μM

Scheme 14.4 Dysidiolide analogs obtained by solid-phase synthesis. The IC_{50} values shown refer to inhibition of Cdc25C.

To determine whether the solid-phase synthesis delivered biologically active natural product analogs with high frequency, 6-*epi*-dysidiolide and compounds **10–16** (Scheme 14.4) were investigated as inhibitors of the protein phosphatase Cdc25C. From the Cdc25 phosphatase family, the Cdc25C protein was chosen because 6-*epi*-dysidiolide (**7**, Scheme 14.3) was previously investigated as an inhibitor of Cdc25A and Cdc25B [44], thus allowing for comparison of data.

The results obtained in the phosphatase assays demonstrate that all dysidiolide analogs inhibit Cdc25C in the low micromolar range, with IC_{50} values varying by a factor of 20. The IC_{50} of 5.1 μM determined for inhibition of Cdc25C by 6-*epi*-dysidiolide (**7**) is considerably lower than the values recorded for the inhibition of Cdc25A (13 μM) and Cdc25B (18 μM). Furthermore, the most active compound in this enzyme assay, ketone **11** (Scheme 14.4), exhibited an IC_{50} value in the high nanomolar range (800 nM) and was 6.4 times as active as **7**. These results indicate that dysidiolide analogs and their derivatives can differentiate selectively between different types of phosphatases and, conceivably, among the three Cdc25 family members. The data also indicate that substantial variation of the precise structural details of the natural product itself is tolerated and leads to inhibitors with significantly enhanced potency. Hence, replacement of the hydroxyethyl bridge between the annelated core ring system and the hydroxybutenolide in compound **7** by an unsaturated three-carbon unit (**14**) or introduction of a keto group (**11** and **12**) lead to more potent Cdc25C inhibitors.

14.5
Development of Natural Product-based Small-molecule Binders to Proteins with Low Sequence Homology yet Exhibiting the Same Fold

It has often been found that proteins with statistically unrelated sequences and/or which play different physiological roles with a different arrangement of binding-site residues have similar folds, evolved from the same ancestors, and can still bind similar ligands [30, 45]. Since sequence homology is sometimes weak or not recognizable, the detection of such distant relatives is not necessarily straightforward. The reason why the most divergent homologs are usually missed in sequence similarity searches is that the respective programs are based on amino acid similarity matrices usually derived from evolutionary models or homology alignments. The instances of divergent evolution with no detectable sequence similarity can therefore be revealed only by comparing the proteins' spatial structures since these are typically more conserved in evolution than are amino acid sequences [46].

14.5.1
Development of Leukotriene A_4 Hydrolase Inhibitors

Leukotriene A_4 hydrolase/aminopeptidase (LTA_4H) catalyzes the final step in LTB_4 biosynthesis. LTB_4 is a potent chemoattractant and immune-modulating lipid modulator involved in inflammation, immune responses, host defense against infection, and platelet activating factor (PAF)-induced shock. The critical role of LTA_4H in LTB_4 generation makes it an attractive drug target. LTA_4H is a bifunctional enzyme whose aminopeptidase functionality is combined with an additional function, namely, the vinylogous hydrolysis of the leukotriene epoxide LTA_4 into LTB_4. Both reactions are catalyzed in the same Zn^{2+}-containing active site [47]. The zinc ion serves as a Lewis acid, polarizes the epoxide ring or the amide carbonyl,

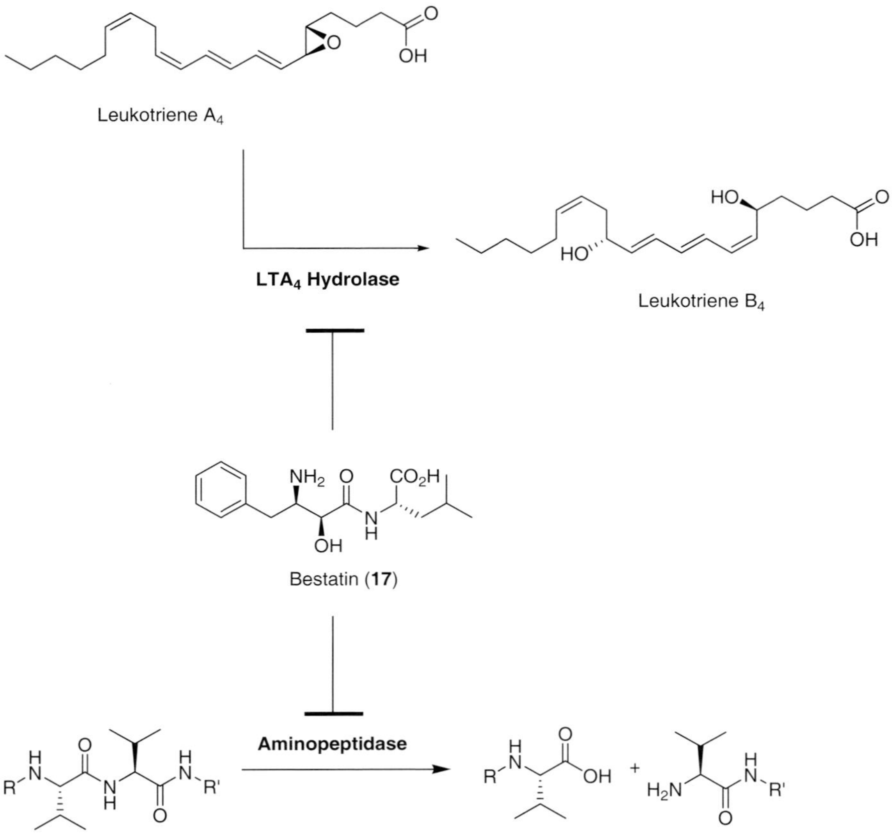

Leukotriene A_4

LTA_4 Hydrolase

Leukotriene B_4

Bestatin (**17**)

Aminopeptidase

Scheme 14.5 LTA_4 hydrolase and aminopeptidases are inhibited by the natural product bestatin, but they catalyze two different reactions.

and stabilizes the negative charge occurring in the transition state. In LTA_4H the presence of the zinc-binding motif (HEXXH-X_{18}-E) was sufficient to prompt investigations of the relationship of this enzyme to zinc-binding metallopeptidases [48]. The evolutionary relationship of the LTA_4H fold to metallopeptidases would have immediately suggested searching for peptidase inhibitors as potential ligands, and indeed, the aminopeptidase inhibitor bestatin (**17**) also inhibits LTB_4 biosynthesis (Scheme 14.5). This finding, and the related observation that the angiotensin-converting enzyme (classified as an M2 metallopeptidase) inhibitor captopril (**18**, Scheme 14.6) also inhibits LTA_4H [48], have inspired the combinatorial variation of these lead structures, which led to the syntheses of potent inhibitors of the peptidase and epoxide hydrolase activity of LTA_4H that proved to be selective for LTA_4H when compared with the inhibitory effect toward other aminopeptidases (compounds **20** and **21**, Scheme 14.6) [49–53].

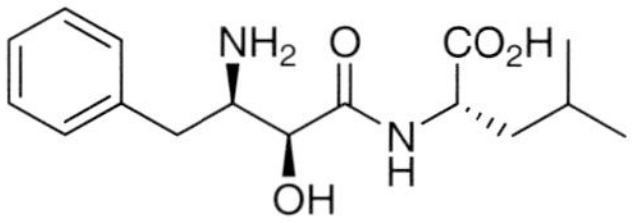

Bestatin (**17**)

K_i (peptidase activity) = 0.2 μM
IC_{50} (epoxide hydrolase activity) = 4 μM

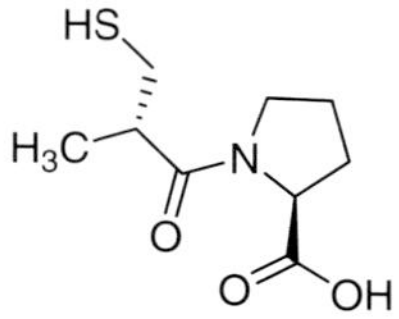

Captopril (**18**)

K_i (peptidase activity) = 0.1 μM
IC_{50} (epoxide hydrolase activity) = 14 μM

Library of Analogs

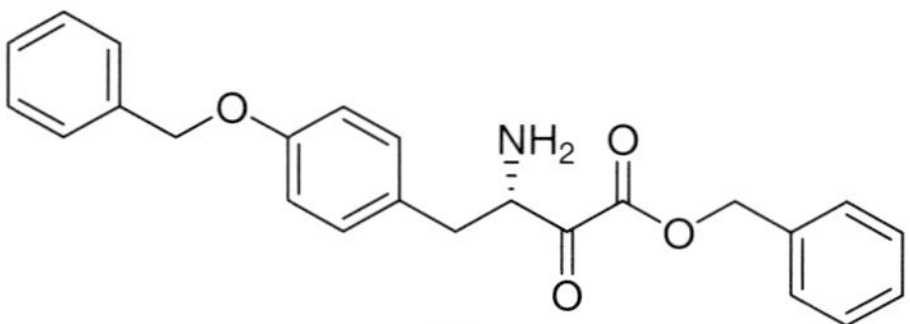

19

K_i (peptidase activity) = 0.046 μM

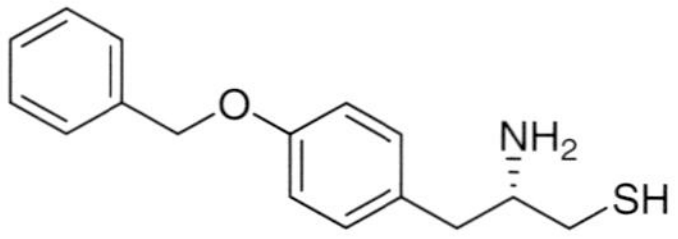

20

K_i (peptidase activity) = 0.018 μM
IC_{50} (epoxide hydrolase activity) = 0.2 μM

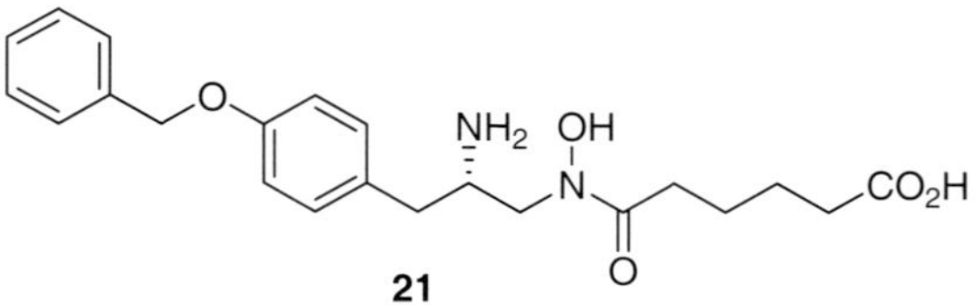

21

K_i (peptidase activity) = 0.002 μM
IC_{50} (epoxide hydrolase activity) = 0.15 μM

Scheme 14.6 Bestatin- and captopril-derived inhibitors of LTA_4 hydrolase.

A comparison of the subsequently determined crystal structure of LTA_4H, classified as a member of the M1 metallopeptidase family, with the recently solved structure of human angiotensin-converting enzyme, a member of the M2 family, and with the structure of thermolysin, which belongs to the M4 family (and shares only 7% sequence identity with LTA_4H) revealed that the catalytic domains of all three enzymes exhibit significant structural homology (Figure 14.2).

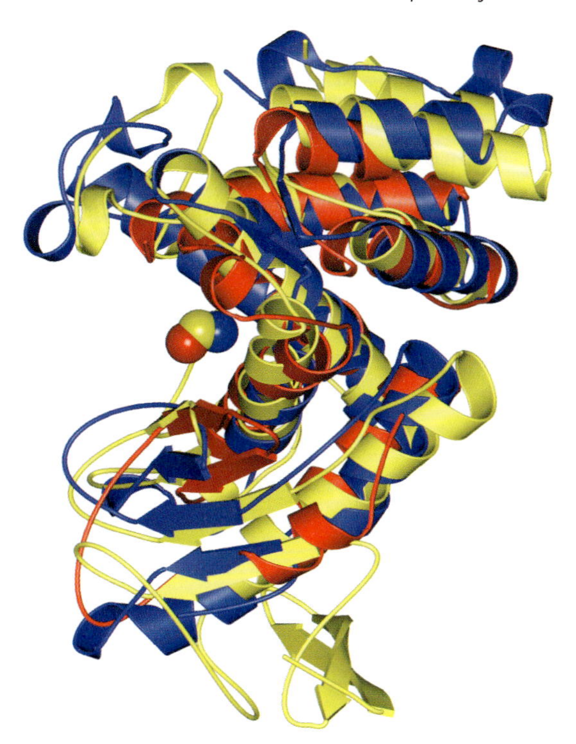

Figure 14.2 Superposed X-ray structures of the catalytic domains of LTA_4H (blue), angiotensin-converting enzyme (red), and thermolysin (yellow), each with bound zinc ion (colored accordingly).

14.5.2
Development of Sulfotransferase Inhibitors

Schultz and coworkers developed a purine scaffold-based compound library synthesized on solid support to target cyclin-dependent kinases (CDKs), using the natural product olomoucine (**22**, Scheme 14.7) as a lead structure [54]. This library afforded a moderately potent inhibitor of CDK2. Further development of this library by synthesizing several hundred 2,6,9-trisubstituted purine derivatives, using solid- and solution-phase chemistry, yielded more-potent CDK inhibitors (CDK1/CDK2), such as **23** [55, 56]. Screening this representative library of purines against recombinant inositol-1,4,5-trisphosphate-3-kinase (IP3K) led to the discovery of inhibitors of IP3K (**24**, Scheme 14.7) [57].

For the combinatorial synthesis of the olomoucine-based library, Schultz and coworkers developed a traceless linkage strategy (Scheme 14.8) [58], the major advantage of which was that anchoring the purine ring did not require that one substituent had to be kept invariant. To achieve linkage to the solid support, primary amines were coupled by reductive amination using sodium triacetoxyborohydride

Olomoucine (**22**)
CDK1/cyclin B: IC_{50} = 4.3 μM
CDK2/cyclin A: IC_{50} = 7.0 μM

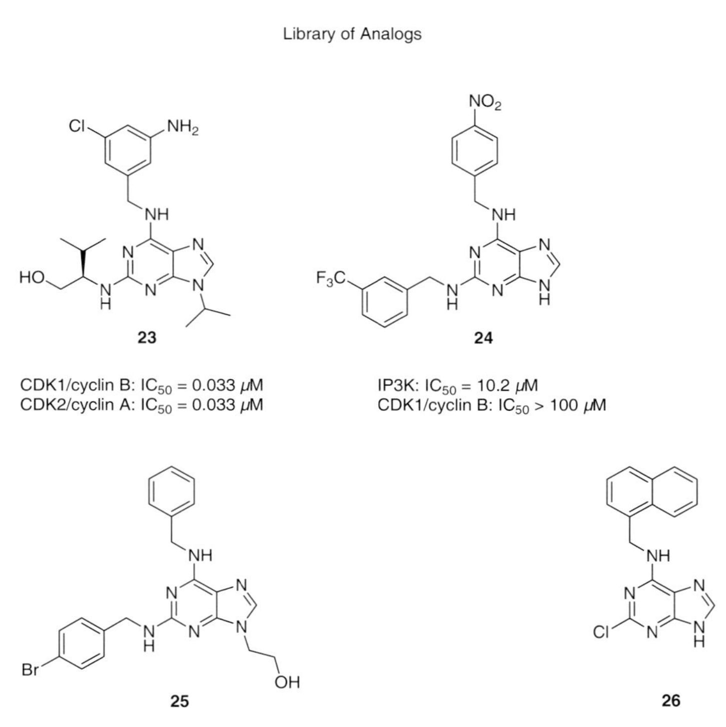

25

NodH: IC_{50} = 20 μM
other carbohydrate sulfotransferases: no inhibition at 200 μM
keratan sulfate sulfotransferase: no inhibition at 200 μM
CDK2/cyclin A: IC_{50} = 15 - 40 μM

26

β-AST-IV: K_i = 0.096 μM
NodH: no inhibtion at 10 μM
protein kinase G: no inhibtion at 10 μM
hexokinase: no inhibtion at 10 μM
pyruvate kinase: no inhibtion at 10 μM

Scheme 14.7 Representatives of olomoucine-based libraries of inhibitors targeting kinases and sulfotransferases.

to a 4-formyl-3-5-dimethoxyphenoxymethyl-functionalized polystyrene resin (PAL) (**27**) [59]. The purine ring (**30**) was then captured at the C2 position by reacting the PAL-amine resin (**28**) with the crude N9-alkylated 2-fluoro-6-phenylsulfenylpurine (**30**) and diisopropylethylamine in *n*-butanol at 80 °C. The C6 position could then be substituted after oxidation-activation of the thioether to the sulfone (**32**).

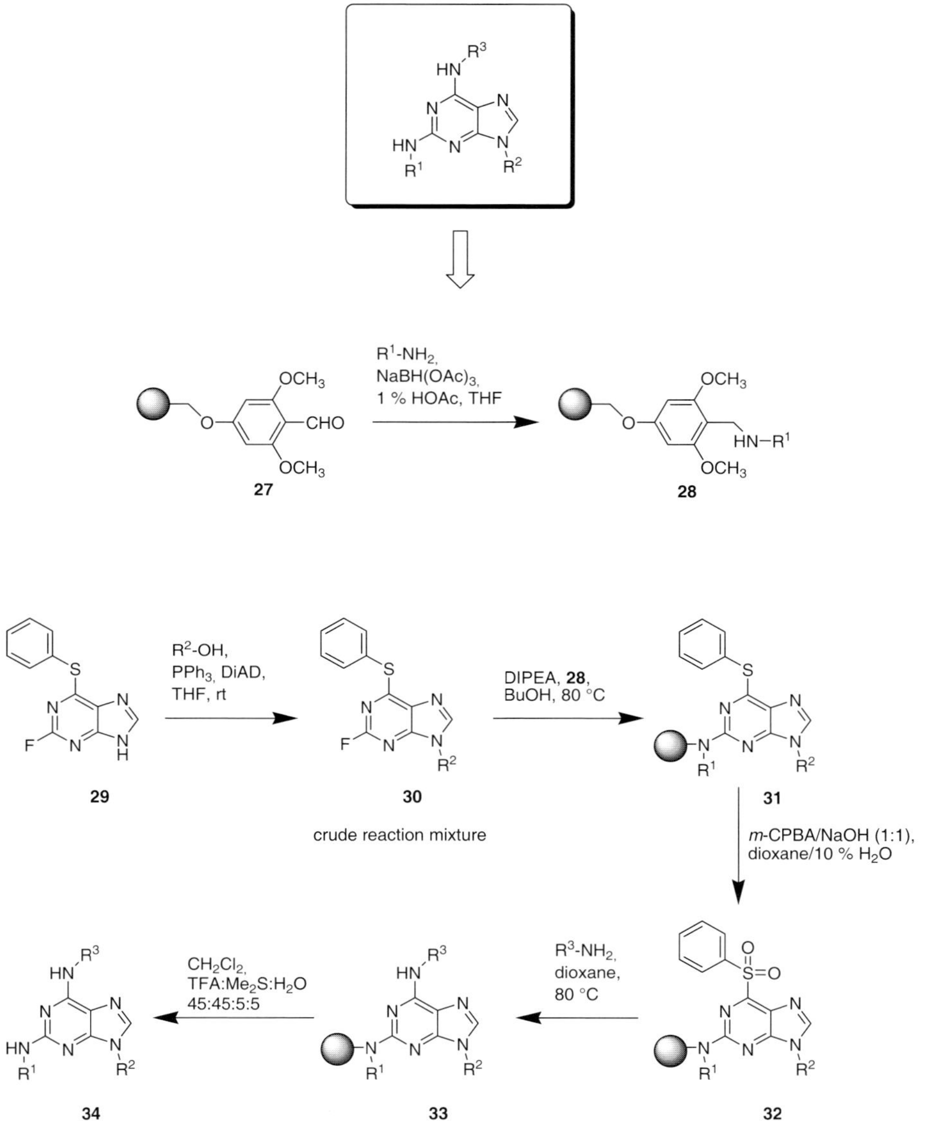

Scheme 14.8 Traceless solid-phase synthesis of olomoucine analogs developed by Schultz and coworkers [58].

Kinases and sulfotransferases utilize similar substrates and catalyze similar reactions. Both transfer anionic groups (Scheme 14.9). Both enzyme classes are capable of binding adenosine-based substrates. Sulfotransferases bind 3′-phosphoadenosine-5′-phosphosulfate (PAPS) (**35**) as a sulfate donor and kinases bind adenosine-5′-triphosphate (ATP) (**36**) as a phosphoryl donor.

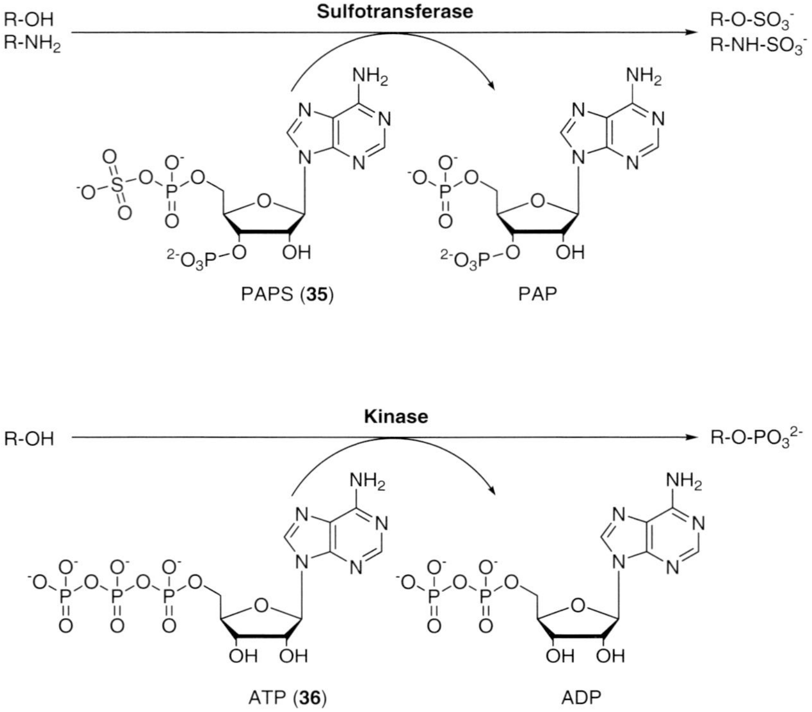

R-OH, R-NH$_2$ —**Sulfotransferase**→ R-O-SO$_3^-$, R-NH-SO$_3^-$

PAPS (**35**) → PAP

R-OH —**Kinase**→ R-O-PO$_3^{2-}$

ATP (**36**) → ADP

Scheme 14.9 Reactions catalyzed by sulfotransferases and kinases. The cofactors are structurally similar. The catalytic mechanisms of nucleotide kinases and sulfotransferases are also similar.

Moreover, the close structural resemblance between sulfotransferases and nucleotide kinases, as shown by the superposition of yeast uridylate kinase (yUK) with murine estrogen sulfotransferase (mEST) (Figure 14.3), is intriguing. It is all the more astonishing when we consider that the catalytic domains exhibit little or no sequence similarity (sequence identity: 8%). Although the specific sidechain interactions differ, both structures bind their cofactors through backbone amide H-bond interaction utilizing a P-loop motif to bind the penultimate phosphate. In addition, the phosphate on the substrate that is phosphorylated by yUK has the same orientation with respect to the cofactor as the phenolic hydroxy group of 17β-estradiol that is sulfated in the mEST-catalyzed reaction. This suggests that the catalytic mechanism of sulfuryl and phosphoryl transfers may be similar. Despite structural resemblance however, there are only a few conserved amino acids, and the specific residues involved in catalysis derive from different locations in the active site [60, 61].

The similarities concerning the bound cofactors, the reaction mechanism, and the adenine-binding pockets led to a screen of the above-described purine-based library of ATP-competitive inhibitors originally designed to target CDKs for cross-reactivity with the carbohydrate sulfotransferase NodH from *Rhizobium meliloti*.

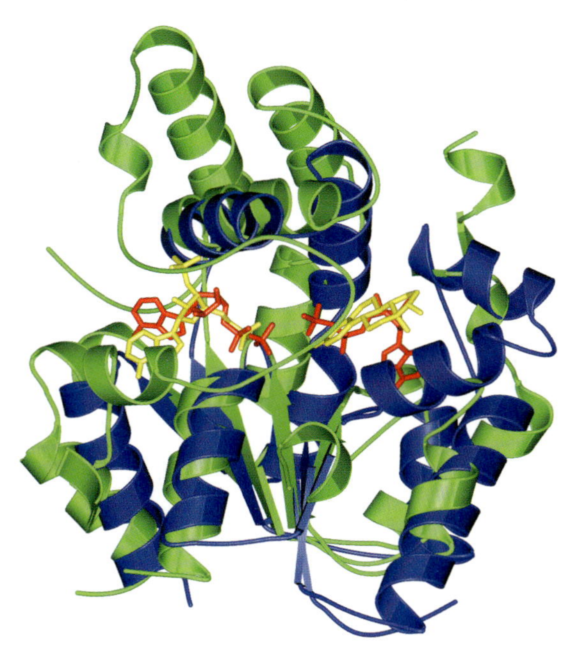

Figure 14.3 Superposed X-ray structures of estrogen sulfotransferase, uridylate kinase, and their cofactors and substrates. Estrogen sulfotransferase: green with consumed cofactor (PAP) and substrate (17β-estradiol) in yellow; uridylate kinase: blue with consumed cofactor (ADP) and substrate analog (ADP) in red.

PAPS-competitive NodH inhibitors (**25**, Scheme 14.7) with modest inhibitory activity (IC_{50} values ranging from 20 to 40 μM) were found that showed selectivity among several tested sulfotransferases. They all displayed inhibitory activity in the micromolar range against several kinases [62].

The library was also tested with murine estrogen sulfotransferase (mEST). This screen afforded a purine-based inhibitor of mEST with nanomolar potency that displayed weak activity against several CDKs and selectivity for mEST when tested with representative members of the carbohydrate sulfotransferase family [63]. Finally, a screen of the library against β-arylsulfotransferase-IV (β-AST-IV) led to the discovery of a potent, highly selective inhibitor (**26**, Scheme 14.7) of β-AST-IV ($K_i = 96$ nM). This compound was also screened against a variety of nucleotide-binding proteins (several kinases, sulfotransferases, and others) and proved to be selective [64].

14.5.3
Development of Nuclear Hormone Receptor Modulators

Nuclear receptors (NRs) are ligand-inducible transcription factors consisting of a ligand-binding domain (LBD) and a DNA-binding domain (DBD). NRs are phylogenetically related proteins that evolved through divergent evolution and therefore are clustered into a large superfamily. Structural comparison of the moderately conserved NR LBDs reveals that these domains exhibit a canonical fold consisting of 12 α-helices, which is better conserved than the primary sequence. The fully buried ligands bind in the hydrophobic core of the LBD. NRs comprise receptors for hydrophobic molecules such as steroid hormones (estrogens, glucocorticoids, progesterone, mineralocorticoids, androgens, vitamin D_3, ecdysone, oxysterols, bile acids, and so on), retinoic acids (all-*trans* and 9-*cis* isoforms), thyroid hormones, fatty acids, leukotrienes, and prostaglandins [65]. Since NRs are naturally switched on and off by small-molecule hormones having physicochemical properties that are very similar to those of therapeutic chemical entities, NRs are one of the most promising target families in terms of therapeutic applications. NRs therefore represent intrinsically very attractive protein targets for the prevention and treatment of diverse diseases. Examples of the current therapeutic exploitation of NRs include the use of estrogen receptor-α (ERα) antagonists (for example, tamoxifen) for the treatment of breast cancer and the clinical use of the structural class of thiazolidinediones or glitazones (agonists of peroxisome proliferator-activated receptor γ (PPARγ) and insulin sensitizers) as antidiabetic drugs [66, 67].

In comparing the natural NR ligands, Bogan and coworkers discovered that they have in common a mean molecular van der Waals volume of 318 ± 53 Å^3 despite their chemical diversity. Their mean molecular weight of 368 ± 110 g mol^{-1} is less conserved. This suggests that coevolution of receptor and ligand took place, leading to the selection of ligands with conserved volumes capable of filling the 3D space of the ligand-binding cavity in the LBD. The canonical LBD fold determines the volume of the binding pocket that the ligand must fill and thus dictates the tolerated range of ligand volumes. Thus, molecular volumes may serve as a valuable tool for judging putative ligands [68].

The plant sterol guggulsterone (GS, 4,17(20)-pregnadiene-3,16-dione (**37**, Scheme 14.10), isolated from an extract of the gum resin of the guggul tree (*Commiphora mukul*), lowers LDL (low-density lipoprotein) cholesterol levels in humans. GS is a highly efficacious antagonist of the farnesoid X receptor (FXR), which is a NR that is activated by bile acids (BAs) such as chenodeoxycholic acid (CDCA, 3α,7α-dihydroxy-5β-cholane-24-acid, **38**, Scheme 14.10). GS competes with CDCA for binding to the LBD of FXR. It also binds to PPARα, whose natural ligand is leukotriene B_4 (**39**, Scheme 14.10), and to the pregnane X receptor (PXR). GS activates PXR approximately 50% as effectively as the specific nonnatural PXR agonist pregnenolone 16α-carbonitrile (PCN, **40**, Scheme 14.10) [69]. Recent results suggest that the physiological effect of GS is due to both inhibition of the bile acid-induced activation of FXR and activation of PXR [70].

Because GS can be seen as a congener of PCN, it is not surprising that it can bind to PXR in addition to FXR. In contrast, its affinity to PPARα, whose natural ligand is leukotriene B_4 (**39**, Scheme 14.10), is quite surprising. This cross-reactivity indicates that combinatorial variation of the structure of a ligand for one NR, irrespective of its scaffold (for example, steroidal or not), also leads to ligands for other NRs.

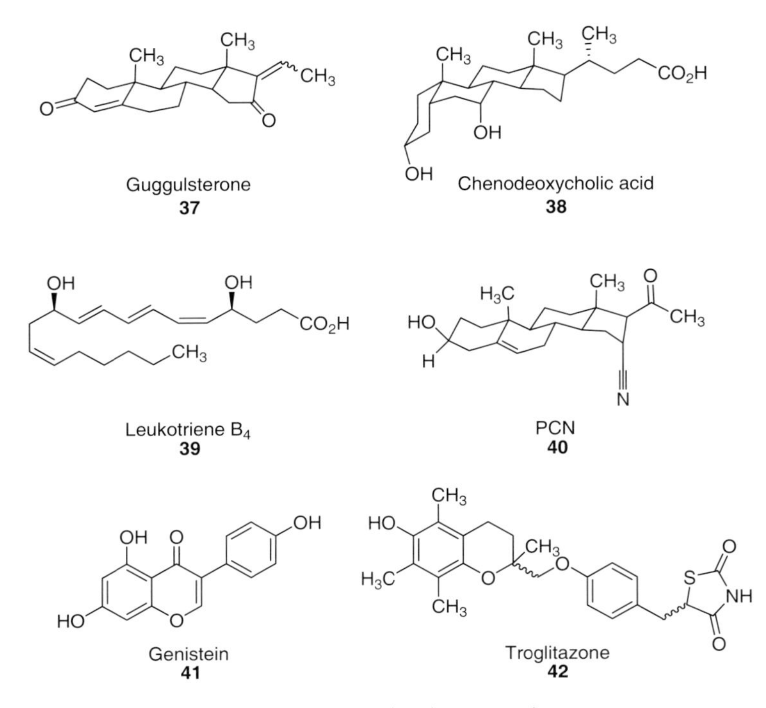

Scheme 14.10 Structurally diverse natural and nonnatural ligands of some nuclear hormone receptors.

As described above, FXR functions as a BA sensor and coordinates cholesterol metabolism. Currently it is hypothesized that FXR senses BA levels and mediates the transcriptional repression of genes responsible for the conversion of excess cholesterol into BAs, as well as the induction of genes necessary for BA transport. This makes FXR an interesting pharmacological target. For further validation of FXR as a potential drug target, it is necessary to understand its physiological role exactly. A selective high-affinity agonist would be helpful as a tool-compound. But where can a biologically validated starting point for the design of a combinatorial library be found? The benzopyran moiety – a privileged motif – occurs in many natural products that cover a broad spectrum of biological activities, such as antitumor, antibacterial, and estrogenic effects, to name but few. Genistein, an isoflavone phytoestrogen (**41**, Scheme 14.10), is found in significant levels in soybeans and soy products. Genistein binds to both estrogen receptor (ER) isoforms α and β with moderate affinity, but exhibits a preference for ERβ as a partial agonist [71]. Additionally, **41** is a ligand of PPARγ and acts as an agonist [72]. A known synthetic PPARγ agonist that also bears the benzopyran core structure is troglitazone (**42**, Scheme 14.10), which was in clinical use as an antidiabetic agent but was withdrawn from the market due to its liver toxicity [73].

The intriguing structural homology of ERβ and PPARγ to FXR despite their low sequence homology (sequence identity: approximately 20%) (Figure 14.4) would

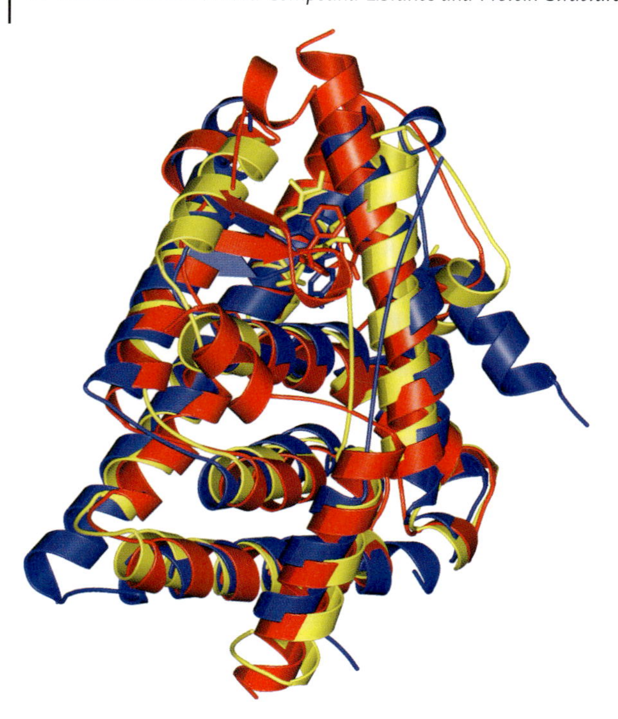

Figure 14.4 Superposed X-ray structures of the ligand-binding domains of ERβ, PPARγ, and FXR, each with bound ligand. ERβ with genistein (**41**, blue), PPARγ with rosiglitazone (red), FXR with **50** (yellow).

have immediately suggested – in light of the structural conservation of the LBD fold – employing the benzopyran moiety, which can be regarded as the core motif of genistein, as the guiding structure for a combinatorial library of potential nonsteroidal FXR agonists.

A combinatorial natural-product-like and diversity-orientated library of 10 000 benzopyran-based small molecules was constructed by Nicolaou and coworkers [74, 75]. They chose a solid-phase approach and an anchoring strategy that does not limit complexity building operations (Scheme 14.11). They used a polystyrene-based selenyl bromide resin (**43**) on which substrates can be immobilized by electrophilic cyclization reactions. Here, *ortho*-prenylated phenol **44** was reacted with the selenyl bromide (**43**) to form the benzopyran scaffold (**45**) via a 6-*endo-trig* cyclization.

The high chemical stability of the selenyl ether bridge through which the benzopyran moiety was linked to the solid support allowed further elaborations at all four possible positions on the aromatic ring (R^1–R^4 of **45**, Schemes 14.11 and 14.12), such as annulations, condensations, aryl/vinyl couplings, glycosidations, and organometallic additions (Scheme 14.12). Finally, the benzopyran analogs were released by oxidation of the selenide, followed by *syn* elimination to produce the benzopyrans **46** (Scheme 14.11).

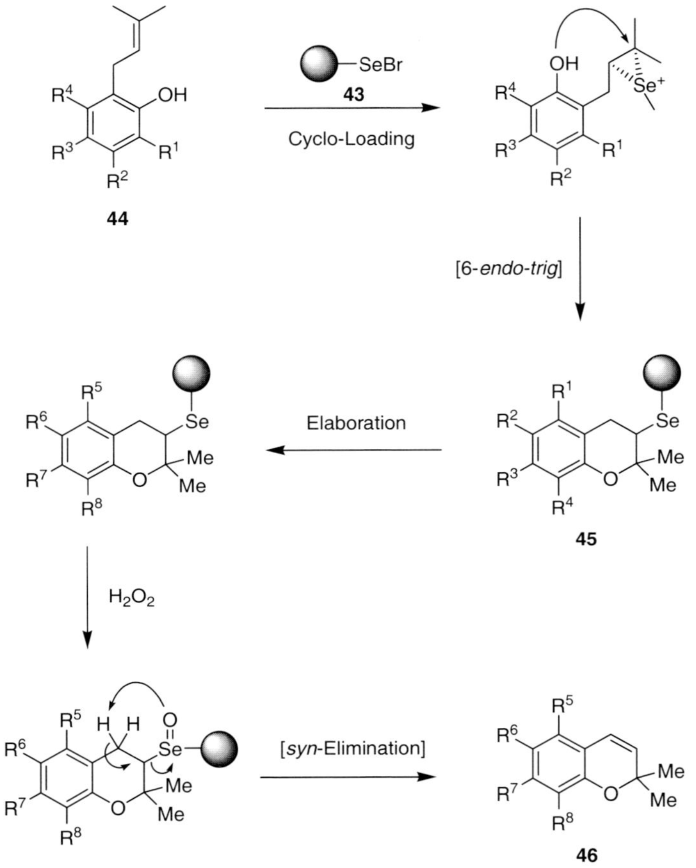

Scheme 14.11 Solid-phase synthesis of a benzopyran-based natural-product-like combinatorial library.

Initial screening of the above-detailed benzopyran library in a cell-based assay for FXR activation afforded several lead compounds (**47** and **48**, Scheme 14.13) with low-micromolar activity (EC_{50} values ranging from 5 to 10 μM). Further elaboration of these lead structures applying a combined solid- and solution-phase approach yielded FXR binders with EC_{50} values in the low-nanomolar range, such as **49** with an EC_{50} value of 188 nM and **50** with an EC_{50} of 25 nM (Scheme 14.13) in which the benzopyran moiety was further deconstructed to the privileged biaryl motif [76, 77].

Condensation
Glycosidation
Annulation
Addition
45
benzopyran scaffold

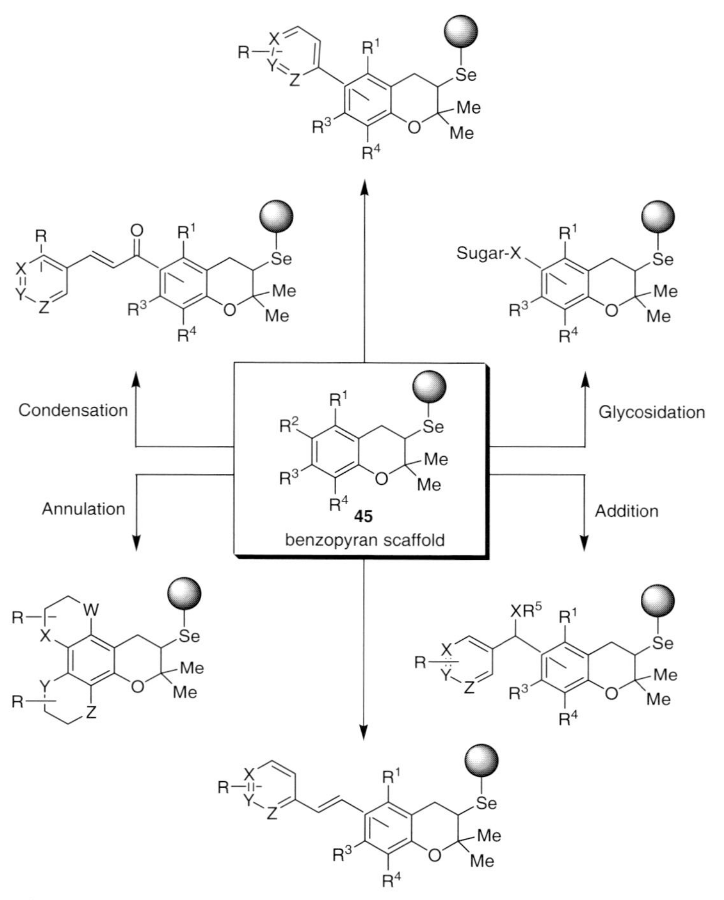

Scheme 14.12 Diversification of the benzopyran scaffold on solid support.

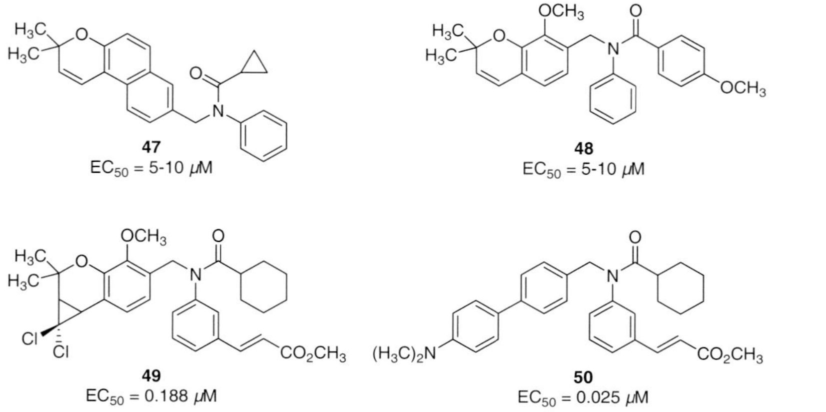

47
EC50 = 5-10 μM
48
EC50 = 5-10 μM
49
EC50 = 0.188 μM
50
EC50 = 0.025 μM

Scheme 14.13 FXR agonists generated by combinatorial solid- and solution-phase chemistry.

14.6
Conclusion: A New Guiding Principle for Chemical Genomics?

In a general sense, chemical genomics can be defined as the genomic response to chemical compounds, that is, chemistry is used to probe a biological system. A more focused, workable definition appears to be the identification of small-molecule lead-like compounds for a member of a gene family product and the subsequent use of these compounds to elucidate the function of other (disease-associated) members of that gene family. Currently, in this approach the gene family products are classified on the basis of sequence alignment and function, that is, into kinases, phosphatases, proteases, and so on (Scheme 14.14). A protein domain-centered approach that considers the domain organization and architecture, however, may provide a new guiding principle for the combinatorial development of compounds, which will pave the way to a new series of proteomics and genomics experiments. Accordingly, a family of gene products (proteins) of interest would be dissected in structural terms, that is, into domains (Scheme 14.14). After domain assignment, structural comparison with known domains/folds would take place, leading to a cluster of structurally related domains that may share little sequence homology. This pool of protein domains with their respective ligands may serve as paradigm for the generation of potent *and* selective small-molecule modulators of protein function. The structures of known ligands of a spatially similar reference domain constitute biologically validated starting points in chemical structural space for the design of focused libraries. Selectivity can be achieved by generating diversity around the ligand core structure, thus taking into account the requirements of the individual binding pockets. This strategy initially reduces complexity and focuses on the 3D similarity of protein domains. This leads to the core structure on which the focused compound library will be based. With natural products, it is postulated that their evolutionarily selected scaffolds represent prevalidated solutions in terms of basic affinity to the protein domains they interact with. Natural products can thus be regarded as inherently promising guiding compounds for the design of domain-selective small-molecule modulators of protein function. But nonnatural, synthetic small-molecule ligands exhibiting selectivity and potency for a specific target protein also can be regarded as valuable starting points, considering that their properties have usually been evolved in an accelerated artificial selection process. Once a biologically relevant structural framework is found, the varying requirements of the different binding sites are addressed by generating diversity around this core structure. At this stage, molecular modeling techniques can be used to plan the substitution patterns required for potency and selectivity. Hence, in the initial step, the overall structural homology of protein domains is employed as a guiding principle for choosing possible ligand scaffolds. In a second step, the structural diversity found in the binding sites of the protein domains is addressed by synthesizing a focused compound library. Ideally, chemical entities are thus evolved that match the diversity found in the binding sites of reference and target domains, yielding selective and potent binders. The advantage of a certain initial indeterminateness in comparing overall domain structures is that predicted and modeled

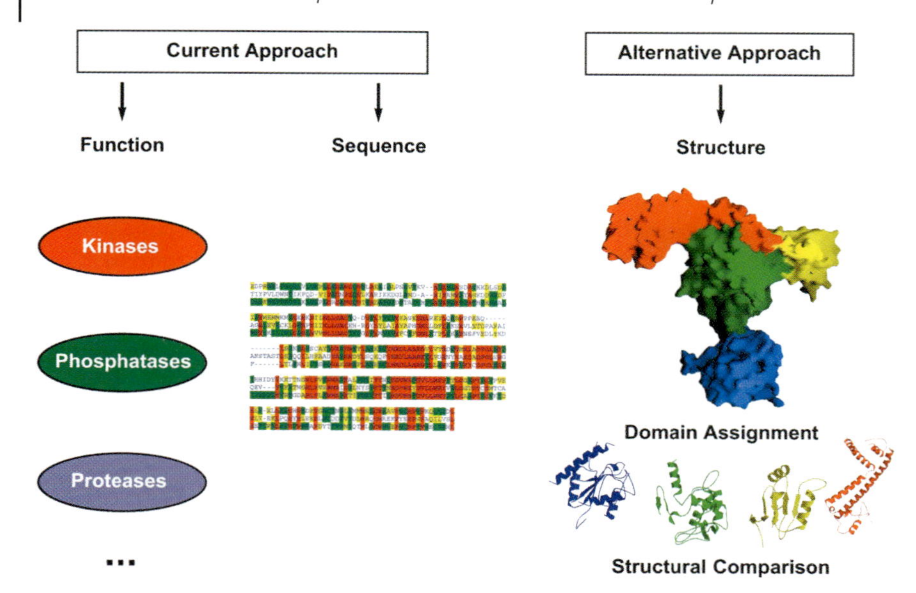

Scheme 14.14 Approaches to the categorization of target proteins. The currently predominating approach in chemical genomics, based on the clustering of target proteins according to their sequence and function, is contrasted to an alternative approach based on structural classification of protein domains.

protein structures having a certain fault tolerance concerning the binding site can also be considered as reference domains because, in the end, the indeterminateness is overcome by the combinatorial approach. A predominantly structure-based approach can be very helpful in the initial stages of screening, when little is known about the function of a newly discovered protein. Here, all existing ligands and their analogs generated by combinatorial synthesis against sequentially and/or structurally related proteins could be tried so as to dissect the physiological role of the protein, in a combined chemical and biological approach, and finally to find new leads for drug discovery.

Acknowledgements

We thank Dr. Ingrid Vetter (Max Planck Institute of Molecular Physiology, Dortmund) for continuing stimulating discussions. This work was supported by the Max-Planck-Gesellschaft, the Deutsche Forschungsgemeinschaft, and the Studienstiftung des deutschen Volkes (doctoral scholarship for M. A. K.).

References

1 B. R. Stockwell, *Nat. Rev. Genetics* **2000**, *1*, 116–125.

2 P. J. Alaimo, M. A. Shogren-Knaak, K. M. Shokat, *Curr. Opin. Chem. Biol.* **2001**, *5*, 60–367.

3 R. S. Bohacek, C. McMartin, W. C. Guida, *Med. Res. Rev.* **1996**, *16*, 3–50.

4 A. Golebiowski, S. R. Klopfenstein, D. E. Portlock, *Curr. Opin. Chem. Biol.* **2001**, *5*, 273–284.

5 J. S. Mason, M. A. Hermsmeier, *Curr. Opin. Chem. Biol.* **1999**, *3*, 342–349.

6 S. L. Schreiber, *Science* **2000**, *287*, 1964–1969.

7 W. P. Walters, Ajay, M. A. Murcko, *Curr. Opin. Chem. Biol.* **1999**, *3*, 384–387.

8 Ajay, W. P. Walters, M. A. Murcko, *J. Med. Chem.* **1998**, *41*, 3314–3324.

9 J. Sadowski, H. Kubinyi, *J. Med. Chem.* **1998**, *41*, 3325–3329.

10 A. K. Ghose, V. N. Viswanadhan, J. J. Wendoloski, *J. Comb. Chem.* **1999**, *1*, 55–68.

11 M.-L. Lee, G. Schneider, *J. Comb. Chem.* **2001**, *3*, 284–289.

12 R. Breinbauer, I. R. Vetter, H. Waldmann, *Angew. Chem., Int. Ed.* **2002**, *41*, 2878–2890.

13 M. A. Koch, R. Breinbauer, H. Waldmann, *Biol. Chem.* **2003**, *384*, 1265–1272.

14 D. Brohm, S. Metzger, A. Bhargava, O. Müller, F. Lieb, H. Waldmann, *Angew. Chem., Int. Ed.* **2002**, *41*, 307–311.

15 B. E. Evans, K. E. Rittle, M. G. Bock, R. M. DiPardo, R. M. Freidinger, W. L. Whitter, G. F. Lundell, D. F. Veber, P. S. Anderson, R. S. L. Chang, V. J. Lotti, D. J. Cerino, T. B. Chen, P. J. Kling, K. A. Kunkel, J. P. Springer, J. Hirshfield, *J. Med. Chem.* **1988**, *31*, 2235–2246.

16 D. A. Horton, G. T. Bourne, M. L. Smythe, *Chem. Rev.* **2003**, *103*, 893–930.

17 A. N. Lupas, C. P. Ponting, R. B. Russell, *J. Struct. Biol.* **2001**, *134*, 199–203.

18 S. Govindarajan, R. Recabarren, R. A. Goldstein, *Proteins* **1999**, *35*, 408–414.

19 C. P. Ponting, J. Schultz, R. P. Copley, M. A. Andrade, P. Bork, *Adv. Protein Chem.* **2000**, *54*, 185–244.

20 A. G. Murzin, S. E. Brenner, T. Hubbard, C. Chothia, *J. Mol. Biol.* **1995**, *247*, 536–540.

21 J. M. Thornton, D. T. Jones, M. W. MacArthur, C. A. Orengo, M. B. Swindells, *Philos. Trans. R. Soc. London, Ser. B* **1995**, *348*, 71–79.

22 Y. I. Wolf, N. V. Grishin, E. V. Koonin, *J. Mol. Biol.* **2000**, *299*, 897–905.

23 E. V. Koonin, Y. I. Wolf, G. P. Karev, *Nature* **2002**, *420*, 218–223.

24 C. Chothia, *Nature* **1992**, *357*, 543–544.

25 P. Green, D. Lipman, L. Hillier, R. Waterston, D. Stobes, J. M. Claverie, *Science* **1993**, *259*, 1711–1716.

26 N. Alexandrov, N. Go, Abstract presented at the Bioinformatics Genome Regulation Structure Conference (Novosibirsk) **1998**.

27 M. Weir, M. Swindells, J. Overington, *Trends Biotechnol.* **2001**, *19*, S61–S66.

28 A. Danchin, *Curr. Opin. Struct. Biol.* **1999**, *9*, 363–367.

29 R. L. Tatusov, D. A. Natale, I. V. Garkavtsev, T. A. Tatusova, U. T. Shankavaram, B. S. Rao, B. Kiryutin, M. Y. Galperin, N. D. Fedorova, E. V. Koonin, *Nucleic Acids Res.* **2001**, *29*, 22–28.

30 W. Anantharaman, L. Aravind, E. V. Koonin, *Curr. Opin. Chem. Biol.* **2003**, *7*, 12–20.

31 J. A. Gerlt, P. C. Babbitt, *Annu. Rev. Biochem.* **2001**, *70*, 209–246.

32 A. E. Todd, C. A. Orengo, J. M. Thornton, *Curr. Opin. Chem. Biol.* **1999**, *3*, 548–556.

33 S. V. Frye, *Chem. Biol.* **2001**, *6*, R3–R7.

34 E. Jacoby, A. Schuffenhauer, P. Floersheim, *Drug News Perspect.* **2003**, *16*, 93–102.

35 P. R. Caron, M. D. Mullican, R. D. Mashal, K. P. Wilson, M. S. Su, M. A. Murcko, *Curr. Opin. Chem. Biol.* **2001**, *5*, 464–470.

36 R. B. Russell, P. D. Sasieni, M. J. E. Sternberg, *J. Mol. Biol.* **1998**, *282*, 903–918.

37 A. G. Murzin, *Curr. Opin. Struct. Biol.* **1998**, *8*, 380–387.

38 P. Stahl, L. Kissau, R. Mazitschek, A. Huwe, P. Furet, A. Giannis, H. Waldmann, *J. Am. Chem. Soc.* **2001**, *123*, 11586–11593.

39 P. Stahl, L. Kissau, R. Mazitschek, A. Giannis, H. Waldmann, *Angew. Chem., Int. Ed.* **2002**, *41*, 1174–1178.

40 D. Brohm, N. Philippe, S. Metzger, A. Bhargava, O. Müller, F. Lieb, H. Waldmann, *J. Am. Chem. Soc.* **2002**, *124*, 13171–13178.

41 D. Brohm, S. Metzger, A. Bhargava, O. Müller, F. Lieb, H. Waldmann, *Angew. Chem., Int. Ed.* **2002**, *41*, 307–311.

42 S. P. Gunasekera, P. J. McCarthy, M. Kelly-Borges, E. Lobkovsky, J. Clardy, *J. Am. Chem. Soc.* **1996**, *118*, 8759–8760.

43 S. Danishefsky, S. R. Magnuson, N. Rosen, Patent WO 99/40079, **1999**.

44 M. Takahashi, K. Dodo, Y. Hashimoto, R. Shirai, *Tetrahedron Lett.* **2000**, *41*, 2111–2114.

45 L. Holm, *Curr. Opin. Struct. Biol.* **1998**, *8*, 372–379.

46 N. V. Grishin, *J. Struct. Biol.* **2001**, *134*, 167–185.

47 M. M. G. M. Thunnissen, P. Nordlund, J. Z. Haeggström, *Nat. Struct. Biol.* **2001**, *8*, 131–135.

48 L. Orning, G. Krivi, F. A. Fitzpatrick, *J. Biol. Chem.* **1991**, *266*, 1375–1378.

49 W. Yuan, B. Munoz, C.-H. Wong, J. Z. Haeggström, A. Wetterholm, B. Samuelsson, *J. Med. Chem.* **1993**, *36*, 211–220.

50 I. R. Ollmann, J. H. Hogg, B. Munoz, J. Z. Haeggström, B. Samuelsson, C.-H. Wong, *Bioorg. Med. Chem.* **1995**, *3*, 969–995.

51 J. H. Hogg, I. R. Ollmann, J. Z. Haeggström, A. Wetterholm, B. Samuelsson, C.-H. Wong, *Bioorg. Med. Chem.* **1995**, *3*, 1405–1415.

52 M.-Q. Zhang, *Curr. Med. Chem.* **1997**, *4*, 67–78.

53 M. M. G. M. Thunnissen, B. Andersson, B. Samuelsson, C.-H. Wong, J. Z. Haeggström, *FASEB J.* **2002**, *16*, 1648–1650.

54 T. C. Norman, N. S. Gray, J. T. Koh, P. G. Schultz, *J. Am. Chem. Soc.* **1996**, *118*, 7430–7431.

55 N. S. Gray, L. Wodicka, A.-M. W. H. Thunnissen, T. C. Norman, S. Kwon, F. H. Espinoza, D. O. Morgan, G. Barnes, S. LeClerc, L. Meijer, S.-H. Kim, D. J. Lockhart, P. G. Schultz, *Science* **1998**, *281*, 533–538.

56 Y.-T. Chang, N. S. Gray, G. R. Rosania, D. P. Sutherlin, S. Kwon, T. C. Norman, R. Sarohia, M. Leost, L. Meijer, P. G. Schultz, *Chem. Biol.* **1999**, *6*, 361–375.

57 Y.-T. Chang, G. Choi, Y.-S. Bae, M. Burdett, H.-S. Moon, J. W. Lee, N. S. Gray, P. G. Schultz, L. Meijer, S.-K. Chung, K. Y. Choi, P.-G. Suh, S. H. Ryu, *ChemBioChem* **2002**, *9*, 897–901.

58 S. Ding, N. S. Gray, Q. Ding, P. G. Schultz, *J. Org. Chem.* **2001**, *66*, 8273–8276.

59 J. Jin, T. L. Graybill, M. A. Wang, L. D. Davis, M. L. Moore, *J. Comb. Chem.* **2001**, *3*, 97–101.

60 Y. Kakuta, L. G. Pedersen, C. W. Carter, M. Negishi, L. C. Pedersen, *Nat. Struct. Biol.* **1997**, *4*, 904–908.

61 Y. Kakuta, E. V. Petrotchenko, L. C. Pedersen, M. Negishi, *J. Biol. Chem.* **1998**, *273*, 27325–27330.

62 J. I. Armstrong, A. R. Portley, Y.-T. Chang, D. M. Nierengarten, B. N. Cook, K. G. Bowman, A. Bishop, N. S. Gray, K. M. Shokat, P. G. Schultz, C. R. Bertozzi, *Angew. Chem., Int. Ed.* **2000**, *39*, 1303–1306.

63 D. E. Verdugo, M. T. Cancilla, X. Ge, N. S. Gray, Y.-T. Chang, P. G. Schultz, M. Negishi, J. A. Leary, C. R. Bertozzi, *J. Med. Chem.* **2001**, *44*, 2683–2686.

64 E. Chapman, S. Ding, P. G. Schultz, C.-H. Wong, *J. Am. Chem. Soc.* **2002**, *124*, 14524–14525.

65 M. Robinson-Rechavi, H. Escriva Garcia, V. Laudet, *J. Cell Sci.* **2003**, *116*, 585–586.

66 M. Schapira, *Curr. Cancer Drug Targets* **2002**, *2*, 243–256.

67 M. Schapira, B. M. Raaka, H. H. Samuels, R. Abagyan, *Proc. Natl. Acad. Sci. USA* **2000**, *97*, 1008–1013.

68 A. A. Bogan, F. E. Cohen, T. S. Scanlan, *Nat. Struct. Biol.* **1998**, *5*, 679–681.

69 N. L. Urizar, A. B. Liverman, D. T. Dodds, F. V. Silva, P. Ordentlich, Y. Yan, F. J. Gonzalez, R. A. Heyman, D. J. Mangelsdorf, D. D. Moore, *Science* **2002**, *296*, 1703–1706.

70 E. Owsley, J. Y. Chiang, *Biochem. Biophys. Res. Commun.* **2003**, *304*, 191–195.

71 A. C. W. Pike, A. M. Brzozowski, R. E. Hubbard, T. Bonn, A.-G. Thorsell, O. Engström, J. Ljunggren, J.-Å. Gustafsson, M. Carlquist, *EMBO J.* **1999**, *18*, 4608–4618.

72 Z.-C. Dang, V. Audinot, S. E. Papapoulos, J. A. Boutin, C. W. G. M Lowik, *J. Biol. Chem.* **2003**, *278*, 962–967.

73 L. Van Gaal, A. J. Scheen, *Diabetes Metab. Res. Rev.* **2002**, *18*, S1–S4.

74 K. C. Nicolaou, J. A. Pfefferkorn, A. J. Roecker, G.-Q. Cao, S. Barluenga, H. J. Mitchell, *J. Am. Chem. Soc.* **2000**, *122*, 9939–9953.

75 K. C. Nicolaou, J. A. Pfefferkorn, H. J. Mitchell, A. J. Roecker, S. Barluenga, G.-Q. Cao, R. L. Affleck, J. E. Lillig, *J. Am. Chem. Soc.* **2000**, *122*, 9954–9967.

76 K. C. Nicolaou, R. M. Evans, A. J. Roecker, R. Hughes, M. Downes, J. A. Pfefferkorn, *Org. Biomol. Chem.* **2003**, *1*, 908–920.

77 M. Downes, M. A. Verdecia, A. J. Roecker, R. Hughes, J. B. Hogenesch, H. R. Kast-Woelbern, M. E. Bowman, J.-L. Ferrer, A. M. Anisfeld, P. A. Edwards, J. M. Rosenfeld, J. G. Alvarez, J. P. Noel, K. C. Nicolaou, R. M. Evans, *Mol. Cell* **2003**, *11*, 1079–1092.

15
Combinatorial Chemistry in the Age of Chemical Genomics

Reni Joseph and Prabhat Arya

15.1
Introduction

The concept of chemical genetics/genomics has emerged recently in the chemical biology community in recognition of a renewed desire to generate small molecules and use them as chemical probes for understanding protein functions [1]. Parallel to genetic approaches, the use of small molecules as highly specific modulators (i.e., inhibitors or activators) of protein functions is a powerful approach and is commonly applied for understanding dynamic processes that involve protein–protein interactions, protein networking, etc. [2]. In general, due to the irreversible effects caused by genetic manipulations, these systems are difficult to study by traditional biological approaches [3].

An excellent viewpoint article by Strausberg and Schreiber [4] discusses the challenges we face today in the post-genomics age, i.e., (1) what is the next step, (2) how to move forward in developing better medicines, (3) how to benefit from knowing the gene(s)/gene products to modulating their functions.

For the success of chemical genetics/genomics-based research programs, rapid access to diverse sets of small molecules is of prime importance, because these derivatives pave the way for dissecting biological processes and are valuable tools as probes for understanding biological events [5]. Over the years, combinatorial chemistry has emerged as an important technology, because it allows efficient synthesis of many compounds in a parallel manner. It has usually been successfully applied to the high-throughput synthesis of simple compounds (i.e., compounds with no stereogenic centers) [6]. With few exceptions, the development of combinatorial methods that allow the high-throughput synthesis of complex, highly functionalized, natural product-like polycyclic derivatives remains a daunting task [7]. The need for these efforts is increasing constantly, due to (1) the rapid rise in new biological targets emerging from genomics and proteomics research, (2) the growing interest in understanding protein–protein interactions in signal transduction, and (3) the need for small molecules that can be used to search for highly specific modulators of protein function.

Chemogenomics in Drug Discovery: A Medicinal Chemistry Perspective.
Edited by Hugo Kubinyi and Gerhard Müller

ISBN: 3-527-30987-X

Over the years, natural products have been widely utilized as small-molecule chemical probes for understanding biological pathways [8]. Many compounds that elicit specific cellular responses have complex, chiral, highly functionalized structures. These properties are valuable in searching for specific binding to protein targets or in differentiating related proteins. Although combinatorial chemistry is well accepted in the medicinal chemistry community for the rapid synthesis of simple molecules, the current challenge is to develop stereo- and enantioselective synthesis (asymmetric synthesis) methods in solution and on solid phase to obtain fast access to complex, functionalized, natural product-like compounds. The development of asymmetric synthesis-derived organic reactions on solid phase allows the synthesis of complex natural products and natural product-like compounds in a high-throughput manner. In recent years, several research groups have taken the challenge of developing stereoselective reaction-based methods on solid phase and are utilizing them for the high-throughput synthesis of complex natural product-like derivatives. Some recent achievements toward this objective are highlighted in this chapter.

The first part of the chapter covers several examples of stereoselective solution and solid-phase approaches to obtaining natural product analogs. The examples discussed here represent focused strategies to library generation that are based on specific bioactive natural products. These libraries have often served as natural product analogs in searches for better biological responses than those exhibited by the parent natural products. The generation of libraries of natural product-like compounds by solution and solid-phase synthesis methods is discussed in the second half.

Here, we cover several examples that included stereoselective diversity-oriented synthesis approaches. The examples covered in this section are indicative of the growing interest in this area and of the need for developing novel approaches leading to fast access to obtaining natural product-like compounds to be used as small-molecule probes.

15.2
Combinatorial Approaches to Natural Product Analogs

Due to their 3D structural architectures, natural products have been a source of inspiration for developing efficient stereo- and enantioselective synthesis methods [9]. There are several examples (see Figure 15.1 for a few bioactive complex natural products: taxol, FK 506, rapamycin, vinblastine) in the literature where the complex architectures of natural products are the key to exhibiting highly specific cellular responses (i.e., highly specific modulations of protein functions). The unique shapes of natural products make them ideal as small-molecule candidates that may selectively bind to enzymes and to other proteins. In combinatorial chemistry, interest in the development of solution or solid-phase methods leading to generation of libraries of bioactive natural products analogs is growing. Several selected examples are discussed in this section.

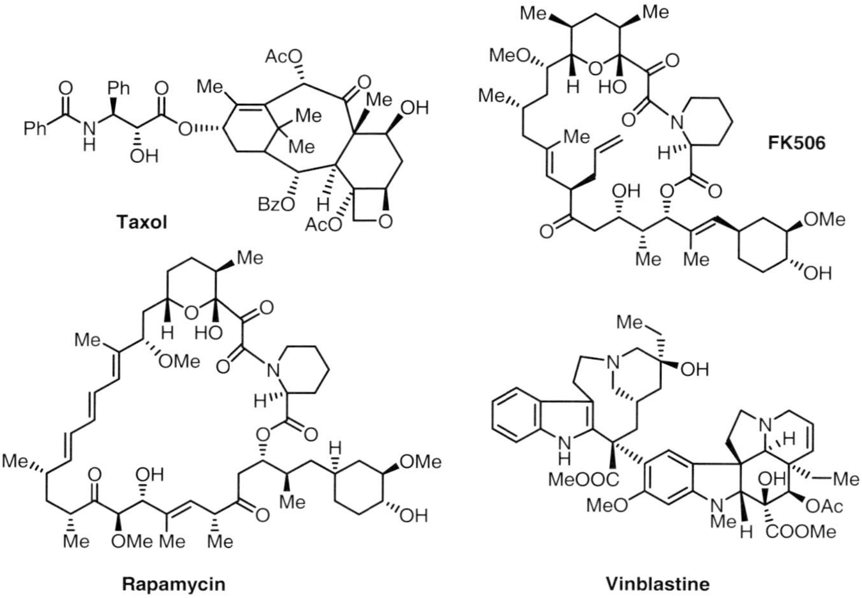

Taxol FK506 Rapamycin Vinblastine

Figure 15.1 Examples of bioactive natural products.

Based on the development of solution-phase synthetic methods [10] for fumiquinazoline alkaloids, Wang and Ganesan [11] developed a solid-phase method and then utilized this approach to generate a library of nonnatural analogs of fumiquinazoline alkaloids. (+)–Glyantrypine was the first target for the development of a solid-phase synthetic method. Commercially available Wang resin loaded with Fmoc-L-Trp **2.1** (Figure 15.2) was deprotected and then coupled with anthranilic acid in the presence of EDC as an activating agent. The next step was acylation of aniline **2.2** with Fmoc-amino acid chloride and pyridine to result in **2.3**. The key dehydrative cyclization of the linear tripeptide **2.3** was then carried out with Ph_3P/I_2/DIPEA, giving the desired product **2.4**. The final step was piperidine-mediated deprotection of the Fmoc and rearrangement of the oxazine **2.4** to the amidine carboxamide **2.5**. The resin was then refluxed in acetonitrile to induce cyclative cleavage, giving the target compound **2.6**. The feasibility of using substituted anthranilic acids was tested by repeating the synthesis with 4-chloroanthranilic acid. Finally, nonnatural fumiquinazoline derivatives were prepared to further examine the scope of the reaction, replacing L-Trp with L-Ala, L-Leu, L-Phe, and performing acylation with a set of five anthranilic acids, followed by final coupling with D or L amino acid chlorides.

Bonnet and Ganesan [12] developed a new route toward the solid-phase synthesis of tetrahydro-β-carboline hydantoins, which appear in a diverse array of biologically active alkaloids. Due to the presence of tetrahydro-β-carbolines and diketopiperazine ring skeletons in several bioactive natural products, these two scaffolds became attractive targets for developing a combinatorial chemistry program.

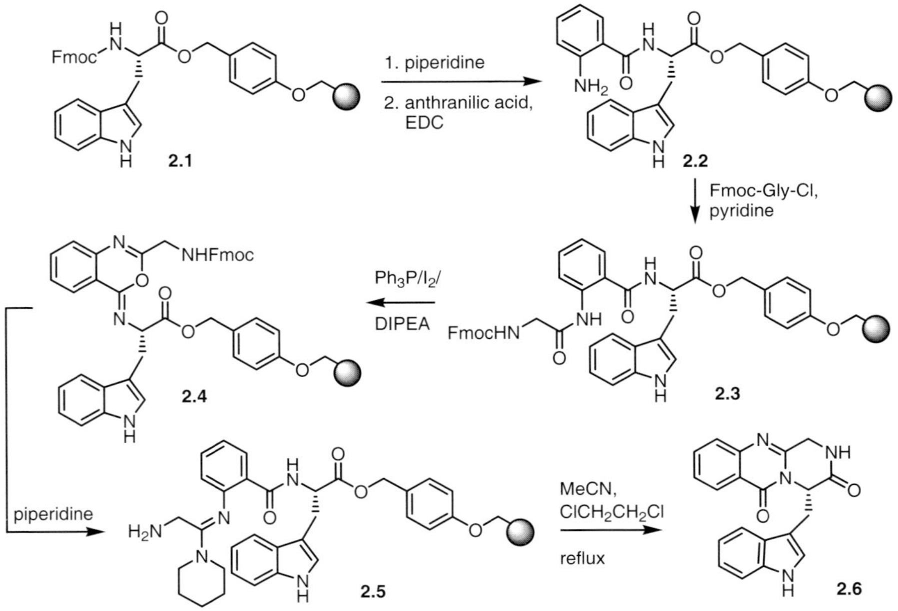

Figure 15.2 Solid-phase synthesis of fumiquinazoline alkaloids by Wang and Ganesan [11].

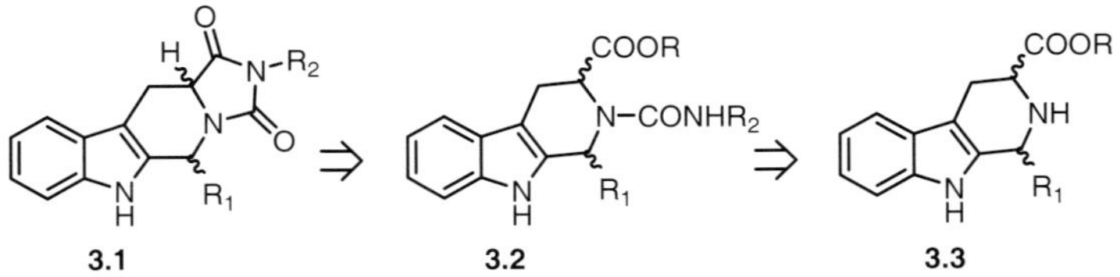

Figure 15.3 Retrosynthetic analysis of tetrahydro-β-carboline hydantoins.

These derivatives could then be tested for small molecules to block eukaryotic cell cycle progression or as chemical probes in mechanistic studies. A series of indole alkaloids having a prenyl functional group are known to interfere with the cell cycle at the G_2/M transition [13]. A retrosynthetic analysis of tetrahydro-β-hydantoins (Figure 15.3) involves construction of the ring skeleton by classical hydantoin synthesis whereby the urea form of **3.2** undergoes intramolecular cyclization. The intermediate urea **3.2** in turn is derived from the reaction of **3.3**, the product of an acid-catalyzed Pictet–Spengler reaction with an isocyanate. In Ganesan's methodology, the urea derivative **3.2** could be obtained by an alternative disconnection approach that involved displacement of an activated carbamate. In a model study, the *cis* and *trans* diastereomers of known tetrahydo-β-carbolines (**4.1** and **4.4** in Figure 15.4) were individually treated with *p*-nitrophenyl chloroformate to give carbamates **4.2** and **4.5**, respectively, which, upon treatment with benzylamine and triethylamine, result in **4.3** and **4.6**. Interestingly, both *cis* (**4.1**) and *trans* (**4.4**) carbolins produced *trans*-hydantoins **4.3** and **4.6**, presumably due to epimerization

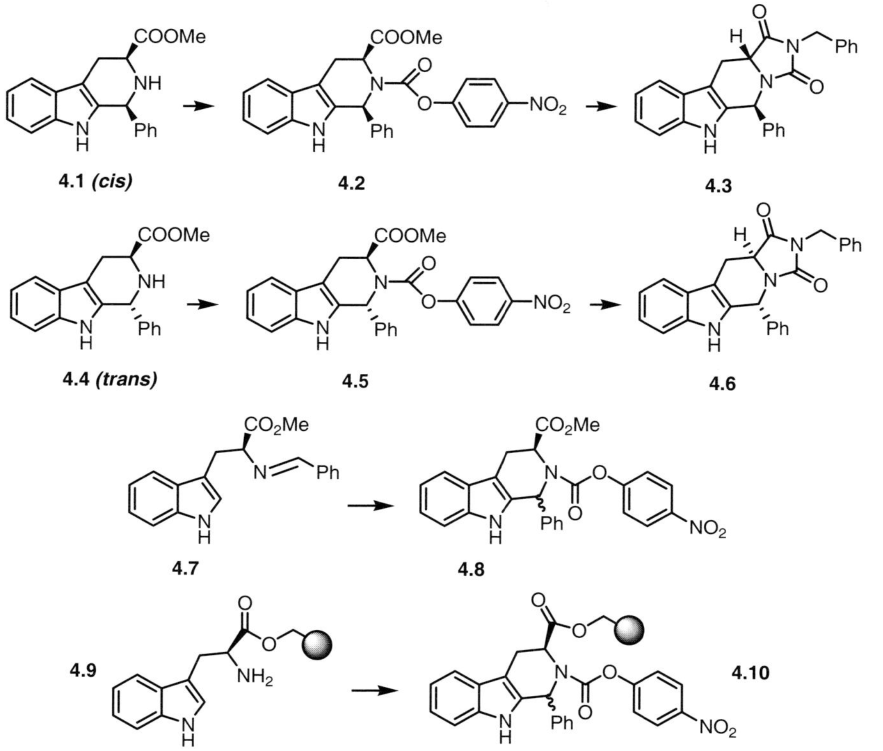

Figure 15.4 Solid-phase approach to tetrahydro-β-carboline hydantoins.

of the *cis* isomer adjacent to the carbonyl functionality to the thermodynamically stable *trans* product. Carbamate **4.8** was also obtained from L-tryptophan methyl ester by an *N*-acyliminium Pictet–Spengler process, in which the tryptophan amine was reacted with *p*-nitrophenyl chloroformate under basic conditions. This reaction was then investigated on a solid support (**4.9** to **4.10**) of polystyrene Wang resin. The scope of this methodology was further demonstrated by the synthesis of a small library on solid support. In summary, this methodology features a new approach to hydantoins that proceeds via amine addition to an activated carbamate and avoids the need for isocyanate building blocks.

Since the goal is to obtain simple analogs, the unique structural architectures of bioactive natural products present tremendous challenges in developing a combinatorial chemistry program. Because natural products have been a major source of lead compounds in drug discovery research, there is growing interest in developing combinatorial chemistry (high-throughput solution or solid phase synthesis) based on their core structures. Curacin A (**5.1** in Figure 15.5) promotes arrest of the cell cycle at the G_2/M checkpoint and inhibits the binding of (^3H)-colchicine to tubulin [14]. This marine natural product has served as the lead compound for combinatorial synthesis of 6-compound mixture libraries [15].

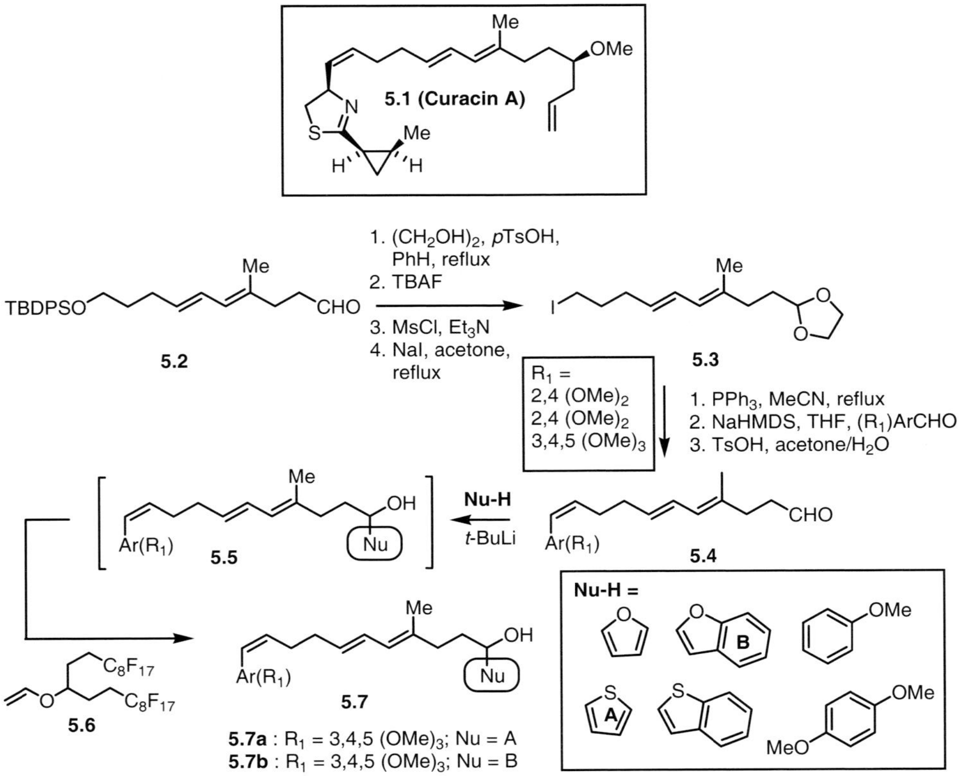

Figure 15.5 Approach to a library based on the antimitotic marine natural product curacin A by Wipf et al. [15].

The solution-phase synthesis achieved by Wipf's group prompted their interest in developing a combinatorial chemistry approach to obtaining fast access to analogs [16]. The goal was to substitute the heterocycle ring of curacin A with electron-rich aromatic moieties and the homoallylic methyl ether terminus with a broad range of hydrophobic benzylic alcohols. Three building blocks of **5.4** were prepared by solution synthesis from the readily available aldehyde **5.2** After protection with ethylene glycol, desilylation, mesylation, and then Finkelstein reaction with sodium iodide, the dioxalane **5.3** was produced. The corresponding Wittig reagent was condensed with 2,4-dimethoxy-, 2,5-dimethoxy-, and 3,4,5-trimethoxy benzaldehydes to give the aldehydes **5.4**. Next, seven mixtures of 6–9 compounds were prepared individually by exposing **5.4** as well as the derivatives having other aryl substituents to a cocktail of 3–6 aryl lithium reagents. The most active biological mixtures were obtained from reactions of 4–6 compounds with 2-lithiated furan, thiophene, benzofuran, benzothiophene, anisole, and 1,4 dimethoxybenzene. The lithiation was performed directly on the mixture. Since a large excess of organolithium was used, the resulting alcohols **5.4** were heavily contaminated with Nu-H. Fluorous quenching with the vinyl ether **5.6** was then utilized to streamline purification of the heterogeneous multicomponent reaction products after the diversification step in the library synthesis and to provide structurally defined mixtures **5.7**. The screening profile of one mixture library was attractive enough to warrant synthesis of the individual components. Several compounds exhibited significant activity in altering the cell cycle as well as the microtubule cytoskeleton in living human carcinoma cells. For example, **5.7a** and **5.7b** inhibited tubulin polymerization with an IC_{50} of ca. 1 μM. Compounds **5.7a** and **5.7b** represent some of the most important curacin A analogs identified to date.

de Frutos and Curran [17] developed an improved radical annulation approach to synthesizing *rac*-mappicine, (*S*)-mappacine (**6.2**), and mappicine ketone (**6.1**) analogs. The mappicine ketone derivative possesses antiviral activity; in particular, it is active against herpes virus and human cytomegalovirus in low micromolar concentrations [18]. Developing a combinatorial chemistry program on the lead bioactive natural product could provide fast access to its analogs. This could also be very useful in understanding the structure–activity relationship of the natural product, and it may also provide interesting simpler analogs with enhanced biological responses. To date, there are very few examples of combinatorial chemistry approaches to polycyclic natural products. The free radical cascade strategy to synthesizing mappacine is summarized in Figure 15.6. The synthesis design permits the combinatorial assembly of three building blocks: BB-1 (**6.3**), BB-2 (**6.4**) and BB-3 (**6.5**). The pyridone D ring (BB-1) was first alkylated with a suitable propargylating agent (BB-2) bearing the B ring substituent, followed by cascade radical annulation with an isonitrile having the A ring substituent (BB-3) to obtain mappicine analogs.

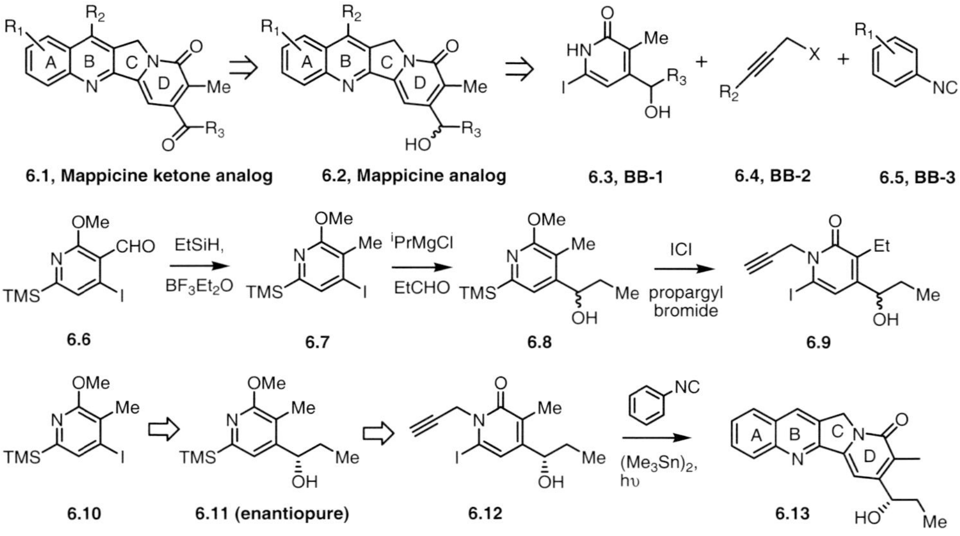

Figure 15.6 Combinatorial approach to mappicine and mappicine ketone analogs of de Frutos and Curran [17].

Mappicine ketones, which exhibit antiviral activity, can be prepared by parallel oxidation of the mappicine analogs. The synthesis begins with the known formyl pyridine **6.6**. Reduction of the aldehyde with Et_3SiH in the presence of BF_3OEt_2 afforded **6.7**, which, after transmetallation reaction with *i*-PrMgCl, generated the heteroaryl Grignard reagent; after quenching this yielded compound **6.8**. TMS–iodine exchange, demethylation, and *N*-propargylation provided the iodopyridone **6.9**. For the synthesis of enantiomerically pure (*S*)-mappicine, a slight modification was made. Quenching of the cuprate reagent with propionyl chloride afforded the ketone, which was subjected to (–)DIP-chloride reduction (i.e., enantioselective reduction), resulting in compound **6.11** as the pure enantiomer. Finally, radical cyclization with **6.12** and phenyl isonitrile in the presence of hexamethylditin in benzene afforded (–)-mappicine, **6.13**. Using four isonitriles (BB-3), four propargyl bromides (BB-2), and iodopyridone (BB-1), respectively, in separate parallel synthesis resulted in a 64-member library of mappicine analogs. Similarly, a 48-member library of mappicine ketone analogs was also synthesized. The researchers also synthesized a library of 560 compounds as mappicine analogs by a fluorous mixture synthesis approach [19].

In 1997, Lavergne et al. [20] reported the solution synthesis of homocamptothecin. This compound is an analog of camptothecin in which the six-membered ring E has been modified to a seven-membered ring derivative. The ring expansion results in improved stability of the compound in human plasma, and this derivative is an interesting lead compound for developing antitumor agents [21]. These researchers proposed that camptothecin and homocamptothecin may not function in a similar manner for topoisomerase-induced DNA cleavage [22]. This observation sparked interest in obtaining further analogs in the homocamptothecin series, with the goal of developing homocamptothecin-based antitumor agents.

Gabarda and Curran [23a] developed a practical, efficient strategy to synthesize the homocamptothecin class of antitumor agents and further accomplished parallel synthesis of 115 homosilatecans. The synthetic pathway to (*rac*)-homosilatecans is shown in Figure 15.7. The synthesis of **7.7** started with 3-formyl-4-iodo-2-methoxy-6-trimethylsilyl pyridine, **7.1** [23b]. Treatment of the iodoformylpyridine **7.1** with $NaBH_4$ afforded the hydroxymethyl pyridine, which, on protection with MOMCl, resulted in **7.2**. Treatment of **7.2** with *i*-PrMgCl followed by addition of CuCN/LiCN and then quenching, by cuprate reagent with propionyl chloride, provided the key intermediate **7.3** [23c]. Aldol condensation between the ketone **7.3** and the enol ether generated from methyl acetate afforded the crude β-hydroxy ester, which, upon treatment with TFA, provided the lactone **7.4**. Iodinative desilylation of TMS-lactone **7.4** was then performed with ICl to afford the iodolactone. Demethylation was accomplished by addition of TMSCl and NaI, providing the iodopyridone **7.5**. The next step involved parallel *N*-alkylation of the iodopyridone **7.5** with several propargyl bromides to yield **7.6**, followed by parallel annulation with ditrimethyltin of the *N*-propargylated iodopyridones with a collection of isonitriles to result in **7.7**. By using this strategy, 115 new analogs were prepared, and biological evaluation of these compounds is under investigation.

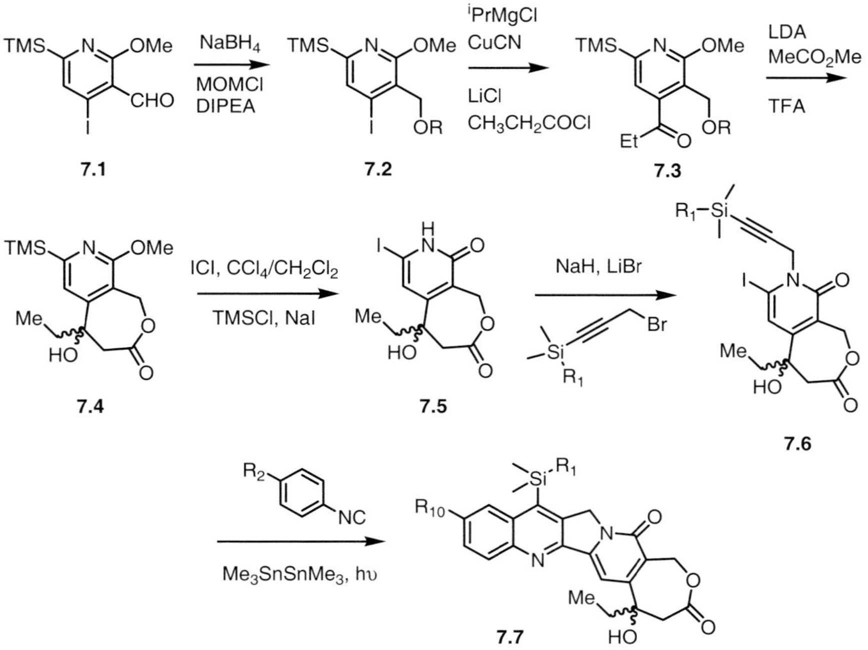

Figure 15.7 Parallel synthesis of homosilatecan analogs by Gabarda and Curran [23a].

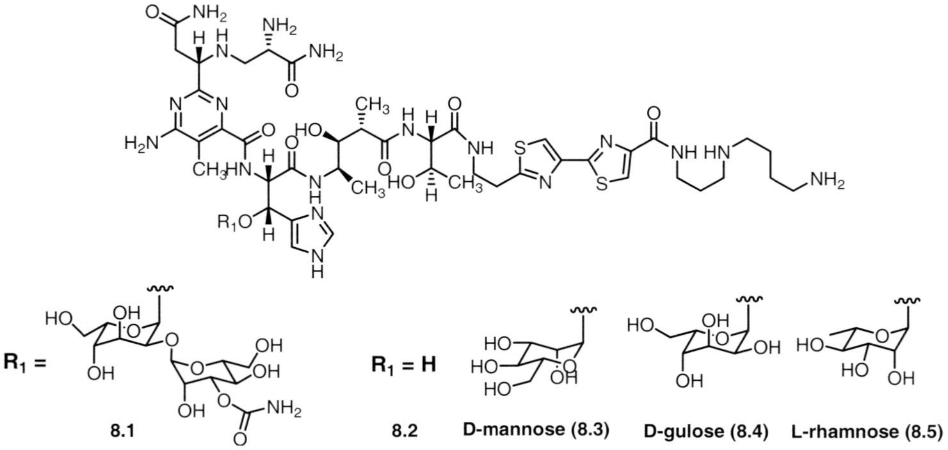

Figure 15.8 Bleomycin A_5 analogs.

The bleomycins (BLMs), exemplified by BLM A_5 (Figure 15.8), are anticancer antibiotics that affect single and double-strand DNA cleavage [24] and RNA cleavage [25] in the presence of metal cofactor and oxygen [26]. Although the deglyco-BLM analogs have provided insights into the function of the amino acid constituents of BLM, the role of the carbohydrate moiety is not well understood.

To explore the role of the glycosides on bleomycin, Boger and coworkers [27, 28] prepared BLM A_2 derivatives containing the monosaccharides α-D-mannose and 2-*O*-methyl-α-L-gulose. These derivatives retained the selective cleavage of DNA, however, only the 2-*O*-methyl-α-L-gulose BLM A_2 derivative cleaved DNA efficiently. The carbohydrate moiety apparently has little influence on the sequence selectivity of DNA cleavage by BLM [29], and no study has defined the effect of the carbohydrate moiety on RNA cleavage.

Thomas et al. [30] undertook a study of solid-phase synthesis of BLM A_5 containing α-mannose, α-L-gulose, and α-L-rhamnose as glycoside moieties. Solid-phase synthesis of BLM A_5 and its monosaccharide derivatives was carried out by analogy with the synthesis of deglyco-BLM. The synthesis utilized Boc- or NBS-(2-nitrobenzenesulfonyl)-protected spermidine resin **9.1** (Figure 15.9). The addition of bithiazole intermediates **9.2** [31] to free amine was accomplished by using HBTU and DIPEA. Fmoc was removed with 20% piperidine in DMF. *N*-α-Fmoc-(*S*)-threonine **9.4** was then coupled by using HBTU and DIPEA in DMF.

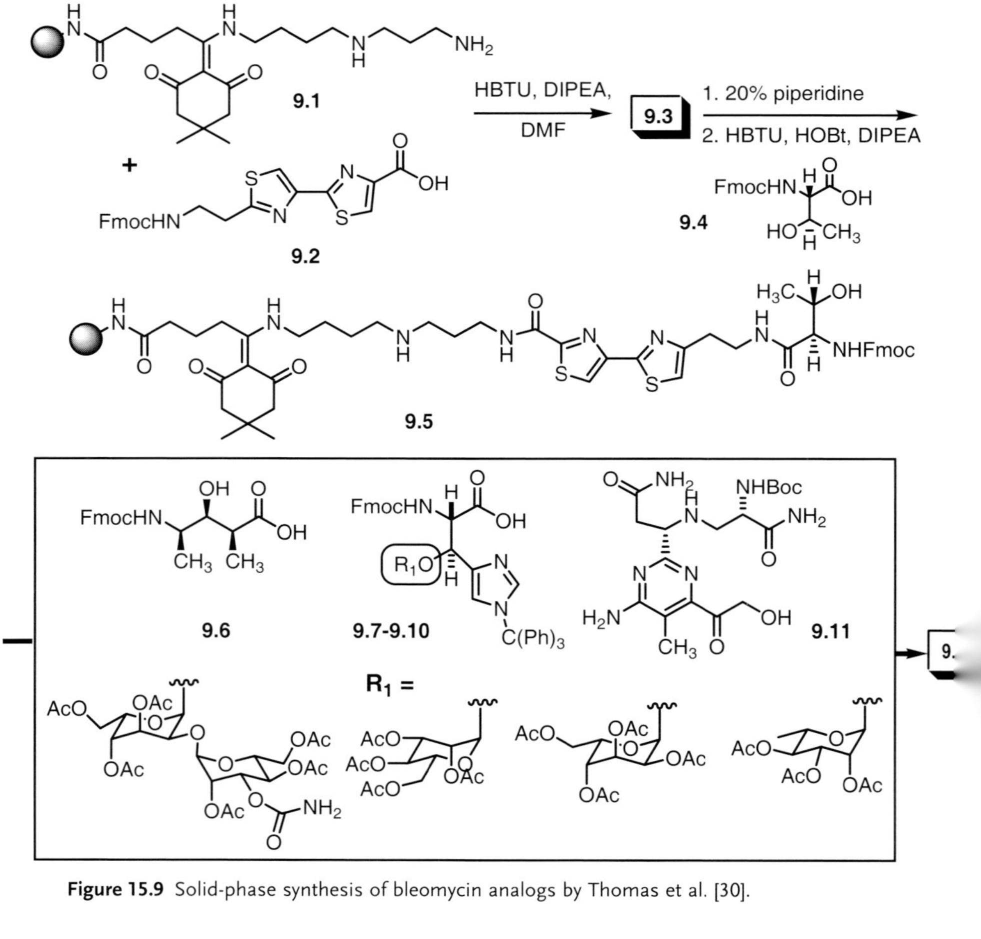

Figure 15.9 Solid-phase synthesis of bleomycin analogs by Thomas et al. [30].

Again following Fmoc deprotection, *N*-α-Fmoc-(2*S*,3*S*,4*R*)-4-amino-3-hydroxy-2-methylvalerate **9.6** was conjugated to the resin in the same fashion. Fmoc- and trityl-protected histidine analogs **9.7–9.9** and **9.10** containing the disaccharide and each of the three monosaccharide derivatives were then coupled. The Boc-protected pyrimidoblamic acid **9.11** [32] was then coupled to the resin-bound pentapeptide. The resin bound fully functionalized BLM A_5 and each of the monosaccharide analogs was then deprotected and cleaved from the resin, affording BLM A_5 derivatives **9.12–9.15**. The α-L-gulosyl BLM A_5 analog affected single- and double-stranded DNA cleavage to nearly the same extent as BLM A_5 itself. The α-mannosyl and α-L-rhamnosyl analogs had diminished DNA cleavage efficiencies, comparable to that of deglyco-BLM A_5. Only the α-L-gulosyl analog exhibited a relatively high level of oxidative RNA cleavage. Thus, these experiments provide evidence that the carbohydrate moiety of BLM plays a significant role in defining its competence in DNA and RNA cleavage.

(−)-Indolactam V, the core structure of tumor-promoting teleocidins [33], has generated strong interest as the key compound for generating its synthetic analogs (Figure 15.10). Waldmann's laboratory [34, 35] reported an efficient synthesis of a library of indolactam analogs based on using a combination of solution and solid-phase approaches. For the design of the library, they took into consideration the substituents at R_1, and R_2 (see **10.3** in Figure 15.10). These two sites influence the

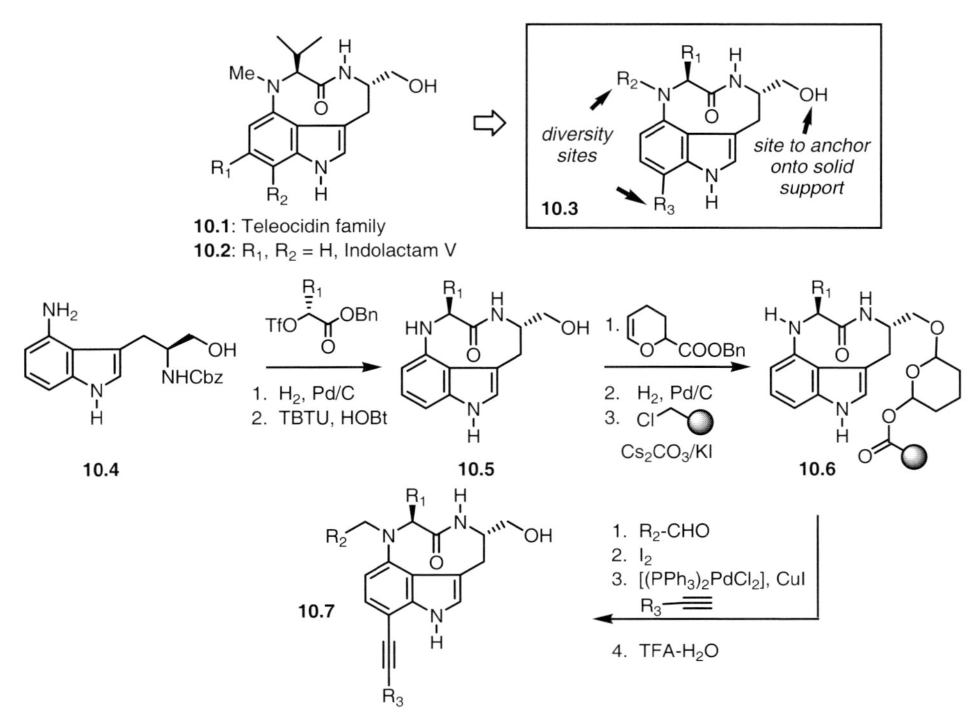

Figure 15.10 Solid-phase synthesis of indolactam library by Waldmann's laboratory [34, 35].

conformation of the nine-membered lactam ring. Furthermore, substitution at R_3 facilitates membrane-binding, and the hydroxyl group in the lactam ring is critical for biological activity. Hence, indolactam analogs **10.3** (with diversity at R_1, R_2, and R_3) may lead to the identification of more potent PKC modulators. Diversity at R_1 was introduced by alkylation of the aromatic amine with three different α-hydroxy acid ester triflates. The triflate derivatives were displaced by aminoindole **10.4** in dichloromethane, and removal of the Cbz group and amide bond formation using TBTU as the coupling reagent gave the indolactam derivative **10.5**. Attachment of indolactams **10.5** to the solid support was achieved with a tetrahydropyran linker by converting **10.5** to the corresponding acetals. Removal of the benzyl ester group followed by coupling to chloromethyl polystyrene beads by esterification with $CsCO_3$ resulted in substrate–linker–resins **10.6**. This resin-bound indole was *N*-alkylated by reductive amination (aldehydes and $NaHB(OAc)_3$) for incorporation of R_2. Diversity was then introduced at R_3 by a Sonogashira coupling reaction with acetylenes to obtain the alkyne derivatives. Finally, the resin was cleaved (TFA/H_2O) to give compounds **10.7** in overall yields ranging from 10% to 65%.

The saframycins are structurally complex alkaloids, constituting a series of natural antiproliferative agents containing a cyanopiperazine core, or its functionalized equivalent, with a complex polycyclic framework. Saframycin A is the most potent antiproliferative saframycin. Myers and coworkers [36] developed a solid-supported, enantioselective synthesis procedure suitable for rapid preparation of large numbers of diverse structural analogs of (–)saframycin A [37]. Several other members of this family are also used in the treatment of solid tumors, which has sparked interest in obtaining analogs showing enhanced pharmacological properties [38]. Myers and Plowright [39] found that a bishydroquinone derivative of saframycin A showed enhanced antiproliferative activity in a few cell lines. Several analogs of bishydroquinone derivatives of saframycin A were synthesized, and in some instances, the compounds exhibited greater than 20-fold activity in antiproliferative assays. Although these two types of compounds are structurally different, it was suggested that they exhibited a similar biological function by a common mechanism [40].

The solid-phase synthesis to obtain fast access to saframycin analogs began with anchoring the antimorpholino nitrile [41] (**11.1**, Figure 15.11) to a solid support by silyl ether formation with 4-(chlorodiisopropyl-silyl)polystyrene, providing the first resin-bound intermediate (**11.2**). Selective deprotection of the *tert*-butyldimethylsilyl ether group occurred upon exposure to tetrabutylammonium fluoride buffered with acetic acid. Subsequent treatment with piperidine in DMF unmasked the amino terminus, affording the phenolic amine. This, upon treatment with N-protected α-amino aldehyde (**X**), provided the corresponding resin-supported imine. Warming the imine in a saturated solution of anhydrous lithium bromide in 1,2-dimethoxyethane induced a stereoselective Pictet–Spengler cyclization reaction, affording the *cis*-tetrahydroisoquinoline derivative **11.3**. The secondary amine group of **11.3** was then reductively methylated. Subsequent deprotection of the phenol and primary amino groups of the resulting *N*-alkylated product produced **11.4**, which was then subjected to a Pictet–Spengler cyclization upon exposure to N-Fmoc glycinal. The resulting bis-tetrahydroisoquinoline derivative **11.5** was formed in quantitative yield with the required *cis* stereochemistry in the newly formed ring. The next key step, the bis-tetrahydroisoquinoline intermediate **11.5**, was subjected to cyclization autorelease by warming in the presence of $ZnCl_2$ at 55 °C for 1.5 h. This transformation was presumed to involve reversible morpholinium-ion formation through expulsion of cyanide, internal capture by cyclization of the secondary amino group formed in the final Pictet–Spengler cyclization, and subsequent extrusion of the resin-bound morpholine dual linker through its secondary point of attachment (i.e., the amino group). The newly liberated iminium ion is presumably captured in solution by the cyanide, providing the saframycin analog **11.6**. A key feature of this reaction is the distereospecificity. Additional diversity was introduced by employing a range of alkyl bromides in *N*-alkylation of the resin **11.4** and simultaneously varying the aldehyde reactant in the second Pictet–Spengler cyclization, thus resulting in a 16-member library of saframycin A analogs prepared by parallel synthesis.

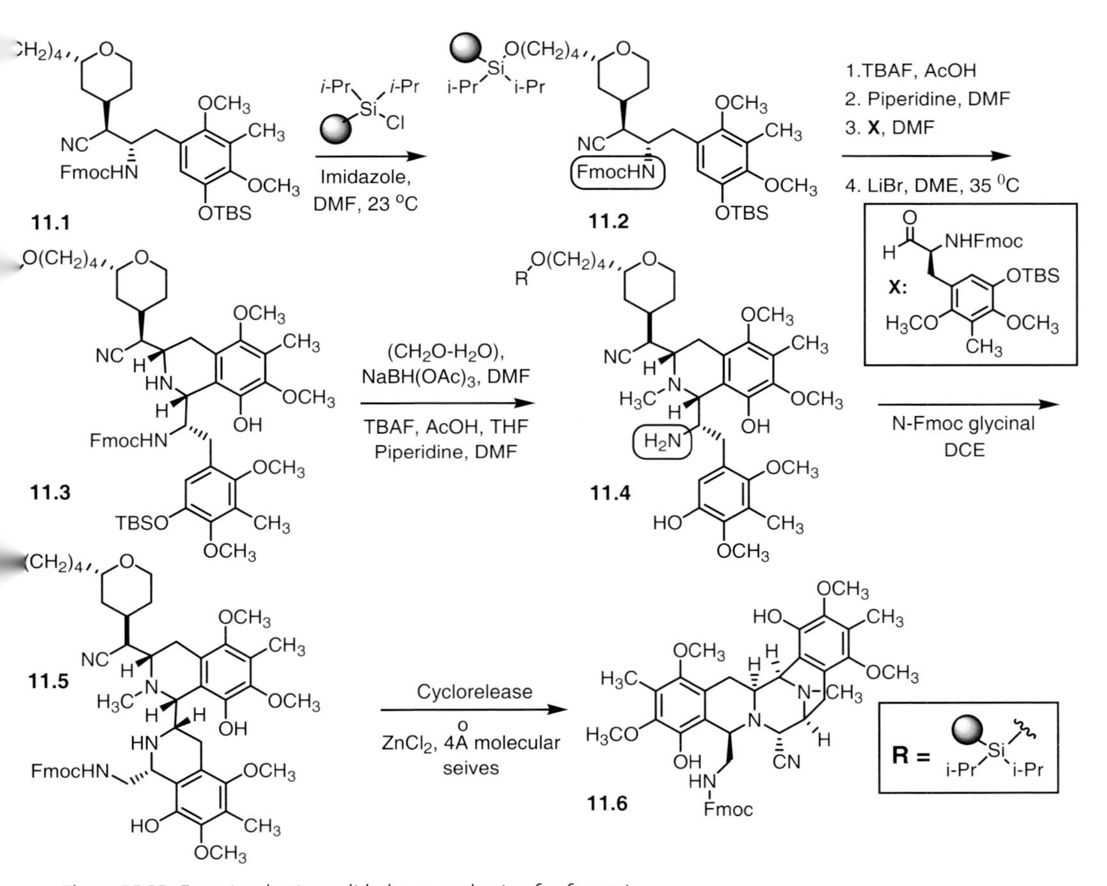

Figure 15.11 Enantioselective solid-phase synthesis of saframycin analogs by the Myers group [36, 39, 40].

15.3
Diversity-oriented Synthesis of Natural-product-like Libraries

The term, 'diversity-oriented synthesis (DOS)' was coined by Schreiber [42a]; the purpose of this approach is to build complex, natural-product-like architectures in a high-throughput manner so as to use these skeletons in library generation. Unlike the traditional combinatorial approaches that focus on generating libraries of aromatic and heterocyclic products, DOS focuses on building 3D structural complexity by exploring stereo- and enantioselective reactions on solid phase. In contrast to the classical analog-based approaches discussed earlier, the libraries generated by DOS are utilized as small-molecule chemical probes for understanding cellular processes. In general, these libraries are not biased toward a given biological target. Although this is a new research activity within the combinatorial community and most of the work reported to date is from Schreiber and coworkers, we have tried to cover DOS-related work from other research groups too.

In 1998, Tan et al. [43a] developed a highly efficient, multistep synthesis process for obtaining enantiomerically pure template **12.4** (Figure 15.12) from shikimic acid **12.1**. Shikimic acid was first converted into both enantiomers of the epoxycyclohexenol carboxylic acid derivative **12.2**, which was then coupled to a photocleavable linker immobilized onto Tentagel-S-NH_2 poly(ethyleneglycol)–polystyrene copolymer. Treatment of the resin-bound epoxycyclohexenol derivative **12.3** with various nitrone carboxylic acids under esterification conditions, followed by an intramolecular cycloaddition reaction, yielded the tetracyclic scaffold **12.4** with complete regio- and stereoselectivity, via a tandem acylation/stereoselective 1,3-dipolar cycloaddition reaction. The tetracyclic derivative **12.4** is rigid and densely functionalized, allowing it to undergo a variety of organic transformations without the use of protecting groups. Upon treatment with a variety of organic and organometallic reagents, this template can be used to obtain highly functionalized bicyclic and tricyclic derivatives, as indicated in Figure 15.12.

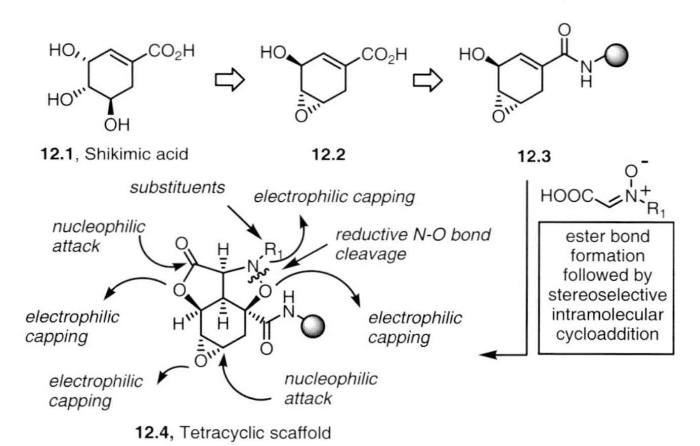

Figure 15.12 Stereoselective diversity-oriented synthesis (DOS) of natural-product-like polycyclics by Tan et al. [43].

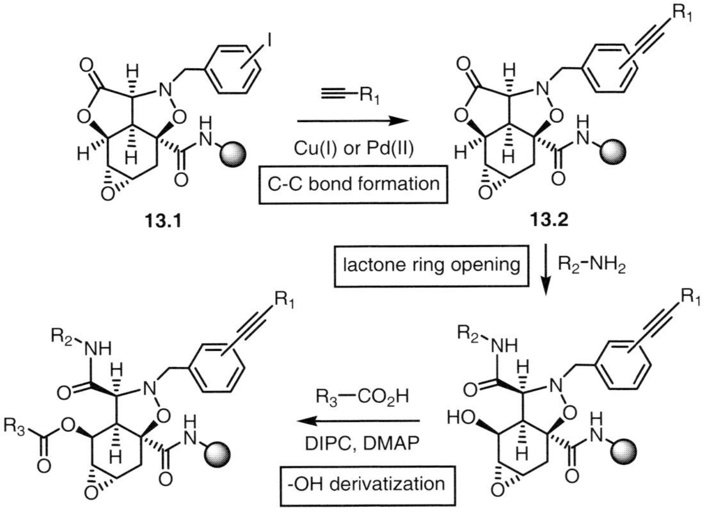

Figure 15.13 Development of specific natural-product-like polycyclics by DOS.

In the example shown in Figure 15.13, the tetracyclic template anchored to a solid support, **13.1**, was subjected to a Cu(I)- or Pd(II)-mediated carbon–carbon bond-forming reaction with an iodoaryl moiety of the template, giving product **13.2**. Upon reaction with various amines, the lactone ring was opened, giving the carboxylamide derivative **13.3**, which has a free hydroxyl group. The hydroxyl derivative **13.3** was further acylated under standard acylation reactions to obtain compound **13.4** in high yield. Several key reactions were well optimized before any library synthesis was undertaken. In a split-and-mix solid-phase synthesis method, the tetracyclic template **13.4** was utilized to develop a method for stereoselective synthesis of a library exceeding 2 million compounds. It is interesting to note that the library synthesis plan did not incorporate functional group protection and deprotection, which is usually a daunting task during total synthesis of complex molecules.

The library was then tested in several miniaturized cell-based assays to search for cell-permeable small-molecule protein-binding agents. For example, several members of this library were found to activate a reporter gene in mink lung cells. The driving force in the library synthesis plan was to develop a method for stereoselective synthesis of the highly functionalized tetracyclic derivative **12.4**, which could undergo several simple nucleophilic and acylation reactions. This is the first example of synthesis of a complex natural-product-based library, in which an enantiomerically pure template was developed by utilizing a highly functionalized chiral starting material.

Pelish et al. [44] demonstrated the use of a biomimetic-based, diversity-oriented synthesis method to produce galanthamine-like molecules (Figure 15.14) with biological properties superior to those of natural product, galanthamine. As first

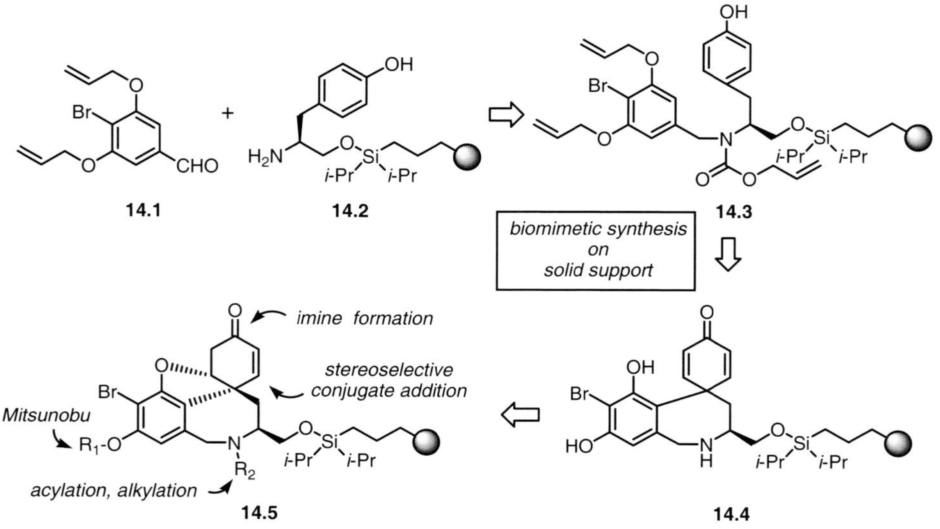

Figure 15.14 Biomimetic solid-phase synthesis of galanthamine-like compounds by Pelish et al. [44].

reported by Barton and Kirby [45], a single precursor, norbelladine, is converted via specific oxidative phenolic coupling pathways to an entire class of natural products, including the crimines, galanthamines, lycoranes, and pretazzetines. Each compound is structurally different and elicits a different biological response.

Shair and coworkers utilized these characteristics in developing a biomimetic synthesis method on solid phase to obtain a chiral template for diverse libraries of complex molecules. Figure 15.14 outlines the procedure for synthesizing amaryllidaceae like alkaloids by mimicking the oxidative phenolic coupling reaction that takes place in Nature with a hypervalent iodine reagent. By using a simple orthogonal protecting-group strategy, a common dienone intermediate was then directed to cyclize at a nitrogen to generate crimine- or galanthamine-type structures after selective liberation of the phenolic moiety. This was followed by split-pool-derived organic synthesis on two core systems to generate a structurally rich amaryllidaceae alkaloid-based library. Library synthesis commenced with attachment of tyrosine derivatives to 500–600-μm high-capacity polystyrene beads through a Si-O bond to generate derivative **14.5**. Reductive amination, followed by protecting-group adjustments, yielded compound **14.3** anchored to a solid support, which, upon exposure to $PhI(OAc)_2$, afforded the quinone derivative **14.4**. Then **14.4** was subjected to Pd-mediated deprotection and a spontaneous intramolecular hetero-Michael-type reaction, giving the cyclic derivative **14.5**. The diversity steps were accomplished by (1) phenolic hydroxyl group alkylation, (2) an intermolecular Michael-type reaction with thiols in the presence of *n*-BuLi, (3) imine formation from the carbonyl group, and (4) secondary amine alkylation or acylation. It is interesting to note that the intermolecular Michael-type reaction with various thiol-based nucleophiles is highly

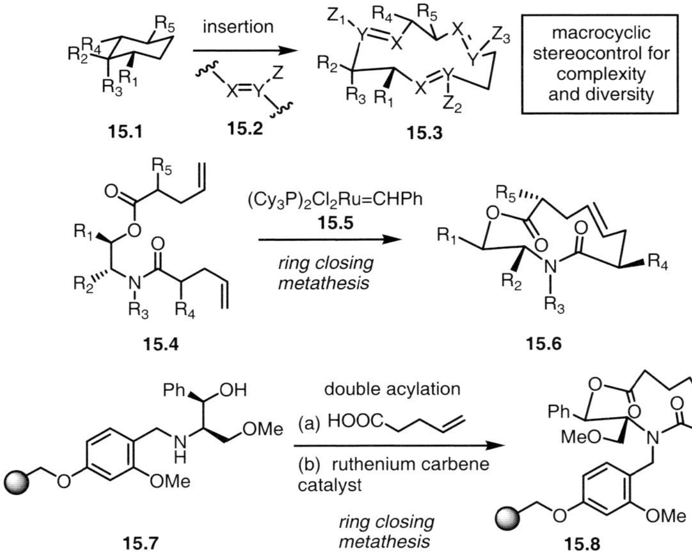

Figure 15.15 DOS approach for synthesizing macrocyclic compounds by Lee et al. [46].

diastereoselective, giving a single diastereomer as the product. The product was finally cleaved from the solid support by using HF-pyridine, and the library was then screened with a cell-based phenotypic assay. A new natural-product-like derivative was identified as a potent inhibitor of VSVG-GFP movement from Golgi apparatus to the plasma membrane, even though galanthamine itself has no observed effects on this secretory pathway.

Another novel method from Schreiber's group [46] was a strategy of macrocyclic ring closure and functionalization aimed at split-pool synthesis. The ring-closure substrate, **15.4** (Figure 15.15), was prepared by simultaneous or sequential acylation of a 1,2-aminoalcohol derivative with 4-pentenoic acid or its 2-substituted derivatives. The ring-closing metathesis reaction was then performed in the presence of Grubbs catalyst to produce compound **15.6** in moderate to excellent yields.

To further illustrate the use of bifurcating reaction pathways to produce different backbone scaffolds, Schreiber and coworkers went on to further functionalize reactions under macrocyclic stereocontrol and to ring-permutation reactions. Functional groups such as olefins and carbonyl groups present in the macrocycle could then undergo various macrocyclic-based stereocontrolled reactions (i.e., epoxidation, enol ether reactions, etc). Finally, to demonstrate the feasibility of performing the ring-closing metathesis reaction sequence on a solid support, a traceless linker was used to synthesize the macrocyclic derivative **15.8** with excellent purity following cleavage with 10% TFA.

Micalizio and Schreiber [47] developed a key reaction, the transesterification of unsaturated boronic esters with allylic esters or propargylic alcohols. This reaction transiently provided mixed organoboronic esters that could be trapped by using

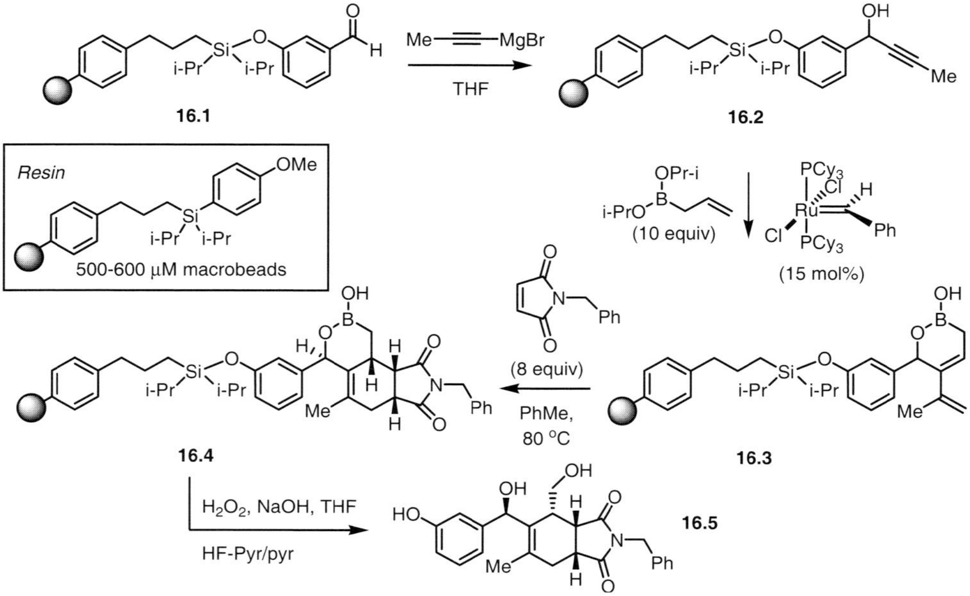

Figure 15.16 Boronic ester annulation, including a Diels–Alder reaction, developed by Micalizio and Schreiber [47].

ring-closing metathesis to afford cyclic boronic esters. The cyclic boronic esters provided access to a diverse family of boron-containing heterocycles whose unsaturation facilitated functionalization reactions (Figure 15.16). 3-Hydroxybenzaldehye loaded onto 500–600-μm polystyrene alkylsilyl-derivatized macrobeads [48] **16.1** was subjected to Grignard reaction with 1-propynylmagnesium bromide to yield **16.2.** This, upon treatment with the unsaturated boronic ester, resulted in a mixed organoboronic ester, which was then trapped by ring-closing metathesis, to afford cyclic boronic ester **16.3**. A Diels–Alder reaction on **16.3** resulted in stereoselective production of polycyclic heterocycle **16.4**. Finally, the resin was cleaved with HF-pyridine to yield **16.5**. This boronic ester approach to diversity-oriented synthesis yields complex structures containing multiple rings, four stereogenic centers, and unsaturated units in just four steps.

Kubota et al. [49] developed an efficient, stereoselective syntheses of tricyclic compounds by exploring Ferrier and Pauson–Khand [50] reactions on a glycal template. This methodology was further utilized to develop a method for stereoselective synthesis of a library of 2500 compounds. Solid-phase synthesis was performed on 500–600-μm polystyrene alkylsilyl-derivatized macrobeads. The initial loading element was synthesized according to the reaction shown in Figure 15.17. A Ferrier reaction of 3,4,6-tri-O-acetyl-D-glucal **17.1** with (*S*)-1[tert-butyldiphenylsilyl)oxy]-3-buytyn-2-ol gave the pseudoglucal as the α anomer, whereas a Ferrier reaction with the (*R*) isomer gave a mixture of the α and β isomers in a ratio of 5:1, from which the α anomer could be isolated by column chromatography. Removal

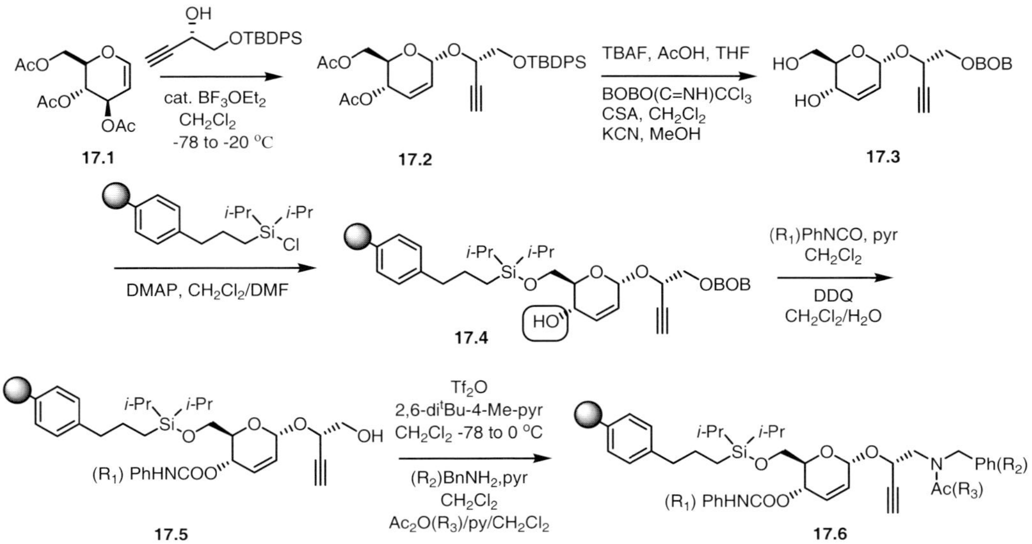

Figure 15.17 Stereoselective DOS-based approach to glycals by Kubota et al. [49].

of the TBDMS protecting group followed by protection of the primary hydroxyl functionality as the 4-butyloxybenzyl (BOB) ether, a compatible protecting group that can be easily removed by DDQ without affecting the solid-phase silyl ether-based linking element, resulted in **17.3**. This was then loaded onto 500–600-μm polystyrene alkylsilyl-derivatized macrobeads, giving **17.4**. The first solid-phase diversity step (R_1) was functionalization of the 4-hydroxy group of the pseudoglucal. Phenylisocyanate reacted quantitatively to afford the carbamate. Deprotection of the BOB group resulted in the alcohol **17.5**, which was the second diversity position after triflation followed by S_N2 reaction with a primary amine. Reaction of the resulting secondary amine with various acylation agents resulted in the third diversity point, **17.6**. Performing a Pauson–Khand reaction on **18.1** (Figure 15.18) resulted in the tricyclic α,β-unsaturated ketone **18.2**, which was further subjected to a hetero-Michael reaction to result in the fourth diversity point, **18.3**. Finally, treatment of the macrobeads with HF-pyridine resulted in the tricyclic compound **18.4** with four diversity points. This methodology resulted in a library of 2500 compounds.

Chen et al. [51] recently developed a procedure to enhance the use of enantioselective 1,3-dipolar cycloadditions of azomethine ylides [52] with electron-deficient olefins. The reaction is of interest because its stereospecificity enables stereochemical diversification of up to four tetrahedral centers on a pyrrolidine ring skeleton. A commercial catalyst, (*S*)-QUINAP, in combination with Ag(I) acetate, was used to carry out the enantioselective cycloaddition reaction (Figure 15.19). Both enantiomers of the new catalyst system are easily prepared from commercially available reagents. 4-Hydroxybenzaldehyde was loaded onto 500–600-μm polystyrene alkylsilyl-derivatized macrobeads to result in **19.1**, which was then subjected

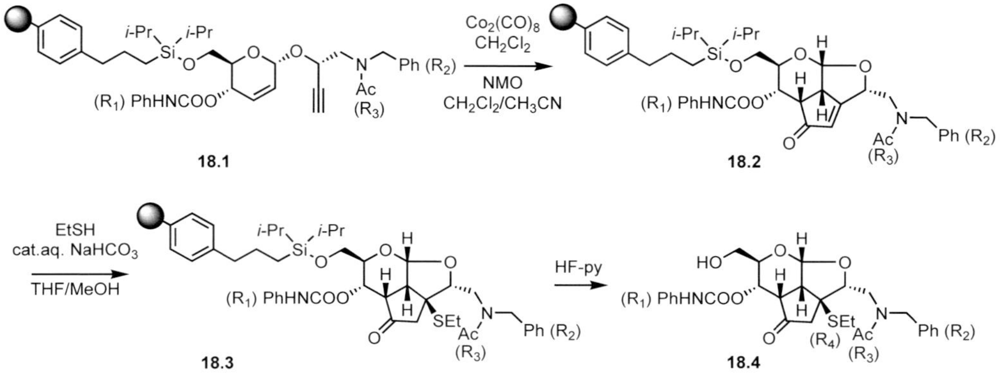

Figure 15.18 Kubota et al. [49] approach (continued).

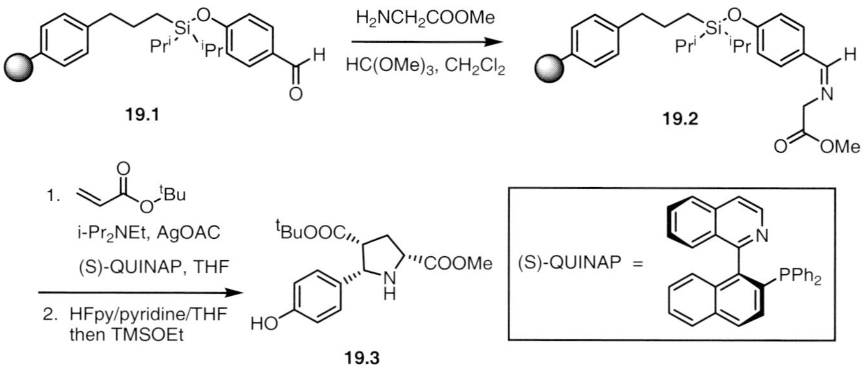

Figure 15.19 Stereoselective DOS-based approach to azomethine cycloaddition by Chen et al. [51].

to condensation with methyl glycinate. The iminoester **19.2** was then reacted with *tert*-butylacrylate using 10 mol% Ag(1) acetate/(*S*)-QUINAP at –45 °C for 40 h, followed by cleavage with HF-py and a TMSOEt quench, producing the pyrrolidine **19.3** in 79% yield and 90% *ee* in just three steps. Thus, these reactions enable introduction of up to four consecutive stereogenic centers in the (3+2) azomethine ylide cycloaddition. This methodology would thus be a powerful tool for the synthesis of stereochemically diverse alkaloid-like compounds.

Su et al. [53] used allylsilanes having C-centered chirality and a distannoxane transesterification catalyst [54] in a sequence of transesterification reactions to rapidly assemble a set of stereochemically diverse macrodiolides reminiscent of polyketide-derivative natural products. Figure 15.20 summarizes the synthesis of stereochemically well defined 14- and 16-member macrodiolides **20.4** and **20.5**, resembling known polyketide-derived natural products, from hydroxyl esters **20.2** and **20.3**. The feasibility of cyclodimeriztion was studied using different solvents and variable concentrations. Reactions were affected by the choice of the solvent.

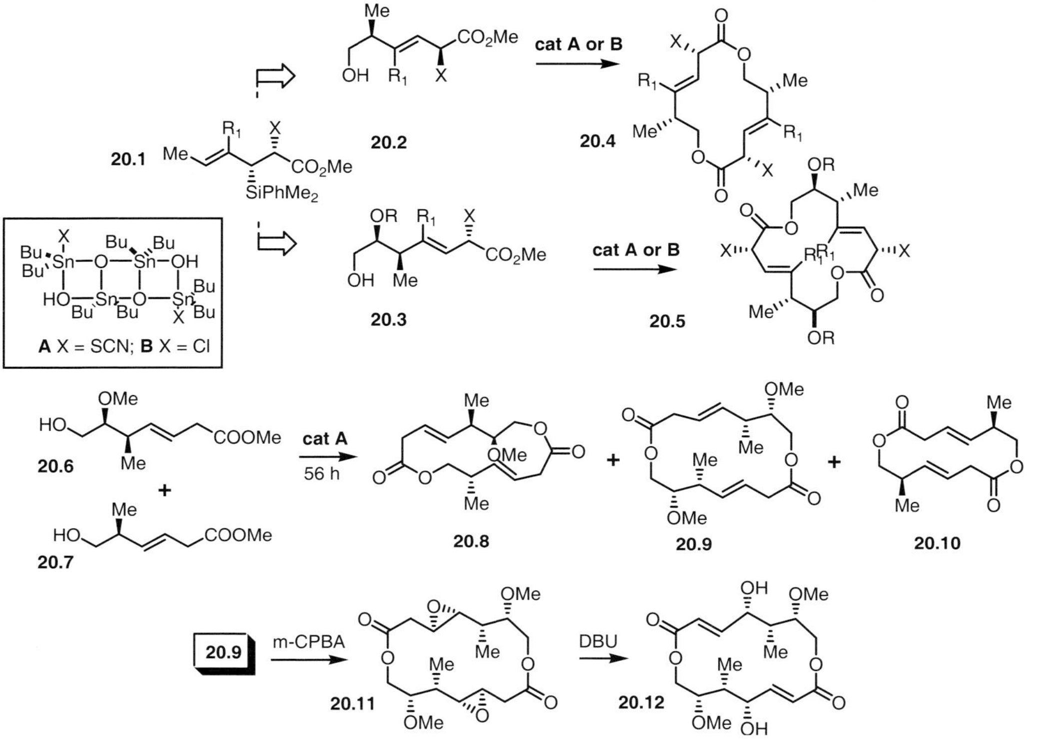

Figure 15.20 Cyclodimerization approach to functionalized macrocycles of Su et al. [53].

High dilutions reduced the amount of oligomers formed. Preliminary experiments on enantio-enriched hydroxyl esters **20.6** and **20.7**, using distannoxane transesterification catalyst, produced stereochemically diverse homo- and heterodimers **20.8–20.10**. Functionalization of the macrodiolides was investigated in an effort to create additional structural diversity. Electrophilic epoxidation of macrodiolide **20.9** afforded bis-epoxide **20.11**. Further diversification was achieved by treating the macrodiolide bis-epoxide with DBU, which resulted in epoxide ring opening to afford α,β-unsaturated macrolide **20.12**.

Several natural products are known that posses indole and indoline scaffolds, and a number of these derivatives exhibit a wide range of biological activities [55]. For these reasons, Arya et al. [56] initiated a program aimed at developing solid-phase synthesis methods to produce a variety of complex polycyclic derivatives. We developed the synthesis route so as to obtain the hydroxyindolinol derivative **21.1** (Figure 15.21). A Mitsunobu reaction-based strategy for the synthesis of hydroxy-indoline-derived tricyclic derivatives **21.4** was achieved by solution and solid-phase synthesis [56]. A 10-step solid-phase synthesis was then successfully utilized to generate a libraries of 16 and 100 derivatives in yields ranging from 20%–35% using the IRORI split-and-mix type approach. Compound **21.1** was anchored to

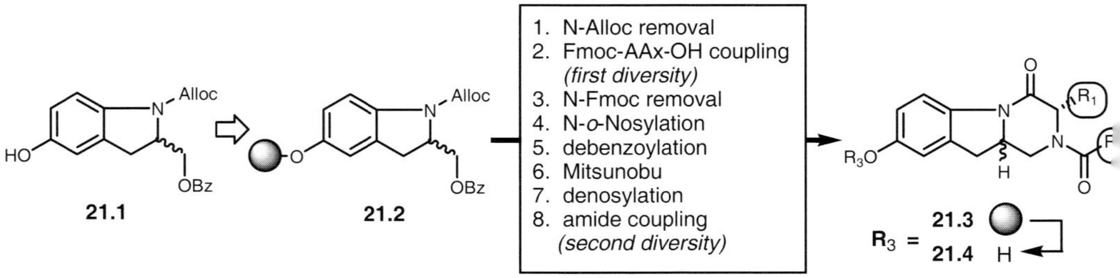

Figure 15.21 Intramolecular Mitsunobu approach to indoline-based polycyclics of Arya et al. [56].

the solid support by using (bromomethyl)phenoxymethyl polystyrene (loading 1.3 mmol g^{-1}). The *N*-alloc group was removed by treating the resin with $Pd(PPh_3)_4$ in CH_2Cl_2 in the presence of acetic acid and *N*-methylmorpholine. The free amine derivative was then coupled with Fmoc-protected phenylalanine under standard amide-coupling reaction conditions (DIC, HOBt). This introduced the first diversity site in an overall yield of 80% in three steps after cleavage from the resin. Following Fmoc removal, the free amine was further protected as an *o*-nosyl derivative by reaction with *o*-nitrobenzenesulfonyl chloride, a requirement for exploring the intramolecular Mitsunobu reaction. The free primary hydroxyl group was generated by debenzoylation (NaOMe, MeOH/THF), giving the free hydroxyl derivative (after step 5, overall yield of 65% in six steps after cleavage from the support). This was then subjected to Mitsunobu reaction conditions (EtOOC-N=N-COOEt, Ph_3P). To complete the synthesis on solid phase, the *o*-nosyl group was removed without any difficulty and the free amine was then coupled with *p*-tolyl acetic acid under standard amide-coupling condition. Cleavage from the support by treatment with 10% TFA in CH_2Cl_2 provided the desired indoline-derived tricyclic derivative **21.4** in an overall yield of 35% in 10 steps. Further work is in progress to explore the use of these indoline-derived tricyclic derivatives as small-molecule probes to study cellular processes. For example, the library is being tested in a search for small-molecule inhibitors of eukaryotic protein synthesis and the findings will be reported when they become available [57].

In another approach, Arya et al. [58] describe solution and solid-phase synthesis of two polycyclic derivatives, **22.2** and **22.3** (Figure 15.22), from enantiopure

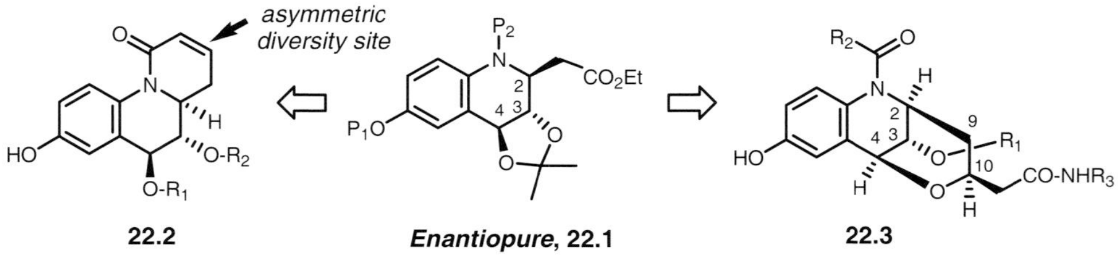

Figure 15.22 Stereoselective DOS approach to tetrahydroquinoline-based natural product-like polycyclics by Arya et al. [58].

tetrahydroquinoline-based β-amino acid **22.1**. The broad usefulness of quinoline- and tetrahydroquinoline-based natural products prompted us to develop a diversity-oriented strategy for synthesizing natural-product-like polycyclic derivatives having this privileged structure.

Central to this idea is the development of an efficient solution method to obtain the enantiopure tetrahydroquinoline-based β-amino acid **22.1**, which is a versatile building block. For model studies, the synthesis of enantiopure tetrahydroquinoline β-amino acid **23.5** (Figure 15.23) was carried out as follows. 2-Nitropiperonal was converted to the unsaturated carboxyl ester by the Wittig reaction (95%) and then subjected to the Sharpless dihydroxylation reaction, giving the enantiopure dihydroxyl derivative **23.2** (88%, > 90% *ee*, determined by chiral HPLC). After acetonide protection, the carboxyl ester was then reduced by lithium borohydride to **23.3**. Compound **23.4** was then obtained from **23.3**. This was then subjected to nitro group reduction, and then treatment with LDA or NaH, to obtain the hetero-Michael product **23.5** as a single diastereomer. The stereochemistry of the new stereogenic center in **23.5** was assigned by NOE (H-2 and H-4). The reaction seems to be independent of the choice of base and provides easy access to enantiopure β-amino acid on a large scale. It appears that acetonide protection of vicinal hydroxyls at C_3 and C_4 is an important factor (see **23.7**) in the asymmetric hetero-Michael reaction. Tetrahydroquinoline β-amino acid contains several important features: (1) vicinal hydroxyls at C_3, C_4, and (2) a phenolic moiety that can be utilized as an anchor site in solid-phase synthesis.

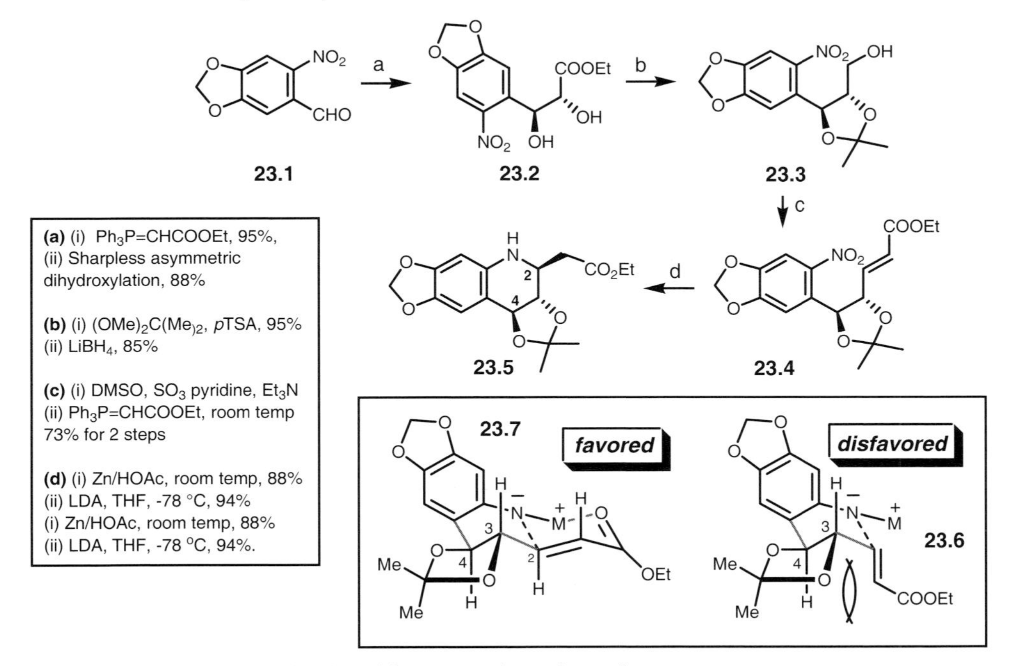

Figure 15.23 Tetrahydroquinoline-based β-amino acids synthesized by an asymmetric hetero-Michael approach.

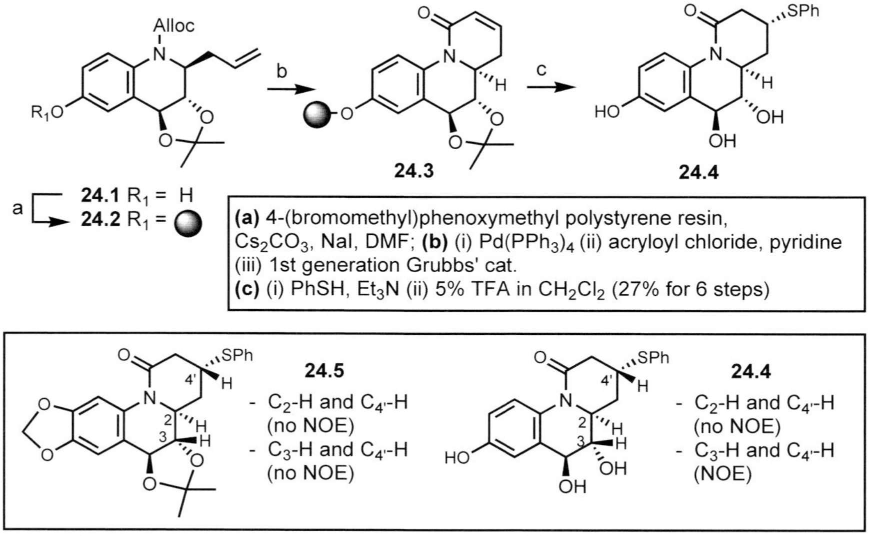

(a) 4-(bromomethyl)phenoxymethyl polystyrene resin, Cs_2CO_3, NaI, DMF; **(b)** (i) $Pd(PPh_3)_4$ (ii) acryloyl chloride, pyridine (iii) 1st generation Grubbs' cat.
(c) (i) PhSH, Et_3N (ii) 5% TFA in CH_2Cl_2 (27% for 6 steps)

24.5
- C_2-H and $C_{4'}$-H (no NOE)
- C_3-H and $C_{4'}$-H (no NOE)

24.4
- C_2-H and $C_{4'}$-H (no NOE)
- C_3-H and $C_{4'}$-H (NOE)

Figure 15.24 RCM-based stereocontrolled DOS approach.

The solid-phase synthesis of tetrahydroquinoline-based tricyclic derivative **24.3** having an enamide functional group is shown in Figure 15.24. Compound **24.1** was obtained from hydroxynitrobenzaldehyde and then anchored to a solid support by using 4-(bromomethyl)phenoxymethyl polystyrene resin (loading 93%). After alloc removal and acryloylation, the ring-closing metathesis reaction gave the cyclic enamide product **24.3**. As observed in solution studies, compound **24.4** was obtained as a single diastereomer (attack from the α face) upon reaction with PhSH after cleavage from the solid support (27% overall yield for six steps). NMR studies of compound **24.4** showed NOE between C_3-H and $C_{4'}$-H.

In a model study of the regio- and stereoselective hetero-Michael approach (Figure 15.25), enantiopure tetrahydroquinoline β-amino acid **25.1** was converted into the free dihydroxyl derivative **25.2**. With compound **25.2** as the starting material, the stage was now set to explore the asymmetric hetero-Michael reaction. We were pleased to note that this reaction proceeded very smoothly and gave a single diastereomer in high yield (84%) [58]. The tetrahydroquinoline-based tricyclic derivative **25.3** was well characterized by MS and NMR. As observed earlier, this reaction seems to be independent of the choice of the base and is an excellent example of a highly regio- and stereoselective (reaction with benzylic-OH at C_4 only) hetero-Michael reaction. Based on extensive NMR studies (compound 25.3) that showed no NOE between C_2-H and C_4-H (except for compound **25.2**, which did show NOE between C_2-H and C_4-H), we propose a boat structure for the newly formed pyran ring, due to a boat-shaped transition state. The regio- and stereoselective outcome could be envisioned by a pseudo-axial occupation of functional groups at C-2, C-3, and C-4, allowing selective facial attack of the oxygen nucleophile at the Michael site.

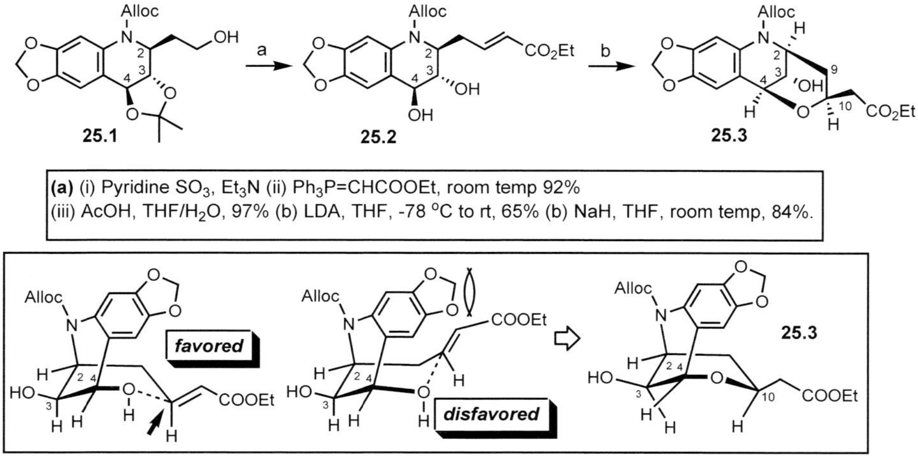

25.1 → (a) → 25.2 → (b) → 25.3

(a) (i) Pyridine SO_3, Et_3N (ii) $Ph_3P{=}CHCOOEt$, room temp 92%
(iii) AcOH, THF/H_2O, 97% (b) LDA, THF, -78 °C to rt, 65% (b) NaH, THF, room temp, 84%.

favored

disfavored

Figure 15.25 Model studies: regio and stereoselective hetero-Michael approach to tetrahydroquinoline-based polycyclics.

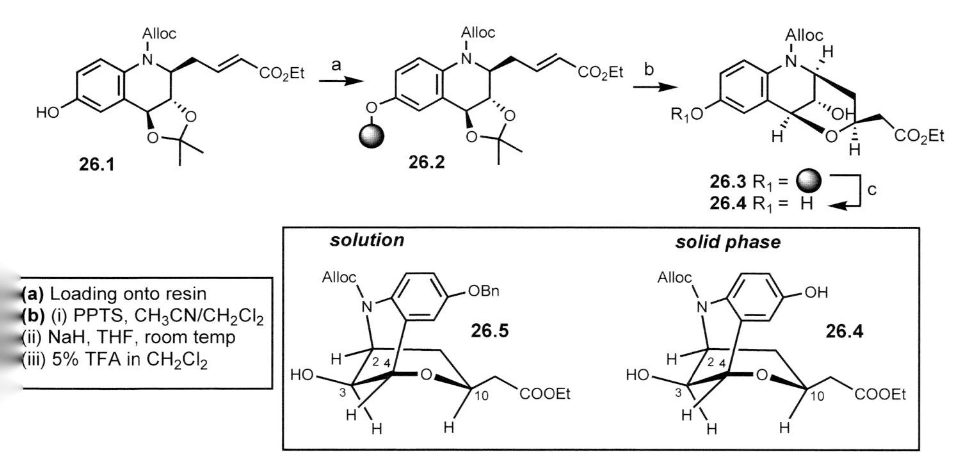

Figure 15.26 Stereocontrolled solid-phase synthesis of tetrahydroquinoline-based natural-product-like polycyclics.

For solid-phase synthesis of **26.4** (Figure 15.26), compound **26.1** was immobilized on the resin as for the previous example (loading 86%, **26.2**). The free hydroxyl derivative obtained after the acetonide removal was subjected to crucial hetero-Michael reaction. The use of NaH as a base at room temperature provided the expected product **26.3**. After cleavage from the support, the crude sample was purified giving product **26.4** (25% overall yield in four steps), which was further assigned by NMR. It was interesting to note that this unusual regio- and stereoselective hetero-Michael reaction worked in a similar manner as in solution synthesis. For comparison purposes, compound **26.5** was also synthesized in solution in a similar manner.

15.4 Conclusions

The early approach of combinatorial chemistry, when library generation based on simple heterocyclic compounds was the main thrust, is now challenged by the growing need for developing efficient high-throughput synthesis methods leading to libraries of natural-product analogs or natural-product-like complex polycyclics. An evolving area of stereo- and enantioselective diversity-oriented synthesis and the development of novel methods leading to 3D skeletal diversity are likely to play important roles in populating the required chemical space so as to provide rapid access to natural-product-like compounds as highly valuable small-molecule probes. The recent combinatorial chemistry literature provides a very good reflection of the growing interest in DOS and in the generation of libraries of natural-product-like polycyclics. An efficient mapping of the desired chemical space and its appropriate function in developing highly effective therapeutics will be a topic of discussions for years to come. Whatever the outcome of this exercise, the synthetic community is well positioned to undertake some of these challenges.

Acknowledgements

We thank the NRC Genomics program, VP Research Funds for Chemical Biology, and NCI Canada for financial support. We thank Dr. Zhonghong Gan for proof-reading the article.

References

1 (a) S. L. Schreiber, *Bioorg. Med. Chem.* **1998**, *6*, 1127–1152; (b) C. M. Crews, R. Mohan, *Curr. Opin. Chem. Biol.* **2000**, *4*, 47–53; (c) C. M. Crews, U. Splittgerber, *Trends Biochem. Sci.* **1999**, *24*, 317–320; (d) B. R. Stockwell, *TIBTECH* **2000**, *18*, 449–455; (e) B. R. Stockwell, *Nature Reviews Genetics* **2000**, *1*, 116–125; (f) C. M. Crews, *Chem. Biol.* **1996**, *3*, 961–965; (g) T. U. Mayer, *Trends Cell Biol.* **2003**, *13*, 270–277.

2 (a) S. L. Schreiber, *Chem. Eng. News* **2003**, March 3, 51–59; (b) J. R. Peterson, T. J. Mitchison, *Chem. Biol.* **2002**, *9*, 1275–1285; (c) D. L. Boger, J. Desharnais, K. Capps, *Angew. Chem., Int. Ed.* **2003**, 42, 4138–4176.

3 T. Gura, *Nature* **2000**, *407*, 282–284.

4 R. L. Strausberg, S. L. Schreiber, *Science* **2003**, *300*, 294–295.

5 M. D. Burke, E. M. Berger, S. L. Schreiber, *Science* **2003**, *302*, 613–618.

6 (a) R. E. Dolle, *J. Comb. Chem.* **2002**, *4*, 369–418; (b) D. G. Hall, S. Manku, F. Wang, *J. Comb. Chem.* **2001**, *3*, 125–150; (c) L. A. Wessjohann, *Curr. Opin. Chem. Biol.* **2000**, *4*, 303–309; (d) L. Weber, *Curr. Opin. Chem. Biol.* **2000**, *4*, 295–302.

7 (a) P. Arya, D. T.-H. Chou, M.-G. Baek, *Angew. Chem. Int. Ed.* **2001**, *40*, 339–446; (b) P. Arya, M.-G. Baek, *Curr. Opin. Chem. Biol.* **2001**, *5*, 292–301; (c) P. Arya, R. Joseph, D. T.-H. Chou, *Chem. Biol.* **2002**, *9*, 145–156.

8 (a) D. J. Newman, G. M. Cragg, K. M. Snader, *Nat. Prod. Rep.* **2000**, *17*, 215–234; (b) J. Mann, *Nat. Prod. Rep.* **2001**, *18*, 417–430.

9 K. C. Nicolaou, D. Vourloumis, N. Winssinger, P. S. Baran, *Angew. Chem., Int. Ed.* **2000**, 39, 44–122.

10 H. Wang, A. Ganesan, *J. Org. Chem.* **1998**, *63*, 2432–2433.

11 H. Wang, A. Ganesan, *J. Comb. Chem.* **2000**, *2*, 186–194.

12 D. Bonnet, A. Ganesan, *J. Comb. Chem.* **2002**, *4*, 546–548.

13 M. Kondoh, T. Ushui, T. Mayumi, H. Osada, *J. Antibiot.* **1998**, *51*, 801–804.

14 (a) A. Jordan, J. A. Hadfield, N. J. Lawrence, A. T. McGowyn, *Med. Res. Rev.* **1998**, *18*, 259–296; (b) S. J. Haggarty, T. U. Mayer, D. T. Miyamoto, R. Fathi, R. W. King, T. J. Mitchison, S. L. Schreiber, *Chem. Biol.* **2000**, *7*, 275–286; (c) P. Verdier-Pinard, J.-Y. Lai, H.-D. Yoo, J. Yu, B. Marquez, D. G. Nagle, W. Nambu, J. D. White, J. R. Falck, W. H. Gerwick, B. W. Day, E. Hamel, *Mol. Pharmacol.* **1998**, *53*, 62–67.

15 (a) P. Wipf, J. T. Reeves, R. Balachandran, K. A. Giuliano, E. Hamel, B. W. Day, *J. Am. Chem. Soc.* **2000**, *122*, 9391–9395; (b) P. Wipf, J. T. Reeves, R. Balachandran, B. W. Day, *J. Med. Chem.* **2002**, *45*, 1901–1917; (c) D. P. Curram, *Angew. Chem. Int. Ed.* **1998**, *37*, 1174–1196.

16 P. Wipf, W. Xu, *J. Org. Chem.* **1996**, *61*, 6556–6562.

17 O. de Frutos, D. P. Curran, *J. Comb. Chem.* **2000**, *2*, 639–649.

18 (a) I. Pendrak, S. Barney, R. Wittrock, D. M. Lambert, W. D. Kingsbury, *J. Org. Chem.* **1994**, *59*, 2623–2625; (b) I. Pendrak, R. Wittrock, W. D. Kingsbury, *J. Org. Chem.* **1995**, *60*, 2912–2915.

19 W. Zhang, Z. Luo, C. H.-T. Chen, D. P. Curran, *J. Am. Chem.* Soc. **2002**, *124*, 10443–10450.

20 O. Lavergne, L. Lesueur-Ginot, F. Pla Rodas, D. C. H. Bigg, *Bioorg. Med. Chem. Lett.* **1997**, *7*, 2235–2238.

21 O. Lavergne, D. Demarquay, C. Bailly, C. Lanco, A. Rolland, M. Huchet, H. Coulomb, N. Muller, N. Baroggi, J. Camara, C. Le Breton, E. Manginot, J.-B. Cazaux, D. C. Bigg, *J. Med. Chem.* **2000**, *43*, 2285–2289.

22 C. Bailly, A. Lansiaux, L. Dassonneville, D. Demarquay, H. Coulomb, M. Huchet, O. Lavergne, D. C. H. Bigg, *Biochemistry* **1999**, *38*, 15556–15563.

23 (a) A. E. Gabarda, D. P. Curran, *J. Comb. Chem.* **2003**, *5*, 617–624; (b) H. Josien, S.-B. Ko, D. Bom, D. P. Curran, *Chem. Eur. J.* **1998**, *4*, 67–83; (c) L. Boymond, M. Tottländer, G. Cahiez, P. Knochel, *Angew. Chem., Int. Ed. Engl.* **1998**, *37*, 1701–1703.

24 (a) J. Stubbe, J. W. Kozarich, *Chem. Rev.* **1987**, *87*, 1107–1136; (b) S. Kane, S. M. Hecht, *Prog. Nucleic Acid Res. Mol. Biol.* **1994**, *49*, 313–352.

25 C. E. Holmes, B. J. Carter, S. M. Hecht, *Biochemistry* **1993**, *32*, 4293–4307.

26 H. Kuramochi, K. Takahashi, T. Takita, H. Umezawa, *J. Antibiot.* **1981**, *34*, 576–582.

27 D. L. Boger, H. Cai, *Angew. Chem. Int. Ed.* **1999**, *38*, 448–476.

28 D. L. Boger, S. Teramoto, J. Zhou, *J. Am. Chem. Soc.* **1995**, *117*, 7344–7356.

29 H. Sugiyama, G. M. Ehrenfeld, J. B. Shipley, R. E. Kilkuskie, L. H. Chang, S. M. Hecht, *J. Nat. Prod.* **1985**, *48*, 869–877.

30 C. J. Thomas, A. O. Chizhov, C. J. Leitheiser, M. J. Rishel, K. Konishi, Z.-F. Tao, S. M. Hecht, *J. Am. Chem. Soc.* **2002**, *124*, 12926–12927.

31 K. L. Smith, Z.-F. Tao, S. Hashimoto, C. J. Leitheiser, X. Wu, S. M. Hecht, *Org. Lett.* **2002**, *4*, 1079–1082.

32 D. L. Boger, T. Honda, Q. Dang, *J. Am. Chem. Soc.* **1994**, *116*, 5619–5630.

33 Y. Nishizuka, *Science* **1986**, *233*, 305–312.

34 B. Meseguer, D. Alonso-Díaz, N. Griebenow, T. Herget, H. Waldmann, *Angew. Chem., Int. Ed.* **1999**, *38*, 2902–2906.

35 B. Meseguer, D. Alonso-Díaz, N. Griebenow, T. Herget, H. Waldmann, *Chem. Eur. J.* **2000**, *6*, 3943–3957.

36 A. G. Myers, B. A. Lanman, *J. Am. Chem. Soc.* **2002**, *124*, 12969–12971.

37 (a) T. Arai, A. Kubo, in *The Alkaloids*, A. Brossi (Ed.), Academic Press, New York **1983**, Vol. 21, chapter 3; (b) W. A. Remers, *The Chemistry of Antitumor Antibiotics*, Wiley-Interscience, New York **1988**, Vol. 2, chapter 3.

38 E. J. Martinez, T. Owa, S. L. Schreiber, E. J. Corey, *Proc. Natl. Acad. Sci. USA* **1999**, 96, 3496–3501.

39 A. G. Myers, A. T. Plowright, *J. Am. Chem. Soc.* **2001**, *123*, 5114–5115.

40 A. T. Plowright, S. E. Schaus, A. G. Myers, *Chem. Biol.* **2002**, *9*, 607–618.

41 (a) J. H. Van Maarseveen, *Comb. Chem. High Throughput Screening* **1998**, *1*, 185–214; (b) O. Seitz, *Nachr. Chem.* **2001**, *49*, 912–916.

42 (a) S. L. Schreiber, *Science* **2000**, *287*, 1964–1969; (b) M. D. Burke, S. L. Schreiber, *Angew. Chem., Int. Ed.* **2004**, *43*, 46–58.

43 (a) D. S. Tan, M. A. Foley, M. D. Shair, S. L. Schreiber, *J. Am. Chem. Soc.* **1998**, *120*, 8565–8566; (b) D. S. Tan, M. A. Foley, B. R. Stockwell, M. D. Shair, S. L. Schreiber, *J. Am. Chem. Soc.* **1999**, *121*, 9073–9087.

44 H. E. Pelish, N. J. Westwood, Y. Feng, T. Kirchhausen, M. D. Shair, *J. Am. Chem. Soc.* **2001**, *123*, 6740–641.

45 D. H. R. Barton, G. W. Kirby, *J. Chem. Soc.* **1962**, 806–817.

46 D. Lee, J. K. Sello, S. L. Schreiber, *J. Am. Chem. Soc.* **1999**, *121*, 10648–10649.

47 (a) G. C. Micalizio, S. L. Schreiber, *Angew. Chem., Int. Ed.* **2002**, *1*, 152–154; (b) G. C. Micalizio, S. L. Schreiber, *Angew. Chem., Int. Ed.* **2002**, *1*, 3272–3276.

48 J. A. Tallarico, K. M. Depew, H. E. Pelish, N. J. Westwood, C. W. Lindsley, M. D. Shair, S. L. Schreiber, M. A. Foley, *J. Comb. Chem.* **2001**, *3*, 312–318.

49 H. Kubota, J. Lim, K. M. Depew, S. L. Schreiber, *Chem. Biol.* **2002**, *9*, 265–276.

50 (a) O. Geis, H.-G. Schmalz, *Angew. Chem., Int. Ed.* **1998**, *37*, 911–914; (b) K. M. Brummond, J. L. Kent, *Tetrahedron* **2000**, *56*, 3263–3283; (c) A. J. Fletcher, S. D. R. Christie, *J. Chem. Soc. Perkin Trans. 1* **2000**, 1657–1668.

51 C. Chen, X. Li, S. L. Schreiber, *J. Am. Chem. Soc.* **2003**, *125*, 10174–10175.

52 (a) S. Kanemasa, *Synlett* **2002**, 1371–1387; (b) R. Grigg, *Chem. Soc. Rev.* **1987**, *16*, 89–121; (c) *Synthetic Applications of 1,3-Dipolar Cycloaddition Chemistry toward Heterocycles and Natural Products* (Eds. A. Padwa, W. Pearson), Wiley-VCH, Weinheim **2002**, pp. 169–252.

53 Q. Su, A. B. Beeler, E. Lobkovsky, J. A. Porco Jr., J. S. Panek, *Org. Lett.* **2003**, *5*, 2149–2152.

54 J. S. Panek, M. Yang, I. Muler, *J. Org. Chem.* **1992**, *57*, 4063–4064.

55 P. M. Dewick, Chapter 6 in *Medicinal Natural Products: A Biosynthetic Approach*, Wiley-Interscience, New York **2002**.

56 P. Arya, C.-Q. Wei, M. L. Barnes, M. Daroszewska, *J. Comb. Chem.* **2004**, 6, 65–72.

57 P. Arya, C.-Q. Wei, M. L. Barnes, J. Pelletier, *unpublished results* **2003**.

58 P. Arya, P. Durieux, Z.-X. Chen, R. Joseph, D. M. Leek, *J. Comb. Chem.* **2004**, 6, 54–64.

Subject Index

Chemogenomics in Drug Discovery: A Medicinal Chemistry Perspective.
Edited by Hugo Kubinyi and Gerhard Müller

ISBN: 3-527-30987-X

c

e

h

j

k

q